The Materials Science of Semiconductors

T0180118

Angus Rockett

The Materials Science
of Semiconductors

 Springer

Angus Rockett
University of Illinois
201a Materials Science and Engineering Building
1304 West Green Street
Urbana, IL 61801
USA

ISBN 978-1-4419-3818-3   e-ISBN 978-0-387-68650-9

Printed on acid-free paper.

9 8 7 6 5 4 3 2 1

springer.com

Dedication

To my parents, for their teaching,

To my family for their love and encouragement,

and to the students in my class for their suggestions, comments

questions, and of course, their enthusiasm.

PREFACE

OBJECTIVES

The primary purpose of this book is to convey insight into why semiconductors are the way they are, either because of how their atoms bond with one another, because of mistakes in their structure, or because of how they are produced or processed. The approach is to explore both the science of how atoms interact and to connect the results to real materials properties, and to show the engineering concepts that can be used to produce or improve a semiconductor by design. Along with this I hope to show some applications for the topics under discussion so that one may see how the concepts are applied in the laboratory.

The intended audience of this book is senior undergraduate students and graduate students early in their careers or with limited background in the subject. I intend this book to be equally useful to those teaching in electrical engineering, materials science, or even chemical engineering or physics curricula, although the book is written for a materials science audience primarily. To try to maintain the focus on materials concepts the details of many of the derivations and equations are left out of the book. Likewise I have not delved into the details of electrical engineering topics in as much detail as an electrical engineer might wish. It is assumed that students are familiar with these topics from earlier courses.

The core prerequisite subjects assumed for use of this book are basic chemistry, physics, and electrical circuits. The most essential topics from an intermediate level in these subjects are reviewed. Students taking my class are assumed to have had a condensed matter physics course and a semiconductor device theory course with significantly more detail than is covered in Chapters 2 and 3. Furthermore they are assumed to have had some organic chemistry (at least at the Freshman undergraduate level) and general materials science courses with significantly more information than I provide in the review in Chapter 4. In spite of these expectations my audience usually includes graduate students lacking background in at least one of these topics.

My background that applies to the material in this book includes more than 25 years doing research in semiconductor materials science and processing. My primary research over the past 18 years has concerned the fundamental materials science of the semiconductor $CuInSe_2$ and related compounds. This material is a fascinating study in all of the topics of materials science rolled into a single field. As such, the results appear from time to time in the book as illustrations. Given rising concerns about energy world wide, the book also makes reference to solar cells, properly known as photovoltaic devices, in several application sections. This also reflects my long study of that field.

TOPICS AND USE OF THE BOOK

I have taught a class based on the material that is presented in this book for many years. I get through most of the topics, excluding detailed discussions of the applications, in one semester. Each chapter typically gets about three hours of lecture, although Chapters 6, 7, 9, and 12 often receive four to five lectures and Chapter 1 gets one hour. My students have generally been a mixture of senior undergraduate and graduate students. One of the common features of these students is that many, especially the graduate students, lack a strong background in some one of the underlying prerequisite subjects such as condensed matter physics, principles of electronic devices, or materials science. Therefore the book includes a brief review of these topics in Chapters 2-4 and I cover these subjects very quickly in the class. Even for the students with a background in all of these areas I usually find it helpful to review selected topics from these chapters. In presenting review topics I hope that the book can be useful to a student body that completely lacks a detailed background if the instructor is willing to go through this material.

One of the challenges in writing the book is that its intended audience does not have much background in quantum mechanics. Therefore, I cannot practically go into details about how the wave function interactions in Chapter 5 are calculated that lead to the matrix element values in the LCAO matrix (or into details of corrections to the LCAO that are used in real calculations today. I have attempted to present the material in such a way that a student without this background can understand the important take-home messages of the subject even if they are unfamiliar with how the values are derived. Of course, if one is doing tight binding theory these matrix elements may be considered fitting parameters, so what I am doing in Chapter 5 is not so much less sophisticated than tight binding theory anyway. One may also note that I have taken two approaches to the question of how atomic orbitals contribute to semiconductor band structures in Chapter 5, one almost purely visual – the Harrison diagram approach – and one semiquantitative, covering the basic approach of LCAO theory. Hence, hopefully the student will have different ways of remembering the relationships among orbitals, bonds, and bands. It is useful when covering the material to also make connections to the material in Chapter 2 on nearly free electron behaviors.

Chapters 6 and 7 are the core of the materials science in the book. They talk about how to engineer a semiconductor material to achieve given properties through modification of its structure and chemistry. Hence, they spend a considerable time on defects and how and why they form. Most of the preceding chapters are building up to these two in hopes that when the student gets here they will say to themselves "of course it should work this way, considering how the atoms are changing and how the bonds are forming."

Chapters 8 and 9 cover less commonly applied materials of great value and interest in applications. Especially in the case of Chapter 9 the organic materials appear likely to increase greatly in range of application in the near future. Amorphous inorganic materials described in Chapter 8 are valueable to understand and preview of what one might expect for the organics in some respects. Many of the issues such as Coulomb blockades at contacts are the same.

Finally Chapters 10-12 cover some of the major methods of processing thin films. When I teach this class this material is covered between the material in Chapter 4 and the material in Chapter 5 but it could just as well be covered last. I put it in the middle because it provides a change of pace in the type of discussion I am presenting. When I was first developing this course together with a second course on dielectrics, metalizations, and other materials and processes used above the semi-conductor surface I tried to separate the course into a materials course and a processing course. This did not work well because the two subjects are so heavily intertwined. It is hard to see why there would be problems in the materials without understanding their processing but it is equally hard to see why one should bother with complex processes unless one has an idea of how the material one is producing responds to defects. Likewise, it is difficult to see why a non-equilibrium process such as evaporation should be important to deposition of an alloy unless one sees how phase separation and defects affect the alloy property.

Of course the biggest problem with writing a book of this type is that the field, especially the organic materials topics, are progressing so rapidly that the book will be somewhat obsolete as soon as it is in print. To improve the longevity of the material covered I have attempted to stick with fundamental concepts and not deal with what I perceive to be passing fads. With luck I will have reason to add some of these topics to future editions.

ACKNOWLEDGEMENTS

I wish to thank the many students who have suffered through the use of this book in draft form in my class and for all of their input. It has been an invaluable help. I also wish to thank the various colleages whose work I use to illustrate concepts in the book and whose contributions have been some of the most important in the field. I have tried to recognize them by name in the presentation of their figures. Finally, I wish to thank the staff at Kluwer who have been so patient. I hope their long wait pays off.

Angus Rockett

University of Illinois, 2007

TABLE OF CONTENTS

3. OVERVIEW OF ELECTRONIC DEVICES   73

4. ASPECTS OF MATERIALS SCIENCE   141

9. ORGANIC SEMICONDUCTORS 395

10. THIN FILM GROWTH PROCESSES 455

LIST OF TABLES

Chapter 1

AN ENVIRONMENT OF CHALLENGES

1.1 OVERVIEW

Modern electronic materials and devices arguably are built upon nearly the entire periodic table (excluding only the actinides and a few other unusual or unstable elements). These diverse materials are required to meet the intense challenges which electronic device applications present. In their full extent, electronic applications range from simple copper wires, to high-performance magnetic materials for computer disks, to semiconductors for state-of-the-art microelectronic devices, and many more. Likewise, the critical properties of the materials range from electronic conductivity, to optical transmission, to diffusion-resistance or mechanical properties. It is not reasonable, nor is it particularly desirable, to cover all aspects of electronic materials in a single text. Consequently, the materials discussed here relate primarily to the most challenging applications, particularly with reference to microelectronic and optical devices. This volume is further restricted to the semiconducting materials used in active devices and leaves the metals, dielectrics, and other materials used in microelectronic processes to other texts. A wide range is considered including some traditional materials, such as silicon, and some in their infancy, such as organic semiconductors. Readers may also wish to consider books on epitaxial growth, and other processes relevant to microelectronics manufacturing as supplements to this text.

The competitive nature of manufacturing of microelectronics has meant that completely new generations of devices and completely new electronic materials and processes have been required on a time scale of months rather than years or decades. One consequence of this situation is that a book covering materials specific to the current generation of devices would be out of date by the time it was written, let alone published and distributed. A more practical approach is to cover fundamental properties of classes of materials and processing methods that are relevant to both old and new device generations. This book attempts to cover these fundamentals but uses illustrations from current technology when reasonable.

As we will see, electronic applications have only one feature in common – they place higher demands for performance on their materials than any other class of products. Working with these materials is truly working in an environment of challenges.

This chapter considers some of the issues currently facing the microelectronics engineer. It refers to constituents of the circuits without explanation, as this would disrupt the flow of the discussion. For the reader who is unsure of the terminology, Chapter 2 presents a review of some of the basic physics of semiconductors. Selected active devices that are currently joined to produce an integrated circuit are described briefly in Chapter 3. Finally, Chapter 4 presents a brief review of concepts from Materials Science that are important in the discussions in later chapters. The details of the terminology are relatively unimportant for purposes of this chapter.

1.2 A HISTORY OF MODERN ELECTRONIC DEVICES

Modern electronic devices build on a long history of invention, discovery, and basic scientific research. The most critical devices are the control circuit elements – the diode and the switching devices. The latter began as triode vacuum tubes and are now generally transistors. Diodes pass current easily in only one direction. The original diode was invented in 1905 by J. Ambrose Flemming based on observations made in the laboratories of Edison Electric. This diode vacuum tube contained a hot filament, which emits electrons, and a metallic plate collector. Electrons flow only from filament to collector. The following year Lee DeForest created the triode vacuum tube and the electronic revolution was launched.

The vacuum triode consists of a heated cathode, an intervening wire grid and a plate or anode, and functions as a diode modified by the control grid. A small change in current at the grid produces a large change in current from cathode to anode. Therefore, the triode allows amplification of weak signals. This ability to amplify is the essential element of both analog and digital circuits. Between 1906 and the mid 1950's, vacuum tubes were developed and adapted to more and more specialized applications with more and more sophisticated internal structures to modify the electron current. Unfortunately, tubes, like incandescent light bulbs, have a very limited lifetime and consume large amounts of electricity, producing a large amount of heat. Even the development of miniature tubes could not overcome these problems. Vacuum tubes are still used in rare applications such as television picture tubes, very high-power amplifier stages in radio transmitters, and in environments where damage to transistors would occur and degrade their performance much faster than a vacuum-tube-based circuit would degrade. As devices included more and more tubes, their lifetime between servicing and their overall reliability decreased dramatically. This situation led researchers to seek new ways of producing a switching effect.

The solution to the tube problem was found in the bipolar junction transistor, created in 1947 at the Bell Telephone Laboratories by John Bardeen, Walter H. Brattain, and William Shockley. The original device was produced from a lump of germanium and worked by virtue of diffusion of metals from the contacts into the Ge crystal. The device was found to control current effectively and yielded amplification as with vacuum tubes but contained no heated filament and consumed relatively little power. As designs progressed the performance improved markedly. While the Ge transistor was revolutionary, it was not a practical solution for the long term. Ge has a relatively low energy gap, making it relatively conductive at room temperature. This allows current to leak backward through a device that is supposed to be turned off. Such leakage causes the entire circuit to consume large amounts of power at all times

Selected Significant Events in the Development of Semiconductor Microelectronics, 1900-2000

1905   Vacuum tube diode invented by J. Ambrose Flemming
1906   Triode vacuum tube invented by Lee DeForest
1916   Czochralski crystal growth technique invented by Jan Czochralski
1935   First patent issued on a field-effect transistor (Oskar Heil)
1938   Early reports of Si rectifiers by Hans Hollmann and Jürgen Rottgardt
1947   Transistor invented by Bardeen, Brattain, and Shockley
1951   First practical field effect transistor
1952   Single crystal Si produced
1954 SiO_2 mask process developed
1958   First integrated circuit invented by Jack Kilby
1959   Planar processing methods, precursors of modern integrated circuit
        fabrication methods, created by Noyce and Moore
1960   First practical metal-oxide-silicon transistor
1960   First patent on a light emitting diode (J.W. Allen and P.E. Gibbons)
1962   Transistor-transistor logic
1962   First practical visible light emitting diode
1962   First laser diode
1963   Complementary metal-oxide-silicon transistors provide lower power
        switching devices
1968   Metal-oxide-semiconductor memory circuits introduced
1971   First microprocessor
1978   First continuously operating laser diode at room temperature
1987   Polymer-based light emitters
1992   Er-doped fiber amplifier
1997   Introduction of Cu-based interconnects

and lowers the gain of the amplification that can be obtained. The solution to this dilemma was to exchange Ge for Si. The bipolar transistor has, following a switch from Ge to Si, become a ubiquitous element in modern circuits.

Another such device, the field effect transistor, was created at about the same time. While patents on field-effect switching devices were filed as early as 1930, the first practical device was produced in 1951. The current control was based on the depletion of charge resulting from the presence of a reverse biased diode junction. The field-effect transistor has a much higher control electrode resistance (that of a reverse biased diode or capacitor) compared to the bipolar transistor. This was a particular advantage because vacuum tubes can be configured to have a very high input resistance. Thus, the field-effect transistor provided a better replacement for vacuum tubes in many applications – for example, in amplifiers for low-power signals.

A switch to Si-based devices occurred rapidly as the technology for its production improved. Silicon has a larger energy gap and consequently pure Si is less conductive at room temperature than Ge. This dramatically lowers reverse leakage current and circuit power. However, the major reasons why Si has remained the most popular semiconductor are the performance, stability and reproducibility of insulating layers and contacts that can be produced on it.

The developments needed for the use of Si in microelectronic devices included two major process improvements – methods for purification of the material, in particular the removal of problem impurities; and methods for growing large single crystals. The latter had been solved as early as 1916 with the creation of the, now ubiquitous, Czochralski method for bulk crystal growth, although many improvements and adaptations were needed before large Si crystals could be grown. As time has progressed and more has become known about the basic science of the Czochralski method, wafer sizes have increased from ~25 mm in the 1960's to 300 mm today.

Purification presented a much greater problem. In the early 1950's The Siemens Company developed a method based on reaction of Si with HCl to produce dichlorosilane, SiH_2Cl_2, an easily evaporated liquid. The dichlorosilane is fractionally distilled and subsequently reduced in a reverse reaction to produce pure Si. (See also Chapter 4.) In 1952, W.G. Pfann created the zone refining method for further reduction of impurities, which was subsequently improved to reduce handling of the material between process steps. This method was used through the 1970's. Around that time, improvements in the initial purification process made zone refining largely unnecessary.

Miniaturization of electronic circuits took another major step forward with Jack Kilby's invention of the integrated circuit in 1958 at Texas Instruments and through the contributions of Robert Noyce and Gordon Moore at Fairchild Semiconductor (later founders of Intel Corporation). Together they developed methods for producing and interconnecting all of the basic elements of a circuit on a single piece of Si. This resulted in single electronic packages containing far more functionality in a compact form than could be obtained from discrete devices. It has proven much less expensive, much more reliable, and much faster to produce complex circuits from standard but individually complex integrated circuits than to wire the devices from discrete components. Today, complete heterodyne radio receivers, video and audio signal processors, amplifiers, and computing circuits are available on single integrated circuit chips. The power and complexity of these circuits has grown amazingly rapidly since the 1950's, as discussed in the next section.

Not all of microelectronics is focused on integrated circuits, although that sector certainly drives much of the field and receives most of the press attention. Optical devices are getting increasing notice as the information age drives a greater and greater need to transfer data. More mundanely, everywhere one looks; one finds small colored dots of light glowing to indicate readiness of electronic devices to serve our needs. These are primarily light-emitting diodes or LED's. The first patent on such devices was in 1955 by R. Braunstein describing electroluminescence from various semiconductors. The first practical visible light emitting diode had to wait for 1962 when N. Holonyak created a GaAs-GaP alloy or "Ga(As,P)" device. Likewise, the first patent on a semiconductor laser was in 1962, but a continuously operating laser diode that would run at room temperature waited until development of advanced heterojunction structures in the early 1970's. The progress in efficiency of light emitting devices is shown in Figure 1.1.

The use of laser and light-emitting diodes has grown explosively in recent years with the current market estimated to be over 30 billion dollars. This growth has been fueled recently by the development of blue and green light emitters. These devices now permit the complete spectrum of colors to be generated, and hence allow full-color emissive displays to be produced. Early blue and green light emitters were based on the II-VI family of compounds including materials such as ZnS and CdS. These proved unsatisfactory for a number of reasons. Several attempts were made to develop SiC-based devices. However, the true breakthrough was the discovery of a method for producing both p and n-type group III nitrides (GaN and related materials) by Nakamura and the demonstration of high-intensity blue and green light emitting devices in 1992.

Laser diodes (based on GaAs and related compounds and primarily emitting in the infrared region) have proven essential in high capacity fiber-optic communications and compact-disk data storage systems. These devices are expensive, and require complex driving circuitry. One of the most recent breakthroughs in optical materials that greatly improved the performance of long-distance fiber-optic communication systems is the Er-doped silica amplifier. This device, at its root, is constructed simply by doping a section of optical fiber with Er atoms, providing higher energy light waves as a pump to excite the system, and allowing stimulated emission as

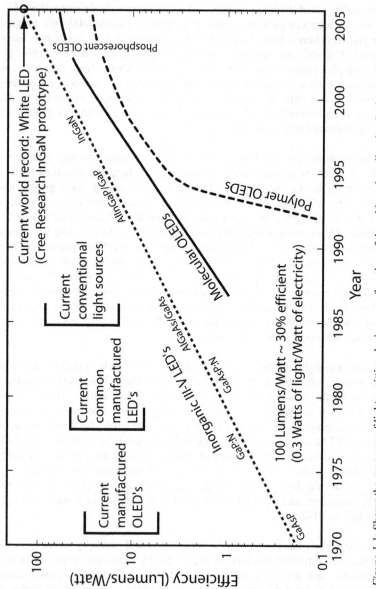

Figure 1.1: Shows the progress of light-emitting devices as a function of time. Note that all technologies are converging to the efficiency of flourescent lamps because the recent devices use a high-energy emitter to drive a phosphorescent material that emits the visible light. Figure based on data in Sheats [1] with additional data from Forrest [2], Andrade and Brown [3] and Cree Research press release [4].

electrons relax back onto the Er atoms. When a light pulse passes through the doped region it is amplified by gain involving the Er atom defect states.

The above are only a tiny fraction of the many remarkable discoveries and innovations that have sustained the recent explosive growth in microelectronics.

Developments in both electronic and optical devices continue at a staggering rate. Most of these are based on the fundamental science of the materials and processing methods described here and in other texts.

1.3 AN ISSUE OF SCALE

As noted above, electronic devices have improved remarkably in recent years. In 1965, Gordon Moore observed that the number of transistors per square inch had doubled every year since the integrated circuit was invented in 1958. His projection that this trend would continue has come to be known as Moore's Law and it has hung like the Sword of Damocles over the minds of most microelectronics engineers, who wonder how long they can maintain it. Despite many firm predictions of its imminent end, Moore's Law has proven correct for over 40 years, as shown in Figure 1.2. The early days of integrated circuits saw only a few devices in the circuit. The number of devices grew to hundreds in the 1960's, thousands to tens of thousands in the 1970's, to the current numbers of millions to tens of millions. The doubling time for transistors per device has been closer to 18 months than one year since the statement of the Law, although recent gains have, remarkably, been faster. There is no obvious reason why it should not be possible to continue the Law until devices shrink well below the tenth-micron scale (individual devices with critical lengths less than 10^{-5} cm). Beyond that it is hard to see how the trend can continue, and a new device paradigm will be needed. Multilayer structures are possible even based on current technology and whole new concepts in device fabrication are anticipated, although none has yet shown a likelihood of replacing the conventional structures.

Along with this staggering increase in device density, performance of the devices has improved dramatically while the price has remained roughly constant, fueling a vigorous market for upgraded computers. To maintain this market and encourage people to purchase new computers, it is necessary for each generation of circuitry to improve sufficiently to justify upgrading hardware. The inevitable failure to improve performance will presumably change the entire nature of microelectronics from a research and development driven field to a commodity driven field. It is our challenge to both stave-off this change as long as possible based on the current device designs and to create new technology fuel for the microelectronics industry based on concepts not yet conceived.

The enormous improvements in device performance have come at the cost of intense and ongoing research and development efforts. These have resulted in creation of new materials and processes, and dramatic improvements in our understanding of the materials involved. There is no doubt that electronic grade Si single crystals are the most-perfect and most highly studied materials ever produced. Further improvements are becoming harder to achieve and the challenges faced are becoming more fundamental.

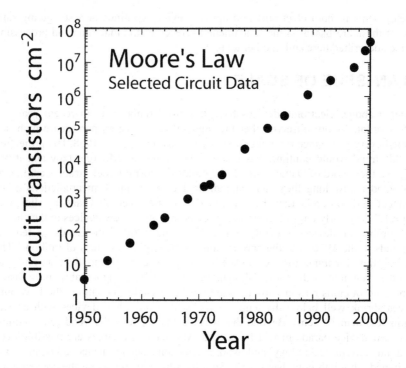

Figure 1.2: Shows the trend known as Moore's Law for microprocessor circuit density. The minimum feature size has decreased proportionate to the square root of the circuit density. Each plotted point corresponds to a marketed product. [Figure based on data on the Intel® web site: http://www.intel.com/technology/mooreslaw/index.htm, 2006.]

Advanced production device dimensions have now shrunk to roughly 0.1 μm minimum feature sizes (see Figure 1.3). The challenges of producing a device of this scale become more obvious when one considers doping of the semiconductor and the electron current density passing through the device. A transistor with a control region length and width of ~ 0.1 μm (10^{-5} cm) and a thickness of 50 nm (5×10^{-6} cm) is current technology. At this scale, the control volume is 5×10^{-16} cm^3. The atom density of silicon is 5×10^{22} atoms cm^{-3}. This suggests that the critical volume of current transistors within which the device is switched on or off contains only 25 million atoms.

Doping the semiconductor with one part per million impurity atoms (5×10^{16} cm^{-3}), a typical value for older devices, means the control region would contain only 25 impurity atoms. The removal of a single dopant atom would therefore correspond to a 4% change in doping level. Such doping levels do not provide adequate conductivity or reproducibility. Consequently, small devices have doping levels closer to 1×10^{19} cm^{-3} or 0.02% (at or near the solubility limit for the dopant). Even at this doping level the control volume contains only 5000 dopant atoms. Reduction in scale of another factor of five in the lateral dimensions is easily foreseen in current technology. With no reduction of thickness, the control volume would contain only 1000 dopant atoms at the higher doping level. A variation of only 10 dopant atoms (well within what one might expect for fluctuations in doping sources) corresponds to a 1% change in doping level. Such changes can affect the resulting performance. Device designs must therefore be more capable of accommodating materials variability. The challenge will grow further if the smallest transistors to date with control volumes of the scale of $2x10^{-18}$ cm^{-3} (10,000 atoms) are to be manufactured.

Data in integrated circuits is carried by small bursts of electrons. Typical devices now operate at $>10^9$ Hz (1 GHz) with some as high as $5x10^{11}$ Hz. Current flowing for the corresponding cycle time (10^{-9} s) transfers only ~$2x10^9$ electrons per amp of current. A current of 1 nA transfers only one or two electrons in a nanosecond. While "single electron" transistors have been produced that can be turned on and off with single electrons, devices responding to such small amounts of charge are not yet practical. A more reasonable number of electrons to activate a device or store a data

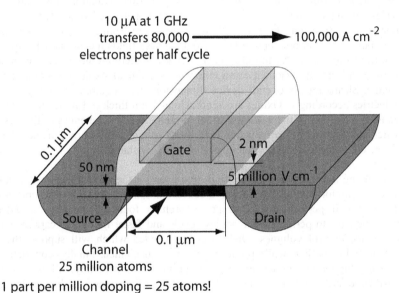

Figure 1.3: A schematic diagram of a state-of-the-art field-effect transistor such as that discussed here. The current densities (10^5 A cm^{-1}) are sufficient to cause conductors to fail, the fields across the gate dielectric ($5x10^6$ V cm^{-1}) are barely supportable by even the nearly perfect SiO$_2$ gate dielectric, the number of dopant atoms in the channel limits the practical dopant concentrations to parts per thousand typically, and the total number of electrons transferred through the device is so small that noise becomes significant.

value is of the order of a few thousand, corresponding to a required current of a few microamps in each cycle. A simple analysis of the 0.1 micron device above would suggest that a current of 10 μA (10^{-5} A) would correspond to a current density through the 5×10^{-11} cm^2 cross section of the device of 2×10^5 A cm^{-2}. Devices operating at 500 GHz require even higher current densities for practical operation, even though the number of carriers flowing in a cycle is reduced.

Current densities of this magnitude flowing through conducting wires produce an "electron wind" with sufficient momentum to literally push atoms along the conductor in materials where atoms diffuse easily. This phenomenon of "electromigration" has been one of the long-standing causes of device failures. It is the primary reason driving the transition from aluminum to copper conductors in current generation devices and has been considered, in some cases, *the* major issue ultimately limiting the lifetime of operating integrated circuits.

At the same time that current densities are rising to unacceptable levels, threatening to melt devices during operation, and total atom numbers are forcing restrictions on the minimum doping levels in device control volumes, insulators are beginning to fail. The best insulator known is SiO_2 grown by thermal oxidation of Si wafers. Such oxides can support fields of up to 10 million volts per cm. Do not depend upon air supporting such a field by placing your finger 1 cm from a 10 MV power source! Even bulk SiO_2 can not support such a field. These fields mean that a 1 V potential requires a minimum of 1 nm (10^{-7} cm) of oxide if there are no defects or thickness changes present. In practical terms, more like 2 nm of oxide are required at this voltage. Shrinking the overall device dimensions has required shrinking the oxide dielectrics accordingly. Oxides are approaching 1 nm thick (~3 molecules). This has required reduction of the voltages, which produce dramatic changes in the design of switching transistors. The reduction in allowed voltages has spurred the development of new dielectric materials with higher dielectric constants.

Considering the issues above, one finds that even for the current technology devices, there are major restrictions facing electrical engineers that prevent optimization of the circuit performance. Current research is focusing on ways to alter the semiconductor to permit higher doping levels and faster motion of carriers through the device control volumes. Dielectrics are needed which will support the same electric fields as SiO_2 while producing higher capacitance. Finally, conductors that can carry higher current densities without failure are under study. All of these issues must be overcome before the next generation of devices can be produced. To continue on Moore's Law, two such generations of device, with their incumbent changes in materials, will have to have been produced between the time it was written and the time you read this book. Imagine the exciting challenges that will face you!

1.4 DEFINING ELECTRONIC MATERIALS

Electronic materials span such a wide range of properties that it is hard to define them. Figure 1.3 illustrates the elements used in electronic applications. Clearly, this involves most of the periodic table. Why is it that electronic devices make use of such a wide variety of elements while most applications such as automobiles make use of only a much more modest subset? Consider the possible types of materials one might select. Electronic devices and device processing methods use solids, liquids, gases, and even gas plasmas frequently. This variety is needed to achieve the level of control required to manufacture the current-generation technology. Of the solids, elemental, alloy, and compound materials are used with no class of solids more common than any other. Among compounds, it is common to require a very specific compound. For example, for advanced dielectrics TiO_2 has been considered. However, this specific compound must be produced rather than any of the other oxides of titanium. Thus, the particular material required is often very well defined and only that material will do. Apparently, the general class of material does not define an electronic material.

We can see in Figure 1.4 that the position of an element on the periodic table does not matter greatly. The properties of elements in the various columns of the periodic table tend to be closely related and hence these elements have similar applications. The inert gases are typically unreactive, but compounds such as xenon fluorides do occur and are used occasionally in microelectronic production processes. Inert gases are largely used for etching materials by physical impacts (sputtering, see Chapter 11). The group VIIa elements (halogens) are highly reactive, tend to form volatile compounds with many elements, and are used in etching. In rare applications, ionic compounds involving halogens such as CuBr are applied directly as device materials. Group VIa elements produce strongly bound compounds. In the case of oxides, these are generally used as dielectrics, while the elements below oxygen form the chalcogenide compounds, most of which are semiconductors.

The most common elements used in microelectronics are in groups IIIa to Va. Group Va elements are largely used in compound semiconductor form, and as dopants in the group IV semiconductors. Group IVa elements include the common semiconductors Si and Ge as well as C. The latter is used in the forms of diamond, graphite, organic compounds or fullerene molecules. The Group IIIa elements include the excellent electrical conductor Al, elements found in semiconducting compounds with group Va elements, and as dopants in group IVa semiconductors.

Group IIb elements form compound semiconductors with the chalcogenides and have a variety of other uses. The most common conductors are the group Ib elements Cu and Au with Ag also occasionally used. Transition metals are commonly used in compound form either as silicides or nitrides, primarily as stable contact materials bridging between Si and a highly conductive metal, or as diffusion barriers. The rare earths are not used extensively. This is gradually changing and elements such as hafnium, erbium and gadolinium are making increasingly common appearances.

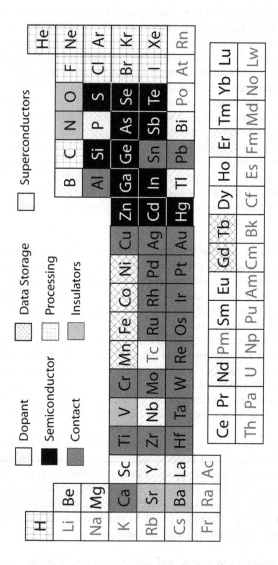

Figure 1.4: Shows the primary application of the various elements in the periodic table in microelectronics. Elements left blank are used only in very rare applications or are not used at all. Many elements such as Al have a number of applications. In the case of Al, these include as a contact/metallization, a semiconductor component, a dopant, and in insulators. Element symbols shown in gray are very rarely used.

In some cases the IIa elements such as calcium are used as conductors or contacts, although their reactivity makes them too unstable in many cases. Finally, the group Ia alkali metals are rarely used because of their reactivity and rapid diffusion rates in many materials, although these too are becoming more common. Both group Ia and IIa elements are increasingly used, for example, in organic electronic devices.

The choice of elements is thus made based on application and properties and is driven by the resulting performance of devices. Microelectronic applications generally require relatively small amounts of material, making their availability in large quantities less important. The competitive performance-driven market provides an impetus for spending more for starting materials in order to achieve superior results. The cost of the few milligrams of paladium or platinum that might be used is a small part of the final cost of the completed device. This suggests on its face that price per unit volume of an element is not critical. If price and availability are not generally important, one might then ask, what is critical?

1.5 PURITY

The fundamental property that most clearly links electronic materials is purity. All of the various classes of compounds that are used are required to contain as few other elements as possible. In many cases extreme measures are taken to prevent contamination. Consider the following examples of the scale of contamination problems that must be dealt with in microelectronic applications.

Aluminum oxide has often been used as the packaging material for military-specification computer chips because of its outstanding resistance to penetration by contaminants. The original packages of this type were made from standard ceramic-process-grade aluminum oxide powder produced directly from bauxite ore. While this material is relatively pure, there is typically a very low level of uranium oxide contamination. This tiny amount uranium was found to be sufficient that radioactive decay in the oxide led to false data in the information stored in the chips. To avoid this, the aluminum oxide must be decomposed electrochemically to aluminum and oxygen. The aluminum is then converted to a vapor compound and purified by fractional distillation. Finally, the compound is reacted with purified oxygen to produce electronic-grade aluminum oxide. It is a lot of trouble to go through to get rid of a few uranium atoms, but it turns out to be necessary.

Electronic grade Si is a second example, and represents the greatest challenge in materials purity in any application. Virtually all contaminants must be eliminated to part per million levels with some impurities, such as transition metals, having detectable effects on devices to the part per trillion level. To visualize the purity requirement for Si, imagine that maple trees represent Si atoms and that pine trees represent Fe atoms. The area of the contiguous United States is about 10 million square kilometers or 10 trillion square meters. If one could plant an extremely dense forest of maple trees, one per 10 m^2 area, then you could have no more than one pine tree in all of the United States. Now that's a weed problem! One trillion Si atoms fill a cube ~ 3 μm on a side. No more than one Fe atom can be allowed in such a volume of typical Si. This is more than 1000 times the purity requirement of other applications.

The actual situation in submicron devices is even worse than this already extreme value. It was estimated above that a 0.1 μm-scale device control volume contains 25 million atoms. Within this volume, a single atom has a bulk atom fraction of 40 parts per billion. This indicates that a single Fe atom could ruin the device in whose control volume it occurs. If a single impurity atom can ruin a device, and if one can only tolerate one bad device in 1000 to achieve a working circuit, then only one in 2×10^{18} atoms could be a particularly destructive impurity such as Fe. The fact is that current circuits are designed with redundancy such that single failed transistors generally do not ruin the entire chip. Even so, it is necessary to account for the possibility of impurities such as Fe in the design of the devices and to achieve and maintain a low level of contamination in every step of the circuit fabrication.

We can now see why raw material price is not an issue with most elements used in semiconductor devices. The cost of purification is generally the bulk of the cost of the material. This rule of thumb does not apply to all materials and devices. Because price is directly related to supply and demand for elements, certain rare elements can be prohibitively expensive for large-scale processes. In general, however, a higher price stimulates greater production, keeping the price roughly constant. On the other hand, some devices such as solar cells and lighting products must be as inexpensive as possible. In these devices, even small amounts of expensive material can be a problem.

1.6 PERFORMANCE

Purity is typically only part of the equation determining the usefulness of an electronic material – most of the rest is performance. Performance can have many aspects including electronic properties of the material such as conductivity, free carrier mobility, etc., and physical and chemical properties such as mechanical strength, stability against diffusional mixing or reaction with adjacent materials, and many more. Electronic and optical properties are related to the way in which electrons interact with the atomic structure of the material. Chemical properties depend upon the atomic bonding and the possible reactions that can take place between one material and others that it touches.

The performance of electronic materials can affect the lifetime, speed, efficiency, or other aspects of the device behavior. Part lifetimes are generally limited by chemical reactions or motions of atoms over time. Current technology devices are becoming so small that motion of atoms over even a few atomic distances could cause the device to become inoperative. This tolerance is nowhere more evident than in the gate capacitor of a metal-oxide field-effect transistor (MOSFET). Current generation devices with tenth-micron minimum gate lengths have $SiO_2+Si_3N_4$ gate insulator (dielectric) thicknesses of 2.5 nm (~7 layers of molecules) or less. Atoms diffusing into this oxide can cause defect states that can disrupt the performance of the insulator or cause the transistors in the device to latch in the on or off state. Part of the solution to this situation is to purify the materials to such an extent that impurities that might move are eliminated. This is the reason why buildings that house integrated circuit fabrication lines ("fabs") cost hundreds of millions of dollars. Purification does not work when a material one wishes to use intentionally as part of the device is intrinsically inclined to move and cause trouble. Such is the reason why it has taken years to switch from Al to Cu as the metal connecting devices in integrated circuits.

Copper diffuses rapidly and causes very large problems if it gets into the active device regions. The solution has been to design exceptional diffusion barrier materials with which to surround the Cu conductors to prevent Cu escape. The performance of Cu is poor in terms of chemical stability but its electrical performance is sufficiently good that it outweighs other considerations. Furthermore, the chemical properties can be mitigated by good materials design.

The debate over performance of a material relative to potential problems it may cause goes on in current technology. The high dielectric constant materials based on Ba compounds are attractive as potential replacements for Si-based dielectrics. Ba, like Cu, can cause significant problems in the wrong parts of the circuit. However, it provides potentially substantial improvements in capacitors used for data storage. Manufacturers are now gradually beginning to introduce Ba compounds into their processes but only for selected devices. Other materials, more attractive than Ba compounds are also making their appearance, making the motivation to pursue Ba compounds smaller.

Performance is most obvious in semiconductors, where electronic transport pheno-mena and optical properties are critical. For example, transistor switching-speeds can be limited by the time necessary for an electron to transit the control volume. The probability that an electron will fill a hole at lower energy and give off the excess energy in the form of light is likewise essential to the performance of light emitting devices. Optical detection and photovoltaic systems rely on a high probability of the reverse process, producing free electrons by light absorption. Great efforts have been made over the years to explore the periodic table in search of new semiconductors. However, it is essential to consider all aspects of performance in a material, as the GaAs integrated circuit community has discovered over the years. Theoretically, electrons can be more easily accelerated in GaAs than in Si and live a shorter time. Both of these contribute to faster device speeds. Why then has GaAs not replaced

Si in normal microprocessors? There are many reasons but the primary ones are the lack of a good insulator and good contacts, and the relative fragility of GaAs. These problems have never been solved, while all of the major problems facing application of Si, except its inability to emit light, have been overcome. Thus, when one considers the performance of an electronic material, one must consider that material in the context of a given application and include all aspects of performance in the analysis.

1.7 SUMMARY POINTS

- Microelectronics traces its roots to vacuum tube electronics of the early 20th century.
- Practical transistors, invented in the late 1940's and early 1950's form the basis of current circuit technologies.
- Significant problems arise as device dimensions shrink concerning the current density carried in a device or in interconnects, the scale of dielectric layers in gate capacitors, and in the number of dopant atoms present in the control volume of a device.
- Microelectronic devices have contained increasing numbers of devices over time following Moore's Law.
- Electronic materials are defined primarily by their purity and performance rather than by the elements that make them up.
- Raw materials cost is generally a small part of the price of electronic-grade materials due to the high cost of purification.
- Impurity concentrations of lower than parts per billion are sometimes required in semiconductors.
- The performance of a material in a semiconductor device includes its optoelectronic, mechanical, and chemical properties. Compatibility with surrounding materials is essential.

1.8 HOMEWORK PROBLEMS

1. Using resources at a local library, write a one-paragraph summary of an invention, material, or process used in current microelectronic technology.

2. A single atomic layer of oxygen atoms ($\sim5\times10^{14}$ atoms cm^{-2}) collects on the surface of a 1 cm^2 wafer of silicon. The silicon is then melted and all of the oxygen dissolves in it. Upon cooling silicon freezes into a single crystal again. Assuming a uniform final oxygen concentration in the silicon of 0.1 parts per million, how thick would the original silicon wafer have to have been? The atomic density of silicon is 5×10^{22} atoms cm^{-3}.

3. Suppose that all circuits shown in Figure 1.2. for Moore's Law correspond to 1 cm^2 chip areas. Thus, 10^6 devices would correspond to a device area of 10 microns x 10 microns. Assuming Moore's Law continues at its current rate of change of circuit density with time, in approximately what year will the average device area reach 1 nm x 1 nm (about 16 atoms area)?

4. What is the lowest doping density in units of cm^3 that can be obtained in the 1 nm x 1 nm device if the thickness of the device is 4 atomic layers (1 nm) as well?

5. Assume that electronic grade Pt can be purchased for $3000 g^{-1} and given that the atomic weight of Pt is 195 g mol^{-1} and its density is 21.4 g cm^{-3}. Estimate the cost of a 30 nm layer of Pt used in a 1 cm^2 microprocessor assuming that it coats the entire area of the device.

1.9 SUGGESTED READINGS & REFERENCES

Suggested Readings:

Ball, Philip, *Made to Measure: new materials for the 21st century*, Princeton: Princeton University Press, 1997.

Braun, Ernest and Macdonald, Stuart, *Revolution in Miniature*, 2nd ed. Cambridge: Cambridge University Press, 1982.

Cahn, R.W., *The Coming of Materials Science*, Amsterdam: Pergamon, 2001.

Dummer, G.W.A., *Electronic Inventions and Discoveries: Electronics from its earliest beginnings to the present day*, 3rd ed. Oxford: Pergamon, 1983.

Seitz, Frederick and Enspruch, Norman G., *Electronic Genie: The Tangled History of Silicon*. Urbana: University of Illinois Press, 1998.

References:

[1] Sheats, J.; "Organic electroluminescent devices". *Science,* 1996; 273: 884-8.

[2] Forrest, Stephen R.; "The road to high efficiency organic light emitting devices", *Organic Electronics* 2003; 4: 45-48.

[3] D'Andrade, Brian and Brown, Julie J., "White phosphorescent organic light emitting devices for display applications", in *Defense, Security, Cockpit and Future Displays II,* Byrd, James C; Desjardins, Daniel D; Forsythe, Eric W.; and Girolamo, Henry J., *Proc. of the SPIE* 2006; 6225: 622514-1.

[4] Cree Research result reported by Whitaker, Tim on optics.org news for June 22, 2006.

Chapter 2

THE PHYSICS OF SOLIDS

Before beginning a general discussion of electronic devices and the more complex aspects of semiconductors and other electronic materials, it is helpful to have an idea of their physics, especially their electronic structure. This chapter provides a partial review of the physics of solids. The nature of materials is determined by the interaction of their valence electrons with their charged nuclei and core electrons. This determines how elements react with each other, what structure the solid prefers, its optoelectronic properties and all other aspects of the material. The following sections describe the general method for understanding and modeling the energies of bands of electronic states in solids. A more detailed discussion of semiconductor bonding is provided in Chapter 5.

2.1 ELECTRONIC BAND STRUCTURES OF SOLIDS

There are two approaches taken when considering how the weakly bound (valence) electrons interact with the positively charged atomic cores (everything about the atom except the valence electrons) and with other valence electrons in a solid. We will consider first the direct approach of solutions to the differential equations that describe the motion of electrons in their simplest form and the consequences of this behavior. This requires many simplifying assumptions but gives a general idea for the least complex problems. The second approach is to follow the electronic orbitals of the atoms as they mix themselves into molecular states and then join to form bonding and antibonding combinations and finally bands of states, which we leave for Chapter 5. To see why it should be possible to mix electronic states in linear combinations, it is useful to consider some direct solutions of the Schrödinger equation, which governs the motion of electrons in an arbitrary potential. As we will see, this is a second-order linear differential equation. It therefore has particular solutions that can be constructed from linear combinations of any set of solutions of the general equation. Consequently, we may suspect that, at least in some cases, a linear combination of atomic orbitals should describe the general problem of a solid. In the following discussion we will assume that the material is a periodic crystal. Some of the results turn out to be applicable in most respects to aperiodic (amorphous) structures as well.

Electrons behave as both waves and particles. The consequences of their wave and particle nature are derived through the formalism of quantum mechanics. The requirement for conservation of energy and momentum forces the electrons to select specific states described by "quantum numbers," analogous to resonant vibrations of a string on a musical instrument. The "resonant" states associated with each set of quantum numbers results in a set of "wave functions" which describe the probability of finding an electron around a given location at a given time. The wave functions of the resonant states are found as follows.

The total energy, E_{tot}, of an electron is the sum of its potential and kinetic energies. In classical terms one could express this relationship as

$$E_{tot} = \mathbf{p}^2/2m + U(\mathbf{r}), \qquad\qquad 2.1$$

where $U(\mathbf{r})$ describes the local potential energy of the particle at position $\mathbf{r}$ and the kinetic energy is given by the classical expression $E_{kin} = \mathbf{p}^2/2m$ in which m is the particle mass and $\mathbf{p}$ is the momentum. This equation applies to any classical body. When the object in question is of a scale small enough that quantum mechanical behaviors become dominant, we need to rewrite Equation 2.1 in quantum mechanical terms. Under such conditions, the exact energy (E) at any specific time (t) can not be described to within an accuracy better than $\Delta E \Delta t = \hbar$ where $\hbar$ is Plank's constant divided by 2π. To account for this uncertainty the particle must be described by a probability distribution (its wave function) rather than by indicating a specific position. Its total energy is given by the change in the wave function, $\Psi(\mathbf{r},t)$, per unit time multipled by $i\hbar$, where $i^2 = -1$. The momentum of a quantum particle is, likewise, the spatial derivative of the wave function multiplied by $i\hbar$. Based on the classical behavior and using the mathematics of operators, the kinetic energy, $\mathbf{p}^2/2m$ becomes the second derivative of the wave function times $-\hbar^2/2m$. Similarly, the potential energy represents a weighted average potential using Ψ as the weighting function. Substituting these expressions in Equation 2.1 yields:

$$i\hbar\frac{d\Psi}{dt} = -\frac{\hbar^2}{2m}\nabla^2\Psi + U(\vec{r})\Psi, \qquad\qquad 2.2$$

where ∇^2 is referred to as the Laplacian and is the second spatial derivative of the function it operates on (in this case the wave function). Equation 2.2 is the full time-dependent Schrödinger Equation and describes not only the steady-state behavior of an electron but also the way in which the electron changes energy as a function of time. Whenever the potential that the electron experiences does not change with time, the time variable can be separated from the space variable. In this case, the energy of the particle cannot change with time, and the spatial-portion of the Schrödinger Equation becomes:

$$E \psi(\mathbf{r}) = (-\hbar^2/2m) \nabla^2 \psi(\mathbf{r}) + U(\mathbf{r})\psi(\mathbf{r}), \qquad 2.3$$

where $\psi(\mathbf{r})$ is the time-independent wave function. The electronic structure of solids is derived by the solution of this equation under the boundary conditions appropriate to the solid being modeled. We will now consider some solutions to Equation 2.3.

2.1.1 Free electrons in solids

The simplest form of Equation 2.3 is the special case of $U(\mathbf{r})=0$, where there is no potential affecting the motion of electrons. For simplicity we will make the further restriction of considering only a one-dimensional problem. In this case,

$$E\psi(x) + \frac{\hbar^2}{2m} \frac{d^2\psi}{dx^2} = 0. \qquad 2.4$$

The general solution to this equation, obtained by Fourier transform methods, is a linear combination of two waves moving in the positive and negative x directions:

$$\psi(x) = A_+ e^{ikx} \pm A_- e^{-ikx}, \qquad 2.5$$

where A_+ and A_- are the amplitudes of the two waves, and $k = \dfrac{2\pi}{\lambda}$ is the wavenumber of the waves (electrons) with wavelength λ. The energies of these waves are determined by substituting Equation 2.5 into 2.4. The second derivative of $\psi(x)$ from Equation 2.5 is $d^2\psi(x)/dx^2 = k^2\psi(x)$. Thus,

$$-\frac{\hbar^2 k^2}{2m} \psi(x) + E\,\psi(x) = 0 \qquad 2.6$$

from which $E = \dfrac{\hbar^2 k^2}{2m}$. This holds for any A_+ and A_- but requires $A_+{}^2 + A_-{}^2 = 1$ for a wave with unit amplitude. The $\psi(x)$ are known as eigenvectors of Equation 2.4 and the energies are the eigenvalues. The momentum of this wave is $p = \hbar k$, thus k represents the electron momentum to within a factor of $\hbar$. This energy vs. wavenumber ["E(k)"] relationship is illustrated in Figure 2.1.

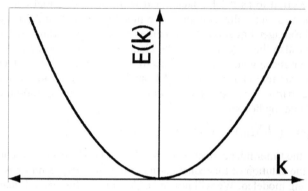

Figure 2.1: The energy vs. momentum diagram for a free electron in the absence of a periodic potential.

2.1.2 Free electrons in a periodic potential

In a solid there is a regular spacing of atoms. In a crystal, this spacing is defined by the translation vectors of the Bravais lattice. (See Chapter 4 for a description of semiconductor crystal lattices.) In an amorphous material the spacing is the average distance from one atom to its nearest neighbors. The result is the same – there is an imposed periodicity on the wave functions. The wave function must have the same value at equivalent positions in the solid. These positions are separated by lattice translation vectors **R**, thus $\psi(\mathbf{r}) = \psi(\mathbf{r+R})$ in a crystal. For a wave of the form given in Equation 2.5, this imposes an additional condition that $\psi(x)=\psi(x+L_x)$ where L_x is the lattice spacing along the x direction. Likewise, the potential energy of a particle will be periodic (at least locally) such that $U(\mathbf{r}) = U(\mathbf{r+R})$. The periodicity of ψ requires that:

$$\psi(\vec{r}) = e^{i\vec{k}\cdot\vec{r}} c_k(\vec{r}),$$

                                                                                                       2.7

where $c_{nk}(\mathbf{r})=c_{nk}(\mathbf{r+R})$ are the Fourier components for wave vector **k** of the wave function $\psi(\mathbf{r})$ [the proof is called "Bloch's Theorem"]. It can further be shown that the electron wave vector is given by $\mathbf{k}=\mathbf{b}/N$ where **b** is a reciprocal lattice vector of the crystal lattice (see Chapter 4) and N is the number of unit cells in the [real-space] lattice. Therefore, any change in **k** must be by a unit vector of the reciprocal lattice, as in any diffraction problem. Electron waves in a solid are susceptible to scattering as one would have for x-rays and the problem can be represented with, for example, an Ewald sphere construction as for normal diffraction of x-rays in a periodic crystal.

The primary consequence of Equation 2.7 is that electron wave behaviors are reproduced whenever the wave vector **k** is changed by a translation vector of the reciprocal lattice, $2\pi/a$, where a is the one dimensional lattice constant along a given direction. This means we can replicate Figure 2.1 every $2\pi/a$ units along the wave

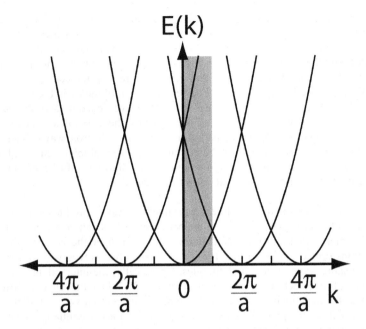

Figure 2.2: The periodic structure of the free electron energy vs. wave number in a periodic solid. The minimum section of the plot needed to provide a complete description of the relationship of E to k is shaded gray.

vector axis as shown in Figure 2.2. Note that if you consider the symmetry of this plot you will find that all of the necessary information is contained within the space between $0 < \mathbf{k} < \pi/a$. Thus, we will represent electron energy vs. wave vector plots in this reduced zone of $\mathbf{k}$ values hereafter.

2.1.3 Nearly free electrons

A somewhat more realistic picture is the case where $U(\mathbf{r})$ is not zero or constant but varies weakly with position. This is the case referred to as "nearly free" electron behavior. The solutions to the Schödinger equation can be constructed from the same set of plane waves we had in Equation 2.7 with the proviso that $U(\mathbf{r})$ is not too large. In this case, a general solution to Equation 2.3 is still given, at least approximately, by Equation 2.7 but now the c_{nk} must account for the effect of the periodic potential. This imposes the following constraint on the coefficients:

$$\left[\frac{\hbar^2}{2m}k^2 - E\right]c_k + \sum_{k' \neq k} U_{k'} c_{k'} = 0, \qquad 2.8$$

where the U_k terms are the Fourier components of the potential U for wavevector k. The free electron behavior is what is given inside the square brackets of Equation 2.8. Because the periodic potential is assumed to be small, the individual U_k terms are modest and the second term in Equation 2.8 represents only a minor perturbation on the result. Furthermore, because the c_k terms are Fourier coefficients, the c_k components can be obtained from the Fourier expansion of the U_k in the free electron plane waves. In other words, because both the wave function and the periodic potential can be expanded in the same Fourier terms, the c_k terms are related to the U_k terms via Equation 2.8. To construct a nearly free electron wave function, as modified by the U_k, from purely free electron waves we multiply each possible free electron wave by c_k and add the results to produce the new wave function solution in the presence of the periodic potential, as in Equation 2.7.

The U_k terms serve to mix the free electron plane waves producing interference effects. The closer the plane wave is to the periodicity of the lattice the more strongly it will interact with the crystal and, likewise, the stronger the component of the Fourier transform of U_k. Consider the interaction of two waves with the same or nearly the same wave vectors and energies. Graphically, the interactions occur near the points of intersection of curves in Figure 2.2. When the energy difference between different branches of the E(k) diagram (different curves in Figure 2.2) is large on a scale of the potential energy, then the behavior is essentially free electron like. However, near the intersection of two curves the energies are modified. Approximating the periodic potential with only its first Fourier component, then Equation 2.8 yields two equations for the two curves, which can be represented in matrix form as:

$$\begin{vmatrix} E - E_1 & -U \\ -U^* & E - E_2 \end{vmatrix} = 0 , \qquad\qquad 2.9$$

where $E_1 = \dfrac{\hbar^2 k_1^2}{2m}$ and $E_2 = \dfrac{\hbar^2 k_2^2}{2m}$ are the free-electron-like behaviors for the two curves (subscripts 1 and 2) near the meeting point, E is the energy at the meeting point, and U is the first Fourier component of the periodic potential. Note that at the intersection, $k_1 = k_2$ so that $E_1 = E_2$. This condition has the solution

$$E = \dfrac{\hbar^2 k^2}{2m} \pm |U| . \qquad\qquad 2.10$$

See Figure 2.3 for an illustration of this situation. The result has two implications that are important:

- Waves interact with each other to raise or lower their combined energies. Graphically, when curves on the E(k) diagram intersect they may interfere resulting in local changes in their energies. Note that

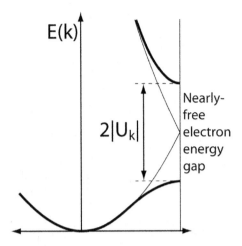

Figure 2.3: A schematic diagram showing the modification of the E(k) relationship near the zone boundary. (Gap is exaggerated to enhance visibility.)

waves which are perpendicular to each other can not interact and the branches of the E(k) diagram cross without modification.

- When the periodicity of the electron wave matches the periodicity of the lattice, the electrons are diffracted by the lattice (the waves interact strongly with the periodic potential and the Fourier component of U is likewise large).

The energy perturbation given in Equation 2.10 technically applies only at the boundary, $k_1=k_2$. As one moves away from the boundary, the magnitude of this interaction decreases quadratically with k. This is the same behavior as at the bottom of the free-electron curve. Another way to view this behavior is that near any extremum (maximum or minimum) of an arbitrary function, a power-law expansion of that function is always quadratic. Thus, near enough to any local maximum or minimum of an E(k) diagram, the behavior of an electron will always appear free-electron like. This provides a partial, if circular, justification of the approximations made above. A rigorous justification is provided by quantum mechanical perturbation theory and may be found in most quantum mechanics texts.

A somewhat more visual representation of the wave functions of the nearly free electron model is shown in Figure 2.4. The wave vector **k** of the electron wave $\psi(\mathbf{k})$ exactly matches the periodicity of the reciprocal lattice at any diffraction condition in a given direction (i.e.: it is a translation vector of the reciprocal lattice). As the wave approaches resonance with the lattice, the electrons interact increasingly strongly with the lattice potential. The magnitude of the lattice potential, U, then becomes

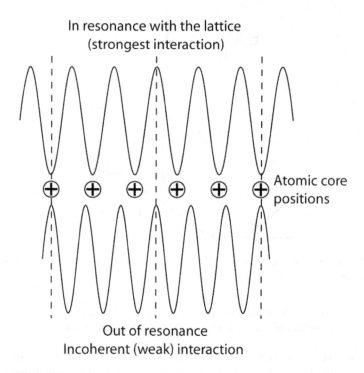

Figure 2.4: A schematic of the amplitudes of the wave functions of two waves. The top wave function shows the resonance of the zone boundary. The bottom wave function has a 90% shorter wavelength. The interaction of the wave with the atomic potential changes the electron energy.

critical. A strong potential affects the energy of the electron waves more, forcing a greater change in wave vector (or momentum) for a given change in kinetic energy. Thus, the parabolic band stretching away from the resonance curves more gradually. Electrons in such a material give the appearance of having a higher mass.

2.1.4 Energy vs. momentum in 3d

So far, we have considered only the nature of electron waves in a simple one-dimensional periodic lattice. Representing the three dimensional behavior of waves is significantly more difficult than representing one-dimensional wave behaviors, although the mathematics is not different. The first problem is to decide what are the wavelengths for diffraction from the lattice in multiple directions. This is solved by the Brillouin zone representation of the lattice.

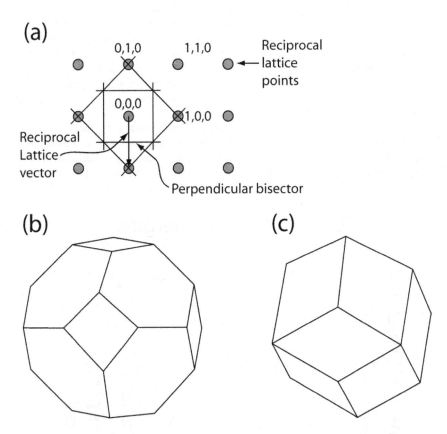

Figure 2.5: (a) The method for constructing the Brillouin zone for a square planar lattice and the first Brillouin zone for (b) a face-centered cubic crystal and (c) a body-centered cubic crystal. For the fcc crystal the diamond-shaped faces of the Brillouin zone are along cube axes, [100]-type directions, while the hexagonal faces are along [111] cube diagonals. For discussion of the [100], [111], and other crystal indices, see Chapter 4.

To construct this representation, one begins by calculating the reciprocal lattice of a given crystal structure (see Section 4.1.2). As described above, the reciprocal lattice gives a picture of the resonant wavelengths of the lattice in given directions. The second step is to determine the volume of reciprocal space closest to a given reciprocal lattice point. This begins by drawing lines (reciprocal lattice vectors) from the point at the origin of reciprocal space to all other points (Figure 2.5a). Each line is then bisected by a plane perpendicular to it. The volume of a given unit cell of reciprocal space is defined by the smallest volume contained within any combination of planes. This volume is referred to as the first Brillouin zone.

Two examples of first Brillouin zone shapes are shown in Figures 2.5b and c. The zone boundary points along primary symmetry directions for each crystal structure are traditionally labeled with letters. The point of zero crystal momentum is at the center of the Brillouin zone and is labeled Γ. The typical representations of energy band structures in three dimensions present the major trends along some of these symmetry directions but not over all of three-dimensional space.

If one carries out the free electron calculation for three dimensions, the complexity of the situation becomes clear, see Figure 2.6. The drawing shows slices through the first Brillouin zone of an fcc crystal along selected symmetry directions. These directions provide a relatively complete view of what the bands are like, although

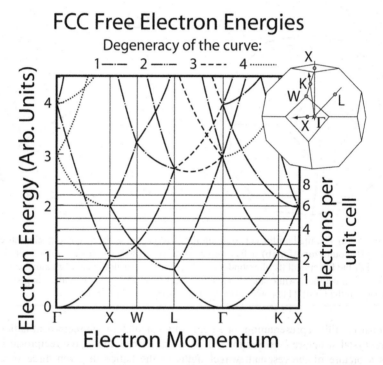

Figure 2.6: Shows the three-dimensional free-electron behavior plotted for a face-centered cubic lattice. Note that the X point is the zone boundary along [100], the L point along [111] and the K point along [110]. Γ is the zero momentum point at the center of the Brillouin zone where the electron wave has no momentum from the perspective of the crystal lattice (it is a standing wave with the lattice periodicity). Figure from Reference 10, Herman, F. in *An Atomistic Approach to the Nature and Properties of Materials* Pask, J.A., Editor New York: Wiley, 1967. Copyright 1967, John Wiley, used by permission.

this does not cover all of three-dimensional reciprocal space. Two typical energy band structures for the real semiconductors GaAs and Ge in three dimensions are shown in Figure 2.7.

(a)

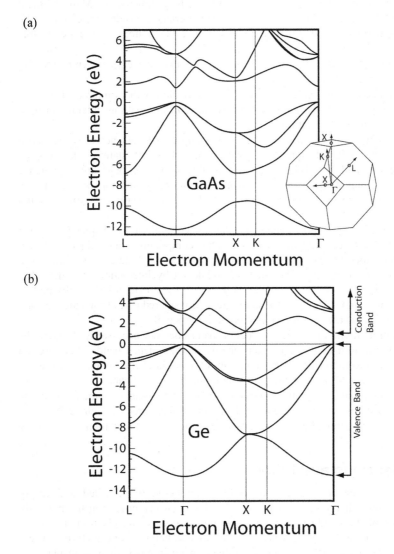

(b)

Figure 2.7: Energy vs. momentum diagram in three dimensions for (a) GaAs and (b) Ge. The directions in the Brillouin zone are shown in the inset and the letters indicate the position in momentum space. Figures adapted with permission from Chelikowsky, J.R. and Cohen, M.L. *Phys. Rev. B,* **14**, 556-582 (1976). Copyright 1976, American Physical Society.

The primary points to notice about the three-dimensional structures compared to the one dimensional structure (Figure 2.3.) are as follows:

- There are many branches to the diagrams in three dimensions. Many of them appear free electron like locally for the real materials. Consider, for example, the lowest-energy state in the real materials and the bottom of the nearly-free electron diagram.
- Behaviors along different directions vary because the crystal symmetry and resonant distances change.
- Some branches cross each other without interaction. Others interact, leaving a gap in the states at that point. Branches that do cross without interaction have perpendicular wave functions.
- Electrons fill states from the lowest energy up to a level known as the Fermi energy. In semiconductors, the lowest empty band of states makes up the conduction band. The filled states nearest the conduction band are the valence band and contain the valence electrons. Traditionally, the top of the valence band defines the zero energy, although band structure energies are often plotted relative to the vacuum level.
- A band of states (due to valence s-orbitals as we shall see in Chapter 5) lies just below the valence band in GaAs and other compound semiconductors. When this band does not connect to the upper valence band states it is formally a shallow core level, even though it is broadened by interaction with the other valence electrons. When a connection does occur as in Ge, it is formally part of the valence band. Usually these states are all considered part of the valence band whether there is a gap or not as the separation is small.
- The minimum energy in the conduction band does not necessarily occur at the same momentum as the maximum energy in the valence band. Therefore, an electron at the conduction band minimum may not have the same momentum as the lowest-energy electron vacancy (a hole) in the valence band.
- When the conduction band minimum occurs at the same momentum as the valence band maximum the semiconductor is termed "direct" (Figure 2.7a)
- An "indirect" semiconductor has these extrema at different momenta (Figure 2.7b).

2.1.5 Electrons and holes

In a semiconductor or insulator, there is a gap between the filled and empty electronic states. Charge can be carried either by electrons moving in the normally empty states of the conduction band or by the absence of an electron (a "hole") moving in the normally filled states of the valence band. If one pictures current flowing in these bands as equivalent to water flowing in pipes, then an insulator or semiconductor can be viewed as two sealed pipes, one completely full of water and one completely empty. Even if one tips the pipes, no water flows. However, either droplets of water in the empty pipe or bubbles in the full pipe, if present, can allow net water transport when the pipes are tilted. The behaviors that apply to electrons also apply to holes (bubbles). Holes are referred to as "quasi-particles" because they do not exist in isolation. They are ghosts of electrons that are only known by the electron's absence in the solid. Nonetheless, in a solid, holes appear to have charge and mass and to be scattered as if they were real Fermions. For our purposes we will treat holes as if they were real particles.

To understand the transport of current it is important to keep in mind the fundamental nature of electrons and holes. Two of the more significant consequences of quantum theory are that electrons and other Fermions have a "spin" quantum number of $\pm 1/2$, and that no two spin $\pm 1/2$ particles can occupy the same state (can have the same set of values for their quantum numbers) at a time (the Pauli exclusion principle). Thus, filling a set of states begins with the lowest energy and adds electrons into higher and higher energy states as the more desirable levels become filled. This is not the case for integer spin "Bosons" such as photons, deuterium atoms, and helium 4 atoms. Bosons can all fill the same low energy state. In superconductors electrons form pairs having integral spin. These "Cooper" pairs act as Bosons with remarkable effects on the electronic properties of the material. Holes behave as do electrons but prefer states with higher energies (from the perspective of electrons) as do bubbles in water.

Electrons in solids are not scattered by each other without some external mediating force that allows one electron to lose energy or change direction by transfer of that energy or momentum to another electron. The possible transitions are limited by availability of empty and filled states (because electrons are Fermions), and thus so are the scattering events. Clearly then it will be important to know something about the distribution of these filled and empty states. The energy and momentum needed for scattering events is stored in the form of heat in the solid and is exchanged by absorption or emission of phonons (lattice vibrations) and photons. The probability of obtaining a given amount of energy from the remainder of the solid decreases exponentially as the energy needed increases. Likewise, the availability of phonons varies with temperature.

As noted above, at zero Kelvin, electronic states are occupied beginning with the lowest energy state and continuing upward until all electrons in the solid are accounted for. Abruptly, one would go from all states being filled to all being empty in this case. At higher temperatures, heat in the lattice puts random amounts of energy into individual electrons. Some of these electrons may have extremely high energies at times. The probability of finding an electron as a function of energy is given by the Fermi function:

$$f(E) = \frac{1}{1 + e^{(E - E_f)/k_B T}} , \qquad\qquad 2.11$$

where the electron has an energy E relative to a reference state, the "Fermi energy" E_f. k_B is the Boltzman constant, and T is the temperature in Kelvin. There are several important facts about the Fermi function that are observable with brief inspection. First, when $E = E_f$ then the exponential is always 1 and the function value is 1/2. This is, in fact, the means of defining and determining the value of E_f. The Fermi energy is that energy for which $f(E_f)=0.5$. When E is very different from E_f then Equation 2.11 can be approximated by a Boltzmann distribution:

$$f(E) \approx e^{-(E-E_f)/k_BT}$$

2.12

It is common to use this distribution in all cases. However, one should bear in mind that this is an approximation and that for some situations its use may not be appropriate. The development of a distribution consistent with the Fermi function requires very little time (typically picoseconds) as electrons rapidly exchange photons and phonons with the remainder of the solid until this distribution is established.

One might ask how one can account for a situation (common in most semiconductors and insulators) in which the Fermi energy is within the energy gap. In this case, there are no states at the Fermi energy itself. The Fermi function remains valid and well defined and the probability of finding an electron drops exponentially moving away from E_f. Whatever remains of the function value above zero for electrons (or below unity for holes) describes the number of electrons/holes that are present in states in the bands at given energies, temperature, and Fermi level. If the distribution of states were the same in the conduction and valence bands and no states existed within the energy gap, then charge neutrality would require that the Fermi energy be near the gap center, as discussed in Section 2.2.1.

To establish the dynamic equilibrium distribution defined by Equation 2.11, electrons are constantly being transferred from the valence band to the conduction band by energy/momentum absorption. This is known as generation of free carriers. Carriers are also constantly falling back into the holes created by the earlier band-to-band transfer. This is recombination. Generation occurs by absorption of light or heat, while recombination releases light, in so-called "radiative" recombination, or heat (non-radiative recombination).

Generation by absorption of energy from the solid is "thermal" generation and from absorption of external light shining on the material is "optical" generation. The thermal generation rate, g_{th}, increases approximately as a Boltzmann distribution with the usual temperature dependence and an activation energy equal to that of the energy gap, E_g, of the material,

$$g_{th} = Ce^{-E_g/k_BT},$$

2.13

where C is a proportionality constant. The constant may be determined based on the requirement that recombination must equal thermal generation if Equation 2.11 is to hold. For optical generation,

$$g_{op} = \int_0^\infty \frac{G(v)Q(v)}{hv}dv,$$

2.14

where $G(\nu)$ is the optical intensity at frequency ν, and $Q(\nu)$ contains the quantum-mechanical conservation rules and other aspects related to the band structure of the material and the optical absorption coefficient. Because optical generation does not depend upon temperature, a carrier distribution determined primarily by optical rather than thermal generation is not described by Equation 2.11.

Recombination depends upon the product of electron and hole densities and a factor giving the probability of a given energy release mechanism (for example, radiative or non-radiative). This probability includes the same quantum-mechanics-determined factors as in Equation 2.13 or 2.14, and incorporates the nature of the band structure. The recombination rate can be expressed as

$$r_{rec} = Q'np, \qquad\qquad 2.15$$

where Q' is the a recombination rate per electron-hole pair including all possible recombination processes. Recombination occurs rapidly in direct-gap semiconductors (Q' large) and slowly in indirect-gap semiconductors (Q' small).

The details of the calculation of the factors C, Q, and Q' above are complex and beyond what we will need here. However, they can be found described in texts on optical generation and recombination. See, for example recommended readings by Bube or Ashcroft and Mermin.

2.1.6 Direct and indirect semiconductors

The "direct" and "indirect" behaviors mentioned above are sufficiently important that they deserve special mention. The critical aspects of the energy band structures of these two types of semiconductor are shown schematically in Figure 2.8. The minimum energy of the conduction band in indirect materials is at a different momentum than that of the maximum energy of the valence band. Electrons in the conduction band rapidly relax to the minimum band energy. Holes equally rapidly move to the maximum energy of the valence band. Therefore, electrons and holes do not normally have the same momentum in an indirect semiconductor while in a direct-gap material these momenta are equal. This has consequences for the minority carrier lifetimes and optical properties of semiconductors.

Optical absorption/emission involves absorption/creation of a photon with a consequent change in energy of an electron, usually resulting in the transfer of that electron to/from the conduction band from/to the valence band. However, because photons have almost no momentum, only vertical transitions on an E(k) band diagram are allowed in purely optical processes. The only alternative to this is the rare case when a phonon is present together with a photon. Phonons have a large momentum (they involve collective motion of massive atoms) and relatively low energy. (See Section 2.3.3.1 for more discussion of phonons.) For practical purposes one can assume that photons lead to vertical transitions among bands on the E(k) diagram while phonons make horizontal transitions (see Figure 2.9.)

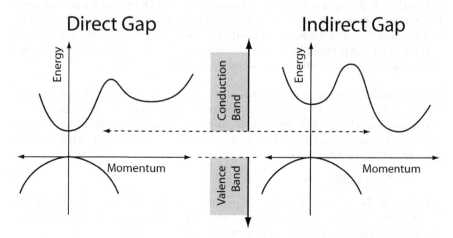

Figure 2.8: A schematic of the difference between a direct gap and an indirect gap semiconductor.

To make a diagonal indirect transition, both an appropriate energy photon and an appropriate momentum phonon must be present together with the electron. This is referred to as a three-body interaction because there are three particles (electron, phonon, and photon) participating. Such collisions are over 1000 times less likely than a simple electron-photon interaction at common temperatures. This means that electrons and holes of different momenta do not recombine rapidly. Typically, electrons and holes in pure direct-gap semiconductors last no more than ~10^{-8} s,

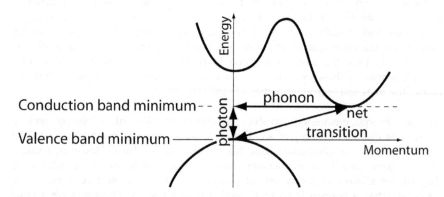

Figure 2.9: A schematic of the transitions produced by a phonon (horizontal) and a photon (vertical) on an E(k) diagram. Indirect transitions require both and thus are less likely.

while in indirect-gap materials the lifetime for free carriers can be 1000 times longer. Furthermore, when recombination does occur in indirect-gap materials, most of the energy is usually emitted in the form of heat in the solid rather than as light. This is why most light-emitting devices use direct-gap materials. Generation of carriers in indirect-gap materials is equally difficult so indirect materials tend to have low optical absorption coefficients below the point where direct-transitions become possible.

Because in indirect-gap semiconductors generation and recombination of carriers are equally difficult, even in indirect-gap materials the distribution of electrons and holes is governed by the Fermi function. Because the Fermi energy can never be farther than half the energy gap from one band edge or the other, the density of carriers in an indirect-gap material is still determined by the *minimum energy gap* in spite of the difference in momenta of the band minima in indirect-gap materials.

The consequences of the need for a phonon to permit an electron and a hole to interact in indirect-gap materials include:

- The maximum recombination time for electrons and holes is much longer (of the order of microseconds) in an indirect-gap material than in a direct gap semi-conductor (nanoseconds).
- The probability of a photon emission or absorption event for energies between the minimum gap energy and the minimum direct gap energy (the lowest energy for a vertical transition) is far below that at or above the minimum direct gap energy.

2.1.7 Effective mass

The nearly-free electron model of solids shows that electrons are strongly affected by the lattice in which they move. The interaction of the electron with a solid modifies the relationship between energy and momentum. Recall that the total energy of an electron is given by the sum of kinetic and potential energies. Because the interactions of an electron with the lattice potential and the lattice spacing along a given direction vary from material to material, the change in kinetic energy with wavelength changes from one material to another. In some materials the energy gain is small for a given momentum increase. In others, it is larger for the same added momentum. Given the classical relationship between energy and momentum, $E=p^2/2m$, the change in proportionality constant between energy and momentum appears as a change in the effective mass of the electron. The exact form of this relationship can be developed from the above energy band relationship as follows.

The energy of a free electron Bloch wave is related to the wave vector as:

$$E = \frac{\hbar^2 k^2}{2m^*}$$

2.16

where m^* is the effective electron mass. Differentiating both sides with respect to k,

$$\frac{k}{m^*} = \frac{1}{\hbar^2}\frac{dE}{dk}$$ 2.17

and differentiating a second time,

$$\frac{1}{m^*} = \frac{1}{\hbar^2}\frac{d^2E}{dk^2}.$$ 2.18

In other words, the effective mass of an electron Bloch wave is inversely related to the curvature of the E(k) diagram. Strong curvatures mean small effective masses and vice versa. The effective mass is generally expressed in units of electron rest mass in a vacuum, i.e. m^*/m_0.

The effective mass concept is only defined near a band maximum or minimum where the energy can be approximated as a quadratic function of wave vector, k. Note that near a minimum in energy bands the E(k) function is concave up and $1/m^*>0$ while near a maximum, $1/m^*<0$. The negative effective mass near a maximum indicates that charge is being carried by holes rather than electrons.

2.1.8 Density of states

One of the most fundamental properties of a material, which determines many of its properties, is its density of states. This refers to the number of states per unit energy in the band structure. To put this in more visual terms, if one takes a thin horizontal slice through an energy band structure diagram such as those shown in Figure 2.7, the "blackness" of the slice (the amount of band line that occurs in that slice) is the density of states. The density of states for a complex band structure can be computed and is normally developed as part of calculations describing a semiconductor (or other material). However, it is not straightforward to present a simple formula for the density of states of such a real system. We will have to be content with a derivation of the density of states for a free electron.

For a free electron in three dimensions the energy is given by

$$E = \frac{\hbar^2}{2m^*}\vec{k}^2 = \frac{\hbar^2}{2m^*}\left(k_x^2 + k_y^2 + k_z^2\right)$$ 2.19

where k_x, k_y, and k_z are the components of the wave vector **k** along the three coordinate directions. If one thinks of wave vectors as resonant states of the crystal, the longest wavelength or lowest energy state would be a half wavelength across the entire width of the crystal, L. Higher energy modes are integer multiples of this state. Therefore, the separation of states along any given direction, i, is π/L_i and the wave vector is then $k_i=n_i\pi/L_i$. Therefore, the energy can be rewritten (with $h = 2\pi\hbar$) as:

$$E = \frac{h^2}{8mL^2}\left(n_x^2 + n_y^2 + n_z^2\right) = k_F^2 E_0 \qquad\qquad 2.20$$

where n_x, n_y and n_z are the indices of the reciprocal lattice points inside a sphere of radius $k_F = \sqrt{n_x^2 + n_y^2 + n_z^2}$. $E_0 = \dfrac{h^2}{8m^*L^2}$ is the lowest energy state of the system. The number of electrons that can be accommodated in states with energy E or less is:

$$N(E) = 2\left(\frac{1}{8}\right)\left(\frac{4\pi}{3}k_F^3\right) = \frac{\pi}{3}\left(\frac{E}{E_0}\right)^{\!3/2} \qquad\qquad 2.21$$

For any process to occur, an electron must be present in an initial state with energy E_i. The higher the density of states at this energy, the more electrons can participate in the process. The electron is changing energy by an amount ΔE. Therefore, there must be an empty state with energy $E_f = E_i + \Delta E$ to receive it. In addition, momentum must be conserved. The process begins from any initial state E_i such that the rest of the above criteria are met. The overall rate of the process is given by an integral over all initial states i containing electrons.

Here the factor of two is because each state has two possible electron spins and the 1/8 is because we must take only positive values of n_x, n_y, and n_z. The density of these states g(E) per unit volume of reciprocal space in an energy interval dE is given by (1/V)(dN/dE), where $V=L^3$ is the crystal volume (for a cube-shaped solid), or:

$$g(E) = \frac{1}{V}\frac{dN}{dE} = \frac{\pi}{2L^3}E_0^{-3/2}E^{1/2} = \frac{\pi}{2}\left(\frac{8m^*}{h^2}\right)^{\!3/2}E^{1/2} \qquad\qquad 2.22$$

This density of states will apply to any band extremum where the band can be approximated at least locally with a quadratic dependence of energy on momentum (free-electron-like). Since all band edges will have this general behavior, Equation 2.22 provides an approximate picture of the number of states per unit energy near a band edge.

The importance of the density of states may be found in the calculation of the rate of any process in a solid, from scattering of an electron off a defect or another electron to absorption and emission of light. The rate of such a process is given in its most general form by "Fermi's Golden Rule". Mathematically, the rate of a process (**H**) moving an electron from state ψ_i to state ψ_f may be written in symbols as follows:

$$r_{i \to f} = \int \left(g(E_i)f(E_i)\right)\left(g(E_f)(1 - f(E_f))\right)\left\langle \psi_i \,|\, \mathbf{H} \,|\, \psi_f \right\rangle \delta(E_i - E_f + \Delta E)dE_i, \qquad 2.23$$

where E_i and E_f are the initial and final state energies, ΔE is the change in energy of the process, g_i and g_f are the density of states at energies E_i and E_f, $f(E_i)$ and $f(E_f)$ are the corresponding Fermi functions, and $\delta(E_i\text{-}E_f\text{+}\Delta E)$ is a Kroniker delta function which enforces conservation of energy. The expression $\langle\psi_i|\mathbf{H}|\psi_f\rangle$ is a mathematical function which enforces the conservation of momentum of the electrons and other quantum mechanical selection rules in process $\mathbf{H}$. In words this formula states the following.

Equation 2.23 is used in calculating quantities such as $Q(\nu)$ in Equation 2.14 and many other values. It is fundamental to detailed analysis of the physics of solids. Therefore, the density of states is an essential element of understanding the general behaviors of materials. We will encounter it regularly in Chapter 3 where the effective density of states at the band edge is fundamental to analysis of the operation of electronic devices.

2.2 INTRINSIC AND EXTRINSIC SEMICONDUCTORS

Semiconductors are called semiconductors because their ability to conduct electricity is neither very great nor very small. Indeed, the conductivity of a typical semiconductor can be controlled by temperature to such an extent that most can appear to be insulators at sufficiently low temperatures and metals at sufficiently high temperatures. Impurities can also play a role. A typical conductivity behavior for a semiconductor containing added impurities is shown schematically in Figure 2.10. A semiconductor free of significant impurities is termed "intrinsic" while those doped with impurities are termed "extrinsic". Both of these behaviors are essential to the operation of microelectronic devices.

2.2.1 Intrinsic semiconductors

An intrinsic semiconductor has no impurities and the number of electrons, n, in the conduction band exactly matches the number of holes, p, in the valence band, n=p. The number of electrons or holes is named the intrinsic carrier concentration n_i. This carrier concentration is given by the probability that a state at energy E is filled (as given by the Fermi function, Equation 2.11) multiplied by the density of states at that energy (Equation 2.22 for free electrons) integrated over all energies at or above the conduction band edge:

$$n_i = \int_{E_C}^{\infty} f(E)g(E)dE \qquad\qquad 2.24$$

Substituting from Equations 2.11 and 2.22, Equation 2.24 becomes:

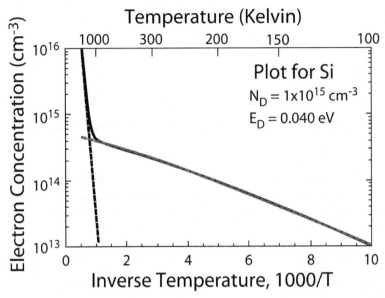

Figure 2.10: A plot of the electron concentration in a piece of silicon doped with 10^{15} cm^{-3} donor atoms having an ionization energy of 0.04 eV. The steep slope at high temperature (low inverse temperature) corresponds to the intrinsic behavior for carriers crossing the energy gap. The lower temperature behavior occurs in the presence of the 10^{15} cm^{-3} electron donors. The slopes of the two curves correspond to the 1.1 eV energy gap and the 0.04 eV donor ionization energies, respectively.

$$n_i = \frac{\pi}{2}\left(\frac{8m_e}{h^2}\right)^{3/2} \int_{E_C}^{\infty} \frac{E^{1/2}\,dE}{1+e^{(E-E_F)/k_BT}} \qquad 2.25$$

This equation can be simplified by noting that the band edge energy is typically far from the Fermi energy in units of k_BT, in an intrinsic material. Therefore, the Boltzmann approximation (Equation 2.12) can be used. With the further definition:

$$N_c = 2\left(\frac{2\pi m_e^* k_B T}{h^2}\right)^{3/2}, \qquad 2.26$$

which is the effective density of states at the conduction band edge, after some algebra, we find:

$$n_i = N_c e^{-(E_C-E_F)/k_BT} \qquad 2.27$$

We can circumvent the issue of the Fermi energy by noting that a similar definition can be made based n_i=p, and with a similar definition for the effective valence-band density of states, N_V:

$$n_i = N_V e^{-(E_F - E_V)/k_B T},$$

2.28

from which, multiplying Equations 2.27 and 2.28,

$$n_i^2 = N_C N_V e^{-(E_C - E_V)/k_B T}.$$

2.29

Taking the square-root gives a value for n_i which does not depend upon E_F:

$$n_i = \sqrt{N_C N_V} e^{-E_{gap}/2k_B T}$$

2.30

The most important point to note about Equation 2.30 is that n_i depends exponentially with temperature on *half* of the semiconductor energy gap. Narrow-gap semiconductors will have large intrinsic carrier concentrations while wide-gap materials will have fewer mobile carriers at a given temperature.

As a final point, note that if one knows that $n=p=n_i$, then from Equations 2.28 and 2.30, the Fermi energy of an instrinsic semiconductor (also known as the intrinsic energy E_i) can be derived. With some algebra, this can be shown to be:

$$E_F = E_i = \frac{E_{gap}}{2} + \frac{3k_B T}{4} \ln\left(\frac{m_h^*}{m_e^*}\right).$$

2.31

In other words, the Fermi energy is near the middle of the energy gap and deviates from the exact center by a factor that depends logarithmically on the ratio of effective masses of the two bands and linearly on temperature.

2.2.2 Extrinsic semiconductors

When impurities are added to a semiconductor the bonding pattern in that semiconductor is modified. This will be discussed extensively in Chapter 7. In the mean time, it is useful to have a general idea of the effect of impurities on carrier concentrations. Impurities that are added intentionally to control the carrier concentrations are called "dopants". When a dopant has the same basic electronic structure as the atom it replaces (for example, partially-filled "s" and "p" orbitals for dopants in Si), the bonding behavior is substantially unchanged. As we will see, this usually results only in small changes in bonding and primarily affects only one band edge region. If the atom then has one more or one fewer electron that the atom it replaces, this state has an extra electron or hole in it, not present in the host semiconductor. The extra electron or hole can often escape the impurity atom and move freely through the semiconductor causing a change in conductivity.

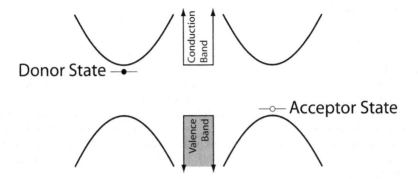

Figure 2.11: Momentum space diagrams of dopant states in the energy gap of a direct semiconductor. Note that the dopant states are somewhat extended in momentum space (they are not points) because they are somewhat localized in real space. Usually one represents the bands as ranges of energies where states are present, as shown at the center of the figure and momentum is largely ignored.

A single impurity atom in a direct gap semiconductor might produce a dopant state such as those shown in Figure 2.11. If the atom has one fewer electron than the host, it is likely to accept an electron from the relatively electron-rich semiconductor. Consequently, it is referred to as an acceptor dopant. If the atom has one extra electron it is generally a donor.

Electrons or holes in these dopant states can be easily released into the bands if the energy of the state is within a small number of thermal energy units (i.e. $k_B T$) of the appropriate band edge (the valence band for acceptors, the conduction band for donors). The probability of finding a carrier is calculated based on the Fermi function and the separation of the dopant state from the band edge. Thus, for donors:

$$n = \frac{N_D}{e^{(E_C - E_D)/k_B T} + 1} \approx N_D e^{-(E_C - E_D)/k_B T} . \qquad 2.32$$

A similar equation can be defined for holes,

$$p = \frac{N_A}{e^{(E_A - E_V)/k_B T} + 1} \approx N_A e^{-(E_A - E_V)/k_B T} . \qquad 2.33$$

In these equations, E_D and E_A are the donor and acceptor energies, E_C and E_V are the conduction and valence band edge energies, and N_D and N_A are the concentrations of donors and acceptors, respectively. The energy differences are the energies necessary to transfer a carrier from the dopant state to the band edge. These equations assume that the Fermi energy is at E_A or E_D, the temperature is moderate, and that the doping concentration is low.

It is typical to assume that all dopant atoms are ionized in a semiconductor at room temperature. However, this is not the case. From Equations 2.32 or 2.33, when the dopant level lies $k_B T$ from the band edge, only 27% of dopant atoms are typically ionized. When the energy difference is half of $k_B T$, only 38% are ionized. Because it is uncommon to find isolated dopant states much closer than $k_B T$ from the band edges, even the best dopants are rarely fully ionized. This becomes significant when working with compensated semiconductors where both donors and acceptors are present together. In these materials donors will spontaneously transfer electrons to acceptors with lower energy states, leading to full ionization of either donors or acceptors, whichever is the less common.

When one adds dopant to a pure semiconductor the Fermi level shifts, approaching the doping state energy. A brief examination of Equation 2.27 shows that if the Fermi energy rises in the band gap, the concentration of holes in the valence band must shrink. While the intrinsic condition n=p no longer holds, the condition

$$np=n_i^2 \qquad\qquad\qquad 2.34$$

is still true. Therefore, doping a semiconductor with an electron donor (making it "n-type") increases the concentration of free electrons and decreases the concentration of free holes by the same factor. In this case, electrons are called "majority carriers" and holes are "minority carriers". The same behaviors hold when holes are the majority carrier due to acceptor doping.

2.3 PROPERTIES AND THE BAND STRUCTURE

Our objective in designing materials is to understand what determines their useful properties and to change them such that they better meet our needs. The preceding sections have described some of the basic physics that underlies useful properties. This section describes in more detail how these properties arise. In later portions of the book we will see how specific properties are obtained by design in specific real materials. While real phenomena are more complex than the simple models presented here, real behaviors can generally be understood based on the concepts outlined.

2.3.1 Resistance, capacitance, and inductance

Resistance, R, capacitance, C, and inductance, L, are all observable phenomena used in circuit elements. They result from the motion of electrons in solids and their interaction with each other and with the atoms surrounding them. Each also depends upon the geometry of the circuit element producing the effect. The basic materials properties, however, are dependent only upon the electronic structure in that material and not on the geometry. These underlying properties are the resistivity, ρ, of a conducting material in a resistor, the dielectric constant, ε, of the insulating material in a capacitor, and the permeability, μ_B, in an inductor. The relationships are as follows.

$$R = \frac{\rho\ell}{A},$$ 2.35

where ℓ is the length of the conductor and A is its cross-sectional area. The resistivity, ρ, of a conductor is inversely related to its conductivity, σ as ρ=1/σ. Likewise, C and ε are related through

$$C = \frac{\varepsilon A}{d}.$$ 2.36

Here A is the area of the capacitor plates and d is their separation or the thickness of the insulating (dielectric) material between them.

Much of the interaction of electrons is through their electric and magnetic fields. The latter is responsible for the phenomenon of inductance. It results from the flow of current around a loop or coil of wire. The magnetic field produced depends upon the permeability of the material inside the wire coil. The greater the energy stored in the magnetic field, the higher the inductance of the coil. The inductance of a solenoid depends upon the permeability as:

$$L = \mu_B \frac{N^2 A}{\ell},$$ 2.37

where N is the number of turns of wire in the coil, A is the coil area, and ℓ is the coil length.

As we will see below, resistivity is directly related to the mobility of charges in the material, the dielectric response to its polarizability, and the permeability to the magnetic moment of the constituent atoms.

2.3.1.1 Mobility and electrical conductivity

Conduction in solids can occur by motion of electrons or charged atoms (ions). Most of this section is devoted to electronic conduction but a brief mention of ionic conduction is in order.

Ionic conduction occurs by diffusion ionized atoms, generally at high temperatures, although we will see it again in organic materials at room temperature. Inorganic solids with strong ionic bonding and high ionic diffusivity conduct charge well at sufficient temperatures. These materials usually include dopant ions that induce formation of charged vacancies on the same sublattice to compensate for dopant charge. Either the vacancies or the charged ions may then diffuse, carrying current.

Fast ionic conductors are used as solid electrolytes in fuel cells and sensors. The search for fast ion conductors operating near room temperature is a matter of current electronic materials research. The practical small-scale application of some types of fuel cells as replacements for batteries may hinge on success in this search.

Much more common and more relevant to most electronic applications is electronic conductivity. When a material is placed in an electric field, $\mathbf{E}$, a current density, $\mathbf{J}$, given by $\mathbf{J}=\sigma\mathbf{E}$, is induced. Here σ is the electrical conductivity:

$$\sigma = q(\mu_p p - \mu_n n),$$

2.38

where q is the moving carrier charge, μ_n and μ_p are the electron and hole mobilities, and n and p are the electron and hole mobile carrier concentrations, respectively. Both the number of carriers and their mobility depend upon the energy band structure of the solid.

The number of carriers moving in a metal depends upon the density of carriers near the Fermi energy. To accelerate a carrier there must also be an empty state to move into with higher energy. The majority of carriers in the material come from or accelerate into states within $\sim k_B T$ of the Fermi energy. The effective density of mobile carriers in a metal is therefore approximately the number within $k_B T$ of the Fermi energy, of the order of one per atom or $>>10^{22}$ cm^{-3}.

When an electron is placed in an electric field, it accelerates continuously until it bumps into something. The acceleration, $\mathbf{a}$, is given by $\mathbf{a} = q\mathbf{E}/m$ where m is the mass of the moving carrier. If the average time between collisions is τ, the "drift" velocity achieved during the acceleration time is $\mathbf{v}_d = \mathbf{a}\tau$. Classically one would expect $\mathbf{v}_d = \mathbf{a}\tau/2$ but if one conducts a detailed calculation doing the average over all trajectories after true times, not average times, then the velocity is increased by a factor of two. Substituting from the formula relating $\mathbf{a}$ with $\mathbf{E}$,

$$\mathbf{v}_d = \frac{q\tau}{m^*}\mathbf{E}.$$

2.39

The quantity $q\tau/m^*$ is the carrier mobility, μ_n or μ_p.

The two variable components of mobility are the mean time between collisions and the effective mass of the moving charge. The mean time between scattering events depends upon how many things there are to bump into in the solid and how often each gives rise to a collision. The three primary scattering centers are other electrons, the atoms of the solid, and defects in a crystal. The first two are generally determined by fundamental properties of the material, while the latter is controlled by the density of imperfections. Defect cross-sections (the probability that a carrier passing through a given area containing the defect will be scattered) are the calculated quantities determining how effectively a defect will scatter a carrier.

Scattering of electrons off other electrons may be estimated from Fermi's Golden Rule (Equation 2.23). In words, one finds that the more electrons there are to scatter the moving charge and the larger the number of states the moving carrier can scatter into, the higher the scattering rate and the shorter the mean time between scattering events. These factors can be related in metals to the density of states at the Fermi energy, $g(E_f)$. When $g(E_f)$ is high there are many carriers to scatter from with energies just below E_f and many states to scatter into just above E_f. Metals such as Mo have low conductivities because they have high densities of states around E_f. Metals such as Cu, Ag, and Au are ideal compromises where there are high enough carrier densities for good conduction but low enough densities of states to keep the mean scattering times long. In metals, the mobility of electrons is generally low (less than 1 $cm^2V^{-1}s^{-1}$). In some semiconductor multilayer structures at low temperatures, by contrast, mobilities may exceed 10^6 cm^2 $V^{-1}s^{-1}$. In spite of this high mobility, the conductivity of these structures is modest because the carrier density is relatively low. Organic conductors tend to have both low mobilities ($<10^{-2}$ cm^2 $V^{-1}s^{-1}$) and low carrier concentrations. Thus, they are very poor conductors in most cases.

Scattering off atoms in the solid can be due to the presence of anomalously charged elements (dopants) that disrupt electrical periodicity. Scattering can also result from transfer of kinetic energy to or from the lattice in the form of phonons.

The details of scattering phenomena are beyond the scope of this book. Additional information may be found in the suggested readings.

As we saw in Equation 2.39, scattering is only part of the picture. Changes in carrier effective mass can also have an effect. Scattering rates often change by orders of magnitude as temperature or defect density change, while effective mass only varies by up to one order of magnitude. Therefore, the effective mass is usually less important than the scattering rate effect. Nonetheless, effective mass differences contribute to determining intrinsic mobility differences from one semiconductor to another and differences between electron and hole conduction within a given semiconductor. The effective mass is relatively easily determined, once one has a calculated band structure, and can be measured experimentally with good accuracy. Therefore, it is generally much better known than the scattering terms. Furthermore, the effective mass is a constant for a given material while the scattering rate is not.

The mobility can be connected to the diffusivity of a carrier. Diffusion depends upon the distance a carrier travels between collisions, as does mobility. Einstein developed a quantitative relation between the two quantities, showing that the mobility multiplied by the thermal energy (k_BT) was equal to the charge multiplied by the diffusivity. For example,

$$\mu_n k_B T = q D_n \qquad\qquad 2.40$$

for electrons. This relationship is most useful in balancing the diffusion current of electrons or holes in a concentration gradient against a drift current due to an electric field and is essential to understanding the behavior of diodes.

2.3.1.2 Dielectric constant, piezoelectric response, and permeability

The dielectric constant and permeability of materials are directly related to their ability to be polarized or magnetized, respectively. In general terms, polarizability is related to the magnitude of the motion of charge in a material resulting from the application of an external electric field. In such a field, electrons would naturally tend to move. Likewise, positive and negative ions would move and in opposite directions. The electrons and ions cannot move far without generating a large electric field of their own. Their motion results in the accumulation of negative charge in the material at the positive end of the external electric field (see Figure 2.12) and positive charge at the negative end, and reduces or cancels the applied field within the material. Thus, one converts the single long dipole associated with the external electric field into two smaller dipoles at each end of the dielectric. This increases the field strength outside and decreases it inside the dielectric. The increased field strength around the edges of the dielectric increases charge accumulation in surrounding materials, resulting in more capacitance. A similar phenomenon occurs in magnetic materials. Rather than charge accumulation, atomic magnetic moments align resulting in cancellation of the magnetic field within the material and increased magnetic fields around the edges. We will return to magnetic response of materials at the end of this section. For now we will consider the dielectric response in more detail.

The optical and dielectric properties of semiconductors are direct results of the dielectric constant, $\varepsilon = \varepsilon_0 \varepsilon_r$ (where $\varepsilon_0 = 8.85 \times 10^{-14}$ F cm^{-1} is the permittivity of free space and ε_r is the relative dielectric constant of the material in question), or the index of

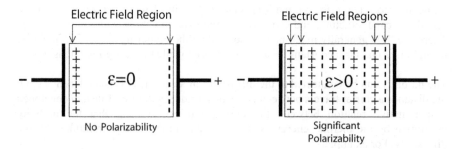

Figure 2.12: A schematic diagram showing the response of a dielectric material to an applied electric field.

refraction, $\eta = \sqrt{\varepsilon}$. To see how this relationship occurs, note that the redistribution of charge resulting from the applied field, E, results in a "displacement" field, D within a material of:

$$D = E + 4\pi P = \varepsilon E ,\qquad\qquad 2.41$$

where P is the polarizability of the material. The dielectric constant, in turn, can be written in terms of the polarizability of the solid as:

$$\varepsilon = (1 + 4\pi P / E).\qquad\qquad 2.42$$

The total polarizability is most conveniently represented in a crystalline material in terms of the polarization, p, of a unit cell of the lattice of volume v, as:

$$P = \frac{p}{v}.\qquad\qquad 2.43$$

The polarization of the unit cell depends upon the applied field. More field produces more polarization. The useful the field-independent materials property is the polarizability per unit cell, α, giving $p = \alpha E_{local}$. Here E_{local} is the electric field inside the material that produces the polarization of the unit cell.

The polarizability, α, is due to two contributions, atomic polarizability and displacement polarizability. The atomic polarizability is caused by the motion of the *electron* cloud in the material with respect to the ionic cores of the atoms on lattice sites. The displacement polarizability is similar but involves motions of charged *atoms* with respect to one another. Consequently, it requires that there be two or more types of atoms in the lattice having a net polarity or difference in electric charge (i.e. at least partial ionic bonding) and results from motion of the negative *atoms* with respect to the positive *atoms*. These two polarization responses are shown schematically in Figure 2.13.

The polarization behavior resulting from atomic polarizability can be estimated by representing the electron cloud as a negatively-charged shell connected to a positive core by springs (Figure 2.14). The very heavy mass of the atomic core relative to the electron cloud means that the electrons can be assumed to be the only particles moving. The spring constant in the model is related to an observable resonant frequency ω_0, which is typically consistent with the frequency of high-energy photons. Thus, for all normal electronic devices, we can assume that the atomic polarizability is constant. However, in optical applications this polarizability can sometimes vary significantly as one passes through the resonant frequencies of the system. Atomic polarizabilities resulting from the displacement of the electron cloud with respect to the atom cores are of the order of 10^{-24} cm^{-2} V^{-1}.

The displacement polarizability results from motion of heavy atoms with respect to each other. These particles, being much more massive than electrons, respond to fields much more slowly. The same general model of masses connected by springs

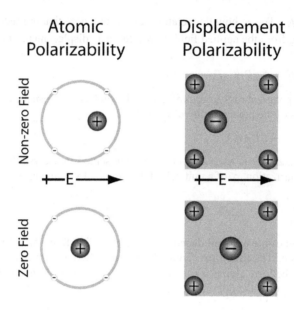

Figure 2.13: A schematic diagram showing the two major components to polarizability. In atomic polarizability, the electron cloud is displaced with respect to the remaining positive components of the atoms (the protons and the non-valence electrons). In displacement polarizability, atoms of different charge (positive or negative) with respect to their average charge.

can be applied, resulting in a harmonic oscillator response driven by any oscillations in the applied field. Such oscillations can induce very large changes in dielectric response with frequency. The displacement polarizability resulting from this behavior is

$$\alpha_{displacement} = \frac{Z^2 q^2}{(\overline{\omega}^2 - \omega^2)}\left(\frac{1}{M^+} + \frac{1}{M^-}\right),$$

2.44

where M^+ and M^- are the masses of the positively and negatively-charged ions, Z is the average ionic charge in units of electron charge (q), ω is the frequency of the applied electric field, and $\overline{\omega}$ is the resonant frequency of the system. Clearly, when $\omega \sim \overline{\omega}$ the system has a very large polarizability and consequently near this frequency there is a large change in dielectric constant. For normal materials the resonant frequency is near the "Debye frequency" which is related to the vibrational modes of the solid (see Section 2.3.3.). The resonant energy of these vibrations is typically 10-100 meV, or of the order of $k_B T$ at room temperature. Thus, the displacement polarizability of partially ionic crystals becomes frequency-dependent for photons in the infrared portion of the optical spectrum. The low-frequency dielectric constants

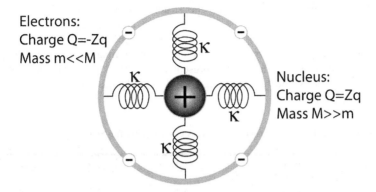

Electrons:
Charge Q=-Zq
Mass m<<M

Nucleus:
Charge Q=Zq
Mass M>>m

Figure 2.14: A model for picturing the response of the electron cloud of a solid to an applied electric field as a simple harmonic oscillator system.

are essentially the static dielectric constants resulting from the nearly constant atomic and displacement polarizabilities for fixed electric fields or very low frequencies.

The above considerations lead to the following general trends in dielectric response of materials.

The atomic polarizability:

- increases with the number of valence electrons;
- increases with decreasing binding energy for valence electrons around atoms (this changes the "spring constant" and lowers the resonant frequency).

The displacement polarizability:

- requires charge transfer among atoms in the unit cell (atomic solids and those without charge transfer show no displacement polarizability) so more ionic materials have a larger polarizability;
- larger displacements and "softer" crystals increase displacement dielectric constant.

Piezoelectric response is related to ionic displacement dielectric response. In a heteropolar (partially ionic) material that lacks a center of inversion symmetry, displacement of atoms of one polarity with respect to atoms of another polarity results in a change in shape of the material. A relationship between shape and applied electric field is termed a "piezoelectric" response. When the unit cell of the lattice includes inversion symmetry such a displacement moves charge but does not change the shape. Consequently, such materials are not piezoelectric. An example of how a material can lack an inversion center is found in all zincblende-structure materials. In these materials, a cation and anion lie at opposite ends of each bond and the structure is not symmetric around this bond. Furthermore, all bond pairs are oriented in the same way. Therefore, the structure lacks an inversion symmetry. Piezoelectric behavior can work either way – an applied field induces a shape change, but a shape change can produce an electric field. Most compound semiconductors exhibit a piezoelectric response. Typical magnitudes of this response for some common semiconductors are presented in Table 2.1. The permeability of a material is tied to its ability to be magnetized as dielectric response is tied to the ability to be polarized.

Table 2.1: **Typical Piezoelectric Stress Coefficients for Selected Materials**

Material	e_{31} (C/m^2)	e_{33} (C/m^2)	Material	e_{31} (C/m^2)	e_{33} (C/m^2)
Calculated values from Reference [1]					
ZnO (h)	-0.51	0.89	AlN (h)	-0.6	-0.47
BeO (h)	-0.02	0.02	GaN (h)	-0.49	-0.84
			InN h)	-0.57	-0.88
CdTe (c)	-0.01	0.03	AlAs (c)	0.01	-0.01
ZnS (c)	-0.05	0.10	GaAs (c)	0.06	-0.12
ZnSe (c)	-0.02	0.04	InAs (c)	0.01	-0.03
AlP (c)	-0.02	0.04	AlSb (c)	0.02	-0.04
GaP (c)	0.03	-0.07	GaSb (c)	0.06	-0.12
InP (c)	-0.02	0.04	InSb (c)	0.03	-0.06
Experimental values from Reference [2]					
ZnO (h)	-0.46 †	1.27	GaAs (c)	-0.16 *	
AlN (h)	-0.48 †	1.55	CdS (h)	-0.21 †	0.44
a-SiC (h)	0.08 †	0.2	$Sr_5Ba_5Nb_2O_6$	5.19 †	9.81
			$LiNbO_3$	3.64 †	1.65

Note: The change in polarization δP_i along direction i is given by

$$\delta P_i = \sum_j e_{ij}\varepsilon_j \quad \text{where } \varepsilon_j \text{ is the strain along direction } j.$$

h: hexagonal   c: cubic   † e_{15} value   * e_{14} value

The magnetic response of a material depends upon the arrangement of electrons in the atoms of that material. Electrons have a "spin" of $\hbar/2$ which is related to their intrinsic magnetic moment of $q\hbar/2m$. When the spin vectors of the electrons in an atom align, the atom has a higher magnetic moment. The ability of electron spins to align is limited by quantum state availability. Half-filled states allow any combination of spins and high magnetic response, while completely filled states allow only one combination of antiparallel spins. When the spins of separate atoms align parallel to each other the system develops a net magnetization in local domains. A sufficiently large applied field can align the magnetic fields of the domains with the external field resulting in a net magnetization. If this field persists after removal of the applied field the material is a ferromagnet.

When the spins within atoms or molecules do not align spontaneously from atom/molecule to atom/molecule but where such alignment can be induced by an applied field, the material is called a paramagnet. In the least magnetic materials, diamagnets, the magnetic moments within an atom or molecule cancel exactly. In these materials, only small magnetic responses can be produced. Finally, it is possible for spins among atoms to spontaneously align opposite to one another. This behavior is called antiferromagnetic and resists magnetization. The permeability of ferromagnets is of the order of hundreds to hundreds of thousands, paramagnets have values greater than unity by 10-10,000 parts per million, while diamagnets have permeabilities smaller than unity by one to 200 parts per million.

2.3.2 Optical properties

The absorption, transmission, and reflection of light all depend critically on the density of states and Fermi energy in a material. Transparent materials have a gap in the density of states around the Fermi energy such that there are no states available into which an electron absorbing a photon (of small-enough energy) could move. A schematic of a typical density of states in a metal is shown schematically in Figure 2.15. A large density of electrons and states available near E_f, allows electrons to move in response to electromagnetic waves causing their reflection. If this occurs uniformly across photon energies, the material appears silvery or white. An exceptionally large density of states at specific energies can produce selective absorption, and consequently the material appears colored (as in Cu or Au). When the material has an energy gap (or very low density of states) between filled and empty levels, it allows light with energy below the gap energy to pass through the material. Transparent materials have very large energy gaps. Examples of transparent materials include silicon dioxide (as in window glass) and aluminum oxide (which is colored by impurities to make ruby and several other gemstones).

The reverse process, emission of light, can also occur when a material has a large density of electrons in a band with high energy (for example the conduction band) and holes at a lower energy (such as in the valence band). This unstable situation is resolved when electrons lose energy to fill the holes. The liberated energy may be emitted in the form of light.

The dielectric constant described above, and in particular the optical dielectric constant resulting from atomic polarizability, has a strong effect on optical properties including reflectivity of the material. Using this effect, even transparent materials can affect reflectivity when coated on another material. This is the basis of antireflection coatings on lens surfaces and window glass. Refraction effects resulting from the change optical dielectric properties of materials across an interface are also fundamental to the operation of lenses, optical fibers, and similar devices.

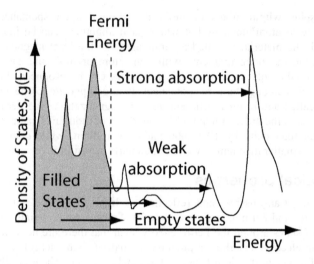

Figure 2.15: A schematic representation of the density of states of a metal which would lead to strong absorption in a limited range of energies and weak absorption at other energies.

2.3.3 Thermal properties

There are a variety of thermal properties of materials such as heat capacity, thermal conductivity, thermal expansion coefficient, and many more. For purposes of this text we will focus on the thermal conductivity as it is critical to many electronic devices. The reader is referred to the suggested readings for details of the full spectrum of thermal properties and for details not described here. The thermal conductivity, κ, of a material results from transport of energy via electrons or via lattice vibrations (phonons). The total thermal conductivity can thus be written simply as:

$$\kappa = \kappa_{phonons} + \kappa_{electrons}. \qquad 2.45$$

Typical thermal conductivities for a range of materials are given in Table 2.2. Let us consider the two contributions to thermal conductivity separately.

2.3.3.1 Heat conduction by phonons

"Phonon" is the name given to a cooperative vibration of atoms in a crystalline lattice much as a wave at sea is a cooperative motion of many water molecules. We saw above that phonons are essential to making transitions between minimum-energy states of indirect-gap semiconductors. They are also the primary reservoir of thermal energy in most solids and therefore contribute to thermal conduction. For these and other reasons, it is important to know more about phonons.

Table 2.2: Thermal Conductivities of Selected Materials

Material	K_{Th} (W/cm K)	Material	K_{Th} (W/cm K)	Material	K_{Th} (W/cm K)
C [3]	20	AlN [5]	2.85	AlAs [4]	0.08
Si [3]	1.56	GaN [5]	1.3	GaAs [4]	0.54
Ge [3]	0.6	InN [5]	0.45	InAs [4]	0.26
β-SiC [3]	5	AlP [4]	0.9	AlSb [4]	0.56
ZnO [6]	1.1	GaP [4]	1.1	GaSb [4]	0.33
Al_2O_3 [7]	0.39	InP [4]	0.7	InSb [4]	0.18

Source citation numbers for values are given parentheses.

One can calculate energy-momentum relationships for phonons just as was done above for electrons. The equations are not the same but the ideas are similar. They are the solutions for the wave equation for atoms rather than electrons in a regular periodic solid. Phonons are sound waves with a wavelength of the order of the lattice constant in the solid. The number of phonon modes is limited by the physical dimensions of the solid and the number of atoms that make it up, and well-defined energy momentum dispersion relations exist for phonons as for electrons. A typical phonon dispersion relation for a one-dimensional solid consisting of alternating heavy and light atoms is shown in Figure 2.16.

When the atoms all tend to move together in one direction or the other over many atom spacings the vibration wavelength is long (small k) and the energy is low. These modes are called acoustic phonons because audible sound vibrations have wavelengths many times the interatomic spacing. When the wavelength approaches the lattice spacing the resonance with the lattice causes a gap in the states, exactly as in electron band structures. In solids where there are two or more distinguishable atoms, not necessarily of different chemistry, additional states occur with short wavelengths. These are referred to as "optical" phonons and result in the upper branch of the dispersion relation in Figure 2.16. Note that the energy is linearly related to momentum for low energy acoustic phonons. Thus, sound waves traveling in a solid have energies that are linearly related to their momenta, as one would expect. As with electrons, the situation in three dimensions is much more complex than the simple one-dimensional behavior shown in Figure 2.16, but the same general results follow.

Full three dimensional phonon dispersion curves are shown for Si and GaN in Figure 2.17.[5] The frequency of atoms moving in solids can be $\sim10^{13}$ s^{-1} for optical phonons in some solids. Among other points, this frequency defines the upper end of the range of attempt frequencies for atomic transport processes such as diffusion and for many reactions.

We can see from Figures 2.16 and 2.17 that lattice vibrations can exist with a range of energies and momenta. The energy of the average phonon in a solid at equilibrium is the average thermal energy of an atom in that solid, $k_B T$. Therefore, one can estimate the distribution of phonons based on the mode energies (see Fig 2.17) and the temperature of the solid. Phonons travel at the speed of sound. This is roughly constant over a wide range of temperatures. To conduct heat by phonon motion, the phonons scatter in the solid and come to equilibrium with the local temperature of the lattice. These phonons can then spread out through the solid as a wave moves across the ocean, carrying heat until they scatter again.

The phonon (lattice vibration) contribution to thermal conductivity is well approximated by:

$$\kappa_{phonons} = \frac{1}{3} C_V \bar{v} \Lambda \qquad\qquad 2.46$$

where C_V is the heat capacity of the solid at constant volume, $\bar{v}$ is the average speed of sound, and Λ is the mean free path of phonons between scattering events. The mean free path for phonons decreases as the number of phonons present in the solid increases due to a higher scattering probability.

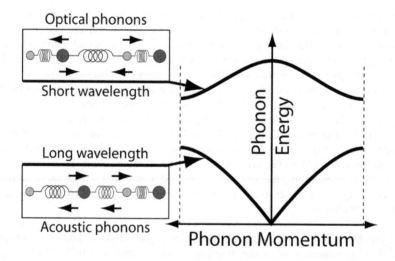

Figure 2.16: A typical phonon dispersion relation for a one-dimensional lattice of balls connected by springs. On the long-wavelength "acoustic" branch atoms move as groups in one direction or another with the direction varying over relatively long distances. For the higher-energy "optical" branch atoms move in opposite directions over very short distances.

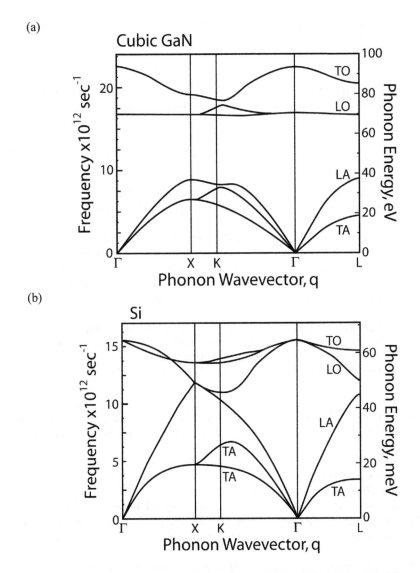

Figure 2.17: The phonon dispersion relations for (a) GaN and (b) Si. TA, LA, LO, and TO refer to transverse acoustic, longitudinal acoustic, longitudinal optical and transverse optical phonons, respectively. Each of these represents a particular vibrational mode. Longitudinal modes run along bonds as in Figure 2.16, while for transverse modes the vibration velocity is perpendicular to the bonds. There are two transverse modes because there are two axes perpendicular to a bond direction. Figures after Levinshtein, Rumyantsev, Sergey, and Shur, Reference [5], p. 27 and 184, respectively. This material is used by permission of John Wiley & Sons Inc.

It can be shown that the phonon mean free path in a typical solid can be written approximately as:

$$\Lambda \approx \Lambda_0 e^{\Theta/2T}$$

2.47

where Θ is called the Debye temperature. At low temperatures the heat capacity is also related to the Debye temperature through the Debye approximation:

$$C_v \approx cNk_B \left(\frac{T}{\Theta}\right)^3,$$

2.48

where c is a constant. Thus, there are phonon-related contributions to the thermal conductivity given in Equation 2.46 that enter through Λ and through C_V.

The heat capacity can be calculated exactly if one knows the relationship between energy and momentum for a vibrational mode and how many such modes exist in the material. Such a calculation can be done for a crystalline solid with results such as in Figure 2.17. A full discussion of the phonon modes in solids is beyond that needed for our purposes but may be found in many solid state physics texts.

At higher temperatures and often by room temperature, the heat capacity is simply determined by vibrations of individual atoms. Each atom has three independent vibrational modes, one for each direction in space, and each mode has an average energy k_BT. The heat capacity is, by definition, $\partial E/\partial T$ from which:

$$C_V = 3Nk_B.$$

2.49

N is the number of particles vibrating per unit volume (the atomic density) and the factor of three results from the three independent coordinates of vibration direction.

In insulators, there are no free electrons and consequently only phonons transmit heat. The maximum thermal conductivity of such a material can be calculated using the formulae above. In most cases, however, this turns out to be a relatively low value and most electrical insulators are relatively good thermal insulators. The notable exception is diamond with the highest thermal conductivity of any material. Because it is an electrical insulator, diamond conducts heat via phonons alone.

In common applications, electrical insulators are also used as heat insulators. Thus, one is concerned with the minimizing, rather than the maximizing thermal conductivity. The best thermal insulators are amorphous materials, which have lower thermal conductivities because phonons cannot exist over extended ranges. Essentially the scattering distance for phonons becomes one atomic spacing. In this case the transmission of heat in the material becomes a process of diffusion of energy among loosely coupled harmonic oscillators. In the classical (high temperature) regime, the minimum thermal conductivity has been estimated to be [8]:

$$\kappa_{min} = \frac{k_B N^{2/3} v}{2.48}.$$

2.50

The value for SiO_2 based on this formula is ~0.01 W cm^{-1}K^{-1}, which is close to the measured value. Thus, amorphous silica (glass) is a nearly perfect thermal insulator. The only improvement would be to decrease the density of the glass, as gases and vacuum are better thermal insulators than any solid can be. The heat shielding tiles on the space shuttle, for example, make use of underdense fiberous silica to achieve an exceptionally low thermal conductivity.

2.3.3.2 Heat conduction by electrons

As with phonons electrons come to thermal equilibrium with the lattice when they scatter. Electrons easily move through the lattice of a conductive solid with relatively long mean-free path lengths. Therefore, they transport heat well. As with phonons, the electron contribution to thermal conductivity is:

$$\kappa_{electrons} = \frac{1}{3} C_{elec} v_{elect} \Lambda_{elec}$$

2.51

where C_{elec}, v_{elec}, and Λ_{elec} are the heat capacity, mean velocity, and mean free path of electrons near the Fermi energy in the solid. One can carry out calculations to show the mathematical contributions to the various terms in Equation 2.51. In the end, one can derive an equation for the thermal conductivity of electrons in terms of the temperature, T, and the electrical conductivity, σ, of the material:

$$\kappa_{elec} = L\sigma T,$$

2.52

which is known as the Wiedemann-Franz law. L is a constant. For a free electron solid L~2.45x10^{-8} WΩ K^{-2}. In real metals, L ranges from ~2.2 to 3.0 x10^{-8} WΩ K^{-2}. In general, the thermal conductivity of electrons is much greater than for phonons.

2.4 QUANTUM WELLS AND CONFINED CARRIERS

In Section 2.1. we examined solution to the Schrödinger equation in free space and in a periodic potential and found families of wave functions having a simple, quadratic relationship between wave energy and momentum over wide ranges of energy values. Because there are a very large number of such states in a solid, it is normally impossible to distinguish one individual state from another. In this Section, we will examine, briefly, the impact of artificial potential structures on the wave solutions. For a complete discussion of such effects, the reader is referred to books on quantum mechanics.

Artificial-potential-barrier structures appear in a number of important microelectronic device applications. The best known is probably the laser diode. These ubiquitous devices are deceptive in their outward simplicity – a single small chip of material

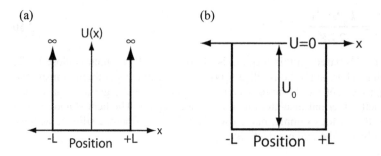

Figure 2.18: (a) a potential well with infinite walls. (b) a well of depth, U_0. Both wells have a width of 2L.

connected by two wires to a switch and batteries, or to a power supply. The trick is in the atomic-scale design of the chip of material. Laser diodes consist of a series of layers that are designed to collect electrons and holes. [A more detailed description of laser diodes may be found in Chapter 3.] Recombination of the free carriers in the trap leads to exceptionally efficient laser light emission. The trap is a potential well resulting from the sandwiching of a narrow-gap semiconductor between two layers of wider-gap material. Discontinuities in the valence and conduction band edges produce a lower-potential region where the carriers tend to collect. Because these wells are of the size scale of an electron wave function, the quantum nature of the system determines its properties. Hence, such structures are known as quantum wells. We will now examine the solutions of Schrödinger's equation in and near such wells.

The traditional beginning is to consider a potential well with infinite side walls [see Figure 2.18(a)]. The potential function in a one-dimensional version of the time-independent Schrödinger's equation (Equation 2.3) then becomes U=0 for $|x| < L$ and U=∞ for $|x| \geq L$. The solution to Equation 2.3 for $|x| \geq L$ must be $\psi(x) = 0$. For $|x| < L$ the equation is (in one dimension):

$$\frac{d^2\psi}{dx^2} + \kappa^2\psi = 0,$$
2.53

where

$$\kappa^2 = \frac{2mE}{\hbar^2}.$$
2.54

The solutions to Equation 2.53 are proportional to $\sin(\kappa x)$ and $\cos(\kappa x)$. Since the wave function is zero outside the well and since the wave function inside and outside must match at the boundaries, the values of κ are forced to be $\kappa L = n\pi$, where n=1,2,3,... for the $\sin(\kappa x)$ solution and $\kappa L = (n-1/2)\pi$ for $\cos(\kappa x)$ to produce nodes at

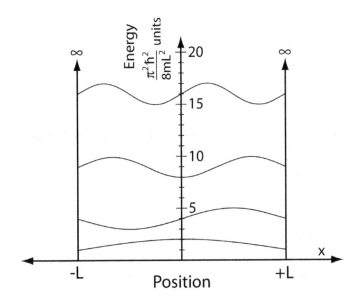

Figure 2.19: Electron states in an infinite-depth quantum well. Note that these are simply harmonics of the fundamental mode with nodes at the edges of the well.

the walls of the well. The allowed energies are then determined from Equation 2.54. Solutions for the infinite well are of the form shown in Figure 2.19.

The situation shown in Figure 2.18 (a) is, of course, not realistic. Infinite potential barriers cannot be produced. A more realistic solution would involve a finite well such as shown in Figure 2.18 (b). Let us consider the solutions of the Schrödinger equation for particles with energies below the energy of the edges of the well. In this case again U=0 for $|x| \geq L$ and U=-U_0 for $|x| < L$. Inside the well and away from the edges, the solutions are roughly the same as one would find in the infinite well with somewhat relaxed boundary conditions. The real part of the most general solution to the Schrödinger Equation for the finite potential well is:

$$\psi(x) = Ce^{\kappa x} \qquad\qquad\qquad x < -L$$
$$\psi(x) = A\cos(kx) + B\sin(kx) \quad -L < x < +L \qquad\qquad 2.55$$
$$\psi(x) = De^{-\kappa x} \qquad\qquad\qquad x > +L$$

The two exponential relationships describe the decay of the wave function outside of the well, while the middle relationship in 2.55 describes the wave function within the well. As in the infinite well, the wave function and its first derivative must be continuous across the well boundaries. The boundary conditions yield four equations (two boundaries, two matching conditions) in four unknowns (A,B,C,D), yielding:

$$\kappa = k\frac{A\sin(kL) - B\cos(kL)}{A\cos(kL) + B\sin(kL)}$$

2.56

$$\kappa = k\frac{A\sin(kL) + B\cos(kL)}{A\cos(kL) - B\sin(kL)}$$

Additional equations for C and D can be developed but will not be reproduced here to save space. Furthermore, one can show that:

$$k^2 = -\frac{2m}{\hbar^2}(U_0 + |E|) > 0$$

$$\kappa = k\tan(kL) \qquad\qquad \text{Even solutions} \qquad\qquad 2.57$$

$$\kappa = -k\cot(kL) \qquad\qquad \text{Odd solutions}$$

The second two formulas in Equation 2.57 are from boundary condition matching. The first part of Equation 2.57 is the bound state requirement and can be rewritten:

$$|E| = -U_0 + \frac{\hbar^2 k^2}{2m}$$

2.58

This is just a statement that the bound state energy ($\hbar^2 k^2/2m$) lies below the well boundary and is harmonic-oscillator-like. If one defines the energy to be zero at the bottom of the well, then the solutions become simple free-electron waves within the well as in Equation 2.16. This is not surprising as the middle relationship in Equation 2.55 for the wave function within the well is a free electron wave behavior. Because the bound states must be standing waves in the well (must have a maximum or a zero at the well center), the values of k are linked to the well half-width, L. This, finally, leads to the constraint:

$$\sqrt{\frac{2mU_0}{k} - 1} = \left\{ \begin{array}{l} \tan(kL) \\ -\cot(kL) \end{array} \right\} \qquad \begin{array}{l} \text{Even solutions} \\ \text{Odd solutions} \end{array} \qquad 2.59$$

Note that Equation 2.59 provides a connection between the wavelength, the well width, and the well depth. Therefore, the well dimensions determine both the wavelength and energy of the states within the well. A lot of algebra and discussion has been skipped in writing these solutions. If you want more details, this problem is treated in any basic quantum mechanics text. The form of these solutions is shown in Figure 2.20.

To reiterate the important points from these equations:

- Equation 2.59 shows that the possible values of the wavelength, k are coupled to the well half width, L, and to the depth of the well.

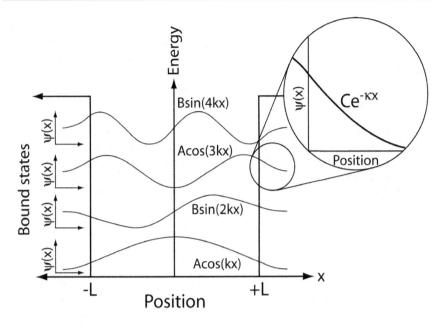

Figure 2.20: Solutions to the Schrödinger equation given by 2.55 and subject to the additional conditions of Equations 2.56-2.59.

- Deeper wells produce more bound states with slightly shorter wavelengths (as the wave function is less inclined to penetrate the well sides).

- Wider wells produce more bound states for a given well depth.

- There are very few states in the well and these have discrete energies.

- There are no states at the bottom of the well as would be judged from the band edges. Therefore, the band gap of the well is greater than the minimum gap of the semiconductor from which it is produced.

- The wave function decays exponentially in the barriers with a decay length κ that increases with decreasing well depth.

- The wave function penetrates the surrounding barriers a significant distance. This gives rise to coupling between adjacent quantum wells.

The final point above deserves some specific attention. Two quantum wells that are close enough to each other for the tails of their wave functions to overlap significantly become coupled together. The interaction of their wave functions alters the energies of states in the wells. This interaction produces bonding-like and antibonding-like pairs of states, one raised in energy, the other lowered. Adding a third quantum well adds a third state in each quantum well. An infinite number of coupled wells would produce a band of states in each well associated with each quantum state for a single well. The widths of the bands are directly related to the well-to-well coupling. Well coupling depends upon the depth of the well and exponentially upon the distance between the wells. This situation is shown schematically in Figure 2.21. [9]

It may be helpful to consider the example in Figure 2.21 in more detail. The discussion below is after Holonyak.[9] Suppose the wells shown have a depth $U_0 = 200$ meV, a well width $2L = 4$ nm, and a well separation $s = 4$ nm. This is typical of a series of quantum wells produced by growth of 4 nm thick $Al_{0.19}Ga_{0.81}As$ layers alternating with 4 nm thick GaAs layers in a superlattice. The GaAs layers produce

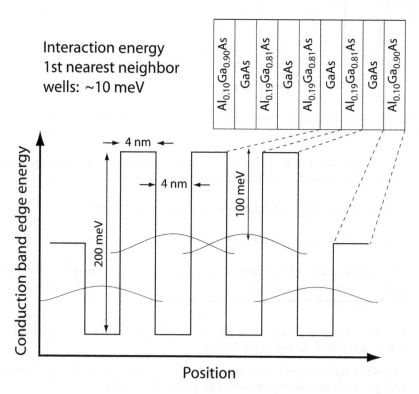

Figure 2.21: An example semiconductor superlattice structure resulting in quantum wells as shown for the conduction band edge. Similar but shallower wells occur in the valence band. The bound states in the wells are also indicated. These states overlap producing splitting and multiple levels in each well (not shown in this sketch). (Courtesy N. Holonyak [9])

the wells and the $Al_{0.19}Ga_{0.81}As$ provides the barriers. The states in the quantum wells couple with each other such that the first-nearest neighbor wells have two states rather than the single state that a single well would have. These states change their energies by ~10 meV relative to the single well. Second nearest neighbor interactions when three or more wells are present cause additional splitting with a further change of energy of ~4.5 meV, and so on. The result is a series of states in the quantum well at the center of the superlattice with energies as shown in Figure 2.22. The more quantum wells that are coupled together the more states occur in each well and the closer together these states are. Wells toward the edges of the superlattice have fewer wells to interact with and, consequently, fewer bound states.

There are several points to notice about Figure 2.22. First, the states increase in energy more than they decrease because as a state drops in the quantum well, the energy barrier separating those states from their neighbors rises. Consequently, the states decay more rapidly in the barrier and their interaction is weaker from well to well. For an infinite number of wells one gets a band of states with a width of

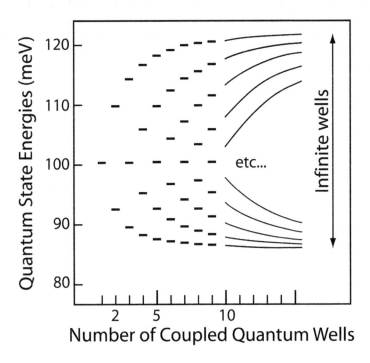

Figure 2.22: A plot of the number of states and their energies in a quantum well which interacts with the specified number of other identical wells. The plot assumes the geometry and energies of Figure 2.21. An essentially infinite superlattice produces a band of states as indicated. (Courtesy N. Holonyak [9])

roughly four times the first-nearest-neighbor interaction energy and with a sinusoidal variation in energy with momentum. Second, for an odd number of quantum wells there is always a state with a binding energy of 100 meV. Third, the result illustrated in Figure 2.22 is for a single bound state. If there had been two states in the well, each would produce a series of states as more wells were added, giving rise to two minibands per state in the isolated well. Such would have been the case for wider wells separated by the same barrier layers. The situation shown in Figures 2.21 and 2.22 is for the conduction band in the example superlattice. A similar set of states would occur in the corresponding wells in the valence band. Because approximately 70% of the band offset (see Chapters 5 and 6 for details) is accommodated in the conduction band, all of the behaviors in Figures 2.12 and 2.22 would hold for the valence band but all barriers would be smaller, leading to more overlap between states and more broadening of the valence band quantum well minibands.

Each state in the quantum well can accommodate electrons or holes and consequently can give rise to absorption and emission of light. In laser diodes we take advantage of single quantum wells to trap electrons and holes to enhance recombination. It is important to use only single quantum wells or widely-separated wells to prevent broadening of the states as in Figure 2.22. Such broadening would give rise to a range of emission wavelengths. In a laser one wants a single emission wavelength (single mode laser) if possible.

2.5 SUMMARY POINTS

- Electrons are Fermions, thus the probability of finding an electron in a given state is given by the Fermi function (Eq. 2.11.).
- The Schrödinger equation solutions describe the electron waves allowed. For a free electron (no potential energy) such solutions result in a quadratic relationship between electron momentum or wave vector and electron energy. In a periodic solid this relationship repeats with a wave vector inversely proportional to the lattice period.
- When a periodic potential is present, energy gaps develop in the quadratic energy/momentum relationship, proportional to the strength of the periodic potential.
- The energy/momentum relationship in three dimensions can include maxima and minima in the energy of given branches. This can lead to indirect or direct energy gaps. This has implications for the optical and electronic properties of the semiconductor.
- The effective mass of the electron is modified from its rest mass in vacuum by the presence of a periodic lattice. The effective mass is inversely related to band curvature in energy/momentum space.
- Density of states per unit volume describes the number of states in a small energy increment around a given energy. The probability of an electron absorbing energy or momentum is generally directly related to the density of states.
- The density of states around any band extremum will be quadratic in energy with respect to the energy of the band edge.
- The density of free carriers in a pure and perfect (intrinsic) semiconductor increases exponentially with temperature with a characteristic energy of half the energy gap.
- The number of holes and electrons exactly matches in an intrinsic material.
- In a doped semiconductor the number of holes and electrons does not balance. The number of majority carriers depends upon the number of dopant atoms and exponentially on the energy separating the impurity state from the appropriate band edge and the temperature. When more than one dopant atom is present, compensation can lower the number of majority carriers.
- Electrical conductivity depends upon the product of carrier density and mobility. Mobility depends upon effective mass and mean time between collisions of electrons causing scattering.
- Dielectric constant depends upon material polarizability. The polarizability consists of atomic polarizability (increases with number of valence electrons and decreasing electron binding energy) and displacement polarizability (increases with polarity of material and magnitude of possible displacements).

- Thermal conductivity depends upon lattice vibration and electronic contributions. The lattice contribution depends upon atomic density while the electronic contribution depends upon conductivity and temperature.
- Lattice vibrations can be described as "phonon" quasiparticles, which carry momentum and heat energy in the lattice. The phonon dispersion relation provides a linear relation between energy and momentum at moderate phonon momenta.
- Quantum wells have states in them whose energy is determined by the well depth and width but is always above the bottom of the well in energy (for electrons).
- Coupled quantum wells result in multiple states in the wells. When enough wells are coupled these become minibands of states in the wells.

2.6 HOMEWORK

1: Energy Band Diagrams
Consider the energy vs. momentum relationships (band diagrams) shown in Figure 2.7.

a) What range of energies does the valence band of GaAs span based on Figure 2.7.?
b) Sketch the most important features of the E(k) diagram for Ge and indicate the minimum energy gap.
c) Which semiconductor (GaAs or Ge) has the indirect energy gap? How do you know?
d) The effective masses of the carriers are given by: $1/m^* = (d^2E/dk^2)$. Which semiconductor, Ge or GaAs has the higher electron effective mass?
e) Sketch the density of states <u>near the top</u> of the valence band. What is the functional form of this curve?
f) If the number of electrons and holes are equal in an intrinsic semiconductor, why is the Fermi Level not exactly in the middle of the energy gap? (Eq. 2.31)

2: Conductivities
a) Could the thermal conductivity of SiO_2 be made lower by changing the material (for example by crystallizing it) but assuming one is not allowed to use alloying, density reduction, etc...?
b) Why is the electrical conductivity of Ag much higher than the electrical conductivity of Mo. Which would you expect to have a higher thermal conductivity? Explain briefly.
c) Explain briefly why it is necessary to have two distinguishable atoms in a crystal unit cell in order to observe the optical branch of the phonon dispersion relationship.

3: Consider phosphorous as an impurity in silicon.
a) Would you expect it to be an electron donor or an acceptor? Explain why.
b) If the ionization energy (the difference in energy between the phosphorous state and the band edge) is 35 meV, calculate the fraction of the phosphorous atoms ionized at (i) 80 K, (ii) 300 K, (iii) 400 K? You may assume that the Fermi energy is at the phosphorous state energy and that electrons escaping into the band need only reach the band edge rather than integrating the product of the Fermi function and the density of states throughout the conduction band.
c) If there are 10^{17} cm^{-3} phosphorous atoms in a sample of Si, calculate the minority carrier concentration at 300K. You may need the following: $N_c=2.8\times10^{19}$ cm^{-3}, $N_v=1.04\times10^{19}cm^{-3}$, and $E_{gap}(300K) = 1.12$ eV for Si.

4: Consider the solutions for the Schrödinger Equation for a quantum well.
   a) Show that Equations 2.55 and 2.56 satisfy the boundary conditions requiring matching of the wavefunctions and their slopes across the boundaries.
   b) Calculate the energies and number of states in quantum wells 85 meV deep and 4 nm wide. Assume the hole effective mass for GaAs. (These are the states that would occur in the valence bands corresponding to the conduction band states in Figure 2.22.)

5: Think about the periodic potential, U(x) discussed in Chapter 2 with which electrons interact.
   a) What is the effect of increasing the periodic potential U on the dispersion of bands in a solid? (Hint: consider Equation 2.10 and Figure 2.3.)
   b) Refer to the free electron dispersion relationship in Figure 2.6 with those for GaAs and Ge. Based on the behaviors where the lower portion of the valence band intercepts the L point (which is in planes most directly bisecting interatomic bonds in these materials) at the Brillouin zone boundary, estimate the magnitude of the first Fourier Coefficient of the atomic potential for these materials. The point in question intercepts the Y axis of Figure 2.6 at roughly 0.8 arbitrary units.

6: Consider the density of states discussion in Section 2.1.8. Between the Brillouin zone boundaries the E(k) relationship turns from concave up to concave down. In this region there is a small area where the energy is roughly linearly related to momentum, E proportional to k.
   a) In this region, derive a formula for the density of states similar to the approach used in developing Equation 2.22.
   b) What is the effective mass of an electron in this region of the energy-momentum relationship in terms of E and k? (Hint: follow the approach by which Equations 2.17 and 2.18 were developed.)

2.7 SUGGESTED READINGS & REFERENCES

Suggested Readings:

Arfken, George, *Mathematical Methods for Physicists.* New York: Academic Press, 1970.

Ashcroft, Neil W. and Mermin, N. David, *Solid State Physics.* Philadelphia, PA: Saunders College, 1976.

Bube, Richard H, *Electrons in Solids: an Introductory Survey.* Boston: Academic Press, 1992.

Cahill, David G, and Pohl, R.O., Lattice Vibrations and Heat Transport in Crystals and Glasses, Ann. Rev. Phys. Chem. 1988; 39: 93-121.

Chelikowsky, J.R. and Cohen, M.L. Nonlocal pseudopotential calculations for the electronic structure of eleven diamond and zinc-blende semiconductors. Phys. Rev. B 1976; 14: 556-582.

Cullity, B.D., *Elements of X-ray Diffraction.* Reading PA: Addison-Wesley, 1978.

Ferry, David K., *Semiconductors.* New York: Macmillan, 1991.

Harrison, Walter A., *Electronic Structure and the Properties of Solids: The Physics of the Chemical Bond.* San Francisco: Freeman, 1980.

Herman, F., *An Atomistic Approach to the Nature and Properties of Materials* Pask, J.A., Editor New York: Wiley, 1967.

Kittel, Charles, *Introduction to Solid State Physics.* New York: Wiley, 1996.

References:

[1] Bernardini, F.; Fiorentini, V.; Vanderbilt, D.; "Spontaneous polarization and piezoelectric constants of III-V nitrides." *Physical Review B,* 1997; 56: R10024-7.

[2] Gualtieri, J.G.; Kosinski, J.A.; Ballato, A.; "Piezoelectric materials for acoustic wave applications." *IEEE Transactions on Ultrasonics Ferroelectrics & Frequency Control,* 1994; 41:53-9.

[3] Eberl, K.; Schmidt, O.G.; and Duschl, R.; "Structural Properties of SiC and SiGeC Alloy Layers on Si," in *Properties of Silicon Germanium and SiGe: Carbon.* Erich Kasper and Klara Lyutovich, eds., London, INSPEC, 2000.

[4] Neuberger, M., *Handbook of Electronic Materials, III-V Semiconducting Compounds.* New York, Plenum, 1971.

[5] Levinshtein, Michael E., Rumyantsev, Sergey L., and Shur, Michael S., *Properties of Advanced Semiconductor Materials.* New York, John Wiley & Sons, 2001.

[6] Florescu, D.I.; Mourokh, L.G.; Pollak, F.H.; Look, D.C.; Cantwell, G.; Li, X.; "High spatial resolution thermal conductivity of bulk ZnO (0001)." *J. Appl. Phys.,* 2002; 91: 890-2.

[7] Kinoshita, H.; Otani, S.; Kamiyama, S.; Amano, H.; Akasaki, I.; Suda, J.; Matsunami, H.; "Zirconium diboride (0001) as an electrically conductive lattice-matched substrate for gallium nitride." *Jpn. J. Appl. Phys.,* 2001; 40: L1280-2.

[8] Cahill, David G.; and Pohl, R.O.; "Lattice Vibrations and Heat Transport in Crystals and Glasses." *Ann. Rev. Phys. Chem.,* 1988; 39: 93-121.

[9] Holonyak, N.; University of Illinois, Department of Electrical and Computer Engineering, private communication.

[10] Herman, F., *An Atomistic Approach to the Nature and Properties of Materials* Pask, J.A., Editor New York: Wiley, 1967.

[11] Chelikowsky, J.R. and Cohen, M.L.; Nonlocal pseudopotential calculations for the electronic structure of eleven diamond and zinc-blende semiconductors. Phys. Rev. B 1976; 14: 556-582.

Chapter 3

OVERVIEW OF ELECTRONIC DEVICES

This Chapter reviews some of the basic physics and operation of selected circuit elements used in microelectronic devices. For a complete review, see the suggested readings. Basic resistance, capacitance, and inductance were covered in Chapter 2. We focus here on diodes, including clas-sic homojunctions, heterojunctions, and Schottky barriers, because they illustrate most of the important issues in microelectronic materials and because both field-effect and bipolar junction transistors are constructed from them. Once we have discussed diodes, a brief review of these two major classes of transistors is provided. Finally, we finish the review with some of the issues unique to light emitting and laser diodes.

The terms "homojunction" and "heterojunction" are used frequently in this chapter. A homojunction is a contact between two of the same semi-conductors (silicon with silicon for example) while a heterojunction is a contact between two dissimilar materials. Heterojunctions, see Sections 3.3-3.4, can include a semiconductor with a semiconductor (Si with Ge for example) or semiconductors with metals (Ni with GaAs for example).

Diodes (solid state devices rather than vacuum tubes) can be any of these junction types and have the property that they pass current easily in only one direction. Understanding diodes requires knowledge of how charges move when junctions are formed and how contact potentials are created. Thus, one must first understand the movement of carriers in solids in response to electric fields or chemical concentration gradients. These motions are typically divided into two categories, diffusion driven by concentration gradients and drift – motion resulting from electric fields.

3.1 DIFFUSION AND DRIFT OF CARRIERS

3.1.1 Chemical potential

Before coming to diffusion and drift, it is useful to review a fundamental concept from thermodynamics that most concisely describes the driving force for particle movement – the chemical potential, the generalized potential energy of a particle at a given position. Real particles (as opposed to pseudoparticles such as holes) always move to lower their chemical potential. Charged particles in an electric field move according to their electrostatic potential. Massive particles in a gravitational field move toward the center of mass of the system. Particles placed in a concentration gradient move from high concentrations to low concentrations (they diffuse). The dominant term in the chemical potential in these three cases results from the electrical, gravitational and entropic energies of the system, respectively. A contribution to the total chemical potential can be defined for any form of energy or entropy. Once the chemical potential of a particle is known, its motion is easily described.

When any two systems are allowed to interact, they will always exchange matter or energy in such a way as to equalize the chemical potentials of the systems. Indeed, equilibrium is defined as equal chemical potential between the two systems. In terms of the concepts described in Chapter 2, the **Fermi energy in a solid gives a measure of the chemical potential of the lowest energy free electron or the highest energy free hole**.

3.1.2 Carrier motion in a chemical potential gradient

The most general way to express the force, F, on a particle is through the corresponding gradient in the chemical potential, μ:

$$F = -C\frac{d\mu}{dx},$$

3.1

where C is a constant. Let us consider a specific type of particle, the electron, for simplicity. The force on an electron due to a chemical potential gradient produces a current density J,

$$J = q\mu_n n\frac{d\mu}{dx}.$$

3.2

Note: keeping as close to the standard symbols used by the community as possible leads to some confusingly similar symbols. Here μ_n is the electron mobility and μ (no subscript) is the chemical potential (of the electron). Electron chemical potential gradients may result from many factors, but for the current discussion, we will consider only electric fields and concentration gradients. The presence of an electric field, E, in a solid corresponds to a voltage (potential) gradient dV/dx and contributes to the chemical potential gradient dμ/dx at constant concentration of electrons as:

$$E = -\frac{dV}{dx} = \frac{d\mu}{dx}.$$ 3.3

Likewise, a change in electron concentration, n, in the absence of an electric field causes a change in chemical potential:

$$\frac{d\mu}{dx} = \frac{k_B T}{qn}\frac{dn}{dx}.$$ 3.4

Therefore, the general chemical potential gradient in the system is given by

$$\frac{d\mu}{dx} = \frac{k_B T}{qn}\frac{dn}{dx} - \frac{dV}{dx}.$$ 3.5

The second term on the right side of Equation 3.5 is the electric field. Substituting for chemical potential gradient in Equation 3.2 gives:

$$J = q\mu_n n \left[\frac{k_B T}{qn}\frac{dn}{dx} + E\right] = \left[\mu_n k_B T\frac{dn}{dx} + q\mu_n nE\right]$$ 3.6

If we prefer, we can use the Einstein relation between mobility and diffusivity $D = \mu_n k_B T/q$ to get the standard form for the current density:

$$J = qD\frac{dn}{dx} + q\mu_n nE,$$ 3.7

where the first term is a current due to a concentration gradient and the second term is due to the electric field. **The first term is called a diffusion current and the latter a drift current.** However, as we have seen, both can be related at their roots to a basic concept, the chemical potential gradient.

3.2 SIMPLE DIODES

Let us now consider the effect of joining two chunks of a semiconductor doped n- and p-type, see Figure 3.1. As we saw in Chapter 2, in the n-type material there is a higher concentration of electrons. With an efficient dopant atom, these electrons have energies close to the conduction band edge (high chemical potential).

Likewise, a p-type material is best doped with atoms that produce empty states near the valence band edge and having a low chemical potential for electrons. When these two materials are brought into contact, electrons are free to move.

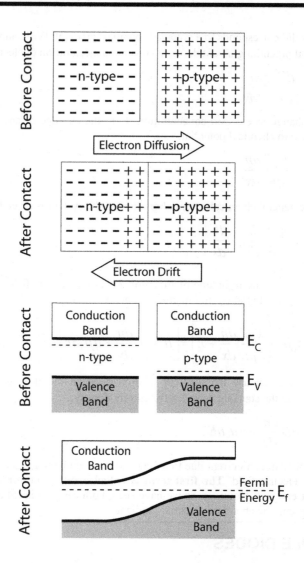

Figure 3.1: Electrons diffuse and drift across a junction between unlike materials. Before contact the Fermi energies (chemical potentials) of electrons in the two materials do not match. After contact, electrons flow until the Fermi energy is matched across the junction. This electron diffusion produces an electric field. At equilibrium the drift current balances the diffusion current and no net electron transfer occurs.

The large electron concentration gradient across the junction causes electrons to diffuse from the n-type to the p-type side, leaving positively-charged donor dopant atoms behind. The residual electric potential resulting from carrier diffusion presents a barrier to further electron motion in one direction and is fundamental to diode behavior.

3.2.1 The junction contact potential

The diffusion of charged species causes an electrostatic potential to build up, which opposes further diffusion of electrons. As charges move, the initial difference in chemical potential of the electrons on the two sides of the junction is gradually reduced and finally eliminated. At this point, diffusion stops. The result is an accumulation of positive charge on the n-type side and negative charge on the p-type side, causing an electric field at the junction. The magnitude of the field can be calculated by setting J=0 in Equations 3.6 or 3.7 and solving for E:

$$\mathcal{E} = -\frac{k_B T}{qn}\frac{dn}{dx},$$

3.8

or, substituting E=-dV/dx, and integrating across all x in the electric field,

$$V_{bi} = \frac{k_B T}{q}\ln\left[n_n/n_p\right],$$

3.9

where V_{bi} is the "built in" voltage (contact potential) across the junction and n_n and n_p are the electron concentrations on the n and p sides of the junction, respectively. For a shallow dopant at high temperature, $n_n \sim N_d$ where N_d is the donor concentration on the n-type side of the junction. Likewise, $n_p = n_i^2/p_p$ or

$$n_p = \frac{N_C N_V}{N_A} e^{-E_{gap}/k_B T}$$

3.10

Substituting $n_n \sim N_D$ and using Equation 3.10 for n_p, we have:

$$\ln\left|\frac{n_n}{n_p}\right| = \ln\left[\frac{N_A N_D}{N_C N_V} e^{E_{gap}/k_B T}\right]$$

3.11

or

$$\ln\left|\frac{n_n}{n_p}\right| = \frac{E_{gap}}{k_b T} + \ln\left[\frac{N_A N_D}{N_C N_V}\right]$$

3.12

from which, substituting Equation 3.12 into 3.9:

$$qV_{bi} = E_{gap} + k_B T\ln\left[\frac{N_A N_D}{N_C N_V}\right].$$

3.13

A similar diffusion of electrons between metals produces the contact potential used in thermocouples. V_{bi} is zero volts if both sides of the junction are undoped and increases with N_A and N_D. The built in voltage can be determined from a diagram of the band edges of a junction at equilibrium. It is simply the amount by which either band edge bends, see Figure 3.2.

The built-in voltage can also be written in terms of the intrinsic carrier concentration n_i. The result after some algebra is:

$$V_{bi} = \frac{k_B T}{q} \ln \frac{N_A N_D}{n_i^2}.$$
3.14

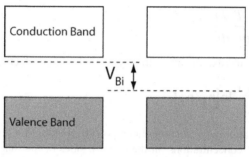

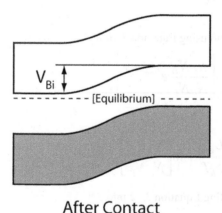

Figure 3.2: The difference in Fermi energies (dashed lines) before contact establishes a band edge bending and a contact potential V_{bi} after contact.

The diffusion of electrons across a junction leads to a distribution of charges as shown in Figure 3.1. The exact distribution can be determined if the dopant concentration depth profile is known. Because the number of free carriers decreases exponentially with the distance of the Fermi energy from the band edge, it is reasonably accurate to assume that a well-defined zone (the "depletion region") around the junction contains no free carriers. The electric field, present in this region, sweeps out the mobile carriers (resulting in a drift current that balances diffusion). The charge giving rise to the built-in voltage on each side of the junction can be approximated based on the number of ionized impurities in the depletion region on that side of the junction. This situation is shown schematically in Figure 3.3.

The electric field E_0 of the junction is found by integration of the charge Q across either side of the depletion region. For example, on the p-type side:

$$E_0 = \int_{E_0}^{0} dE = \int_{-x_p}^{0} Q^-(x)dx = \int_{-x_p}^{0} \frac{-qN_A}{\varepsilon}dx = \frac{-qN_A}{\varepsilon}\int_{-x_p}^{0} dx = \frac{qN_A}{\varepsilon}x_p, \qquad 3.15$$

where x_p is the depletion width on the p-type side, N_A is the ionized acceptor concentration (usually assumed, somewhat inaccurately, to be the same as the acceptor concentration for typical dopants), q is the acceptor charge, and ε is the dielectric constant of the semiconductor. The voltage one accumulates from such a field is the integral of the field over a distance. Graphically, it is the area under the electric field vs. position plot in Figure 3.3, $V_{bi} = -E_0 W/2$. Substituting from Equation 3.15,

$$V_{bi} = \frac{qN_A}{2\varepsilon}x_p W. \qquad 3.16$$

The variables N_A, W and x_p in Equation 3.16 are not independent. Specifically, W = $x_n + x_p$. Furthermore, it is necessary to balance the charge accumulated on the two sides of the junction, from which $x_p N_A = x_n N_D$. Therefore, W = $x_p(1 + N_A/N_D)$. Solving for x_p gives

$$x_p = W\frac{N_D}{N_A + N_D}. \qquad 3.17$$

Substituting for x_p in Equation 3.16 in turn, gives

$$V_{bi} = \frac{qW^2}{2\varepsilon}\frac{N_A N_D}{N_A + N_D}. \qquad 3.18$$

Substituting for V_{bi} from Equation 3.14 and solving for W leads eventually to:

$$W = \left[\frac{2\varepsilon k_B T}{q^2}\left(\ln\frac{N_A N_D}{n_i^2}\right)\left(\frac{1}{N_A} + \frac{1}{N_D}\right)\right]^{1/2}. \qquad 3.19$$

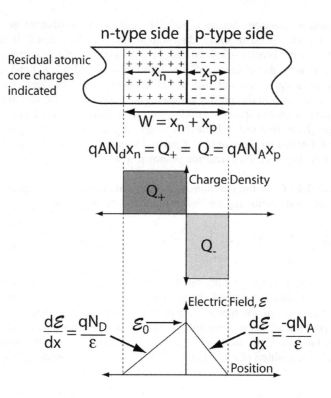

Figure 3.3: The depletion region and the charge associated with it showing how the depletion is primarily on the lightly-doped side of the junction. The charges on the two sides of the junction must balance. The electric field that results is due to the charge on the two sides of the junction. The maximum electric field is directly related to the total charge on each side of the junction.

Examination of this equation shows that most of the depletion lies on the lightly-doped side of the junction.

Equation 3.18 relating voltage to depletion width is actually more general than the discussion above suggests. When a voltage is applied to a p-n junction, the total voltage across the junction is $V = V_{bi} - V_{applied}$. In this case, Equation 3.18 still holds but V_{bi} is replaced by V. Therefore, a voltage that is the opposite sign as the built in voltage increases the depletion width. A voltage of the same sign decreases the depletion width:

$$W = \left[\frac{2\varepsilon(V_{bi} - V_{applied})}{q}\left(\frac{N_A + N_D}{N_A N_D}\right)\right]^{1/2}.$$

3.20

From a brief consideration of the capacitance, C=εA/d, and the fact that the depletion region is the most resistive portion of the diode, it should not be surprising to learn that diode junctions have a significant capacitance. The effective separation of the "capacitor plates" in the junction is the depletion width, W. The resulting capacitance is voltage dependent as can be seen from Equation 3.20 and depends primarily on the concentration of dopant on the lightly-doped side of the junction as

$$\frac{2C_{junction}}{A} = \sqrt{\frac{2q\varepsilon N_{net}}{V_{bi} - V_{applied}}}, \qquad 3.21$$

where N_{net} is the difference between the number of ionized majority carrier dopant atoms and the number of minority carrier dopants on the lightly-doped side of the junction. When the doping is nearly symmetric the width formula from Equation 3.20 must be used. It may not be surprising then to learn that by measuring the capacitance of an asymmetric junction as a function of voltage, the doping concentration on the lightly doped side can be determined as a function of depth. For an applied reverse bias voltage V_R:

$$\frac{1}{N_{net}(V_R)} = \frac{q\varepsilon}{2} \frac{\partial C_{junction}^{-2}}{\partial V_R}. \qquad 3.22$$

This relationship is used in the capacitance-voltage technique for profiling carrier concentrations near diode junctions and is usable for any type of diode in which one side of the junction is much more heavily doped than the other. It can also be used in a transient mode to detect and analyze point defects as they charge and discharge with bias voltage changes.

3.2.2 Biased junctions

To see why diodes behave electrically as they do one needs to consider the biased junction in more detail. Doped semiconductors are more conductive than the depletion region where all of the mobile carriers are swept away by the junction field. Because the voltage applied to a diode is dropped primarily across the most resistive element, the majority of an applied voltage adds to or subtracts from the field in the depletion region, as we saw in developing Equation 3.20. This, in turn, means that the amount of band bending is increased or reduced. The band edge diagrams for biased junctions equivalent to the equilibrium diagram in Figure 3.2 are shown in Figure 3.4.

The current density given in Equation 3.7 must be rewritten to include both carrier types to completely describe the situation in a p-n junction:

$$J = \left(qD_n \frac{dn}{dx} - qD_p \frac{dp}{dx} \right) + q\left(\mu_n n_p + \mu_p p_n \right)E, \qquad 3.23$$

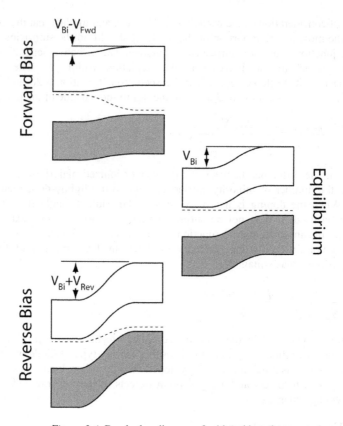

Figure 3.4: Band edge diagrams for biased junctions.

where n_p and p_n are the electron concentration on the p-type side and the hole concentration on the n-type side, respectively. In other words, these are the concentrations of the minority carriers ($n_p \ll p_p$ and $p_n \ll n_n$). The first term is the total diffusion current while the second term is the total drift current. In a biased junction the chemical potential barrier associated with the junction (the junction electric field) is changed. Drift current flows in all cases but diffusion current changes dramatically with the applied bias. A detailed derivation of the drift and diffusion currents, discussed briefly in the next two sections, may be found in the suggested readings.

3.2.2.1 Drift current

The drift current is relatively insensitive to the junction electric field. As noted above, this field sweeps away mobile electrons and holes. However, this sweeping effect does not depend upon how large the field is as long as it is not too small.

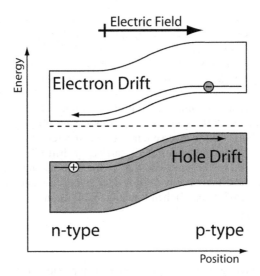

Figure 3.5: A schematic of the motion of minority carriers by drift across a junction. Drift current does not depend significantly on the electric field magnitude.

Therefore, ideally the drift current does not change under reverse bias. In sufficiently high reverse bias other conduction mechanisms set in, as we will see in Section 3.2.3. Although it is not field-dependent, the drift current requires minority carriers to reach the edge of the high-field depletion region. Minority carriers are required because the field must be of the appropriate sign to move them across the junction. The carrier motions for drift current under equilibrium conditions affect minority carriers, as shown in Figure 3.5.

The number of minority carriers reaching the edge of the depletion region per unit area depends upon the equilibrium number of minority carriers on each side of the junction (p_n or n_p for the n- and p-type sides of the junction, respectively) multiplied by the effective velocity at which they reach the edge of the depletion region. The latter depends upon their diffusivity (or mobility) and inversely on the average distance they can diffuse before recombining with a majority carrier, $v_{eff}=D_p/L_p$. The resulting drift minority carrier current density is given by

$$ J_0 = -q\left(\frac{D_p}{L_p} p_n + \frac{D_n}{L_n} n_p \right), \qquad\qquad 3.24 $$

where D_p and D_n are the minority carrier diffusivities and L_p and L_n are the carrier diffusion lengths for holes in the n-type and electrons in the p-type materials, respectively. This is the reverse current in a junction. Equation 3.24 applies to any p-n diode between similar materials (homojunction). We will see in Section 3.3 that a different formula must be applied for heterojunctions. In Section 3.2.3 we will find that when sufficient defects are present in the depletion region, this formula is also modified.

When a forward bias $V_{applied}$ is greater than the built-in voltage, the drift current changes direction as the field at the junction is reversed. This means that the carriers change direction, and that majority carriers would then drift across the junction in addition to diffusing. However, the drift current is so small by comparison with the diffusion current under these conditions that it is undetectable.

Reverse current density is important in determining the power that a device consumes when in the "off" state as it determines the effective resistance of the device in reverse bias. Hence reverse current is also referred to as leakage current. From Equation 3.24, the reverse current is carried by drift of minority carriers from the lightly doped side of the junction (that is the side having the higher minority carrier density) toward the heavily-doped side. If one assumes a p$^+$-n diode (the p-type side is very heavily doped) then Equation 3.24 becomes:

$$J_0 \approx -\frac{qD_p}{L_p} p_n .$$
3.25

The hole current flowing from the lightly-doped n-type side of the junction can then be related to the doping level on the n-type side and the intrinsic carrier concentration through the relation $n_n p_n = n_i^2$ as:

$$J_0 \approx -\frac{qD_p}{L_p} \frac{n_i^2}{n_n} .$$
3.26

Using Equation 2.29 and assuming $n_n \sim N_D$ Equation 3.26 becomes:

$$J_0 \approx -\frac{qD_p N_C N_V}{L_p N_D} e^{-E_{gap}/k_B T} .$$
3.27

From this equation it can be seen that reverse saturation current density increases exponentially with decreasing energy gap. This relationship was the downfall of Ge as a semiconductor for normal devices. The much lower energy gap of Ge produces a far higher reverse current density and consequently much greater "off" currents. In short, Ge devices consume too much power. Indeed, there has been discussion of a renaissance in Ge devices if active cooling to ~77K (liquid nitrogen temperature) was employed. At this temperature $k_B T$ is one quarter of its room temperature value. Consequently, a semiconductor with half the energy gap of silicon (such as Ge) would have lower leakage currents at 77K than Si has at room temperature.

3.2.2.2 Diffusion current

The diffusion current is a majority carrier current. Thus, diffusion current is carried by electrons leaving the n-type side of the junction and by holes leaving the p-type side. It depends upon the potential barrier opposing motion of majority carriers and on the velocity with which these carriers can diffuse into the material of opposite type, and their concentration gradient. When the carriers cross the point where the Fermi energy is at mid-gap, they switch from being in the majority (and hence stable on average) to being in the minority and are therefore more likely to recombine. Because recombination rate depends upon the product of n and p (see Chapter 2), rapid recombination does not usually become significant in a good diode until the carrier reaches the far end of the depletion region where the majority carriers are not depleted. It is common to assume that no recombination occurs in the space-charge region in a good diode.

The situation is most easily described and visualized based on the concept of the quasi-Fermi level. The normal Fermi level or Fermi energy, E_F, describes the equilibrium situation in a junction, where $np=n_i^2$. The Fermi level can be used to describe the concentrations of both electrons and holes as a result. When the junction is forward biased, electrons and holes diffuse into the depletion region faster than they can recombine locally. In this region $np>n_i^2$. A brief analysis will show that, the Fermi level for electrons cannot lie at the same location as for holes in this case. To distinguish these non-equilibrium and bias-dependent levels from the equilibrium Fermi level, they are referred to as "quasi-Fermi levels", F_n and F_p for electrons and holes, respectively. They are defined based on the relationship between the number of carriers and the intrinsic carrier concentration:

$$F_n - E_i = k_B T \ln(n/n_i)$$
$$E_i - F_p = k_B T \ln(p/n_i)$$

3.28

E_i is the intrinsic energy, near midgap, given by Equation 2.31. The quasi-Fermi levels for a forward biased junction are shown schematically in Figure 3.6. The excess carriers that diffuse across the junction eventually recombine with the local majority carriers and the quasi-Fermi levels return to the equilibrium value, E_F.

The number of carriers reaching the far side of the depletion region rises exponentially with the separation of the quasi-Fermi level from E_F. If there is negligible recombination of carriers in the depletion region, then this separation is the forward bias voltage, $V_{applied}$, see Figure 3.6. The current density for the junction is given by the injected carrier densities, which one could calculate from Equation 3.28. This increases exponentially with voltage but is limited, as was drift current, by the rate at which carriers diffuse into the depletion region. An alternate way to see why there should be an exponential relationship between applied voltage and injected current is that to overcome the potential barrier opposing diffusion current an electron must have extra energy. The number of such electrons is governed roughly by the Boltzmann distribution (see Chapter 2), which increases exponentially as the potential barrier decreases. Therefore, as the barrier decreases the number of injected carriers increases exponentially. It is preferable to use the quasi-Fermi level description because it continues to hold for strong forward bias where the barrier to diffusion has vanished but the current continues to increase exponentially.

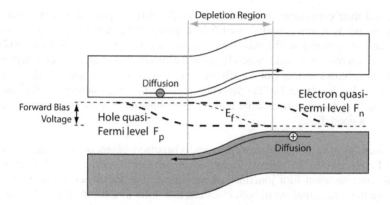

Figure 3.6: Shows the quasi-Fermi levels for a forward biased junction. The deviation of the quasi-Fermi levels from the Fermi level indicates the excess number of injected carriers that changes exponentially with the difference in these levels.

The rate at which carriers diffuse into the depletion region, J_0, is calculated by determining the amount of current at the edge of the depletion region where minority carriers are injected. This is the equilibrium minority carrier population enhanced by injected carriers (as determined by the quasi Fermi level relative to the equilibrium Fermi level). The calculation is conducted in the same way as for drift current and uses the same variables. The resulting prefactor is therefore the same:

$$J_{forward} = q\left(\frac{D_p}{L_p}p_n + \frac{D_n}{L_n}n_p\right)e^{qV_{applied}/k_BT} = J_0 e^{qV_{applied}/k_BT} \qquad 3.29$$

3.2.2.3 Total current in a junction

The total current in a junction is just the sum of the drift current (Equation 3.24) and the diffusion current (Equation 3.29):

$$J = J_0(e^{qV_{applied}/k_BT} - 1), \qquad 3.30$$

where we have taken advantage of the fact that the coefficient of the exponential term in forward bias is just the reverse current J_0. This current/voltage relationship, known as the diode equation, gives rise to an operating behavior as shown in Figure 3.7. The current density in reverse bias is nearly constant, as one would expect from

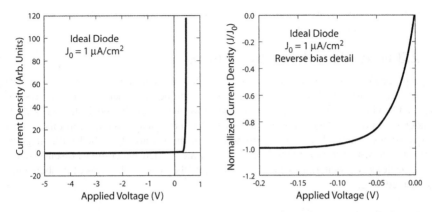

Figure 3.7: The current voltage curve for an ideal diode obeying the diode equation. Note that the left-hand plot is exponential although on this scale it looks square. This shows how dramatic the turn-on behavior of a diode is.

equations 3.24 and 3.30, while it rises exponentially in forward bias. The smaller the reverse saturation current density, the larger the forward bias that must be applied before a given current density is reached (Figure 3.8).

Because the magnitude of both the forward and reverse currents in a homojunction diode depend upon the minority carrier properties, the devices are referred to as minority carrier devices. This is critical to the relationship between diodes and defects in the material as defects affect the minority carriers far more than the

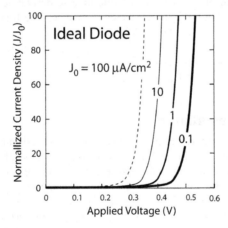

Figure 3.8: Comparison of forward currents as a function of reverse current J_0.

majority carriers. This, in turn, is a major reason why the materials science of semiconductors is important to electronic devices.

3.2.3 Non-ideal diode behaviors

The preceding discussion is for an ideal diode. Many silicon devices operate very nearly as perfect ideal diodes. Our task when engineering materials for diodes and related devices is to understand not only how they work when right, but also to understand how defects in the materials relate to non-ideal behaviors. In this section, some of the major observable malfunctions in diodes are described without going into the causes of the problems extensively. Later, when we come to discussions of specific defects in materials we will consider how they lead to the sorts of misbehaviors described here.

3.2.3.1 Series resistance

One of the simplest forms of poor performance in a diode is series resistance. In reverse bias an ideal diode has an extremely high resistance. In this case, series resistance in the remainder of the device is irrelevant as the diode junction is the most significant resistance in the circuit. In forward bias the resistance drops exponentially as the forward voltage is increased. At some point, the resistance of the diode will drop below that of some other circuit element. In an otherwise perfect diode the dominant series resistance term will usually be the semiconductor on the lightly-doped side of the junction outside of the depletion region. In some cases, contact resistances enter the picture before the resistivity of the doped semiconductor becomes a factor. A series resistance leads to a linear current-voltage curve in forward bias as shown in Figure 3.9. This can be accounted for by determining the resistance of an ideal diode as R=dV/dJ (the inverse derivative of the ideal current voltage curve) and adding this to the series resistance. The final current is then the voltage divided by the net resistance:

$$ J_{total} = \frac{V}{R_{series} + |dV_{diode}/dJ|_{no-series-R}}. \qquad 3.31 $$

3.2.3.2 Shunt resistance

A perfect diode has no short circuits or other unintended resistances from the front to the back contact. However, in some non-ideal cases it is possible for a top contact to punch through the device and make a direct connection to the back contact. This results in a shunt resistance in the diode. Shunt resistances can arise from physical defects such as holes in the device layer, high conductivity paths such as grain boundaries or dislocations (see Chapter 7) or simply poor performing regions of the diode. The shunt resistance allows current to flow in spite of the high resistance of the good portion of the diode in reverse bias. In forward bias the shunt only has an effect at very low voltages where the diode resistance is still high. Figure 3.10 shows a typical behavior for a shunted diode.

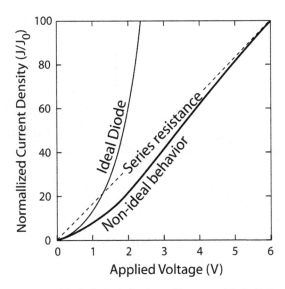

Figure 3.9: Compares an ideal diode behavior with a non-ideal diode exhibiting series resistance.

Shunt resistances can be taken into account easily in the diode equation by addition of a term $J_{shunt}=V_{applied}/R_{shunt}$ Together with the series resistance of Equation 3.31, this yields:

$$J_{total} = \frac{V}{R_{series} + \dfrac{1}{1/R_{shunt} + |dJ_{diode}/dV|_{no-series-R}}}. \qquad 3.32$$

Note that the dJ/dV derivative is the inverse of the derivative shown in Equation 3.31.

3.2.3.3 Recombination in the depletion region

When many defect states are present in the depletion region of the semiconductor, significant recombination can occur in spite of the electric field attempting to sweep away free carriers. Typically recombination in the depletion region is by trapping of moving charges by defect states (see later chapters). This has no obvious effect on reverse (drift) current as the injected carrier begins as a minority carrier and becomes a majority carrier upon transiting the depletion region. If it recombines before it becomes a majority carrier then one effectively reduces J_0. In contrast, however, there is a large effect on forward current, which is increasing exponentially.

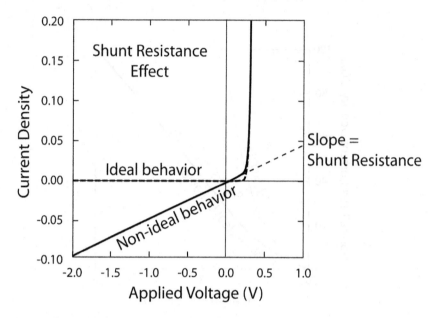

Figure 3.10: A non-ideal behavior due to shunt resistance compared with an ideal diode.

Recombination in the space charge region dramatically increases the voltage needed to achieve a given level of current injection. It can be shown that the exponent in the diode equation is modified to yield:

$$J = J_0(e^{-qV_{applied}/ak_BT} - 1).$$

3.33

In this equation the added term "a" is called the diode ideality factor. The value of the ideality factor ranges from 1 for an ideal diode to at most 2. The latter occurs when the minority carrier diffusion lengths (before recombination) are small relative to the depletion width, W. The consequence of non-ideal diode behavior is shown in Figure 3.11. Effectively, the forward resistance of the diode can be doubled at a given voltage.

As a practical note for those working with poor quality junctions, it is possible to observe current-voltage curves that appear to have ideality factors greater than two. This is generally an indication of problems with the contacts or other voltage-dependent series resistances. Consider, for example, the curve shown in Figure 3.9 for a diode with series resistance. If the series resistance decreases gradually as the voltage is increased then the behavior will appear, at least locally, to be diode-like but with a high ideality factor.

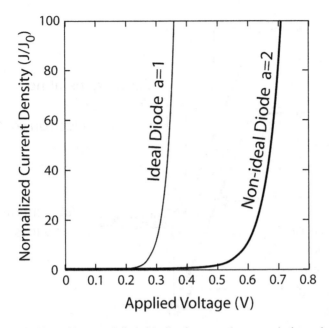

Figure 3.11: The effect of increased diode ideality factor on the current/voltage characteristic.

3.2.3.4 Reverse breakdown

When the diode is reverse biased, the ability of the junction depletion region to resist current flow is limited. Eventually, the dielectric behavior of the reverse-biased junction must fail. There are three forms of breakdown failure in reverse bias: tunneling breakdown, avalanche breakdown, and the soft breakdown associated with current flow through defects. Of these, avalanche breakdown is generally catastrophic because the resistance of the diode drops as the process begins. The resulting increased current flow heats the diode and heating increases the current flowing. Thus, avalanche breakdown is self-amplifying. The other mechanisms tend to be less of a problem because they are generally self-limiting. Indeed, tunneling breakdown is used in Zener diodes to provide voltage control. The three mechanisms above are considered briefly in turn below.

Avalanche breakdown occurs when the depletion width is relatively large (light doping on both sides of the junction). It is most obvious when the semiconductor is of high quality. The basic process is shown in Figure 3.12.

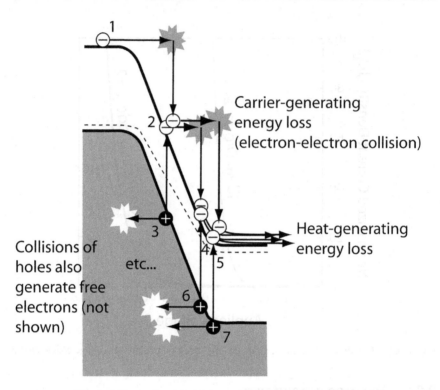

Carrier-generating
energy loss
(electron-electron collision)

Heat-generating
energy loss

Collisions of
holes also
generate free
electrons (not
shown)

etc...

Figure 3.12: A schematic of the avalanche breakdown process.

In the event illustrated, a single (typically thermally-generated) minority carrier (1) enters the depletion region and is accelerated. The carrier gains more kinetic energy than the energy gap before colliding with another electron. As its energy exceeds the gap energy, the number of electrons with which it can interact rises abruptly (think about application of Equation 2.23) and it experiences a collision with a valence band electron, losing its kinetic energy and transferring the valence band electron to the conduction band, resulting in an additional free electron (2) and a free hole (3). In the example shown, both the free electron and the free hole can go on to experience additional ionizing collisions. The original electron (1) and the newly free electron (2) are shown accelerating and undergoing such additional collisions, creating two more electrons (4 and 5) and two more holes (6 and 7). The electrons 4 and 5 are too close to the edge of the depletion region to gain sufficient kinetic energy to cause more ionization. The holes, however, can do so. Thus, beginning from a single electron, a continuously increasing current is created. Note that only a fraction of the minority carriers entering the depletion region will be so efficient in producing additional carriers. Many collisions produce heating but no new carriers. However, as the voltage increases the fraction of collisions producing new carriers increases dramatically.

As the reverse bias rises, at some point the cascade begins. Worse yet is the fact that the additional current and non-ionizing collisions heat the system and increase the rate at which minority carriers are created, increasing the rate of the process and hence the reverse current. Therefore, the resistance continues to drop rapidly as the current increases. Unless some other series resistance limits the current, this leads to the diode acting as a fuse and burning out.

The avalanche breakdown voltage can be estimated as follows. The mobility of an electron is $\mu_e = q\tau_e/m_e^*$, where τ_e is the mean time between scattering events. A similar expression can be written for holes. An electron traveling at a velocity v between scattering events moves a distance $\ell = v\,\tau_{.e}$ After some algebra, it can be shown that $\ell = m_e^*\,v\mu_e/q$. For silicon, the velocity of an electron is $\sim 10^7$ cm s^{-1}, the free electron mass is 9.1×10^{-31} Kg and the relative effective mass is 0.19 (light hole), and the electron mobility is ~ 1000 cm^2 V^{-1}s^{-1}. A typical mean length between scattering events is then $\ell = 1.1\times10^{-6}$ cm (11 nm). The breakdown field $\mathcal{E}_{BR}$ is that field at which an electron accumulates at least the energy gap of energy over a distance of ℓ. The relationship is $q\mathcal{E}_{BR}\,\ell = E_{gap}$. For silicon, the energy gap is ~ 1.1 eV at 300 K. Using the value of ℓ calculated above, $\mathcal{E}_{BR} \sim 1\times10^6$ V cm^{-1}. Note that $\mathcal{E}_{BR} = V_{BR}/W$ where V_{BR} is the breakdown voltage and is a negative value. Therefore, as the doping rises and the depletion region shrinks, the voltage that must be applied decreases to obtain a given field. Thus, the breakdown field is inversely related to the square root of the doping concentration.

Putting the pieces above together,

$$\mathcal{E}_{BR} = \frac{E_{gap}}{m^*\,v\mu_e},$$

3.34

or using Equation 3.20, assuming $V_{BR} \gg V_{bi}$ and noting that $V_{BR} < 0$,

$$-V_{BR} = \frac{2\varepsilon}{q}\left(\frac{E_{gap}}{m^*\,v\mu_e}\right)^2 \frac{N_A + N_D}{N_A N_D}.$$

3.35

The relationship of breakdown voltage to doping is shown schematically in Figure 3.13. The experimental results for Si are very similar to the result obtained from Equation 3.35 using the parameters for Si given above. The fact that the sample calculation was based on the minimum *indirect* energy gap of Si shows that only a bare minimum opportunity for new carrier generation is required for avalanche breakdown to begin.

Tunneling breakdown is far better behaved. This process occurs in heavily-doped junctions where the depletion region is relatively narrow. In this case, before the avalanche breakdown process can begin, electrons can tunnel through the depletion region from filled states into empty adjacent states (see Figure 3.14).

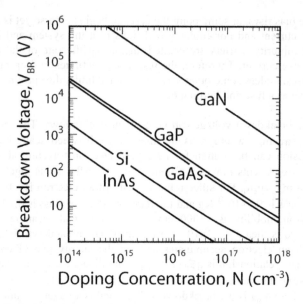

Figure 3.13: Typical breakdown voltages as a function of doping level calculated from Equation 3.34 and 3.20. Note: the curvature at low breakdown voltage shows the correction for non-zero V_{bi}.

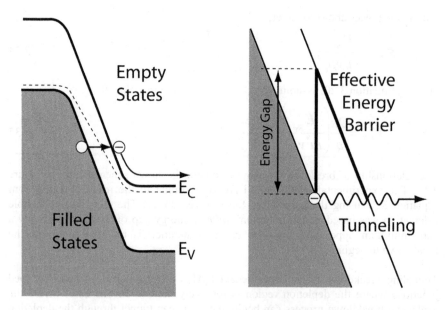

Figure 3.14: A schematic of the tunneling breakdown process.

The tunneling probability, P_{Tu}, through the barrier is given by:

$$P_{Tu} = e^{-C\sqrt{m^* E_{gap}^3}/\mathcal{E}}$$

3.36

where C is a constant, and $\mathcal{E}$ is the electric field of the junction [$\mathcal{E}=(V_{bi} - V_{appl})/W$]. This formula may be developed based on the decay of the electron wavefunction within a potential barrier, similar to the discussion in Section 2.4. The tunneling current is proportional to the tunneling probability and the number of empty states adjacent to filled states. It increases exponentially with applied voltage. Higher energy gap materials have lower tunneling currents, as do junctions with higher depletion widths. Likewise, semiconductors with higher densities of states at the band edges have higher tunneling currents.

One of the important points about tunneling breakdown is that as current begins to flow, other resistances in the circuit drop more of the applied voltage. This reduces the voltage across the diode and consequently limits the tunneling current. Furthermore, the tunneling current does not increase significantly with increasing temperature. These factors combine to make tunneling breakdown a relatively benign process, which is less likely to cause the diode to fail than is avalanche breakdown.

Finally, soft breakdown occurs by generation of carriers through defect states in the energy gap. The sources of these defects are described in detail in Chapter 7. Both the avalanche and tunneling breakdown mechanisms can be mediated by such states. Carriers may tunnel through defect states rather than moving directly band-to-band. Likewise, collisions with energies too low to excite carriers across the entire energy gap can move electrons into states in the middle of the gap or from there on into the higher energy band edge, resulting in avalanche-like breakdown, but with a current limited by the number of defect states participating. As the reverse voltage is increased, more and more defects come into the depletion region. Furthermore, the Fermi energy, which is making a smooth transition from the conduction band edge to the valence band edge in the depletion region, shifts with bias. This leads to changes in the charge state of defects in the energy gap. These effects combine to increase the amount of reverse current flowing in a diode with increasing reverse bias. While this does not lead to the very sudden increases in reverse current which the other two mechanisms produce, soft breakdown can render a device unusable.

The result of the above three breakdown mechanisms on the diode current-voltage curve is shown in Figure 3.15. The avalanche and tunneling mechanisms have very abrupt onsets. Typical results can be modeled roughly as

$$J_{rev} = \frac{J_0}{1-(V_{rev}/V_{BR})^n},$$

3.37

where n normally ranges from 3 to 6. Higher quality materials produce more abrupt onsets and larger values of n.

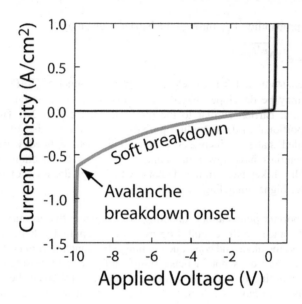

Figure 3.15: Combined effects of soft breakdown and avalanche breakdown on the current/voltage characteristic of a diode. The soft breakdown is shown as might be observed in a very poor diode. Tunneling has a very similar behavior to avalanche breakdown but is not self-amplifying.

3.3 SCHOTTKY BARRIERS AND OHMIC CONTACTS

Not all diode junctions involve contacts between different doping types of the same semiconductor ("homojunctions"). The remainder, junctions between dissimilar materials or "heterojunctions", are generally divided into metal/semiconductor and semiconductor/semiconductor junctions. We consider metal/semiconductor heterojunctions first. A sufficiently degenerate semiconductor behaves essentially as a metal and produces results nearly identical to the metal/semiconductor behavior. Metal/semiconductor junctions turn out to have either linear (ohmic) or diode-like current voltage characteristics and are called ohmic contacts or Schottky diodes, respectively.

3.3.1 Ideal metal/semiconductor junctions

To figure out the consequences of a junction between a metal and a semiconductor we need to know how electrons flow upon making the contact. Consequently, we need to know the chemical potentials (Fermi energies) of electrons in the two materials. The materials property describing the electron chemical potential in metals is the work function, Φ. It measures the position of the Fermi energy with respect to the vacuum level (the lowest energy to which an electron must be raised to escape from the surface). This is a well-defined constant for a given pure metal, although it changes in alloys.

The work function of a semiconductor also measures the position of the Fermi level. However, this value changes with doping and so it does not make a very good general-purpose description such as one might provide in a table of semiconductor properties. The energies of the band edges with respect to the vacuum level do not depend significantly upon doping so they are much better choices. The conduction band energy relative to the vacuum level (the electron affinity, χ_s) is therefore used to quantify the general properties of semiconductors (see Figure 3.16).

Experimental measurements generally determine the work function or other related doping-dependent quantities. Careful conversion to the doping-independent value is therefore needed to determine the electron affinity. The measurements are difficult and many factors can affect the outcome. Therefore, electron affinity values have often been the source of significant debate in the experimental literature. Real device

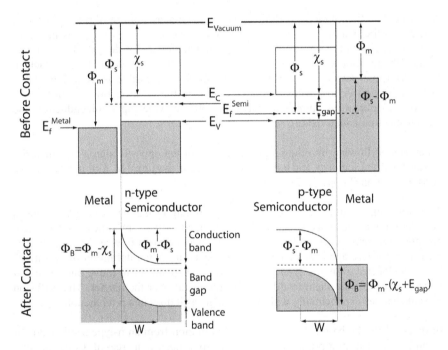

Figure 3.16: Shows the definitions of the electron affinity χ_s of the semiconductor and the work functions of the semiconductor, Φ_s and metal, Φ_m. Also indicated are the depletion width, W, the Fermi levels, band edges, and the Schottky barrier heights, Φ_B. Note that Schottky barriers to both n and p-type semiconductors exist but that to calculate the barrier for the p-type semiconductor you need to compare the valence band edge with the metal work function while for the n-type material it is the conduction band that is compared to the work function.

behaviors depend upon doping, of course, so while dopant-independent values are handy for tables and for device design, doping must be known to understand a given device under observation. Below, we will assume that we know the actual Fermi level of the semiconductor and metal precisely and will leave it to the experimentalists to worry about how this can be managed (see Section 5.5.1).

When contact is made between a metal and a semiconductor, electrons flow from one material to the other as determined by their relative Fermi levels. This flow can lead to depletion or enhancement of the majority carriers. There are four possible situations to consider: a low work function metal in contact with (1) a p-type or (2) an n-type semiconductor, and the same cases for a high work function metal. "High" and "low" work function here simply indicate whether the Fermi level for the metal lies above or below the Fermi level of the semiconductor before contact is made. These four possibilities are shown schematically in Figure 3.17.

When the direction of electron flow is into n-type material or out of p-type material, it increases the majority carrier concentration of the semiconductor. This increases the conductivity of the semiconductor near the junction and is described as "ohmic" because the junction acts only as a resistor. Such junctions are desired for contacts to microelectronic devices. Unfortunately, it is often impossible to find a metal of sufficiently high or low work function to arrange the electron flow in the correct direction to obtain an ohmic contact, especially to wide-gap semiconductors and organic electronic materials.

When charge flowing into the semiconductor is of the opposite sign as the majority carrier type, the majority carriers are depleted near the junction forming a resistive depletion region (Figure 3.17).

An applied bias voltage appears across this relatively insulating depletion region, as in a homojunction. This means that the applied voltage modifies the chemical potential of electrons across the junction and induces current flow as shown in Figure 3.18. For positive applied voltages (forward bias under the definition of "positive" here), the barrier for majority electrons in the semiconductor to move into the metal is decreased, causing them to diffuse into the metal over the barrier. The diffusion current changes exponentially with bias voltage, as it does in a p-n junction.

In reverse bias the barrier height for electron motion from an n-type semiconductor or hole motion from a p-type semiconductor to a metal is increased. Consequently, the flow of electrons/holes in this direction decreases exponentially to near zero. Charge flow from the metal to the semiconductor faces a constant barrier height, Φ_B (Fig 3.18). This height is characteristic of any metal-semiconductor junction independent of the semiconductor doping, and is known as the Schottky barrier height, $\Phi_B = \Phi_M - \chi_S$. Note that in the case of a Schottky barrier on a p-type semiconductor one must compare the metal work function with the valence band edge so $\Phi_B = \chi_S + E_{gap} - \Phi_M$.

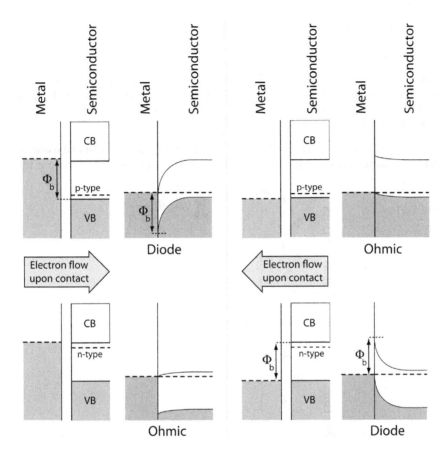

Figure 3.17: The four possible metal-semiconductor cases before and after contact. Two of the four produce ohmic contacts (resistors) while the others produce Schottky diodes.

The current voltage curve for the resulting metal-semiconductor junction has the same form as a homojunction diode:

$$J = J_0 (e^{qV/ak_BT} - 1),$$
                                                                                                3.38

with the same variable definitions as in Equation 3.29; although the reverse saturation current is altered. The reverse saturation current now depends upon a Boltzmann factor describing the probability that an electron will have sufficient energy to overcome the barrier height rather than the semiconductor energy gap. The complete derivation of this current is based on the theory of thermionic emission, which describes the similar process by which electrons are emitted from a hot filament. The final form of the reverse saturation current is:

$$J_0 = ABT^2 e^{-q\Phi_B/k_BT},$$ 3.39

where A and B are constants.

Note that nowhere in Equations 3.38 or 3.39 do minority carrier properties appear. A very significant advantage of this type of diode is that its operation does not depend upon the minority carrier properties of the semiconductor. The only carriers injected across the junction are majority carriers.

Schottky diodes are very useful. However, they can be annoying when one wishes to make an ohmic contact to a semiconductor. It is often impossible to make a true

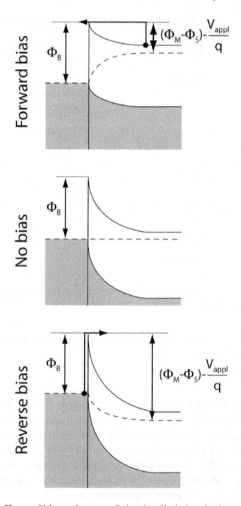

Figure 3.18: The effects of bias voltage on Schottky diode band edges and band bending.

ohmic contact. Fortunately, a Schottky diode can be made to operate as an ohmic contact in many cases by increasing the doping level of the semiconductor near the junction. In this case, the depletion width in the semiconductor becomes very small (and can be estimated from Equation 3.20 taking one of the doping concentrations to be one electron per atom for the metal). Very small depletion regions lead to very low tunneling breakdown voltages. Once breakdown occurs the diode operates as an ohmic contact. The question of whether an ohmic-like behavior can be obtained often revolves around whether the semiconductor can be doped sufficiently strongly.

3.3.2 Real schottky diodes

Unfortunately, the real situation with Schottky diodes is not so good as was presented in the previous section. A hint of such problems was provided in Equation 3.38 where a diode ideality factor appears. Theoretically, one can measure the work function of both the metal and the semiconductor using photoelectron spectroscopy and other techniques. From this and the work function of the metal, one should be able to estimate the Schottky barrier height.

Unfortunately, it is often very difficult to determine the Fermi level of semiconductors because of the dangling bonds at their surfaces. Even if no surface phase forms and no oxide is present, clean surfaces of semiconductors typically have a large number of defect states. Because of the limited doping and carrier concentration of the semiconductor, the Fermi level at the surface is determined by the energy of these defects, rather than by the dopant atoms. Consequently, a surface Fermi level measurement usually gives an unreliable estimate of the barrier height. Indeed, it is generally more reliable to measure the barrier height (i.e. the Schottky barrier reverse current as a function of temperature) to determine the Fermi level of the semiconductor than the reverse. The typical barrier potential in a real Schottky diode is shown schematically in Figure 3.19. Clearly it does not match the shape given in Figures 3.16-3.18.

As an additional complication, thin layers of compounds can form at the junction or short-range diffusion can broaden the effective interface. This spreads out or changes the potential at the boundary and causes a change in the effective barrier. We take advantage of such reactions in forming metal-Si Schottky barriers by reacting the metal with the semiconductor to form a silicide. This gives a very reliable contact because the junction is formed by solid phase reaction within the Si rather than at the original semiconductor surface resulting in few interfacial defects.

Finally, interfacial defects can permit reverse current to flow by tunneling through the resulting states (see Figure 3.19), which can lead to an unexpectedly high current (an anomalously low apparent barrier). The net result is a continuing series of debates about what is the electron affinity of most semiconductors. The debates are most important for wide-gap semiconductors where the barriers are high, and knowing the height is critical to understanding devices. In such semiconductors, surface defects and phases are more important, and obtaining good contacts is difficult.

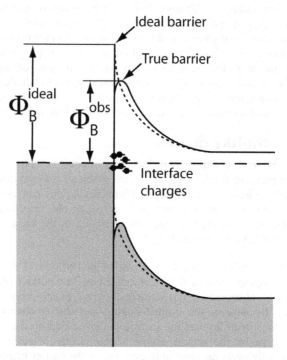

Figure 3.19: A defective Schottky contact, typical of real barriers. The small lines indicate defect states in the semiconductor.

3.4 SEMICONDUCTOR HETEROJUNCTIONS

Heterojunctions are any electrical junctions between dissimilar materials. The term is generally used to refer to semiconductor heterojunctions, although technically it applies to Schottky contacts and even to metal-metal junctions such as thermocouples. Here we will use the conventional terminology and consider semiconductor/ semiconductor junctions.

Semiconductor heterojunctions are very useful because they allow the materials on either side of the junction to be optimized for specific tasks. For example, one semiconductor can be used to emit light while a second can be used as a top contact and/or can be selected for transparency to the emitted light and for high conductivity.

For semiconductor heterojunctions one must consider a wide variety of issues. These include:

- **Optoelectronic Properties** – What are the electronic and optical properties of the two semiconductors under consideration and do they meet the needs of the application?

- **Interface Barriers** – What electrostatic fields develop at the heterojunction? Are these compatible with the desired application? A related issue is the presence and nature of defects at the interface that contribute to and modify the resulting fields.

- **Current Flow** – It is possible with semiconductor heterojunctions to select for current flow by one charge or the other (electrons or holes) across a junction with less dependence on doping than in the homojunction.

- **Epitaxy** – It is possible to grow one semiconductor as a single crystal on another but one must decide upon the best growth method (see Chapters 10-12). A non-epitaxial heterojunction will generally have far more interface states than an epitaxial one.

- **Lattice mismatch** – If epitaxy is possible any difference of lattice size must be accommodated at the interface and may introduce defects.

- **Interface stability** – Do the semiconductors mix across the junction or is the interface thermodynamically stable?

In this section we consider only the electronic properties of semiconductor heterojunctions. Optical properties will not be considered, as they are determined primarily by the individual materials and not by the nature of the heterojunction.

3.4.1 Heterojunctions at equilibrium

The critical points in determining the electronic properties of a semiconductor heterojunction, assuming structurally and chemically perfect materials, are the energy gaps, electron affinities, and doping types and levels. It is easiest to begin by assuming that the Fermi levels of two semiconductors are arranged (through doping or just good fortune) to be at the same energy before contact. This "flat band" condition means that upon contact there is no net flow of electrons across the junction and the band edges remain flat.

The energy gaps and electron affinities of flat band semiconductor heterojunctions lead to three possible band-edge configurations, shown in Figure 3.20. These are referred to as follows. "Straddling" is the situation where the conduction and valence band edges of the narrow-gap semiconductor lie within the energy gap of the wide-gap semiconductor. To obtain a straddling gap one requires one gap to be significantly larger than the other ($E_{gA} > E_{gB}$) and the electron affinity of the smaller gap material to be slightly larger than that of the wider gap material ($E_{gA}-E_{gB} > \chi_B - \chi_A > 0$). A typical example of a straddling gap configuration is the AlAs-GaAs heterojunction. "Offset" refers to a situation where the energy gaps are roughly equal ($E_{gA} \sim E_{gB}$) but the electron affinities are different (eg.: $\chi_A > \chi_B$). Further, that the difference is less than the energy gap of either constituent. This means that both band edges of one semiconductor lie above both band edges of the other but where a portion of the energy gap overlaps across the interface. An example of an offset-gap heterojunction is between InSb and InP. Finally, the "broken gap" semiconductor heterojunctions are composed of semiconductors with such extreme differences in electron affinity that the band gaps do not overlap at all. In this case the valence band

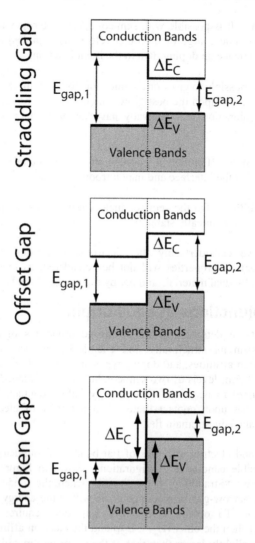

Figure 3.20: Shows schematically the band edge offsets in the flat band condition for the types of semiconductor heterojunctions.

edge of one semiconductor lies above the conduction band edge of the other. An example of this type of junction is between InAs and GaSb. Such semiconductor heterojunctions usually involve at least one relatively narrow-gap material in order to achieve sufficient difference in electron affinities.

What is at issue in these junctions is the energy of the two band edges. There are some simple "rules of thumb" which indicate the trends one can expect in band edges as a function of which elements are changed across the junction. We will see in Chapter 5 that there are very good reasons for these rules, and we will examine their basis in the materials physics of bonding in semiconductors in that chapter. The commonly cited rules of thumb are:

Linearity and Transitivity: Because the intrinsic energy gap and band edge positions do not change in one semiconductor just because it is joined to another, if you know the relative band edge positions for two semiconductors the band edge discontinuity can be determined. Mathematically, the statement of this "linearity" principle is

$$\Delta E_V = E_V(B) - E_V(A),$$ 3.40

where $E_V(A)$ and $E_V(B)$ refer to the valence band edge energies of two semiconductors, A and B and $\Delta E_V = \Delta E_V(A{:}B)$ is the valence band offset. A similar equation may be written for the conduction band. Furthermore, if you know discontinuities between any two semiconductors and a third semiconductor, the discontinuity between the first two can be inferred. This "transitivity" criterion is:

$$\Delta E_V(A:B) + \Delta E_V(B:C) + \Delta E_V(C:A) = 0.$$ 3.41

The Common Anion Rule: When the anion (the electron accepting atom such as As in GaAs and InAs) is in common across a semiconductor heterojunction, the change in the conduction band edge is greater than the change in the valence band edge across the semiconductor heterojunction. Mathematically, $\Delta E_V < \Delta E_C$.

The Common Cation Rule: When the cation (e.g. Ga in GaAs or GaSb) is in common across the junction, the valence band edge energies scale with the anion electronegativities. For example, the valence band edge of phosphide semiconductors will lie below those for arsenides which will lie below those of antimonides. Mathematically, $E_V(CA_1) < E_V(CA_2) < E_V(CA_3)$, where C designates a cation, and A_1, A_2, and A_3 designate three anions with decreasing electronegativities.

When arbitrary doping is allowed in a semiconductor heterojunction or when the Fermi energies of undoped materials do not match spontaneously, electron transfer causes band bending as in homojunctions, bringing the Fermi energies to equilibrium. The following discussion describes how to determine the effect of electron flow on the potentials at the heterojunction. There are many possible configurations of a heterojunction and only a few examples are discussed here.

Data used in drawing this figure:		
Compound	Electron Affinity	Energy Gap
n-AlAs	3.56 eV	2.16 eV
p-GaAs	4.07 eV	1.42 eV
N_A(GaAs) ~ N_D(AlAs)		

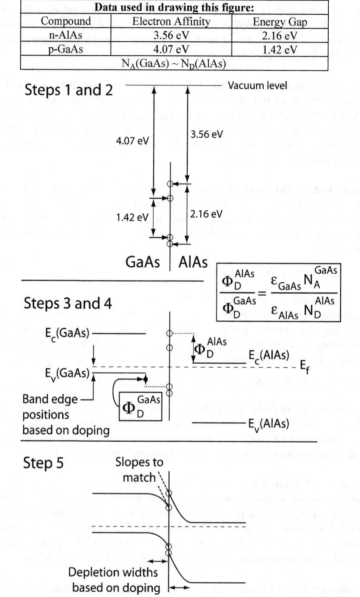

Figure 3.21: Recommended steps in drawing a semiconductor heterojunction. Step 5 shows the complete band edges for this junction.

The best way to determine the effect of electron transport on a semiconductor heterojunction is to sketch a band edge diagram. This can be complicated. Here is a series of steps for making sketches that may help to obtain qualitative ideas of the junction behavior. (See an example in Figure 3.21.)

- Mark a vertical line on a page to indicate the junction and a horizontal line to indicate the flat-band vacuum level.

- Based on a constant vacuum level, mark the band edge positions on the vertical line using the electron affinity and energy gap values for the two materials. These positions are doping-independent and fix the band edges in relation to one another at the junction.

- Draw a horizontal dashed line indicating the Fermi level at equilibrium (see discussion below for where to mark it relative to the points of the band edges). Deciding where to mark this line is the most tricky part of the drawing. How it is marked determines how band bending is distributed in the various semiconductors.

- Far from the junction, mark the band edges relative to the Fermi level as appropriate to the doping in each semiconductor.

- Connect the band edges far from the junction to the band edges you marked in Step 2 as appropriate. Requirements to be met in this step are:

I.   *the slopes of the connecting segments must match* where they intercept the points marked in Step (2) to satisfy the electrostatic continuity equation.
II. *the energy gaps must be constant on each side of the junction* because the gap for a given semiconductor does not change.
III. *the depletion widths are fixed by the doping levels.* These are calculated as in Equation 3.19.

These steps are shown for a p-GaAs/n-AlAs heterojunction in Figure 3.21.

The major question is where to put the Fermi level with respect to the flat-band band edges (step 3). Marking this position will determine on which side of the heterojunction most of the band bending occurs. When one semiconductor is very heavily doped the choice is easy. Mark the Fermi level with respect to the flat band edge of the heavily-doped material. There will be very little band bending in that material (see Figure 3.22).

When the doping levels are nearly equal (see Figure 3.21), the resulting electric field will be distributed roughly equally across the heterojunction and the band bending is nearly symmetric. In this case the Fermi level must be drawn close to the middle of the distribution of flat band points (Figure 3.21). The details of exactly how to distribute the band bending in moderately-doped materials become messy and are beyond the scope of this text. The results are determined by solving the diffusion

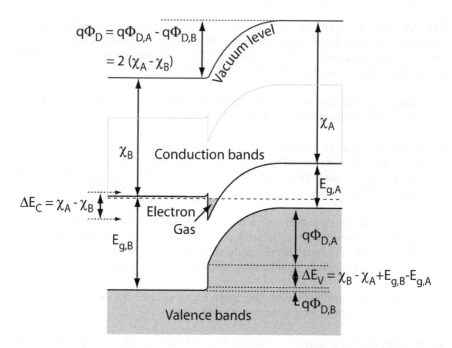

Figure 3.22: Shows the formulas for calculation of various energies in the heterojunction. This case is similar to that shown in Figure 3.21, but with very heavy doping in the AlAs layer to produce the electron gas.

equation (which covers how electrons respond to chemical potential differences) self consistently with the Poisson equation (which determines the resulting electric fields). Here, we will have to be content with an estimate as to how the bands bend. The potentials (as shown in Figure 3.21) are distributed approximately, assuming fully ionized dopants, as:

$$\frac{\Phi_D^B}{\Phi_D^A} \approx \frac{\varepsilon_A N_{dopant}^A}{\varepsilon_B N_{dopant}^B},$$

3.42

where Φ_D indicates the potential energy difference associated with band bending on the designated side of the junction. Furthermore, $\Phi_D^A + \Phi_D^B = \Delta E_F$ or the contact potential (difference in Fermi energies of the semiconductors before contact).

As in the flat band picture there are discontinuities in the band edges at the hetero-junction. These are one of the significant features of a semiconductor heterojunction and have important uses in making devices. The discontinuities result from the fact that not only the doping and the electron density are changing across the interface, but that the atoms themselves are changing. The effective charges on the cores of the atoms making up the lattice on the two sides of the junction change as one goes across the interface. This leads to a step function in electron potential. The magnitudes of the band edge discontinuities can be calculated from the electron affinities and energy gaps of the two materials, as shown in Figure 3.21

One remarkable example of a discontinuity-related effect is the interfacial spike in the conduction band edge, shown in Figure 3.22. The wider gap AlAs is doped n-type in this case. Enough electrons flow into the GaAs that near the heterojunction it is made so n-type that it becomes a metal. With careful design, this metallic layer can be rendered very thin such that the electrons find it very hard to move perpendicular to the heterojunction. In other words, the potential traps them adjacent to the interface. Because they are free to move randomly in this plane and have a moderate to high concentration they are referred to as a two-dimensional electron gas. Because their motion is restricted, it is more difficult for them to scatter since they must remain in the interface plane. The consequence is that the electrons in the interface have a higher mobility than electrons in the bulk semiconductor.

The presence of a spike at the heterojunction relies on the junction being abrupt on the atomic scale. Many billions of dollars worth of research have led to significant enhancements in crystal growth methods and to understanding and controlling the interface abruptness so that phenomena such as spikes at the interfaces can be used. Applications of heterojunction discontinuities are mentioned briefly in Sections 3.5 and 3.6.

Band-edge discontinuities are not always present at heterojunctions. If one intentionally or unintentionally grades the interface between two semiconductors, for example by diffusion, the electric field resulting from the change in atomic core potentials is averaged over the distance of the grading, leading to a smooth transition from one band to another. This will round off discontinuities in the band edges at the interface or eliminate them entirely. A major issue related to whether a junction will be abrupt is the miscibility of the two semiconductors. Thus, two materials that have little or no solid solubility in each other (for example Si and GaAs) will form an abrupt interface. The stability of this interface is based on thermodynamics and will persist over wide ranges of temperature and other experimental conditions. Materials with large solid solubilities such as GaAs and AlAs will not necessarily form an abrupt interface. In this case, the growth conditions must be adjusted to prevent mixing through limiting atomic diffusion by choice of materials and by minimizing the temperature of the junction at all times. Heterojunctions between materials which are both highly miscible and for which atomic diffusivities of constituent atoms are high may not to lead to abrupt interfaces. Interdiffusion may also be enhanced by impurities. This is used selectively in some laser diodes to control where current injection and emission occur.

3.4.2 Heterojunctions as diodes

For now, let us assume a perfect abrupt semiconductor heterojunction, and consider the implications of applied electric fields on the current conduction process. The junction shown in Figure 3.21 (p-GaAs/n-AlAs) will exhibit a diode-like behavior. A "forward" applied bias will reduce the potential between the layers, narrow the depletion region, and lower the barrier for carrier injection (by diffusion) across the junction. However, because of the opposite signs of the band gap discontinuities (one is a step upward, the other a step downward), electrons emitted into the GaAs, in this case, will have a much lower barrier to overcome than will holes injected from the GaAs into the AlAs. Thus, most of the forward diffusion current will be carried by electrons. In reverse bias the currents are simply the thermally-generated carriers in the GaAs which reach the depletion region and are swept into the AlAs. The much larger energy gap of the AlAs means that there will be relatively few minority carriers in this material and thus there will be no significant contribution of minority carrier drift from the AlAs to the GaAs. The barriers at the heterojunction will usually have little effect on the reverse current drift process in high-quality materials, as the carriers will generally gain enough energy to pass over them. The barrier is more likely to have an effect the higher it is. However, in a lightly-doped junction where the depletion region is broad, and where the scattering of carriers is rapid for any reason, the spike in the conduction band can influence the behavior of the device. The minority carrier properties in the GaAs will generally affect this type of device as current injection and recombination processes are very similar to the homojunction case. In other types of heterojunctions, this is less apparent.

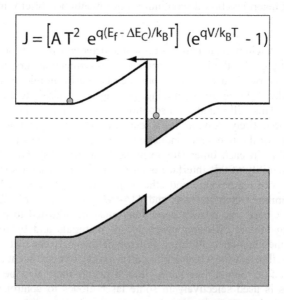

$$J = \left[A\,T^2\ e^{q(E_f - \Delta E_C)/k_BT} \right] (e^{qV/k_BT} - 1)$$

Figure 3.23: An n-n heterojunction with offset gap. This heterojunction may show Schottky diode-like behavior based on injection of electrons across the spike in the conduction band. The larger potential barrier in the valence band reduces hole injection from the narrow gap material into the wider gap material. An applied voltage reduces band bending in both semiconductors.

The behavior of the junction shown in Figure 3.22 responds in general in the same way as described above except that now the AlAs is very heavily doped. This further enhances the dominance of injection of electrons into the GaAs. The injected electrons come from the electron gas induced in the GaAs, and the spike at the interface is truly irrelevant (except that it is responsible for the formation of the electron gas in the first place). The behavior now becomes that of an n^{++}-p GaAs homojunction.

Not all heterojunctions act as one-directional homojunctions. If one has a heterojunction such as shown in Figure 3.23, the current is carried primarily by injection of electrons across the spike. In this case, both semiconductors are n-type but even so, a depletion region occurs on one side of the junction (and an enhancement region on the other side). Electrons moving from the electron gas into the wide-gap material face a barrier that does not change significantly for modest bias voltages. Forward bias reduces the scale of the field causing the electron gas to accumulate and reduces the barrier for electron injection from the wide-gap material across the spike. In reverse bias the behavior is (at least initially) determined by injection of electrons from the electron gas across the spike. Thus, there will be a Schottky-like behavior for at least modest bias voltages.

Various other variations on these themes occur in heterojunctions, including the possibility of barriers to conduction in junctions between materials of the same doping type (Figure 3.23).

3.5 TRANSISTORS

Transistors are the subject of whole books by themselves. This section presents only the bare outline of transistor types and modes of operation, focusing primarily on materials-related issues. Later chapters make reference to these devices and so it is useful to know the general aspects of their structure and operation. However, virtually all important aspects of the science and engineering of electronic materials can be illustrated through discussion of diodes and capacitors. Consequently, the following description is very brief.

3.5.1 Bipolar junction transistors

The bipolar transistor, which began the microelectronics revolution, consists of two diodes joined back to back by a thin common semiconductor layer (the base). Thus, bipolar transistors are referred to as p-n-p or n-p-n depending upon whether the common base layer is n or p type. Consider a typical n-p-n bipolar junction transistor, shown schematically in Figure 3.24 along with the electrical connections for its operation. When turned on the emitter junction is forward biased. This injects electrons into the base and holes into the emitter. The holes injected into the emitter recombine there and are of no value to operation of the device. The electrons emitted into the base may either recombine, contributing to base current, or transit the base. If the base is thin, most of the electrons pass through it without recombining, reach

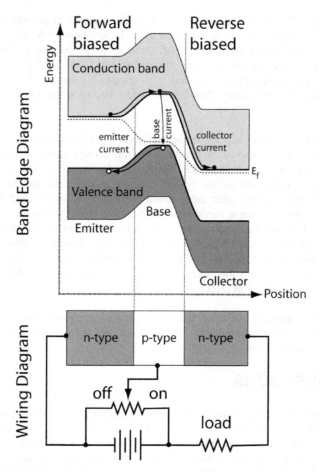

Figure 3.24: An example of the connection of an n-p-n type transistor. The circuit schematic shows the transistor in forward bias. Electrons are injected from emitter to base due to the forward bias. Some electrons flow across the base and reach the collector. A minority are lost through recombination in the base. These are few and far more reach the collector.

the reverse-biased collector junction, and are accelerated into the collector. As long as the time required for the injected electrons to diffuse through the base to the collector is less than the minority carrier lifetime, the transistor will permit more current to flow emitter to collector, $i_{collector}$, than from emitter to base, i_{base}.

The current gain of the transistor is approximately:

$$\frac{i_{collector}}{i_{base}} = \frac{\tau_n}{\tau_t} = gain, \beta \qquad\qquad 3.43$$

where τ_n and τ_t are the electron lifetime in the base region and the transit time required for an electron to pass through the base. If the base is wider than the mean-free-path between collisions, then the carrier must diffuse across the base. This is the normal condition for bipolar junction transistors. When no current can flow from emitter to base, for example by removal of the forward biased emitter junction, there is no injection and therefore no current emitter to collector either.

It is useful to repeat for emphasis that the current emitter to collector is controlled by *current* emitter to base as much as by the *voltage* emitter to base. The collector current increases rapidly with base current but the base current (and therefore collector current) increases exponentially, as in any diode, with base voltage. Therefore the emitter-to-base resistance is moderate and decreases with increasing voltage. In other words, a bipolar junction transistor has a low input impedance (resistance) and is a *current* amplifier by nature.

As with the homojunction diode, bipolar junction transistor (BJT) performances are controlled by the minority carrier properties. Thus, BJT's are known as minority carrier devices. From Equation 3.43 it is hopefully clear that the minority carrier lifetime in the base is critical to achieving high gains. Likewise, to achieve low transit times by diffusion of carriers through the base, a high carrier mobility is needed. Both of these require very good quality material with relatively low doping concentrations and no unintended defects. Some impurities such as Fe in Si cause such significant loss of minority carrier properties that degradation in bipolar transistor performances can be detected at Fe concentrations of parts per trillion. This has been a major driving force for improved purity in Si wafers.

It is because BJT gains and switching speed are controlled by base transit times that many billions of dollars in research has been done on GaAs devices. GaAs has a higher electron mobility than Si, which should lead to higher speed n-p-n devices. Unfortunately, it has turned out to be so difficult to work with very brittle GaAs and to be so difficult to form a good contact to and insulator on it that in the end, Si transistors outperform most GaAs devices. This is not because carriers move faster (they do not) but because the devices can be made smaller and other problems such as contact resistance and surface and interface states are less of an issue. This compensates for the reduced carrier mobility and results in similar net performances. Only with great care and effort can GaAs outperform Si, which makes GaAs devices expensive and their applications limited.

Significant improvements in the performance of bipolar junction transistors can be obtained through using semiconductor alloys, which allow tailoring of the energy gaps and band offsets of the various semiconductors and junctions. Heterojunction bipolar transistors are described in detail in the applications section of Chapter 6. For now, it is sufficient to say that by adding Ge to the base region of the device, the base transit time and emitter injection efficiency can be dramatically improved.

Materials issues in bipolar junction transistors have driven massive materials research programs. Through these, the current bipolar junction transistor performances have been achieved. Much of the discussion of circuit technologies has focused on smaller and smaller transistors. The above examples and discussion hopefully convey that advances can be achieved by other means as well. This is useful as at some point the continuing reduction in device scale must end.

3.5.2 Field-effect transistors

The field-effect transistor (FET) works in a somewhat different manner than the bipolar junction transistor. In these devices, back-to-back p-n junctions are produced and one of the junctions is therefore reverse biased to prevent current flow, as in the BJT. However, current flow is induced by elimination of these junctions to produce a continuous conducting path of one conductivity type, rather than by minority carriers passing through a region. FET operation may be accomplished in a number of ways. A typical structure of the most common implementation is shown in Figure 3.25. This example device uses two heavily n-type regions in a lightly p-type substrate, fabricated in such a way as to leave a short p-type barrier separating the n-type areas. The heavily doped materials are called the "source" (from which electrons are flowing when on) and the "drain" (into which the electrons drain). A gate electrode

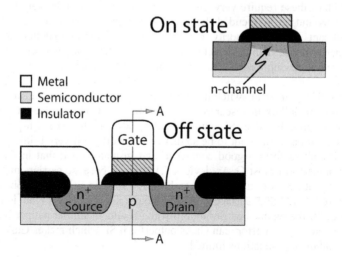

Figure 3.25: A schematic of a typical enhancement-mode n-p-n "metal-insulator-semiconductor" field effect transistor. The device is normally off and conduction is enhanced by application of a negative gate voltage relative to the source, creating an n-type channel. The source-to-drain voltage adds to the gate bias when turned on, which is why the channel is wider at one side. The hatched area under the gate indicates an optional heavily-doped polycrystalline semiconductor region as part of the gate (see also Figure 3.27). One must further prevent current flow to the substrate as this would turn the device on as if it were a bipolar junction transistor and the base were the substrate.

provides the current control by causing majority carrier depletion and ultimately inversion of the channel in the substrate, drawing charge into the channel as in a capacitor. This eliminates the diodes temporarily in a thin area under the gate and allows current to flow easily from source to drain. More bias voltage produces a wider and more conductive channel, leading to a lower resistance from source to drain. The FET is considered a majority carrier device because, unlike in a BJT, the source, drain, and channel all conduct majority carriers at all times that the device is on. Consequently, FETs are sensitive to majority carrier properties *and are less influenced by defects in the materials* than are BJTs.

The type conversion, providing the conducting channel, can be achieved by creating a heterojunction or Schottky contact to the surface of the transistor. The depletion region of such a junction provides type conversion in the channel. This depletion region can be increased or decreased depending upon the gate bias voltage used. These "junction" field effect transistors (JFET's) are used more for GaAs-based devices than for Si-based structures. For GaAs and other compound semiconductors, high quality heterojunction diodes are relatively easy to fabricate but a sufficiently good insulator (other than a wider-gap semiconductor) is difficult to produce. The major problem with the JFET design is that some gate current can occur even in a reverse-biased junction. Therefore, the gate resistance is limited by the conductivity of the reverse biased gate junction.

The most common form of the field-effect transistor uses a metal-insulator-semiconductor (MISFET) structure such as that shown in Figure 3.25. When the semiconductor is Si, the most common insulator is SiO_2 or a silicon oxynitride. These devices are called metal-oxide-semiconductor FET's or MOSFETS when SiO_2 is used for the gate insulator and are by far the most common type of FET. Note that MISFET is a more generic name for the same device and is more correct when the insulator is not a pure oxide. However "MOSFET" has become a somewhat generic name and is often used where "MISFET" would be more appropriate. A properly produced silicon oxynitride gate dielectric can be a remarkable insulator, withstanding fields of more than 10^7 V cm^{-1} with very low leakage currents. Indeed, devices now produced have gate dielectrics less than 5 nm thick in many cases with some in research prototypes as thin as 1 nm. Intensive research is now underway to find an adequate replacement for silicon oxynitrides that would produce more capacitance for a given thickness (ie: have a high dielectric response). The greatest successes to date have been achieved with complex (relative to SiO_2) transition metal oxides and rare-earth oxides but research continues and silicon compounds still dominate the industry. The effects of bias voltage on the electric fields in the gate insulator and the underlying semiconductor are shown in Figure 3.26.

FETs can operate in a number of modes. Enhancement mode devices, such as in Figure 3.25, are normally off. There is no channel under the gate and current does not flow unless a bias voltage is applied to turn conduction on. Depletion-mode devices are normally on, having a pre-existing channel. In these devices a voltage

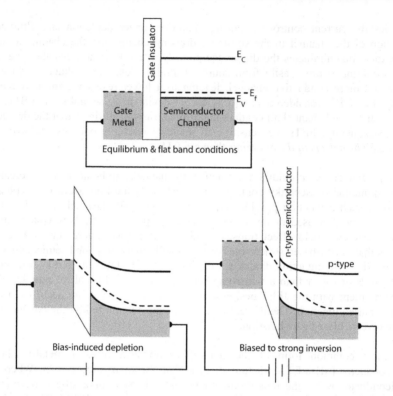

Figure 3.26: Shows the change in carrier concentration in a semiconductor near the gate of a MISFET for three applied fields. Under accumulation conditions the majority carrier concentration of the channel region is enhanced. Depletion conditions reduce the majority carriers without causing the majority carrier type to change. In the inversion regime the majority carrier type near the insulator is reversed.

must be applied to deplete the channel of charge and turn them off. Note that a depletion-mode device must have an underlying insulating layer against which the channel can be closed. Simply changing the p-substrate to an n-substrate in Figure 3.25 will not work as current can simply flow deeper in the substrate.

The majority carrier type of the conducting channel can be either p or n type with the heavily doped regions being the same type. The channel type is chosen depending upon the sign of the charge one wishes to apply to the gate to achieve control.

The speed of a FET is determined by the time it takes for an electron or hole to transit the channel. Consequently, as the channel dimension shrinks, the devices become faster. This is part of the reason smaller devices produce faster computers. In addition to size, electron or hole mobility is important as these determine the number of collisions a carrier makes in transiting the channel. Because electron mobility is higher than hole mobility in general, devices in which electrons transit the channel are faster, all other factors being equal. When the channel length is significantly less than the mean distance between collisions, the electrons or holes in the channel move ballistically (without collisions on average), further increasing the device speed. A modification of the FET that increases carrier mobility in the channel makes use of the formation of a two-dimensional electron or hole gas at a heterojunction, as was discussed above. Carriers confined to two-dimensional motion have fewer options for scattering and consequently have higher mobilities. To use this in a FET, a heterojunction is produced under the gate from materials that will form such a carrier gas. The resulting carrier gas yields the conducting channel just below the gate insulator. Applying a charge to the gate depletes or enhances the carrier gas density and changes the device conductivity. Finally, carrier mobilities increase as the doping concentration decreases. The use of heterojunctions also can allow doping in one material to transfer carriers to another, effectively separating the moving carriers from the dopant atoms that produce them.

An important advantage of MISFETs is the very high resistance between the gate and the remainder of the device. A voltage applied to the gate induces a current source to drain without dc current flow through the gate (although ac current can flow according to the ac impedance of the gate). The high input resistance of MISFET's makes them ideal for amplifiers of low power signals such as from antennas or other sensors as well as allowing FET's to operate as sensors themselves. Indeed, many modifications of FET's have been produced that induce a voltage on the gate without needing a wire connection. For example, adsorption of gasses, absorption of photons, and other methods can produce such a voltage and consequently an output signal. Although an FET has high input impedance, the output impedance can be low, allowing high current flow. This combination makes the FET very versatile and contributes to its high popularity as a circuit element.

A problem with FET's that has spawned much materials research is the "latch up" phenomenon. An examination of Figure 3.25 suggests that such a device contains a bipolar junction transistor with the semiconductor substrate acting as the base. If current can escape through the substrate, it is possible to turn the bipolar transistor on and current will flow from source to drain, independent of gate voltage. The solution to this is simple from an electrical standpoint – isolate the channel electrically such that no current can flow out of it except to the source or drain. In practice this has meant that each transistor must sit in a small insulating well or that it must be isolated by an insulating layer below. A wide variety of techniques for achieving isolation been developed including epitaxial growth of silicon on sapphire ("SOS"), buried oxide layers formed by oxygen implantation and annealing, wafer bonding, and many others. All of these are expensive and complex and can reduce process yields. No perfect solution exists.

From an electrical engineer's perspective, one of the essential variables in design of an FET is the voltage necessary to turn it on. It is important for complex circuit design that all threshold voltages be the same. The exact value of the threshold voltage is generally of less importance, as long as it is well known. Several variables can affect the turn-on behavior of the device and these are crucial to many aspects of the fabrication processes. Bonding defects, as we will see in Chapter 7, produce states in the semiconductor energy gap. These are likely to have a charge associated with them and/or are likely to influence the Fermi energy as if they were doping states. This, in turn, modifies the natural depletion or enhancement of charge in the channel and consequently changes the threshold voltage. In a lightly doped material such as the channel of a FET, even a moderate number of defects can dominate the local Fermi level. This is the root of the success of SiO_2 as a gate dielectric (even more than its dielectric properties) – it can be produced easily with an acceptably low defect state density at the SiO_2/Si interface. Defect states under the gate at the channel/dielectric interface are the cause of most failures in producing compound semiconductor MISFETs. Unintended impurity atoms in or around the gate dielectric also cause changes in the threshold voltage.

Once a threshold voltage can be made reproducible by reducing defect states and unintended impurity atoms, the next question is, how can the threshold voltage be changed and, in particular, how can large turn-on voltages be avoided. Large threshold voltages limit the possibility to reduce gate dielectric thickness, which is a crucial element to scaling the transistor to smaller sizes. Turn-on voltage is directly related to current flow from the gate electrode to the channel when contact is made. This current flow is determined by the relative difference in work function of the gate metal and the channel semiconductor. For example, electron flow from the metal to the semiconductor at equilibrium can induce a depletion region in the semiconductor without an applied bias (Figure 3.27). This can lead to an induced inversion region at equilibrium and a normally on device, requiring an applied gate voltage to turn the device off. Alternatively, current flow could lead to carrier accumulation in the channel of the same type as the substrate doping. This could lead to increased turn on voltages. At high enough accumulation reverse breakdown of the junctions to the source and drain can occur and, again, produce an on state.

The control of threshold voltage has largely been solved by making the gate of a very-heavily-doped semiconductor of the same type as the barrier between source and drain (Figure 3.27). In silicon devices, polycrystalline silicon (poly-Si) can be used. It is polycrystalline because it is deposited on a non-single-crystal substrate (typically amorphous SiO_2) and so there is no template for formation of a single crystal material. Fortunately, the polycrystalline form works well when heavily doped. Using a poly-Si gate, the equilibrium Fermi energy is essentially constant between the gate and the channel (Figure 3.27).

The most significant advantages of poly-Si gates in Si devices are as follows. (1) They can be doped the same type as the underlying material, resulting in very little and very predictable band bending in the adjoining Si channel region for either channel type. (2) Any of a wide variety of metals can be used for contact to the semi-conductor gate material, regardless of the carrier type, because the gate semiconductor is

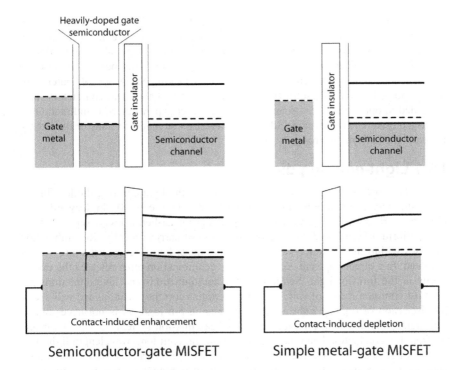

Figure 3.27: A schematic diagram of a typical MISFET structure with a conventional gate metal with spontaneous band bending due to electron flow (right) and the situation with a heavily-doped semiconductor gate using the same semiconductor as in the channel (left). The heavily-doped semiconductor has the same conductivity type as the channel. Contact results in a small enhancement in the carriers in the channel. This has relatively little effect on the threshold voltage. The strong depletion of carriers in the contact to the gate metal now happens at the contact with the doped gate semiconductor. The latter is doped sufficiently that the contact breaks down at essentially zero volts, leading to an effectively ohmic contact.

heavily doped and so the resulting Schottky diode breaks down at very low voltages. (3) Metal contacts to poly-Si are relatively stable and predictable, and the poly-Si/SiO_2 interface is very stable and reproducible. Poly-Si works well for Si-based MISFET's and has been the industry standard gate material. FET's based on other semiconductors may still be susceptible to threshold voltage problems because sufficiently heavily-doped polycrystalline forms of these semiconductors may not be available. In spite of these advantages, other gate materials are being explored and poly-Si is being replaced in some cases.

3.6 LIGHT-EMITTING DEVICES

Solid-state light emitters have become increasingly important in microelectronics and in our daily lives. We encounter light emitting diodes and semiconductor lasers routinely. Optical devices have become critical to long-distance fiber-optic communications systems as both light emitters and detectors. These devices are based on homojunction or, more commonly, heterojunction diodes. Optical devices are the only common application of non-silicon semiconductors. This persists in spite of long and determined research programs aimed at making Si an optical material or making non-Si materials useful for switching devices. Other semiconductors have succeeded in optical devices because they have direct gaps while Si is indirect. Only direct-gap materials can provide adequate interaction strengths to transfer the energy from/to light to/from electrons.

3.6.1 Light-emitting diodes

One of the most basic solid-state optical devices is the light emitting diode. This is, in its simplest form, a p-n homojunction diode (Figure 3.28). In forward bias, minority carriers are injected across the junction. Quasi-Fermi levels separate from the equilibrium Fermi level, indicating an excess of carriers locally. Recombination increases within the depletion region when the quasi-Fermi levels are more widely separated. In a direct-gap semiconductor this recombination often releases the excess energy in the form of light. Semiconductor homojunctions and heterojunctions are the most common forms of solid-state light emitter currently. But, as we will see in the discussion of organic light emitting devices, a p-n junction is not necessary to such light emission. In the erbium-doped silica emission system, in which carriers fall from the conduction band into a defect state and emit long-wavelength light, it is not even necessary to have recombination occur across a valence-band/conduction-band energy gap. It is only necessary that electrons in high-energy states occur at the same location as holes in lower energy states, and that a light-emitting recombination process is sufficiently likely. By adjusting the material in which the electron and hole recombine, the energy, and hence the color, of the emitted photons can be controlled. Thus, the light-emitting diode can be a remarkably simple device to understand. The critical issues relate to the probability of light emission and the possible recombination of carriers through non-radiative processes. A digression into these topics is therefore in order.

3.6.1.1 Radiative recombination

Light emission requires that electrons and holes recombine in a single step, satisfying requisite conservation laws and with allowed quantum-mechanical selection rules. The rate at which radiative recombination, $R_{radiative}$, takes place between free electrons and holes across the energy gap of a semiconductor depends upon the product of the number of electrons and the number of holes in the material. The rate at which these interact depends also upon their thermal velocity v_{th} (the velocity of carriers with a kinetic energy near $k_B T$), the cross section for the recombination process S_R, and a factor due to internal reflection in the material. The final relationship is:

$$R_{radiative} = npv_{th}S_R/2\eta^2$$

3.44

(a)

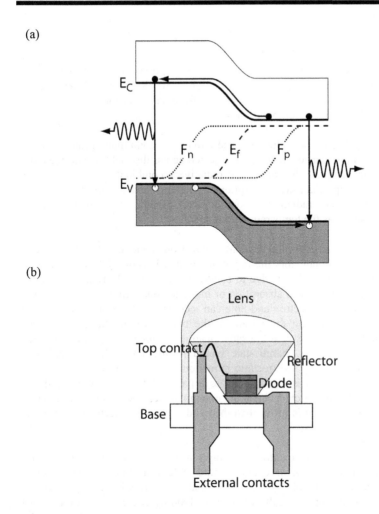

(b)

Figure 3.28: (a) A band edge diagram for light emission in a forward-biased homojunction diode. F_n and F_p indicate the quasi Fermi levels. The difference between these levels and the equilibrium Fermi level E_f indicates the excess hole or electron minority carrier populations. (b) A schematic of a typical light emitting diode in a standard mounting. The reflector and lens focus almost all of the emitted light out of the front of the lens providing high efficiency and brightness in the forward direction.

where η is the index of refraction of the semiconductor. The cross section is the apparent size of an electron to a moving hole, or vice versa, and gives a measure of how effectively one particle is captured by the other.

The recombination cross section can be estimated by detailed balance (see Bube in the recommended readings for a discussion) to be approximately

$$S_R \approx 5 \times 10^{-25} \eta^2 E_{gap}^2 \alpha \left(m_e^* m_h^*\right)^{-3/2} \left(300/T\right)^{5/2} cm^2 \qquad\qquad 3.45$$

where T is the temperature in Kelvin, and α is the absorption coefficient in cm^{-1}. For example, the radiative recombination cross section at 300 K is 1.8×10^{-19} cm^2 for Si and 9.3×10^{-18} cm^{-2} for GaP. Note that using this formula the refractive index terms cancel in the final radiative recombination rate.

Some points to notice are the following: The radiative recombination rate scales as np. Therefore, the more carriers that can be injected across the junction and the more heavily-doped the junction, the higher the radiative recombination rate. Materials with higher energy gaps and lower absorption coefficients have higher cross sections and hence more rapid radiative recombination. Finally, note that radiative recombination goes up as the temperature goes down.

Radiative recombination does not have to occur from band edge to band edge. Indeed, in normal semiconductors the radiation emitted is actually lower in energy than one would expect for a band-edge to band-edge transition. Here the band-edge transition is judged from optical absorption or intrinsic conductivity vs. temperature measurements. Before the electron and hole can recombine (before the electron can drop into the state containing the vacancy), the two wave functions have to be physically close to each other with the same quantum numbers. One would expect, however, that their opposite charge state would result in the electron and hole becoming bound together by electrostatic attraction. This bound pair exists and is called an exciton. The term is also applied to a pairing of positively-charged and negatively-charged bonding defects in an organic molecule, as we will see in Chapter 9. The energy of the exciton is lower than the band gap by the electrostatic binding energy of the pair.

A rough approximation of the exciton, viewing it as a hydrogen atom in a dielectric medium, suggests that the binding energy should scale as a reduced effective mass relative to the masses of the electron and hole, and inversely with the square of the relative dielectric constant. Typically the exciton binding energy is of the order of 10 ± 6 meV for common inorganic semiconductors. In organic materials it can be 1 eV. Thus excitons may break apart due to heat in inorganic semiconductors but dissociate very slowly in organic materials. An interesting consequence of the exciton binding energy is that the emitted photon has a lower energy than the gap of the semiconductor and is therefore not significantly reabsorbed as it leaves the material.

Extrinsic semiconductors contain shallow doping states (near the band edges). Just as exciton binding energies can reduce emitted photon energies, transitions involving donor or acceptor states can also lower recombination energies. Complex situations can arise in which an exciton can bind to the charged core of an impurity atom. All of these situations lead to electron-hole pairs that have a characteristic energy and consequently a characteristic emission wavelength. While many semiconductors, especially semiconductor alloys, often contain enough variability to hide such effects, in a very high quality layer it is possible to observe most or all of the various emissions associated with these and other combinations.

3.6.1.2 Non-radiative recombination

Unfortunately, recombination is not always radiative (it does not result in light emission). Indeed, in many materials radiative recombination is rare, especially when lattice defects are present. Other mechanisms for recombination energy dissipation include Auger processes (the electron recombining with a hole passes its energy to another electron which then gives up the energy to heat), and trap mediated recombination. Auger processes are always possible if permitted by the electronic structure of the solid. Trap-related recombination only occurs when the material contains a defect of some sort. Non-radiative recombination via traps can be shown to depend upon the np product and the total number of traps. It also depends on the trap capture coefficient for electrons or holes (limited by whichever is the smaller). Theoretically, the trap capture coefficient for a defect in a semiconductor should not vary much with the energy of the defect in the gap. In real semiconductors, the depth of the state significantly modifies its bonding and, consequently, modifies the capture cross section. A detailed and predictive theory of such processes is not, however, currently available.

Note that because both radiative and non-radiative processes scale with the np product, driving the light emitting diode harder produces more light but does not produce a significant increase in efficiency. However, a decrease in defects and states producing non-radiative recombination centers has a direct effect upon improving radiative recombination and light emission efficiency. In some cases increasing the np product can induce a change in the charge state of a defect and effectively eliminate it. This is connected with the concept of a "demarcation level" (see Bube for a discussion).

Since their invention, light-emitting diode efficiencies have been significantly improved (see Figure 1.1). This has involved better growth techniques to reduce non-light-emitting processes and improved doping methods, as well as better materials. In some cases surface passivation techniques are used to reduce electron-hole recombination at free surfaces of the device and optimized contact designs reduce recombination there. Efficiencies now exceed those of incandescent lamps and high brightnesses are obtained. These results, combined with the longer lifetime of solid-state light emitters, led to them replacing conventional lamps in many applications. One of the more visible of these changes is the use of LED's in vehicle taillights and traffic signals. Worldwide production has increased along with efficiency and lifetime. Large area displays incorporating hundreds of thousands of red, green, and blue LED's are now seen in many public locations.

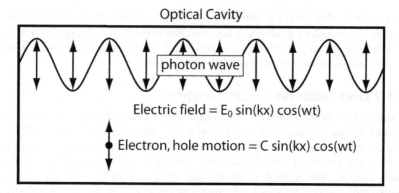

Figure 3.29: A schematic diagram of the electric field portion of the optical standing wave in a cavity and the resulting motion of electrons resulting from it. The electrons must move at the same speed and in the same direction as the electric field, resulting in a very strong coupling of the electrons to the optical wave. Note also that the cavity length constrains the photon wave vector, k. In the figure, w is the photon frequency, and E_0 and C are constants.

3.6.2 Laser diodes

Laser diodes function, in general terms, as light emitting diodes do. Electrons and holes are injected from opposite sides of a p-n junction, and recombine radiatively. What is different in a laser is that a high intensity standing light wave is created in the device. The standing wave is stationary in space, unlike in normal light emitters where emitted waves vary rapidly in both time and space. Diode lasers, as other lasers, consist of a cavity in which the optical wave accumulates, bounded along the standing wave direction ideally by one highly-efficient mirror and one partially-reflective mirror through which some of the optical wave can escape. Other surfaces are also designed to confine the optical wave to the emitting region. Most laser diodes also include mechanisms for channeling injected current into a small area of the device to increase current density locally. The remainder of the laser serves as a heat sink and contacting area for the emitting region.

3.6.2.1 Laser physics

The detailed physics of stimulated emission does not concern us here and is complex to present. We will make do with a brief summary. A complete description may be found in one of many texts on lasers. For example, see Zhao and Yariv in the recommended reading by Kapon. In the laser cavity the electric field portion of the optical wave can be added to the electron potential energy in the time-dependent Schrödinger Equation (Equation 2.2). Through this change in the potential energy the wave interacts with carriers to organize the electrons (and holes) to promote recombination, with the energy released simply amplifying the optical wave, see Figure 3.29. Such amplification may not be surprising if one recalls that the Schrödinger and electromagnetic field equations are linear so any solution may be

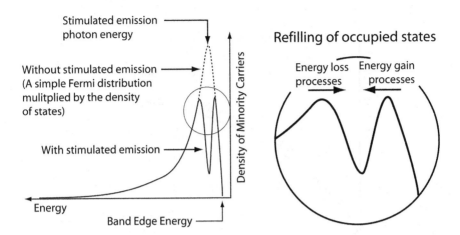

Figure 3.30: The density of minority carriers as a function of energy above the band edge energy in the presence of stimulated emission and without stimulated emission. The dashed curve represents a properly thermalized distribution of carriers in the band edge following normal Fermi statistics. The solid curve shows the reduction in carrier density at a specific energy where emission is occurring. The inset indicates how energy gain and loss mechanisms change the remainder of the electron energies to refill the gap in the density of carriers caused by recombination.

added to any other with given boundary conditions. The results show that the recombination rate in the laser cavity as enhanced by the optical field scales linearly with the photon energy and the non-equilibrium carrier density. Therefore, increasing carrier injection raises the recombination rate and the intensity of the wave.

Because the electromagnetic field of the standing wave must be strong enough to be significant relative to the atomic potentials, such that the electron trajectories are altered meaningfully, there is a threshold optical intensity for lasing. To maintain this threshold, a certain level of recombination, optical emission, and hence a critical injected carrier density are necessary. At this injection level stimulated emission exceeds random spontaneous emission. The more effectively the injected carriers can be gathered together and the more efficiently they can be caused to recombine radiatively, the less current injection (threshold current) is needed to maintain lasing. Therefore, a considerable effort has been expended optimizing the design of solid-state lasers. The results of this design are discussed in the next section.

An interesting result of the complete Schrödinger Equation solutions for the laser cavity is that recombination becomes so efficient that it significantly decreases the carrier density where the minority carrier recombination energy matches the photon energy, producing a "hole" in the electron energy distribution (see Figure 3.30). The resulting carrier distribution does not obey a simple Fermi or quasi-Fermi type

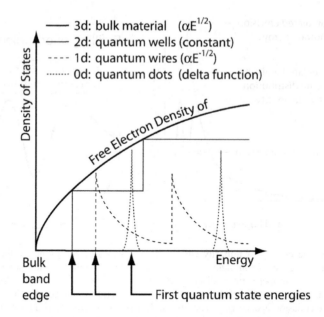

Figure 3.31: The density of states in laser cavities large in three, two, one and zero dimensions. Note that the density of states becomes increasingly restricted in energy range as the number of dimensions decrease (with a corresponding decrease in degrees of freedom and Fourier coefficients for the electron waves). As the distributions become sharper the minimum energy of the states also increases above the free electron band edge.

distribution. Therefore, at steady state the recombination rate depends strongly upon how fast minority carriers with different energies can relax to repopulate the carriers lost to stimulated recombination. One gets an idea of how fast lasing recombination is when one considers that this relaxation of carrier distributions usually occurs on picosecond time scales! The consequence of the reduction in carrier density at the resonant energy is that the recombination rate is actually somewhat lower than might be expected from a linear dependence on injected carrier density. The slower the relaxation to refill the distribution of carriers, the more non-linear the behavior becomes. This relaxation (refilling) rate therefore is a critical element of the equations describing laser diode operation.

Another useful result of the simple modification and solution of the Schrödinger Equation is that one may explore the effect of the dimensionality of the laser cavity. For example, as one restricts the electron wave functions from three dimensions to two, one, or even zero, the density of states in the material is strongly modified (see Figure 3.31) as are the types of waves that can occur in the cavity, their polarization, and how they can interact with the electrons. Into all of this come the symmetries and energies of the electronic states near the band edges of the material and their quantum numbers. The final recombination rates and how they drive and are driven by the optical wave must be summed over the various possible transitions of the system as weighted by the appropriate state occupancies.

All of these considerations select for only a few transitions, and with careful design only one wavelength of light may be emitted. Amplification of that resonant mode causes the laser to become locked into a specific emission frequency. One of the objectives of laser design is to enhance this preference to select a single desired frequency/wavelength. Such a design has the advantage that the emission is optically very pure and no energy is wasted in generating undesirable light waves. A second design objective is to restrict the density of states around the emission energy to enhance the rate of refilling of states emptied by lasing. In general the selection of specific modes and densities of states is accomplished through laser cavity design.

3.6.2.2 Diode laser design

There are two major geometries for semiconductor lasers. Examples are shown in Figure 3.32. The conventional edge-emitting laser is a small chunk of semiconductor cleaved from a larger wafer. The cleaved facets are very smooth and serve as mirrors for the device (each about 30% reflectivity, not ideal but easy). Current flows perpendicular to the optical cavity with holes injected from above and electrons from below, or vice versa. In most devices separate quantum well recombination layer and optical confinement regions are included. The second design is the vertical-cavity surface-emitting laser or VCSEL. Here, mirrors are integrated into the stack of semiconductors carrying injected current and light is emitted from the top of the structure, parallel to current injection. The practical edge-emitting laser is the older design, dating from the 1970's while the VCSEL began to be common roughly 20 years later.

To get an idea of the challenges facing the materials engineer in laser design, let us pause for a moment and consider the awesome amount of power in the small light-emitting region of a normal laser diode in, for example, a laser pointer. The lasers found in these devices are edge-emitting type (see Figure 3.32). Typical dimensions for the recombination region might be 0.5-1 mm long, 2 μm wide, and 10 nm high. The optical confinement region is ~200 times greater in thickness (1-2 μm high) because light cannot be confined in distances less than roughly its wavelength. Under normal operation the device current might be ~5 mA at ~3 V corresponding to an input power of ~15 mW. The output power might be perhaps 5 mW for a 33% efficient (external quantum efficiency) laser. Highly efficient laser diodes can reach 95% efficiency, far exceeding the performance of gas lasers. The input power of normal lasers is not large and is easily dissipated in the semiconductor substrate if mounted on an adequate heat sink. However, typical threshold current densities are 100-200 A cm^{-2}, with operating current densities even higher, corresponding to a power density of 300-600 W cm^{-2} in the carrier recombination zone, a prodigious amount of power in a small space. (For comparison, the intensity of sunlight is roughly 100 mW cm^{-2} on a very sunny day and normal room light is far less still.)

The situation becomes even more critical when one considers where power is dissipated. For normal laser diodes with simple cleaved facet mirrors a significant amount of power is lost due to recombination of carriers at the exposed faces. This recombination has a somewhat surprising effect. Because it decreases the density of carriers (the extent of inversion), it increases the optical absorption near the facet. The carrier recombination and optical absorption near the facets cause heating and melting of the facet face. Typical unpassivated surface facets can withstand an optical power density of 1-5 MW cm^{-2} while a surface passivation that decreases carrier recombination allows operation at 10-20 MW cm^{-2} without facet failure. [1] To put these numbers into a more usable form, one may note that the optical emission area at the end facet of a diode laser is typically a few square microns in area (say ~5 x 10^{-8} cm^2). For a 5 mW power output this would correspond to a power density through the emitting facet of roughly 0.1 MW cm^{-2}. Laser diodes in single-mode operation are considered high power when their output exceeds 50 mW. This

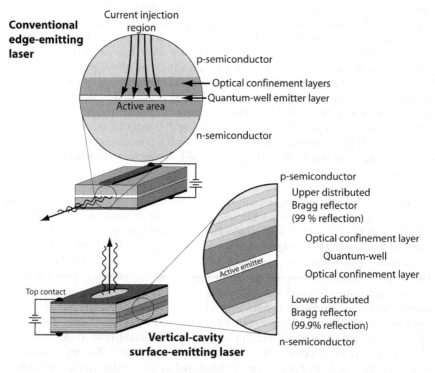

Figure 3.32: Schematic diagrams showing the two major designs for laser diodes. Edge-emitting lasers have a quantum well, restricted to a narrow area, from which light is emitted. The cleaved edges of the semiconductor form the mirrors of the laser cavities. In vertical-cavity surface-emitting lasers (VCSELs) the reflectors are semiconductor superlattices and light is emitted from the top of the device through a transparent contact material.

is the point at which the power density exceeds the ability of the facet to withstand the absorbed heat. Therefore, high power lasers require special treatments of the facets to reduce power dissipation there. For shorter pulse lengths the operating power may be increased significantly. Likewise, operation of the laser in a broad-area multimode condition spreads the optical field across the mirror facets and allows nearly an order of magnitude increase in power output under otherwise similar conditions.

An injected current of 5 mA results in the addition of $\sim 3 \times 10^{16}$ electrons and holes per second to the semiconductor. The electron density in the cavity rises to $\sim 2 \times 10^{18}$ cm^{-3}. Higher injection currents do not further increase this density significantly because more injection leads to more recombination and higher optical power output. The excess (injected) carrier density is above the conduction band-edge density of states. At this point the quasi-Fermi levels will be outside the energy gap, resulting in strong inversion of the carrier distribution, which is a necessary condition for lasing. [Unless the quasi-Fermi levels are outside the gap, spontaneous recombination will dominate over stimulated recombination resulting in a light-emitting diode rather than a laser behavior.] The high injected-carrier density also shows why it is not necessary to dope the active recombination region of the device, as the injected carrier densities overwhelm any doping.

Returning to laser diode design, a few more details on edge-emitting lasers would be helpful to understand the materials issues. In particular, the reader is encouraged to keep the following in mind when reading Chapters 5, 6, and especially Chapter 7, as much of the discussion there is essential to the production of efficient lasers. Current edge-emitting lasers typically incorporate a series of layers as follows (see also Figure 3.33).

1) p and n current injection regions define the diode-like properties of the device and affect its current-voltage curve and contact resistances. These regions are normally produced in relatively wide-gap "cladding" materials. The wider gap semiconductor serves to prevent injected carriers from passing completely through the device and being lost to recombination in the heavily doped regions of opposite type. Cladding layers also provide areas of different dielectric constant for optical confinement. Normally the top doped layer is contacted with a narrow stripe of metal to better confine the injected current to the area where lasing is intended to occur.

2) A thin, undoped, narrow-gap "quantum well" layer, usually a few atomic layers across, separates the doped layers. The quantum well ideally collects and traps all of the injected carriers of both types, producing an inverted population in which a large number of electrons occupy high-energy states while a large number of holes occupy low energy states. In practice, over 90% of the injected carriers can be induced to recombine in the well.

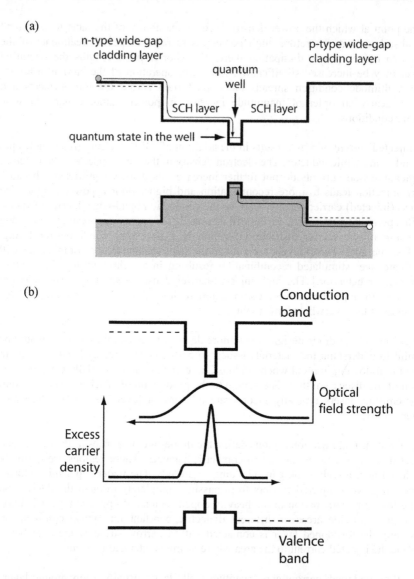

Figure 3.33: Illustrates the three primary component layers of an edge-emitting diode laser. (a) shows the energy gap as a function of distance from top to bottom of the device layer. The lower figure shows how injected carriers and the optical field are confined by the SCH/quantum well structure.

An analysis of the electron states in the well shows that with careful design the well may contain only a single energy level (see Chapter 2), strongly determining the energy of the emission and hence wavelength. The advantages of having emission occur between states in a quantum well include the following:

- The density of states is modified to produce a much sharper distribution.
- The limited energy range of the density of states restricts the energies of free carriers, enhances recombination rate, and reduces the optical emission modes.
- The limited number of optical modes produces a more efficient coupling of the optical energy to the electrons and holes.
- The quantum mechanical selection rules become less of a problem.

Variations on the quantum well design include use of strained layer structures, in which the lattice of the narrow-gap material is distorted in the plane of the structure (see Chapters 7 and 10); barrier structures, where there is an additional potential barrier that improves injected carrier trapping in the well; and multiquantum well structures with several wells in the active region.

3) Because the quantum well is normally much thinner than the wavelength of light, the quantum well does nothing to confine the optical field. For maximum efficiency the optical field should be localized near the emitting carriers. To provide optical confinement, a separate confinement heterostructure (SCH) is added to the diode surrounding the quantum well (Figure 3.33b). Many variations on the SCH design with different energy gap gradings have been tested and produce different confinement behaviors.

As we considered laser efficiencies earlier, it is worth noting that the energy difference between the electron and hole injected in the wide-gap regions and that in the quantum well is lost as heat. Therefore the benefits of the quantum well are slightly offset by a loss of efficiency. Overall, however, the benefits outweight the losses and quantum well structures are a very good idea.

Edge-emitting laser diodes have been designed with continuous operating output powers exceeding 600 mW per device. To achieve such a power density significant effort must be devoted to: passivation of the mirror facets; addition of highly reflecting and antireflective coatings on the back and front facets, respectively; and expansion of the emitting area to reduce power density at the facet faces. In such devices up to 70% of the emitted light escapes the device in the output beam and over 90% of the carriers injected into the device contribute usefully to this output power.

The top contact of the edge-emitting laser indicated in Figure 3.32 defines the location of current injection, along with any patterning of the top of the device into a mesa-type structure. However, it has also been found that eliminating the quantum well and averaging the energy gap of the semiconductor vertically through the device away from the active emitting area can also serve to channel current into a small active region. The method for achieving this takes advantage of a remarkable phenomenon. When certain impurities diffuse through AlAs-GaAs alloy multilayers, the Al and Ga atoms diffuse rapidly along the direction of motion of the impurity. This results in disordering of the layers in the device, producing an alloy with an average energy gap throughout the interdiffused region.

Let us now turn to the vertical-cavity surface-emitting laser (VCSEL) design. In this geometry, mirrors are grown into the structure above and below the emitting region. Current is injected along the axis of the laser cavity, which is perpendicular to the top surface. Light is emitted from the top of the device. The VCSEL mirrors are not based on cleaved facets, but rather on a series of Fresnel cavities (superlattices [a regularly-spaced series] of alternating dielectric constant materials), known as distributed Bragg reflectors (DBRs). The DBR emitting surface mirror is typically 99% reflective while the back mirror is over 99.9% reflective. The high reflectivity of the DBRs increases the optical power in the cavity and thus reduces the threshold current. Recent edge-emitting laser designs have also incorporated DBRs into the sides of the cavity to improve optical confinement and reduce threshold currents in a similar way.

VCSEL laser arrays can be easily integrated onto substrates. As with the edge-emitting diode, the active emitting region is patterned so that not all of the surface area of the device is active. The active recombination volume in an edge-emitting laser is $\sim10^{-11}$ cm^3. By contrast, a VCSEL might emit from a circular region of ~10 microns diameter from a quantum well ~1 nm thick, resulting in a volume of $\sim10^{-13}$ cm^3. The two order of magnitude decrease in active volume in the VCSEL further contributes to reducing the threshold current. At the same time, the optical emitting area in the VCSEL increases by up to a factor of 100 over the edge-emitting laser, reducing the optical density at reflective DBR's. Furthermore, because the DBR's are not free surfaces, carrier loss due to surface recombination at the DBR does not occur and hence optical absorption in VCSEL's is similar to or better than in passivated edge-emitting devices. Finally, distributed optical losses (absorption and conversion to heat) of the optical wave increase with optical cavity volume. The VCSEL has a significantly smaller optical confinement volume than the edge-emitting laser, leading to up to an order of magnitude reduction in distributed power losses.

To summarize, the VCSEL design has a number of significant advantages over edge-emitters, including surface-normal output (for better coupling to optical fibers), ability to pattern large arrays of devices using standard lithographic methods, low beam divergence, low threshold currents, single mode (one wavelength) operation, high speed (due to small volumes), high thermal stability, ease of integration, and simpler manufacturing. At the same time there are some limitations to VCSELs. For example, resistance in the current injection path can lead to significant power losses, especially for current entering through the top DBR mirror, spreading inward from an annular contact into the active region. In addition, methods for defining the electrically active region may limit the heat-sinking capabilities of the devices that can limit the power output. Because of the large emitting area, the ability to focus the emission from VCSELs is limited.

We are primarily concerned with the materials issues in this text so at least a brief mention of this topic with regard to lasers is appropriate before moving on. Many of the basic ideas are the same as for light emitting diodes. Lasers are more efficient when fewer defects are present and are even more sensitive to such defects than are light-emitting diodes. We have referred to changes in the energy gap, dielectric constant, state of strain, and other properties of the materials as well. These will be left for later chapters after we have seen how forming semiconductor alloys may control these properties. However, they are all essential elements to control in fabrication of the devices.

3.7 SUMMARY

The primary electronic materials issues to keep in mind from this chapter are:

- Homojunctions – are contacts between identical materials.
- Heterojunctions – are contacts between different materials.
- Chemical potential of electrons and the Fermi level determine the tendency of carriers to move and determines their lowest-energy configuration.
- Diffusion of carriers occurs along concentration gradients.
- Drift of carriers occurs down electrical potentials.
- Net carrier transport is determined by transport down the chemical potential gradient.
- Band offsets at heterojunctions can be straddling, offset and broken gap types.
- The common cation and common anion rules suggest that the majority of the change in energy band edges when the energy gap changes occurs in the valence and conduction band edges, respectively.
- Steps to draw a heterojunction band edge diagram are shown in Figure 3.21 and related text.
- Defect states in semiconductors affect bipolar junction transistors strongly.
- Interface states at the dielectric affect MISFET devices strongly.
- The wavelength of emission is determined by the energy gap of the semi-conductor from which a LED is produced.
- Radiative recombination depends upon energy gap, absorption coefficient, the product of electron and hole densities, and the thermal velocity.
- Defect states can enhance non-radiative recombination processes.
- Laser diode design relies on control of energy gap and refractive index of the semiconductor constituents.
- Laser operation is affected by the relaxation rate for carriers to states separated by the emitting energy.
- Shrinking the emitting volume of a laser diode to quantum size scales modifies the density of states in which carriers are trapped prior to emission.
- Surface passivation is critical for high-power diode laser design.

In addition, some of the more important device-related points include:

- Diodes are junctions between two semiconductors resulting in a potential barrier allowing current flow only in one direction exponentially increasing with voltage while reverse bias results in relatively little current.
- Contact potential (see Equations 3.9, 3.13, and 3.14).
- Depletion width (see Equations 3.19 and 3.20).
- Reverse saturation current (see Equation 3.24).
- Quasi-Fermi levels (see Equation 3.28).
- Non-ideal diode behaviors (reverse leakage, series and shunt resistances, non-ideal behavior, reverse breakdown) and the ideality factor.

- Reverse breakdown mechanisms (tunneling, impact ionization).
- Schottky barriers occur when majority carriers are depleted when a semiconductor contacts a metal and behave very much as do diodes. They are, however, majority carrier devices with a different form of the reverse current (see Equation 3.39).
- Ohmic contacts occur when majority carriers are enhanced by contact with a metal.
- Heterojunction diodes can behave as homojunction diodes or Schottky barriers. They are used, for example, to control direction of carrier injection, induce electron or hole gas layers, and control energy gap on one side of the junction.
- Bipolar junction transistors consist of three layers of semiconductor with alternating doping type where the center layer of the three is relatively thin. The three layers are, respectively, the emitter, base, and collector. A small current emitter to base allows a large current emitter to collector when properly biased.
- Field-effect transistors have three regions, source, channel, and drain as well as a gate, which controls the conductivity of the channel connecting source to drain.
- Forward biased diodes may be induced to emit light as carriers recombine with proper design.
- LED light emission can be more efficient than other lighting sources.
- Edge emitting and vertical cavity laser designs are the most common for solid-state lasers. Each has specific advantages and disadvantages.
- Laser design is optimized to trap electrons and holes in the same area where the optical wave is confined.

3.8 HOMEWORK PROBLEMS

1. What are the two major causes of reverse bias breakdown in diodes?

2. With reference to Equation 2.23, explain why the rate of ionizing collisions (those which generate additional electron hole pairs) in avalanche breakdown increases rapidly as the accelerated free carrier kinetic energy increases through energy gap. Note the specific portions of the equation that are most strongly affected by this increase.

3. What effect does n_i have on diode performance (ie: the current/voltage curve)?

4. What would cause the current/voltage curve for a diode to become linear in the forward bias region?

5. Consider a p-n junction diode in GaAs doped with 1.6×10^{15} cm^{-3} acceptors on the p-type side and 1×10^{17} cm^{-3} donors on the n-type side at a temperature of 300 K. Assume that GaAs has an energy gap of 1.4 eV, a relative dielectric constant of 13, and effective band-edge densities of states of $N_c = 5 \times 10^{17}$ cm^{-3} and $N_V = 7 \times 10^{18}$ cm^{-3}. Calculate:

    a. the built-in voltage of the device,
    b. the depletion width,
    c. the average electric field in the depletion region.

6. The development group at MegaJoule Industries is working on producing a series of devices that rely on both Schottky Barrier and ohmic contacts. Other members of the device group have selected metal "A" because of its metallurgical properties and phase relationships with constituents of the finished device. You do some studies of metal A and find that the work function is 0.5 eV greater than the electron affinity of Si.

    a. Would you expect this to form an ohmic contact or a Schottky barrier on:
        i)   n-type Si (doped with 10^{15} As cm^{-3})
        ii) p-type Si (doped with 10^{17} B cm^{-3})

    b. Sketch the band edge diagram for each device.
        i)   If the device is an ohmic contact would you expect the semiconductor to become more or less conductive near the metal interface?
        ii) If the device is a Schottky barrier indicate the depletion region and the Schottky barrier height on the above diagram.
        iii) If the device is a Schottky barrier suggest a way of obtaining an ohmic behavior.

c. The engineering team is concerned about the effect of poor minority carrier properties on the devices. Is this a concern? Explain briefly.

7. As part of the MegaJoule Industries device design group, you are asked to help analyze the (abrupt) heterojunction expected for p-type GaAs with n-type AlAs. Given the following data,

	GaAs	AlAs	
E_{gap}	1.41	2.13	eV
Doping	1×10^{15}	1×10^{18}	cm^{-3}
Electron Affinity	3.6	3.1	eV

Assume the Fermi energy is 0.3 eV above the GaAs valence band edge and 0.05 eV below the AlAs conduction band edge.

a. Sketch a band edge diagram for the p-n heterojunction.

Be sure to include:
i) Values for the conduction and valence band offsets
ii) Reasonable band bending for the doping levels

b. Discuss the effect you would expect if the heterojunction is graded rather than abrupt.
c. What type of heterojunction is this (straddling, offset, broken gap?)

8. If the average velocity of an electron in GaAs is $\sim 1\times10^7$ cm s^{-1} at a field of 10^4 V cm^{-1} and the effective mass of the electron is 0.07 m$_e$,

a. estimate the electron mobility (hint: see Equation 2.39),
b. estimate the mean time between electron scattering events,
c. estimate the breakdown voltage of the diode described in Problem 4, above.

9. Suppose you have a tunnel diode made from GaAs and having a doping concentration of 2×10^{18} cm^{-3} donors on the n-type side and 3×10^{19} cm^{-3} on the p-type side. Sketch approximate band-edge diagrams for the following conditions.

a. a moderate negative bias voltage,
b. small positive bias voltage,
c. positive bias voltage of roughly half the energy gap,
d. positive bias voltage of roughly 90% of the energy gap,
e. estimate the current/voltage curve you would expect for this device. (Hint: the diode resistance is negative over a portion of its operating range.)

10. Consider a p-n junction in Si with 10^{17} cm^{-3} As on one side and 10^{15} cm^{-3} B on the other side as dopants. If the device was identical except that the Si was replaced with Ge then given the following data, how would you expect the reverse leakage current density in the diode to change (quantitatively) at 300 K ?

	Si	Ge
Relative dielectric constant ε_r	11.8	16
Energy gap (eV)	1.1	0.65
N_C (cm^{-3})	2.8×10^{19}	1.04×10^{19}
N_V (cm^{-3})	1.04×10^{19}	6.0×10^{18}

Assume that the ratios qD_n/L_n and qD_p/L_p for both semiconductors are 1.0.

11. What is the purpose of the SCH region in a conventional edge-emitting laser diode?

12. Why is surface passivation less important in a VCSEL laser than in a conventional edge-emitting laser?

13. In a light-emitting diode, what is the relationship between light intensity emitted and forward bias voltage assuming no changes in the operation of the diode?

3.9 SUGGESTED READINGS & REFERENCES

Suggested Readings:

Adamson, Arthur W., *Physical Chemistry of Surfaces, 4th edition*, New York: Wiley, 1982.

Bube, Richard H., *Photoelectronic Properties of Semiconductors.* Cambridge: Cambridge University Press, 1992.

Chang, C.Y. and Sze, S.M., editors, *ULSI devices,* New York: John Wiley & Sons, 2000.

Kapon, E., editor. *Semiconductor Lasers I: Fundamentals*, San Diego: Academic Press, 1999.

Kapon, Eli, editor. *Semiconductor Lasers II: Materials and Structures.* San Diego: Academic Press, 1999.

Liu, William, *Fundamentals of III-V Devices: HBT's, MESFETS, and HFETs/HEMTs.* New York: John Wiley & Sons, 1999.

Mahajan, Subash and Harsha, K.S. Sree, *Principles of Growth and Processing of Semiconductors*, Boston: McGraw Hill, 1999.

Matthews, J.W., *Epitaxial Growth*, New York: Academic Press, 1975.

Streetman, Ben G., *Solid State Electronic Devices*, 5th ed. Englewood Cliffs: Prentice Hall, 1999.

Sze, S.M., *Semiconductor devices, physics and technology*, New York: John Wiley & Sons, 1985.

Vossen, John L. and Kern, Werner, editors, *Thin film processes II*, Boston : Academic Press, 1991.

Reference:

[1]   Mehuys, David G. "High-power Semiconductor Lasers", in *Semiconductor Lasers II Materials and Structures*, ed. by Kapon, E., San Diego: Academic, 1999.

Chapter 4

ASPECTS OF MATERIALS SCIENCE

This chapter provides a brief description of materials concepts that may be useful in understanding electronic materials. The review is not exhaustive but is intended to provide a minimum (and rather basic) level of familiarity with important concepts used in other chapters. As elsewhere, the reader is referred to the recommended readings for additional background and details.

4.1 STRUCTURES OF MATERIALS

One of the most fundamental issues in understanding a material is the arrangement of the atoms that make it up. Gases consist of widely spaced atoms distributed randomly in a volume. Because the atoms are spaced, on average, beyond the range of normal electronic interactions, their arrangement is random. Furthermore, for a material to be a gas the bonding interactions that would draw atoms together must be smaller than the thermal interactions that drive them apart, therefore bonds among atoms in a gas must be relatively weak. In a liquid or amorphous solid the atoms are nearly as closely packed as they would be in a crystalline solid (hence the volume change upon freezing is modest) but their interatomic angles are rather random and thus there is no regular long-range order. As we saw in Chapter 2, regular crystal structures are essential determining the properties of most semiconductors, although amorphous semiconductors also exist.

Amorphous solids are much like very viscous liquids. Their organization may be relatively good within the first ring of nearest neighbors relative to a given atom.

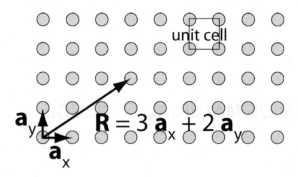

Figure 4.1: A schematic diagram of the periodic arrangement of atoms in a lattice. The basis vectors of the lattice are a_x and a_y. Any lattice point can be written as a linear combination of the two basis vectors.

However, the distance and direction to atoms beyond the first nearest neighbor becomes increasingly uncertain to the point that beyond roughly second-nearest neighbors, the chance of finding an atom in any direction is roughly equal. The simplest way to picture an amorphous solid is to imagine small shapes (triangular pyramids for example) connected through flexible links. The individual shape would be relatively well defined, but the distance and direction from one shape to another would be ill defined and some of the pyramids would be significantly distorted. Many common amorphous solids are based on such a network. Examples of amorphous materials include the semiconductor amorphous silicon and the insulator amorphous silicon dioxide. Amorphous silicon and other amorphous semiconductors are described in more detail in Chapter 8.

4.1.1 Crystal lattices

Crystalline solids form in one of the fourteen basic structural arrangements, known as Bravais lattices. All lattices have one basic property in common – there are a fixed set of translation vectors which will take an observer from any given lattice site to any other site. As we will see below, it is also possible for a lattice site to include more than one atom. Any combination of the basic translation vectors will always end at a lattice site, no matter how far one is from the original position, as long as the crystal contains no defects. Thus, if the translation vectors of a given lattice are a_1, a_2, and a_3, and a reference site is located at coordinate 0,0,0, then all other sites are located at $R = ua_1 + va_2 + wa_3$, where u, v, and w are integers. For example, a square planar lattice, such as the one shown in Figure 4.1, has two translation vectors along the x and y axes. This regularity is important to the collective vibrational motion of atoms in the solid (the phonon modes) and to the nature of electronic states, as we saw in Chapter 2.

In discussing the various directions and planes of atoms in a crystal, it is useful to have a convenient mechanism for naming them. "The planes that run from corner to corner to corner of the cube in a cubic material," would hardly be a helpful name and is imprecise in addition. The labels u, v, and w, by which atom translation vectors were described above, provide a handy index for a direction because any combination of the three defines a direction (as in vector **R**, above) relative to the origin. To indicate a negative direction it is conventional to place a bar over the negative index, rather than using a negative number. This shortens the indexing and keeps the numbers better aligned. Thus, in the index $[1\bar{1}0]$, the $\bar{1}$ indicates "negative one in the second index direction."

To index planes one may take advantage of the fact that any plane has a single direction perpendicular to it. Therefore the same indexing scheme used for directions can index planes as long as we have a notation by which we may know if we are talking about a plane or a direction. Thus, [u, v, w] with square brackets indicates a direction while we use parentheses, as in (h, k, l), for the plane. If the plane and direction are to be perpendicular it is required that $uh+vk+wl=0$ in a cubic system. Typical examples of plane indices are shown in Figure 4.2. A brief examination of the cubic lattice in Figure 4.2 shows that there are groups of similar planes. For example, the (100), (010), and (001) planes are all identical except for rotation of the coordinate system as are the corresponding directions. To indicate any of a set of otherwise identical planes one uses braces – thus {100}, while for a set of identical directions, triangular brackets are used as in <100>. One may also note that the (h, k, l) plane intercepts the x, y, and z axes at 1/h, 1/k, and 1/l within the cube for a cubic system. These indices are known as Miller indices.

The fourteen Bravais lattices include three cubic forms (simple, face-centered or body centered), two tetragonal forms (simple or body-centered), four orthorhombics (simple, body-, base-, or face-centered), rhombohedral, hexagonal, two monoclinic (simple or base-centered), and triclinic. All crystal structures, no matter how complex, can be reduced to one of these basic lattices. The most important structures for semiconductors are face-centered cubic (fcc) and hexagonal, although others such as tetragonal and simple cubic are also common. For drawings of these lattices, see Cullity in the recommended readings or a similar crystallography book.

The majority of simple metals form hexagonal, body-centered or face-centered cubic lattices, although other lattices also occur. Which lattice is formed depends upon the details of the atomic orbitals that give rise to bonding in the solid (see Chapter 5). However, there are few obvious trends when one looks at the distribution of lattices on a periodic table. Group IA metals and those with partially filled d-orbitals often form bcc structures, while the noble metals tend to be fcc, as are the group IB metals. Rare earths and metals with slightly filled d-orbitals are primarily hexagonal. Four elements stand out as somewhat unique. C, Si, and Ge as well as the semiconducting form of Sn can all occur in the diamond-structure, a face-centered cubic structure but with two atoms per lattice site. All of these are semiconductors.

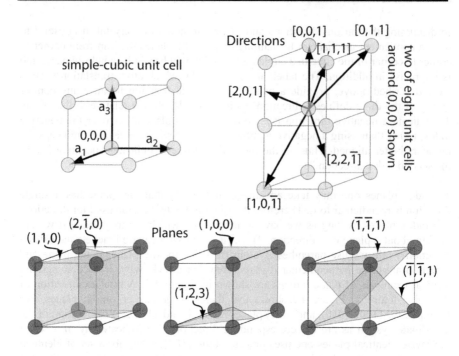

Figure 4.2: Shows the Miller indexing scheme for crystal planes and directions in a simple cubic crystal. Similar indexing methods are used in other Bravais lattices, although the basis vectors are not necessarily parallel to Cartesian coordinate axes. In hexagonal lattices an alternate labeling scheme employing four indices in which only three of the four are independent is often used. Notice that the two ($\bar{1}\,\bar{1}\,1$) planes marked are adjacent and parallel to one another.

Semiconductors can be found having many different crystal structures. Examples include silicon (a face centered cubic [fcc] lattice), cadmium sulfide (hexagonal close packed [hcp]), lead teluride (simple cubic), copper indium diselenide (tetragonal), and amorphous silicon (amorphous). Common semiconductors are built of sp^3 hybrid molecular orbitals with a tetrahedral geometry (for both hcp and fcc structures) as discussed in detail in Chapter 5. Lead Teluride and other IV-VI compound semiconductors are an exception to this behavior. This is not surprising when one considers that the IV-VI semiconductors have an average valence of five, while the common group IV, III-V and II-VII semiconductors have an average valence of four. Thus, the IV-VI semiconductors cannot produce an energy gap through tetrahedral bonding, as sp^3 hybrid orbitals can only accommodate four electrons per atom in bonding states.

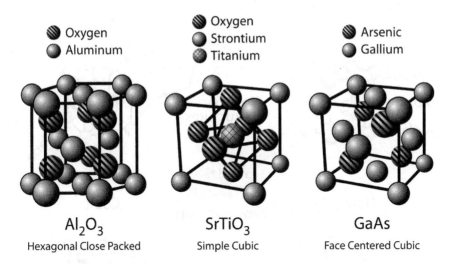

Oxygen
Aluminum

Oxygen
Strontium
Titanium

Arsenic
Gallium

Al_2O_3
Hexagonal Close Packed

$SrTiO_3$
Simple Cubic

GaAs
Face Centered Cubic

Figure 4.3: Shows the crystal structures of three compounds used in microelectronics. Each has a different Bravais lattice. To see the Bravais lattices of each compound, look only at the aluminum, strontium, and gallium atoms, respectively.

Chemical reactions giving rise to compounds place additional constraints on the arrangement of atoms. In general, there will be at least the number of atoms per Bravais-lattice structural-unit as there are atoms in the chemical formula or an integral multiple of the formula. Three compounds used in microelectronics illustrating three crystal structures are shown in Figure 4.3. The advanced dielectric $SrTiO_3$ has a simple cubic structure with Sr atoms at the cube corners, a Ti atom at the cube center, and O atoms at the face centers. In some ways this looks like an fcc structure with an extra atom at the cube body center or like a body centered cubic structure with extra atoms on the face centers. In the end, the only periodic structure is the simple cube because only the simple cube takes you from any atom to another of the same chemistry. Within each cube there are one Sr, one Ti, and three O atoms as expected from the formula. In the common semiconductor GaAs there are two atoms in the compound and two atoms per lattice site. This compound has a fcc Bravais lattice and each cube contains four formula units of the compound. A different unit cell must be defined to encompass only one formula unit. For each Ga atom, there is an As atom displaced by one quarter of the unit cell length along all of the three axes. In Al_2O_3 there are five atoms in the compound and five in the hexagonal unit cell. Because the aluminum planes defining the hexagonal volume of the solid contain a layer of aluminum in the vertical center of the unit cell, the structure is the close packed arrangement rather than the simple hexagonal structure.

Close packed structures are distinct from simple structures because rather than stacking atoms directly on top of each other along a given axis, the atoms stack in the

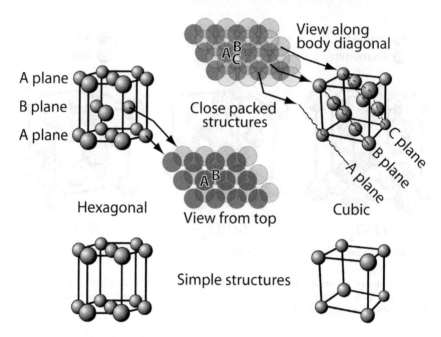

Figure 4.4: A comparison of fcc and hcp structures and their corresponding simple structures. Atoms in the "A" plane correspond to the darkest gray circles in the top view while atoms in the "B" and "C" planes are progressively lighter shades of gray.

hollows between atoms in a lower layer. The close packed face centered cubic (fcc) and and hexagonal close packed (hcp) structures are compared in Figure 4.4 with their non-close-packed cousins. If one considers the hexagonal planes that are the body diagonals of the fcc cube, one finds that there are three different planes displaced with respect to one another. In Figure 4.4 these are labeled A, B, and C. The hexagonal close packed structure consists of two such planes, A and B. The only difference between the fcc and hcp structures is that the fcc is formed by an ABCABC stacking of planes, while the hcp structure is stacked ABABAB with reference to the sites labeled A, B, and C in the figure. In other words, the fcc structure uses both distinct hollows between atoms in a given layer to stack the surrounding layers, while the hcp structure uses only one.

As noted above, a basic feature of all crystal lattices is that beginning from any point on the lattice and moving by any combination of translation vectors of that lattice you arrive at another lattice site. Diamond-structure semiconductors are composed of regularly stacked tetrahedra with atoms at their corners and at their centers. If one examines the possible lattices formed by arrangements of tetrahedra, one rapidly realizes that if one moves from a tetrahedron corner site to a tetrahedron center and then moves a similar distance along the same direction, one does not come back to another atom. This means that bond directions from one atom to the next do not represent translation vectors of a lattice made up of tetrahedral subunits. Rather, the atoms in the corner and the center of the tetrahedra must lie on different sublattices within the Bravais lattice (see the GaAs structure in Figure 4.3). When a crystal structure contains two or more inequivalent atom sites (the two positions are not connected by lattice translation vectors) then the structure is said to have a multi-atom basis. No matter how one reduces the unit cell to its most fundamental repeat unit, one can never have fewer than two atoms in these structures. More complex compounds such as $SrTiO_3$ or Al_2O_3 have even more atoms in their structural unit cells. Note that atoms such as Si that crystallize in the diamond structure have two identical atoms on non-identical lattice sites. (The diamond structure is the same as for GaAs in Figure 4.3 if the Ga and As atoms are replaced by identical atoms such as Si.) The diamond structure is not itself a Bravais lattice. It is an fcc Bravais lattice with, for example, two atoms Si per lattice site. These two atoms have the same orientation with respect to one another at all sites of the fcc lattice. One can also view the structure as two fcc Bravais lattices displaced by 1/4 of a unit cell along each coordinate axis with respect to one another.

A number of semiconductors can exist in both the fcc and hcp lattices. A close examination of the two structures in Figure 4.4 shows that the primary difference is at the second-nearest neighbor distance. In the diamond (as in GaAs) or sapphire (Al_2O_3) structures shown in Figure 4.3 the distinction is even smaller as each lattice site has a pair of atoms associated with it so the distinction between the fcc and hcp lattices is essentially made at the fourth-nearest neighbor distance. Therefore, the energy difference between the two lattices and the difference in bonding will be very small. Of course, a small energy difference summed over a whole crystal can represent a significant energy. So long as the energy difference is greater than the thermal energy (k_BT per atom), the solid will still strongly favor the lower energy structure. However, factors such as temperature, impurities, or the presence of a lattice-distorting substrate can shift the balance and change the resulting crystal structure. Most of the II-VI semiconductors occur in both cubic and hexagonal forms although most have a preference for one or the other at room temperature.

Defects in these crystal structures are essential to determining the properties of the materials. The crystalline defects relevant to semiconductors will be discussed in detail in Chapter 7. Amorphous materials have no regular order so there are no well-defined defects in the material. Nonetheless, we will see in Chapter 8 that the continuum of distortions in the structures of amorphous semiconductors play a key role in determining their properties. Here we will list only the types of atomic-scale (point) defects in crystalline materials and leave more complex structures and detailed discussion to Chapters 7 and 8. Point defects in crystals include vacancies, interstitials, and antisites. Vacancies are missing atoms in the crystal structure. They are essential to the diffusion of atoms among lattice sites in many materials. Interstitials are atoms lying in spaces between atoms in the crystal structure. More open lattices such as the diamond structure accommodate interstitial atoms relatively easily. Finally, in ordered structures one may disorder the atomic sites locally. For example, in GaAs there are a specific set of lattice sites for Ga atoms in an fcc arrangement and a separate set of fcc lattice sites for As. An antisite defect places an As on a Ga lattice site or vice versa. Extended line, planar and volume defects also occur but these disrupt the lattice and are left for description in Chapter 7.

4.1.2 The reciprocal lattice

It is important in understanding electron (and x-ray) diffraction to know how waves interact with solids. The energy band diagrams described in Chapter 2 all refer to electron wave vectors and diffraction of the electron wave off the crystal lattice at the Brillouin zone boundary. To understand diffraction one has to consider the lattice from the perspective of a wave, not from the perspective of points.

Any function, such as a function f($\mathbf{R}$) giving the probability of finding an atom at a given position $\mathbf{R}$ in real space, can be Fourier transformed to provide an indication of the amplitude f($\mathbf{k}$) of waves having wave vector $\mathbf{k}$ which, if summed or integrated, replicate f($\mathbf{R}$). Mathematically, the Fourier transform relationship is:

$$f(\mathbf{R}) = \int f(\mathbf{k})\, e^{i\mathbf{k}\bullet\mathbf{r}} dk.$$ 4.1

The amplitude function f($\mathbf{k}$) is similar to a display of the intensity of sound as a function of wavelength on a graphic equalizer or to the intensities of colors in a spectrum produced by passing a light beam through a prism.

Perfect infinite crystals may be described by a series of harmonics of very well-defined wavelength, while a small crystal has much less capability to specify the waves needed to reproduce it exactly. Therefore, a much wider range of wavelengths is allowed. One can see this difference if we take two patterns of dots overlain upon one another. Twisting one set of dots relative to the other by a small angle is immediately obvious for a large dot array while it is not so detectable for a small array. This comparison is shown in Figure 4.5. The result is that a Fourier transform of a small perfect crystal gives significant intensity over a range of wave vectors, $\mathbf{k}$, while a large crystal yields intensity only for much more specific values of $\mathbf{k}$. The Fourier transform intensity spectrum f($\mathbf{k}$) directly determines the diffraction of waves off the lattice, as we will see shortly. Thus, a small crystal will give a broad diffraction peak in a diffraction spectrum, while a larger or more perfect crystal will give sharper diffraction peaks, as shown in the figure.

The relationship between the intensity spectrum, f($\mathbf{k}$), and diffraction may be seen by considering the conditions necessary for diffraction in more detail. In diffraction an electromagnetic or electron wave with flat wave fronts and wave vector $\mathbf{k}_i$ strikes an array of atoms as shown in Figure 4.6. The interaction of the wave with the electron clouds of the atoms causes them to emit spherical waves of the same wavelength.

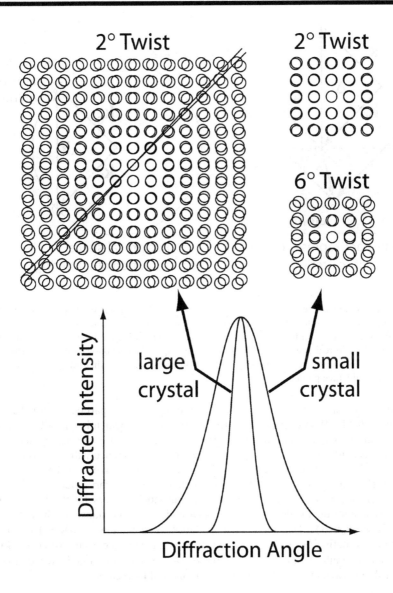

Figure 4.5: Shows the effect of crystal size on sensitivity to angular distortions. A "large" crystal shows an obvious effect of a 2° twist of one layer with respect to another. By contrast, a "small" crystal shows an almost undetectable effect. To achieve a similarly visible distortion in the small crystal a 6° twist is required. The sensitivity to variations in angle produces similar effects on diffraction patterns. Large crystals produce diffraction (constructive interference) only over a very small angular range. Small crystals allow diffracted intensity over a much larger range.

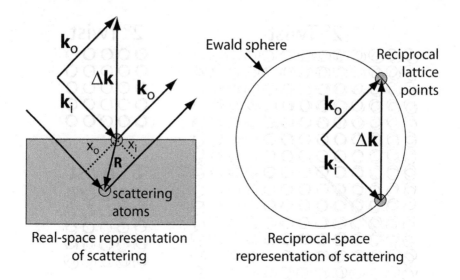

Real-space representation
of scattering

Reciprocal-space
representation of scattering

Figure 4.6: Shows the relationships among the incident, k_i, and scattered, k_o, wave vectors, the change in wave vector, Δk, the extra path lengths, x_i and x_o, of one part of the scattered wave front along the incident and outgoing waves, and the distance between two scattering centers (atoms) in real space, R. Also shown is the equivalent reciprocal space construction based on the Ewald sphere concept showing that diffraction occurs when Δk begins and ends on a reciprocal lattice point.

As these waves interfere, they produce an outgoing wave at a particular angle having a wave vector k_o. If no energy is absorbed in the process (which is typically the case in diffraction), the magnitude of the two wave vectors are the same, but their directions are different. To determine at what angle interference will be constructive one may ask what the extra path length is for the incident wave to strike an atom at position $R = ua_1 + va_2 + wa_3$, relative to an atom at an arbitrary reference position, 0,0,0 where a_1, a_2, and a_3 are the primitive translation vectors for the lattice. The extra distance for the incident wave is $x_i = k_i \cdot R$ and the extra distance for the outgoing wave is $x_o = k_o \cdot (-R)$. The total extra distance is then $x = x_i + x_o = (k_i - k_o) \cdot R$. It is convenient to define $\Delta k = k_i - k_o$. Then, constructive interference occurs when $x = \Delta k \cdot R$ is an integer multiple of 2π. If we write $\Delta k = hb_1 + kb_2 + lb_3$ and if we define a relationship between the a_1, a_2, and a_3 and b_1, b_2, and b_3 vectors as

$$b_1 = (a_2 \times a_3)/(a_1 \bullet a_2 \times a_3)$$
$$b_2 = (a_3 \times a_1)/(a_1 \bullet a_2 \times a_3), \qquad\qquad 4.2$$
$$b_3 = (a_1 \times a_2)/(a_1 \bullet a_2 \times a_3)$$

then the dot product becomes $\Delta \mathbf{k} \cdot \mathbf{R} = 2\pi(uh+vk+wl)$. If u, v, and w are integers based on the definition given above for the crystal lattice atom positions, then this imposes the additional criterion that h, k, and l also must be integers. This implies that $\Delta \mathbf{k}$ must fall upon a lattice defined by points $h\mathbf{b}_1$, $k\mathbf{b}_2$, and $l\mathbf{b}_3$.

> **The reciprocal lattice is defined by the set of points $h\mathbf{b}_1$, $k\mathbf{b}_2$, and $l\mathbf{b}_3$ with the basis vectors defined by Equation 4.2.**

The diffraction condition may also be described by the "Ewald sphere" construction, shown in Figure 4.6. Vectors $\mathbf{k}_i$ and $\mathbf{k}_o$ have the same length and therefore can be presumed to define the surface of a sphere if both originate from the same point. Positioning the end of vector $\mathbf{k}_i$ on a reciprocal lattice point, diffraction will result in any direction in which the sphere passes through another reciprocal lattice point. The angle between $\mathbf{k}_i$ and $\mathbf{k}_o$ intersecting these two reciprocal lattice points, (0,0,0) and (h,k,l) defines the diffraction angle for reflection off the real-space lattice planes (h,k,l).

The conditions described here also define the conditions for diffraction of electron waves at the Brillouin zone boundaries. Likewise the Brillouin zones described in Chapter 2 are reciprocal-space objects with the symmetry of the reciprocal lattice rather than the real-space lattice. The reciprocal lattice points in Figure 2.5, for example, are located at points $h\mathbf{b}_1$, $k\mathbf{b}_2$, and $l\mathbf{b}_3$. The reciprocal lattice for a simple cubic system with basis vectors $\mathbf{a}_1$, $\mathbf{a}_2$, and $\mathbf{a}_3$ has reciprocal lattice vectors parallel to the real space vectors. However, larger distances in real space correspond to shorter distances in reciprocal space. Thus, planes that are widely spaced in real space have closely spaced reciprocal lattice points and vice versa. One may determine by examination of Figure 4.2 that the (100) planes are $\sqrt{3}$ times farther apart than are the (111) planes. In general, the distance, d, between (hkl) lattice planes in a cubic system may be shown to be:

$$\frac{1}{d^2} = \frac{h^2 + k^2 + l^2}{a^2},$$ (4.3)

where a is the cube side length.

4.2 BASIC THERMODYNAMICS OF MATERIALS

An essential issue in materials science is to establish the stability of any of the crystal structures discussed above relative to any other, to determine the energy released or absorbed upon reaction to form or decompose compounds, and to determine how conditions such as temperature and pressure affect the stability of materials. This section describes the basic approach to tackling these questions.

Before calculating the stability of a material, we need to decide how to describe it in a well-defined way. We will make extensive use of the term "phase."

A phase is a chemically and structurally homogeneous material.

Thus, liquids and solids are different phases. Solids of different crystal structures are different phases even if they are chemically the same. Mixtures of two solids are called two-phase mixtures. Alloys are single-phase mixtures because one cannot distinguish isolated regions of one or the other constituent within the alloy. Thus, a phase can have different chemical compositions in different mixtures and still be the same phase as long as a given mixture is homogeneous and no structure change occurs. For example, varying amounts of sugar can be dissolved in water. As the mixture contains more sugar it becomes thicker and more syrupy. However, it is still the same phase (a chemically and structurally homogeneous liquid sugar water solution). What crystal structure forms for a given solid is a question of what phase is most stable.

When a material changes its crystal structure or when a solution of several constituents decomposes into separate parts with different compositions or structures, this is known as a phase change. Phase changes occur with changes in temperature, pressure, or composition as a result of a change of most stable structure. However, a phase change does not generally occur all at once across an entire material. (The martensitic transformation in steel is about the closest thing to an exception to this statement where the phase transformation moves across the material at the speed of sound and requires no long-range transport of atoms.) A phase change requires formation of a distinct interface between initial and final phases. Although the phase change might be favorable if it could be accomplished across the entire solid, one of the primary questions in determining whether a phase change will occur is how hard it is to form such an interface. Furthermore, it is often necessary to redistribute atoms if the product phases have different compositions. This atom movement takes time. Two fundamental fields describing these topics are thermodynamics and kinetics. Thermodynamics answers the question: which state does the system prefer, while kinetics asks: how fast will the system transform from an initial unstable state to any other state.

The stability of any material or interface is determined by the energy (or chemical potential) of the starting state relative to the energies of potential reaction products or other crystal structures. From our perspective, the key components in determining the energy of a material are the heat of the material, the work done in forming it, and any chemical bond energies. These contributions determine the total energy of the solid as a function of system variables such as temperature, pressure, and number of atoms in the system.

Let us examine these components from a thermodynamic perspective. In Chapter 2 (and again in Chapter 5) we consider quantum mechanical methods for determining the energy change upon bond formation and hence the energy of a solid at zero temperature. Knowing the vibrational modes allows us to go on to estimate the energy at a non-zero temperature. This is fine if we have the patience and determination to work out the full quantum mechanical results for all arbitrary configurations of the system to calculate which is most stable. A more traditional and practical method is to look at the response of a material to macroscopic variables such as temperature and pressure and their effects on the energy of the material. From these results we can work out a lot about the energies of the phases.

Thermodynamics begins by relating the energy of a material to the work, W; heat, Q; and bond energy Q_B that are created or absorbed in its formation. Other energies could be included as necessary. For example, the gravitational energy is not normally important to a chemical reaction and is thus generally ignored. Thus, the total energy, E, of a solid may be written

$$E = Q + W + Q_B + \text{other terms as needed.} \quad\quad 4.4$$

Likewise, when a system changes state, its total energy is modified based on heat and work done during the transformation and changes in bond energy as,

$$dE = dQ + dW + dQ_B +... \quad\quad 4.5$$

We will consider the heat, work, and bond energies of forming a solid separately.

The work done in creating or modifying a material can be divided into a number of contributions including mechanical work, electrical work, surface work, etc... The mechanical work done, dW, in any process is

$$dW = -PdV, \quad\quad 4.6$$

where P is the pressure on the material and dV is the change in volume resulting from the process by which the change occurs.

If one considers a collection of atoms widely separated such that they have virtually no interaction energy and one allows the atoms to come together to form a solid, the system converts potential energy to kinetic energy. In other words, to remain at constant temperature it must do work on some outside environment as it comes together to release energy. Typically, the amount of chemical energy produced is the cohesive energy of the solid and is the chemical work of bond formation, dQ_B. The magnitude of the cohesive energy is discussed in more detail in Chapter 5. It can also be expressed as the chemical potential of an atom, μ, multiplied by the number of atoms, N, in the system. Because there is a separate chemical potential for each chemical species in the system, one must sum over all i species present:

$$dQ_b = \sum_i \mu_i dN_i. \quad\quad 4.7$$

The heat of formation of a material dQ results from a change in the number of states accessible to it at a given temperature. The number of states at a given energy is related to a quantity called the entropy of the system, S. By definition, $S = k_B \ln\Omega$, where Ω is the number of states accessible to a system with an energy between E and E+dE.

To get a better sense of entropy, consider a single gas molecule in a large volume. That molecule will have a large number of energetically-equivalent places it can be within the volume and consequently a high entropy. Simply confining the same molecule in a smaller volume without otherwise changing the system reduces the number of places it can be and hence reduces the entropy of the molecule. A characteristic of thermodynamics is that spontaneous processes increase the total entropy of the system. All reversible processes involve no change in the total entropy of a system.

The nature of temperature is also connected to this concept of the number of states accessible to a system. Specifically, the temperature of a system is $T = \partial S/\partial E$ by definition. In other words, the temperature of a system is the change in the number of states accessible to a system resulting from a small change in the energy of the system with all other system variables held constant. One can rewrite this relationship as $dE = TdS$ when other variables are held constant. A quick inspection of Equation 4.5 reveals that holding volume and number of atoms constant $(dV = dQ_B = 0)$, implies that

$$dQ = TdS. \qquad\qquad 4.8$$

from which we can see that the generalized change in energy of a system for a constant number of particles is:

$$dE = TdS - PdV. \qquad\qquad 4.9$$

If we revisit the issue of gas molecules in a volume, we can see that the change in entropy of an isolated system at constant temperature and number of molecules would be directly related to the change in volume as $dS = P\, dV/T$.

Equation 4.9 is fine as long as the number of particles in a system is kept constant. There is an obvious problem if this is not the case. For example, suppose that we keep the volume and energy of a system constant but allow the number of particles to change. If the temperature and pressure are not allowed to change, then the entropy must change as the number of states accessible to the system is modified. Alternately, changing the number of particles at constant volume and entropy must change the energy of the system as each atom added also adds a unit of chemical potential of that species. To account for changes in the number of particles of type i:

$$dE = TdS - PdV + \sum_i \mu_i dN_i. \qquad\qquad 4.10$$

The index i can also designate a series of phases.

Returning to the absolute energies of Equation 4.4, rather than energy changes, one can write

$$E = TS - PV + \sum_i \mu_i N_i .$$ 4.11

Several other energies are useful to define. The enthalpy, H=E+PV describes the energy of the solid excluding mechanical work and describes the energies associated with bond energy and configurational energy. The Helmholtz free energy, F=E-TS removes contributions due to entropy, and the Gibbs free energy is G=F+PV. From these definitions,

$$H = TS + \sum_i \mu_i N_i$$

$$F = -PV + \sum_i \mu_i N_i .$$ 4.12

$$G = \sum_i \mu_i N_i$$

Each of these free energies is most useful for describing certain behaviors of the system or for conditions under which only certain variables are changing. For example, because it is directly related only to the chemical potential of a set of particles, the Gibbs free energy is a broadly useful quantity for determining the chemical stability of a closed system of given size. Furthermore, because systems are in equilibrium when their chemical potentials are equal, the condition of equilibrium is when there is no change in their Gibbs free energies if they exchange particles. An example of a system that can exchange particles is a mixture of two phases in a solid. Phase changes will occur whenever the chemical potential of one phase is lower than the other. Stability is therefore the condition in which the chemical potentials of all phases or particles are equal, as was discussed in Chapter 2.

One final point to note about the energy of a system is that at higher temperatures the entropy term becomes more important than the bond energy terms. Therefore, as one heats a solid one often observes changes in structure because different crystal lattices have different entropies. If one makes mixtures of materials the entropy of mixing often becomes important. It is the desire to mix materials to raise the entropy of the system that produces a corresponding decrease in the melting points of many mixtures. This is how antifreeze works.

4.3 PHASE DIAGRAMS

It will be important in discussions in other parts of this text to be able to describe easily which phases will be present under a variety of conditions and if a combination of phases is in equilibrium or not. For example, at what temperature will an element, alloy or mixture of phases melt? What is the composition of the material that melts at

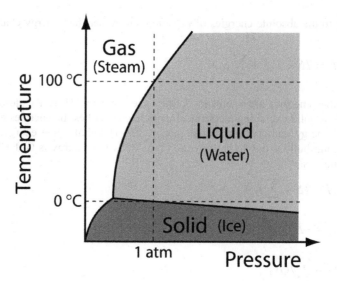

Figure 4.7: The pressure-temperature phase diagram for water. Note the "triple point" which occurs just above 0°C at a fraction of an atmosphere pressure where solid, liquid and vapor can coexist. This must be a point because all degrees of freedom are taken up by the three phases present simultaneously. This can only occur at a single point.

such a temperature? And so forth. These questions can be answered immediately by reference to a phase diagram. This is a map showing which phases are stable as a function of variables such as temperature, pressure, or composition.

It can be shown that the number of degrees of freedom F which a system has available to it is

$$F = C - P + 2, \qquad\qquad 4.13$$

where C is the number of constituents in the system, P is the number of phases, and the 2 refers to the availability of work and heat as additional variables. Normally the laboratory variables we use to affect work and heat are temperature and pressure, although entropy and volume could also be used. If we have a single constituent, such as water, the number of degrees of freedom will be $F=3-P$. Therefore, we can vary temperature and pressure independently in the presence of a single phase or we can change either temperature or pressure if we want to have two phases present simultaneously (one degree of freedom). This means that the boundary between two phases (where they coexist) will be a line on a temperature-pressure diagram. A typical example of a phase diagram is that for water, shown in Figure 4.7. It is even possible to have three phases coexist as long as it is only at one temperature and pressure. This is known as the "triple point."

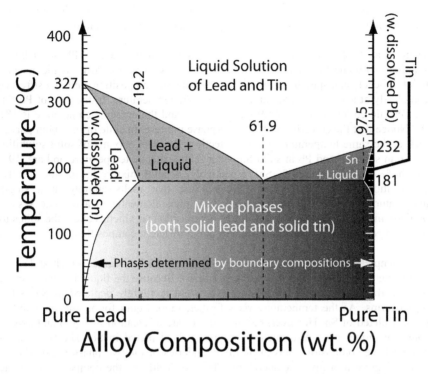

Figure 4.8: The Pb-Sn binary phase diagram. This type of phase diagram is typical of materials with limited solid solubility.

When two components are present, an additional degree of freedom exists. The three degrees of freedom are now temperature, pressure and composition. This means that two phases can be present in a two-dimensional region of phase space. One can, for example, change both the composition of the mixture and the temperature and retain the two phases. It is even possible to have three phases present together. For example, two solid phases and a liquid phase can coexist, but only on a line.

A typical example of a phase diagram with coexisting phases is a binary eutectic system such as that of Pb and Sn, commonly used in solders. The Pb-Sn phase diagram is shown in Figure 4.8. There are a number of features to notice. First, both Pb and Sn can dissolve some of the other element in their solid form, but Pb can dissolve much more Sn and this solubility persists to much lower temperatures than for solubility of Pb in Sn. Second, at temperatures below 181°C (the horizontal line on the figure) the system consists of a mixture of solid phases of Pb and Sn or a solid solution of one or the other of these. The composition of the phases is determined by the left and right boundaries of the two-phase region at the temperature in question.

There is a "triple point" at which a liquid phase (in the case of Pb-Sn, a liquid solution of Pb and Sn), and two solid phases coexist. This is known as the eutectic point. It occurs at a specific composition, the eutectic composition, and melts/freezes completely at a specific temperature, the eutectic temperature. In the case of Pb-Sn, the eutectic composition is ~61.9 weight % Sn, and the eutectic temperature is 181 °C. Notice that the eutectic solid is a two-phase mixture. In spite of this, both phases melt at the same temperature for the eutectic alloy. The maxima of solid solubility for Sn in solid Pb and Pb in solid Sn occur at the eutectic temperature and are, 19.2 weight % and 97.5 weight % Sn, respectively. A solid mixture having the eutectic composition on average is the only mixture that melts entirely at a single temperature. All other compositions (except pure elements) in a eutectic system partially melt leaving a solid with altered composition (if kinetics allow the solids to maintain equilibrium) in contact with a liquid of different composition.

For example, a Pb-Sn alloy containing 19.2 weight % Sn begins to melt at 181°C, producing a liquid with the eutectic composition. Because the liquid has much more Sn than the starting solid composition, the solid is slightly depleted of Sn as the liquid forms. As the temperature rises further, more liquid forms and the solid is more depleted of Sn. However, because of the rate of creation of liquid and loss of solid, the liquid must also become progressively depleted in Sn. The two phases present have compositions given by the boundaries of the two-phase region on the phase diagram at a given temperature. Thus, to decide on the compositions of the liquid and solid in equilibrium after melting the 19.2 weight % Sn alloy one draws a horizontal line across the phase diagram at the temperature in question. Where this line intersects the boundaries of a two phase region determines the phase compositions. For example, at ~263°C the two phases in equilibrium for starting Sn fractions below the eutectic composition will be solid Pb with about 10 at.% Sn dissolved in it and a liquid of about 30 at.% Sn. The entire mixture has melted at ~289°C.

Often mixtures of two or more elements form compounds – in some cases a wide variety of compounds. Taking the Si-Ta phase diagram shown in Figure 4.9 as an example one can observe a wide range of behaviors. First there are seven separate phases represented on this diagram including the liquid phase and the two elemental phases, Si and Ta. There are four intermetallic compounds. Because these compounds require significant space to write as chemical formulas, they are usually designated with small Greek letters. Most of the phase diagram consists of two-phase regions and none of the phases except the liquid can deviate much from a single composition.

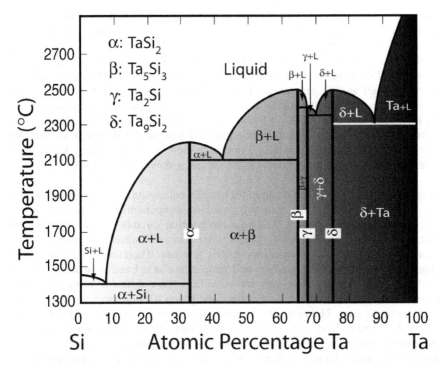

Figure 4.9: The Si-Ta phase diagram. Each vertical line corresponds to a separate compound and is labeled with a Greek letter. Note that between each pair of compounds a simple binary phase diagram can be drawn. Phase diagrams between specific compounds may be referred to as pseudobinary phase diagrams.

Limited solid solubility in the elemental phases is not uncommon when compounds form easily, as the presence of these compounds indicates a driving force in favor of clustering and ordering of the atoms. Between any two compounds one finds a simple binary phase diagram behavior. One may therefore draw a "pseudobinary" phase diagram representing the equilibrium between the compounds bounding each region. Most of the individual equilibria between solids in this phase diagram are eutectic-like. Note that a liquid having exactly the composition of one of the compounds freezes to the single phase but all other liquids form two solid phases when frozen. Also notice that although the solids have very little solubility in each other, the liquid is one phase across the whole composition range. Thus, one would notice a continuous change from liquid Ta to liquid Si in the single phase melt as the composition changed.

The above discussions refer to phase equilibria for unary (water) and binary (Pb-Sn, or Ta-Si) materials. Very few semiconductor alloys exist as simple binaries. An example of one that does is the Si-Ge alloy, discussed in detail in Section 6.4. Most semiconductor alloys contain at least three elements. Therefore, for a phase diagram there should be three possible composition ratios to account for. However, because the sum of all atomic percentages must be 100%, only two free variables exist. Consider, for example, the AlAs-GaAs alloy. Increasing the Ga content at constant As level must decrease the Al in the alloy. The resulting composition space can be plotted effectively as points within an equilateral triangle. Each edge of the triangle represents an individual binary alloy, similar to those shown in Figures 4.8 and 4.9. A ternary phase diagram for the Al-Ga-As system is shown in Figure 4.10.

In Figure 4.7 we had a phase diagram for a single compound. In the plane of the plot we could therefore accommodate two variables, temperature and pressure, which determined the phases present. Adding a composition variable in a two-dimensional plot, such as those in Figures 4.8 and 4.9, we had to eliminate one variable. The most common case is to assume atmospheric pressure. Therefore, the binary phase diagrams have axes of composition and temperature at a fixed pressure. In Figure 4.10 and other ternary diagrams we now have two different compositions to represent – group III element composition, and group III to group V elemental ratio in the example case. Therefore, we have eliminated the last environmental variable. Figure 4.10 had to have been made at a fixed temperature and pressure (normally atmospheric pressure). Fortunately, most ternary systems of interest have a critical or particularly interesting temperature at which, for example, growth of crystals occurs or heat treatment of the solid takes place. Furthermore, the phase diagrams for semiconductors often remain relatively constant over a fairly wide temperature range.

Let us consider the ternary phase diagram in Figure 4.10 in more detail. To begin to evaluate it we note that the interior of the triangle contains no compounds. In other words, no compound such as $AlGaAs_2$ exists which would have a plotted position away from the edges of the triangle. Along each edge one may also search for compounds. Only two exist in this phase diagram, AlAs and GaAs. As both of these compounds have virtually no solubility for point defects, they occur at single compositions and are plotted as points on the phase diagram. There are no Al-Ga intermetallic compounds. Below we will consider a more complex case with phases within the triangle. All together there are five separate phases on this diagram, Al, Ga, As, AlAs, and GaAs. Note that this diagram shows no specific temperature and ignores both liquid metals and gaseous elemental As. These are all likely to exist as Ga, for example, melts just above room temperature. Furthermore, to be in equilibrium at any given temperature above 50% As, the system must be enclosed and under sufficient As pressure to keep the excess As from boiling. Consideration of vapor pressure is essential in any system including a high vapor-pressure compound or element. Therefore, we can expect that the diagram in Figure 4.10 must be somewhat schematic or must imply a closed system and ignore molten phases or

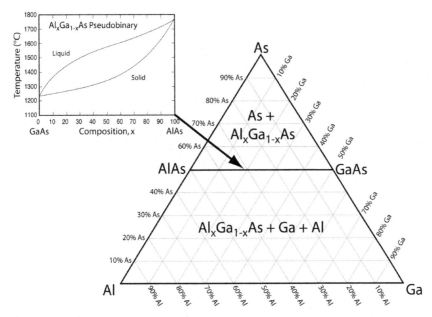

Figure 4.10: The ternary phase diagram for Al-Ga-As. Each light line represents alloys of constant composition as labeled for each line. The dark horizontal line represents the pseudobinary alloy $Al_xGa_{1-x}As$. This is a continuously-soluble alloy as shown in the inset. Figure inset based on data from Reference 1.

must represent a low temperature. It is common to simplify ternary phase diagrams in this way to present the general aspect of the phase relationships.

One can create a mixture of any pair of compounds and/or elements. In general there are three possible results: 1) you get a two phase mixture consisting of distinct regions of the compounds you mixed, 2) the two materials can dissolve in one another forming a single phase alloy, or 3) a reaction may occur, resulting in new phases. All of these situations may be found for Figure 4.10. Mixing AlAs and GaAs results in an $Al_xGa_{1-x}As$ alloy with no other phases (case 2). If one mixes As with GaAs, AlAs, or any alloy thereof, one gets a two-phase structure (case 1). Finally, mixing Ga with AlAs or Al with GaAs can result in reactions producing an $Al_xGa_{1-x}As$ alloy with both Ga and Al metallic second phases. In other words, one begins with two phases and ends with three (case 3).

Ternary phase diagrams have lines drawn connecting any two phases, which, when mixed, result in either case 1 or 2, above. These are referred to as "tie lines". Rephrasing, lines on the phase diagram indicate phases which, when mixed, will be unaffected or will alloy. The two-phase result indicates that the two materials are

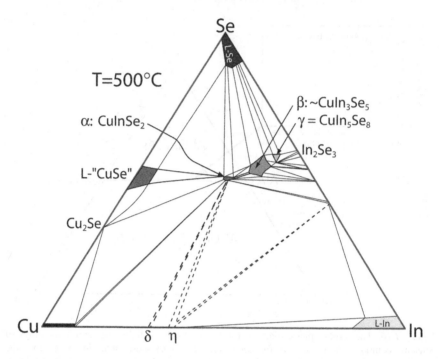

Figure 4.11: A ternary phase diagram for Cu, In, and Se at 500°C based on data in [2]. Additional data available for 750°C in [1]. Uncolored regions on the diagram are two or three phase areas. Numerous In-Se compounds occur which are not labeled to simplify the diagram. The diagram is approximate and most relevant for 500-750°C.

thermodynamically stable with respect to one another. This is desired for producing sharp interfaces such as at semiconductor-metal contacts. Any tie line may be represented as a pseudobinary alloy between two compounds or elements and a phase diagram such as in Figure 4.8 can be created. An example for the $Al_xGa_{1-x}As$ alloy is shown as an inset in Figure 4.10.

To illustrate some of the complexities possible in ternary phase diagrams, consider the Cu-In-Se ternary shown in Figure 4.11. This ternary system contains three semi-conducting phases within the triangle at 500°C, $CuInSe_2$, $CuIn_3Se_5$, and $CuIn_5Se_8$. Note that each of these is represented by a shaded area rather than a point because each can dissolve a significant number of defects and produce the same phase over a range of compositions. Numerous additional phases are present on the perimeter of the diagram including some with extended solid solubility. For example, a very In-rich liquid solution containing some Cu and Se occurs near the In-corner of the diagram. The phases present in a given region of the diagram will be the phases defining the corners of that region. Liquids generally form solutions and are single phase, as occurs in this diagram.

All three phases within the triangle lie roughly along a line connecting Cu_2Se and In_2Se_3. Thus, it is common with this system to discuss a tie line between these two compounds. However, a more precise description would be to refer to a series of tie lines. In some instances such as between $CuIn_3Se_5$ and $CuIn_5Se_8$, rather than a tie line, a volume of the phase space connects the compounds in which a range of compositions occur. Similar behaviors are found in many semiconductor quaternary alloys as will be seen in Chapter 6. The various lines and dashed lines (for estimated boundaries) indicate tie lines connecting phases. Many of these lines show stable mixtures of materials. For example, vacuum growth of Cu-rich $CuInSe_2$ results in the formation of Cu_2Se second phases, usually on the surface of the $CuInSe_2$, in agreement with the diagram. The complexity of ternary phase diagrams can make them difficult to interpret and difficult to plot. However, they are very convenient for determining the thermodynamic stability of interfaces.

Fortunately, when dealing with semiconductors, we are more often faced with phase diagrams such as the Al-Ga-As ternary in Figure 4.10, which can be reduced to a few relevant binaries, rather than complex ternaries such as in Figure 4.11. By contrast, metallic contacts to semiconductors usually require looking at the more complex ternary phase diagrams to determine reaction and stability issues. A very complex situation such as the Au-Ge contact to $Al_xGa_{1-x}As$ would, in principle require a five element phase diagram. This is impossible to represent even with a three dimensional plot, so one must be content with an estimation based on, for example Au-Ga-As, Au-Al-As, Ge-Ga-As and Ge-Al-As ternary diagrams supplemented as possible by experiments.

4.4 KINETICS

Chemical reactions and phase transformations occur between elements or materials because by organizing or reorganizing atoms, the energy of the electrons in the system (or the chemical potential) can be lowered, in the case of an exothermic reaction, or raised if that energy can be obtained from a temperature bath (endothermic reactions). Whether a reaction occurs and what the energy barriers are is a question of thermodynamics. However, even exothermic reactions may never occur because the atoms need to move relatively long distances under conditions where movement is prohibitively slow. The question of how fast a reaction will occur is one of kinetics rather than thermodynamics. A common example of a kinetically-limited reaction is the stability of metals in air. All metals except Au will oxidize exothermically. Al reacts particularly rapidly with oxygen. Indeed, it is the fuel in many solid rocket boosters. What prevents Al soda cans from bursting into flame is kinetics.

Getting a reaction to proceed usually requires the following 1) sufficient reactants to provide the source materials for the reaction, 2) a small enough number of product molecules being present to prevent reverse reactions from dominating in reversible reactions (reversible reactions have little or no entropy change during the reaction), 3) sufficient atomic mobility that the atoms can sample the configurations of the system in which products are formed, 4) the energies of intermediate states between the reactants and the products are not too high, and 5) no parasitic (unintended) reactions occur that will dominate in competition with the desired reaction. All of these can be used to advantage or can represent a problem in a given process.

4.4.1 Reaction kinetics

For reversible reactions, reactant and product concentration effects are often modeled with the Law of Mass Action, which states that the reaction rate is proportional to a ratio of the concentrations of reactants to products, each raised to an appropriate power. Consider the reaction: $2A + BC_2 \rightarrow A_2B + C_2$ where A and BC_2 are the reactants and A_2B and C_2 are the products. Suppose that the concentrations of these species are [A], $[BC_2]$, $[A_2B]$, and $[C_2]$, respectively. The rate of the forward reaction is:

$$R_f = k_f \frac{[A]^2[BC_2]}{[A_2B][C_2]},$$

4.14

where the forward reaction rate constant is

$$k_f = k_0\, e^{-\Delta H/k_B T},$$

4.15

and k_0, and ΔH are the attempt frequency and the energy barrier for the reaction. The energy of the system during a typical exothermic reaction is shown schematically in Figure 4.12. Because the reaction requires two A atoms, it is necessary to bring them together simultaneously. (Otherwise the reaction would have to be broken down into two sequential reactions with one A atom each.) The time required to find two A atoms at adjacent locations at the same time is proportional to $[A]^2$. If we had required three A atoms the concentration would have been cubed. The reverse reaction rate is the inverse of the forward rate, leading to a change of sign of ΔH and the switch of reactants and products. In general the value of k_0 will also change. At high temperatures where the reaction rate constants, k, are all large, the forward and reverse reaction rates are determined primarily by the ratio of reactants to products. If [A] and $[BC_2]$ are much larger than $[A_2B]$ and $[C_2]$, then the forward reaction rate will be high and the reverse rate will be low. When the converse is true, small [A] and $[BC_2]$ relative to $[A_2B]$ and $[C_2]$ gives a predominant reverse reaction. If the temperature is low then the exothermic reaction rate will tend to dominate over the endothermic rate because of the difference in the values of the forward and reverse attempt rates, k_f and k_r.

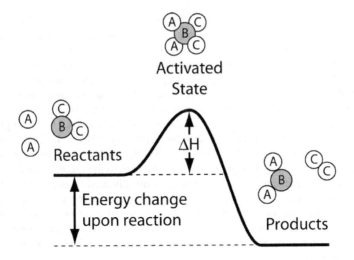

Figure 4.12: The energy of a system of atoms during the course of a reaction such as 2A + $BC_2 \rightarrow A_2B + C_2$. The reactants must be configured for reaction. In the example this might mean attaching two A atoms to the BC_2 molecule prior to reaction. The reverse reaction would cross a higher energy barrier climbing the entire energy hill from the right ("Products") to the activated state energy.

In the event that there are multiple steps in the reaction, the rate constants of the various steps can be included separately in an overall rate. For example, consider a three-step reaction in which two reactants form a complex, the reaction occurs and then the complex decomposes to yield products:

$$A+B \underset{k_{-1}}{\overset{k_1}{\rightleftharpoons}} A{:}B \underset{k_{-2}}{\overset{k_2}{\rightleftharpoons}} C{:}D \underset{k_{-3}}{\overset{k_3}{\rightleftharpoons}} C+D$$

,

where A and B are reactants, C and D are products, the formation step for the A:B complex has a rate constant k_1, calculated based on Equation 4.15, and so on. The forward reaction rate is k_2, and the product complex decomposition rate is k_3. The reverse reaction rates are designated with negative subscripts. The overall forward (k_f) and reverse (k_r) reaction rates for the three step processes are

$$k_f = \frac{k_1}{1+(k_{-1}/k_2)+(k_{-1}k_{-2}/k_2 k_3)}$$

$$k_r = \frac{k_{-3}}{1+(k_3/k_{-2})+(k_2 k_3/k_{-1}k_{-2})}$$

4.16

Note that the final forward and reverse reaction rates have significant differences in their parametric dependencies. Thus, the rate-limiting step for the forward and reverse multistep reactions may be different. For example, the forward reaction may be limited by formation of the reactive complex, while the reverse reaction may be limited by the reverse reaction rate [i.e. k_{-2}]. The overall rate of both k_f and k_r is generally dominated by the slowest step. When all forward reaction rates are much faster than any reverse reaction step then the denominator of the formula for k_f becomes unity and $k_f = k_1$. When one of the reverse steps is fast, for example large k_{-1}, this will lead to a large second term in the denominator if $k_2 < k_3$. In this case, the forward reaction rate becomes $k_f = k_2 (k_1/k_{-1})$. Similar analyses may be made for each of the various possible rate limitations. Various changes to the reaction conditions may supply or remove reactants or products and consequently modify any of the individual rates. In this case each rate constant must be multiplied by the appropriate reactant and product concentrations. Temperature also affects different processes in different ways. Thus, it is often possible to design a process to favor a given reaction over another. This is fortunate because we often want to select a desired event and suppress an undesirable one.

4.4.2 Nucleation

For a synthetic chemist the reaction rate is often the most crucial aspect of kinetics. Synthetic chemistry generally treats reactions in the gas or liquid phase where reactants have high mobilities and where distinct second phases are not normally formed. In semiconductor processing, reaction kinetics such as discussed above is essential to modeling techniques such as chemical vapor deposition (Chapter 12). Solid state chemistry and most materials science of solids is concerned with atomic transport as much as with reactant and product concentrations or thermodynamics. Furthermore, it is often the case that the forward reaction will not occur rapidly until some product is formed. This is normally referred to as a problem of nucleation and is similar to the classic conundrum, "which came first, the chicken or the egg?" In the remainder of this section we will examine nucleation and then turn to diffusion transport kinetics in the following section.

The phenomenon of nucleation is commonly observable in social situations, which may serve to illustrate some of the issues. Consider a party at which everyone is free to take food from a table. The food has been lavishly prepared and all of the dishes look very special. It is common that the guests will hesitate to take any of the food until someone has cut into it and spoiled the perfection of the presentation. Likewise, in a classroom students are often reluctant to ask questions. However, if someone interrupts the lecture with one question, often many more will follow. The same situation occurs in solids.

Suppose that a material exists as a single phase (say, an A + B alloy) at a high temperature. As the material cools, the phase diagram suggests that it should want to change to a two-phase mixture (of A and B phases). With all of the atoms uniformly mixed in the alloy, it would be difficult for the atoms to know what to do to form phase-separated regions. The only chance is to rely on random events to bring a relatively large concentration of A or B atoms together locally. Such a configuration, being more stable, would tend to last relatively long. If it lasted long enough to add additional A or B atoms, then this phase would grow. Conversely, if the cluster of one type of atoms were not sufficiently stable, then it would shrink.

To further complicate the problem in a solid, nucleation of a new phase requires that a small volume of that phase be carved out of the existing materials. This creates a volume of the new phase but also a surface or interface area. If the reaction is favorable, then the energy change associated with the formation of the new volume of material should be negative (exothermic) and scales as the volume of new material created. For a spherical volume, that would be as the radius of the new phase cubed or r^3. Surfaces and interfaces are generally unfavorable (positive energy change) and have an area that rises as r^2. For an interface energy γ_s, a volume energy change ΔH, and assuming that the energies have no size dependence, the total energy of the system (i.e.: Equation 4.11) includes a term:

$$E = \pi r^2 \gamma_s + (4\pi/3)r^3 \Delta H \qquad 4.17$$

Normally, γ_s is positive and ΔH is negative for a favorable reaction with an unfavorable interface energy. For this case the total energy increases roughly as r^2, for small r, reaches a maximum, and then decreases, ultimately converging to an r^3 dependence. This situation is shown schematically in Figure 4.13.

The energy maximum occurs where

$$\frac{\partial \Delta E}{\partial r} = 2\pi\gamma_s r^* + 4\pi\Delta H r^{*2} = 0. \qquad 4.18$$

From which,

$$2\pi\gamma_s = -4\pi\Delta H r^* \qquad 4.19$$

and

$$r^* = -\frac{\gamma_s}{2\Delta H} \qquad 4.20$$

where r^* is referred to as the critical radius of a spherical nucleus. This has a positive value when γ_s is positive and ΔH is negative, as assumed above. Any addition of an atom to the cluster when $r<r_c$ has a net energy cost (the total energy increases), while beyond r^* addition of atoms is exothermic. As the surface energy increases relative to the energy gain from the reaction, the critical radius increases. The energy barrier to nucleation E^* is

$$E^* = \frac{\pi}{12} \frac{\gamma_s^3}{\Delta H^2} \qquad 4.21$$

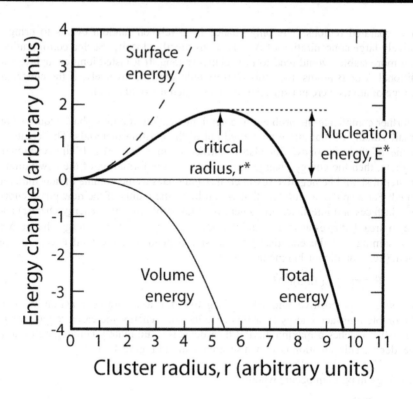

Figure 4.13: Shows the energy of a cluster of atoms during nucleation as a function of cluster size. Surface energy generally increases as r^2, while volume energy decreases as r^3. The energy of the cluster increases with addition of atoms up to a critical size, r^*, beyond which the energy decreases with atom addition. The maximum energy at r^* and is the activation energy barrier for nucleation.

for a spherical nucleus. In general, the nucleation rate, R, will depend exponentially upon the size of this barrier as

$$R = R_0 \, e^{-E^*/k_B T},$$

4.22

where R_0 is an attempt frequency. The attempt frequency will generally be energy dependent as well and may include a diffusion activation energy or some similar term. If E^* limits nucleation then point (4) in the list of considerations for reaction rates is most critical. If R_0 is the most significant limitation to nucleation rate, then point (3) in the same list dominates the overall rate. Because for a product of exponential terms the exponents may be added, the overall temperature dependence of nucleation may include a higher or lower energy than E^*. For R_0 including a diffusion rate $D = D_0 \exp(-E_D/kT)$, where E_D is the activation energy for diffusion of atoms, the nucleation rate may, for example, have an exponential dependence upon temperature consistent with an overall energy $E = E^* + E_D$. In other words, the nucleation and diffusion barrier terms are combined in the final rate expression into a single behavior with the same general form as Equation 4.22.

Nucleation rates can present severe limitations to the speed of processes. In my own work we were coating glass slides with CdS by evaporation but were finding that for moderate exposures there was no detectable coating on the slide. Only a short time later more than one micron of CdS was found. Extrapolation of film thickness vs. deposition time data showed a delay of several minutes before growth began. This is known as a nucleation delay and was the result of a high nucleation barrier. The CdS was landing on the slide at a high rate, but would remain only a very short time in the absence of large particles of CdS to which to stick. Similar nucleation barriers are observable in many systems.

Nucleation is an essential element to most phase transformations and to growth of one material on another (heteroepitaxy). It is not an issue when the two materials are the same (homoepitaxy) or so similar as to make little or no difference. Nucleation often determines the microstructure of thin films grown from the vapor phase as well as bulk materials produced by cooling from a melt.

A final consideration is the presence of unintended or alternate processes. Often reactions or phase changes can operate along a number of different pathways. One of the oldest commonly known cases of this type (not understood in detail until the mid 20th century) was the martensitic transformation in steel. Martensite forms when a blacksmith heats a piece of medium-carbon steel to orange heat (above ~740°C) and quickly quenches it (usually in water). A slow cooling leads to a relatively soft but thermodynamically-stable two-phase material (pearlite), while the sudden quench produces a hard brittle phase known as martensite. The issue of whether pearlite or martensite forms is one of diffusion kinetics for carbon in the steel.

A second familiar example of parasitic reactions is loss of charge or loss of capacity in rechargeable batteries. In the classic case of lead-acid storage batteries, normally used in automobiles, Pb reacts with sulfate ions (SO_4^{-2}) in the sulfuric acid electrolyte solution to produce $PbSO_4$ and electrons. The electrons flow through the outside circuit to the other electrode where they allow PbO to react with the sulfuric acid (H_2SO_4) to produce $PbSO_4$ and H_2O. During charging, supplying excessive voltage can drive a competing reaction producing hydrogen but not resulting in net charge on the battery. A similar reaction in a stored battery can result in hydrogen generation and discharge of the battery without current flow through the external circuit.

Parasitic reactions also occur in electronic device fabrication. Thus, in chemical vapor deposition of AlAs one can use $Al(CH_3)_3$ (trimethyl aluminum) or $Al(C_2H_5)_3$ (triethyl aluminum) as the source gas for Al. The triethyl compound reacts with

AsH$_3$ to produce AlAs. However, the trimethyl compound can also decompose to produce AlC and hydrogen. This parasitic reaction leads to large carbon contents in the resulting films when the trimethyl compound is used.

4.4.3 Atomic transport

The common factor controlling reaction and interaction rates is atomic transport. In the vapor phase and somewhat less in liquids, transport is relatively rapid. Even so, transport often limits reactions. For example, in normal fires fuel is released from the burning material into the gas but cannot burn immediately as all of the oxygen has been used up near the burning surface by previous reactions. Combustion only begins when sufficient oxygen has mixed with the flame to allow the reaction. The flame temperature increases as the fuel-to-oxygen ratio approaches the ideal value, decreasing again as the mixture becomes fuel-poor. Most of this mixing is through convection and turbulence in the gas, but on a smaller scale, diffusion controls the mixing rate. Nonetheless, mixing is rapid enough to allow the fire to burn. Similar mixing processes occur in liquids.

In solids atomic transport is much slower. Particularly in crystalline materials the atoms have well-defined positions with strong forces holding each atom in place. Diffusion in crystalline solids is generally divided into two classes, substitutional and interstitial, depending upon whether the atom is moving primarily from lattice site to lattice site or from interstitial site to interstitial site. There are several factors that must be overcome for diffusion to occur. First, there must be a vacant site for the atom to move into. This is primarily important for substitutional diffusion as in interstitial diffusion most or all of the surrounding interstitial sites will be vacant. However, interstitial diffusion may require transfer of an atom from a lattice site to an interstitial site before diffusion will occur. There are some mechanisms for substitutional diffusion that extend the range over which one can look for a vacant site but such a site is still required to be close to the moving atom. Second, generally even for movements to the nearest-neighbor site in a lattice is it necessary for some of the atoms on surrounding lattice sites to move out of the way of the diffusing atom. This movement is normally a result of thermal vibrations of atoms in the structure and contributes to the temperature-dependence of the diffusion process. Third, as the diffusing atom moves atomic bonds are distorted or broken and reconstructed. This bond distortion contributes most of the energy barrier for interstitial diffusion. In substitutional diffusion the energy required to form vacancies is usually the largest term in the diffusion activation energy. The processes of substitional atomic diffusion in an fcc lattice are shown schematically in Figure 4.14. Interstitial diffusion is nearly the same with the exception that the moving atom starts and ends on an interstitial site.

Because there is energy required in the diffusion process, which must be obtained from the heat of the crystal lattice, the diffusivity (related to the diffusion rate) depends exponentially on temperature following a typical Boltzmann relationship:

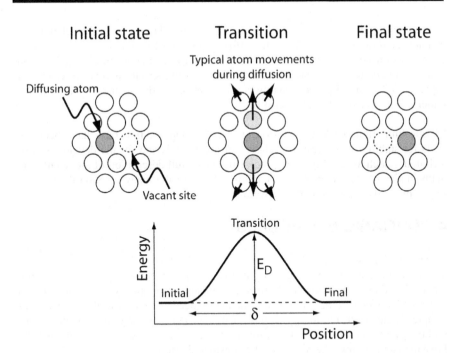

Figure 4.14: A schematic of the atomic movements involved in a typical substitutional diffusion process in a fcc lattice and the change in system energy throughout the process.

$$D = D_0\, e^{-E_D/kT} \qquad\qquad 4.23$$

where E_D is the activation energy. The prefactor $D_0 = v\delta^2$ involves both the attempt frequency, v [which is roughly the lattice vibration (phonon) frequency consistent with the lattice temperature] and the jump distance δ. In general, stronger bonds will give higher vibrational frequencies, higher energy barriers, and slower diffusion.

Diffusional mixing is usually the result of a series of uncorrelated atomic movements of the type described above. Because the movements are not related to one another, any individual atom may move in any direction. However, the net movement of atoms depends upon the concentration gradient for a given species. The result is that the flux of atoms F across a given plane perpendicular to the direction of diffusion, x, may be described by Fick's first law:

$$F = -D\frac{dC}{dx} \qquad\qquad 4.24$$

where C is the concentration of the diffusing species. This is the same relationship that appeared in Equation 3.2 and the diffusion portion of Equation 3.7 for motion of electrons in a potential gradient. One may generalize the concentration gradient in terms of a chemical potential gradient for the diffusing species and arrive exactly at Equation 3.2 with the proportionality constant related to the diffusivity of the atoms rather than to the mobility of electrons.

Kinetic limitations such as diffusion, nucleation, and reaction rates have been the basis for stability of several new classes of compounds and for the development of techniques such as rapid thermal processing. For a full description of the details and mathematics of atomic transport in solids see one of the many texts on the subject, for example, the classic text by Shewmon.

4.5 ORGANIC MOLECULES

When we discuss organic semiconductors in Chapter 9, we will be discussing organic molecules and making some use of organic chemistry concepts. It is not practical to present a summary of organic chemistry or organic synthesis here, nor is a detailed understanding of organic chemistry needed to follow the concepts of Chapter 9. However, familiarity with some nomenclature, definitions and approaches will be useful. Therefore, in this section only a few of the more important items from Freshman and organic chemistry will be considered briefly.

Organic molecules are constructed on carbon backbones. For examples of organic molecules see Figure 4.15. Most of the carbon atoms are sigma-bonded via sp^3 hybrid molecular orbitals (see Chapter 5 for more discussion of bonding). Carbon-carbon bonds define the geometry of most structural units of the molecule. Those containing exclusively carbon backbones with tetrahedral sp^3 single bonds are referred to as alkanes. Familiar members of this family include propane, butane, and so forth. Double carbon-carbon bonds (sp^2-sp^2 σ + p-p π bonds) have triangular planar molecular orbitals in their backbone bonds. Molecules containing double bonds are known as alkenes, such as ethylene and propylene.

One of the most important distinctions of alkenes from alkanes is that the single bonds can rotate while the double bonds cannot. This greatly modifies the physical properties of the molecules. Finally, alkynes include triple bonds (sp-sp σ + 2(p-p) π bonds). These are similar in many respects to the double bonded alkenes and are rigid and linear. All single and multiple carbon-carbon bonds are purely covalent, strong, and have well-defined bond angles and lengths. Interactions among organic molecules are much weaker than within the molecule and are dominated by either by van der Waals bonds or hydrogen bonds. Van der Waals bonds are important between molecules having no charged species (nonpolar) such as polyethylene. When charged subunits are present on the chain the charges may interact electro-statically to produce the stronger hydrogen bonds. The difference between polyethylene

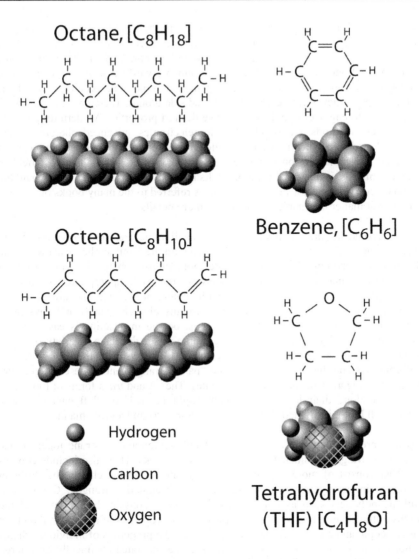

Octane, $[C_8H_{18}]$

Octene, $[C_8H_{10}]$

Benzene, $[C_6H_6]$

Hydrogen

Carbon

Oxygen

Tetrahydrofuran
(THF) $[C_4H_8O]$

Figure 4.15: Chemical formulas and sketches of the molecular structure of selected organic compounds.

and polyamide (for example nylon) is largely due to the hydrogen bonding present between charged amide units (CONH) which are absent in otherwise identical polyethylene.

The simplest organic molecules consist only of carbon chains with the remaining bonds terminated by hydrogen (see Figure 4.15). The properties of the materials are easily modified, however, through replacement of a hydrogen with a larger organic group. A common example might be the difference between ethane (a gas used for cooking and heating) and ethyl alcohol, found in beverages, which differ only in the replacement of one hydrogen with an -OH group. Very effective, if time-consuming, electronic structure calculations exist by which it is possible predict some of the underlying properties of molecules with selected side groups. Therefore, it is increasingly possible to engineer a molecule to have desired properties. Modern drug design utilizes these methods extensively. In microelectronics, advanced photoresists are also developed using such tools. The challenge is often to find an effective means to synthesize the molecule, once the desired structure is predicted. The vast field of organic synthesis deals with this problem. Its methods are fascinating and elegant but far too extensive to review here. The reader is referred to the many books on organic and, for that matter, inorganic chemistry for more details.

Organic molecules are often divided into small molecules and polymers. Small molecules include most of the species that make up the familiar chemicals we use in laboratories every day. They also include more unusual materials synthesized for specific tasks. Small molecules usually dissolve readily in appropriate solvents and have properties that vary significantly with molecular weight. The differences in properties with weight make them easily distinguishable chemically and physically. As a result, they are usually separable. Thus, octane in gasoline is easily distinguished from the lighter hexane molecule. Some molecules have a variety of forms depending upon where groups or atoms are attached. This results in the "cis" or "trans" conformations and the "ortho" "meta" and "para" arrangements. These varieties are illustrated in Figure 4.16 for selected molecules. The *cis* and *trans* forms of molecules often twist along their length either in a "right handed" or "left handed" form. Virtually all biologically-active molecules twist in a "right handed" manner.

Polymers are chains of identical small molecules (monomers), strung together like beads on a string, in branched structures, or in networks. They also include copolymers that consist of more than one type of monomer in the chain. The different monomers in copolymers may be regularly or irregularly arranged. Commercial synthetic polymers such as polyethylene often have considerable variation in the lengths of the chains. This leads to variations in the properties of the material on both a microscopic and a macroscopic scale. However, the properties of individual chains do not normally vary enough to allow them to be separated chemically on a large scale basis. (Electrophoresis and related techniques do manage this separation but are slow and work with small volumes of material.) Natural polymers such as proteins synthesized by living cells, by contrast, frequently have remarkable uniformity. Most natural proteins are complex strings of selected monomer units arranged in extremely specific sequences of extremely specific lengths. This uniformity is what makes the enzymes in our bodies operate effectively.

Many polymer families have similar backbone chains but with different side groups attached to the molecule. Thus, polyethylene is a bare carbon backbone with hydrogens attached, while polystyrene has 1/4 of the hydrogens replaced with six membered phenyl rings. These side chains are often referred to as *functional groups or ligands*. An additional term used in organic chemistry is *moiety*, a part of a molecule. Thus the -CO-NH- unit, which is present in all polyamides, is a moiety. Benzene moieties are often referred to as *aromatic* groups because they determine the smell of many odorous molecules. Examples of molecules with different functional side chains and ligands are shown in Figure 4.17. This figure also shows equivalent structural representations for two of the molecules to illustrate organic chemistry short hand.

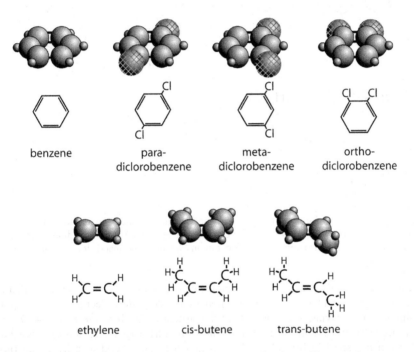

benzene	para- diclorobenzene	meta- diclorobenzene	ortho- diclorobenzene

| ethylene | cis-butene | trans-butene |

Figure 4.16: Shows conformations of several organic molecules to illustrate the *para, meta, ortho, cis,* and *trans* organization of moieties on the central structure of the molecule. If one takes the $H_3CHC=C$ portion of the *cis-* and *trans*-butene molecules as a fixed reference frame one can see that this causes the molecule to turn right or left down its length. The *para, meta,* and *ortho* arrangements of functional groups such as Cl determines the distribution of electric charge on the benzene ring structure. This, in turn, modifies the chemical behavior of the molecules above and beyond changes directly related to the organization of the functional groups themselves.

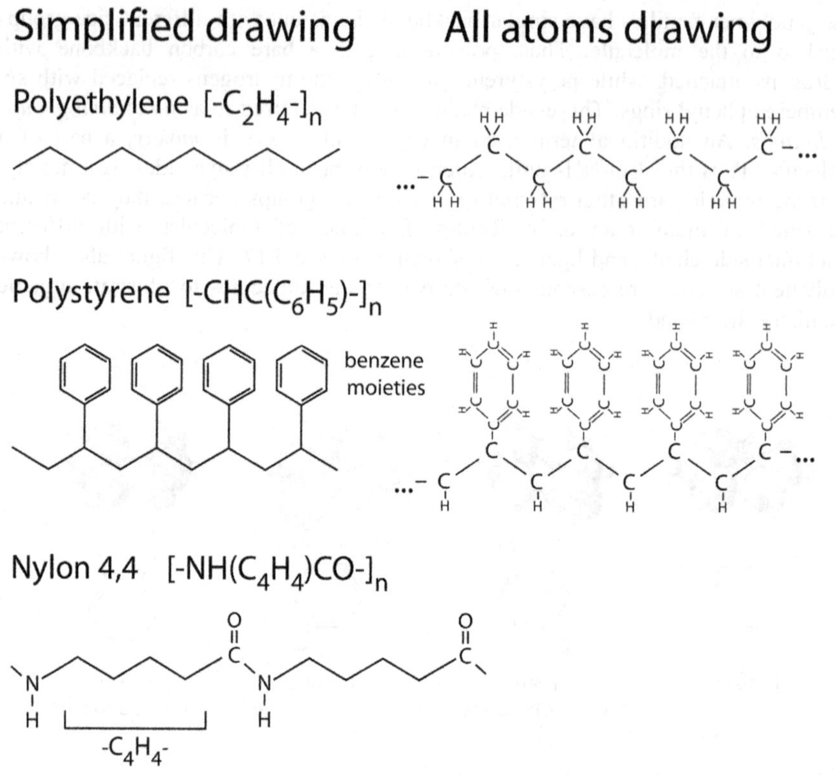

Simplified drawing   # All atoms drawing

Polyethylene $[-C_2H_4-]_n$

Polystyrene $[-CHC(C_6H_5)-]_n$

benzene
moieties

Nylon 4,4 $[-NH(C_4H_4)CO-]_n$

$-C_4H_4-$

Figure 4.17: Chemical formulas for three polymers. Structures are shown in three forms. The basic chemical formula is shown after the common name, the structure including all atoms is shown for two of the structures, and the common chemical shorthand drawing is given for each structure.

Because showing all of the atoms in a molecule produces a very busy drawing that is hard to interpret and because moieties often are turned with respect to the drawing, it is simpler to abbreviate some aspects of the structure. As shown in Figure 4.17, the simplified structures do not include an indication of hydrogen. Any left-over bond to a C atom not specifically indicated is presumed to be a bond to a hydrogen (not shown). A dangling bond is generally indicated by a dot indicating an unpaired electron or two dots if the electrons are paired (see Chapter 9 for examples). C atoms are also not shown explicitly on this diagram. C atoms are presumed to be present at vertices where lines indicating bonds meet if no other atom is shown. Non-C atoms are shown explicitly.

Aspects of functional groups which change the behavior of a molecule include 1) polarity or electronic charge, 2) size or shape, 3) reactivity, etc... Polarity is essential to the interaction of one molecule with another, for example through the formation of hydrogen bonds. This is also evident in solubility of common materials. Polar materials generally dissolve in polar materials and non-polar materials dissolve in non-polar materials, but the two do not usually mix well. Water and alcohols are examples of polar materials, while methane, hexane, and most oils are non-polar. Polymers or small molecules may include both polar and non-polar segments. This controls how they react and interact. In such cases the polar segment dissolves in a polar solvent while the non-polar parts prefer to clump together. This is how micelles form and is also the basis for the cell wall in biological systems.

Polymers are formed by reactions joining small molecules together to create the final structure. When the small molecules have two reactive groups, reactions joining them result in straight chains. Straight chain polymer molecules may be relatively rigid or relatively flexible, depending upon the flexibility of individual bonds in the chain. Ring structures are always inflexible units. Monomers with more than two reactive sites produce networks rather than chains. In this case the entire solid becomes effectively one huge molecule. The extent of interconnection in a polymer results in dramatic changes in its properties and divides polymers into two groups, thermoplastic and thermosetting.

Thermoplastic polymers consist of individual molecules, usually of variable length but relatively uniform chemistry, entangled with each other but not directly bound together by covalent chemical bonds. Weak interchain bonding may further organize the polymer chains into sheets or small crystallites if the polymer has few side groups and is very regular along its length. Because they consist of a jumble of individual molecules, thermoplastic polymers may melt if heated or may dissolve in solvents. Examples of thermoplastic polymers are polyethylene and polystyrene.

Thermosetting polymers consist of networks of monomers. It is common to make thermosetting materials from thermoplastic materials by adding small molecules, which will link the thermoplastic polymer chains together with covalent bonds to form a network. Epoxy is a common example of such a material. The epoxy resin is a thermoplastic polymer with reactive side chains. The hardener is a cross-linking molecule that links the resin molecules together. Because of the strength of the individual C-C bonds, thermosetting materials cannot be melted. Because the entire material is one huge covalently-bonded molecule, they do not dissolve in solvents, although solvents often will intercalate into the network and cause the material to swell. Thermosetting materials tend to be hard and brittle because it is not straight-forward for them to deform.

4.6 APPLICATIONS

4.6.1 A basis for phase transformations

To apply the formulas in Section 4.2 in more detail, one can, for example, calculate the free energy of a mixture of materials, from which the binary phase diagram may be obtained. Consider a mixture of two elements, A and B with composition x. The composition of this mixture would then be designated A_xB_{1-x}. A and B might be Pb and Sn in the case of Figure 4.8 but we will consider them generically here. Under the regular solution model these two elements can form three types of bonds, A-A, B-B, and A-B with distinct energies. It is assumed in this model that individual pairs have well defined energies and that there is no distinction among A-B bonds in which the A atom is also bonded to other A or B atoms. In real materials, especially semiconductors, the other atoms surrounding a given atom have a strong effect. The presence of significant interfacial energies also suggests that the pairwise bond energy assumption is highly approximate. Nonetheless, we will use it here. These considerations will be seen to be important when we discuss semiconductor band gap bowing and miscibility criteria in Chapter 6.

The difference in energy of two A-B bonds (energy=$2\omega_{AB}$) relative to one A-A (ω_{AA}) and one B-B (ω_{BB}) bond, is $\xi = 2\omega_{AB} - \omega_{AA} - \omega_{BB}$. ξ defines the energy gain or loss upon mixing of the elements. In a random alloy the probability of A-B bonds is the fraction of A atoms multiplied by the fraction of B atoms or x(1-x). The average bond energy, Ξ, of the resulting material is thus:

$$\Xi = 2x(1-x)\xi \qquad\qquad 4.25$$

This is also the energy of mixing of A and B atoms as it is the average change in energy per bond of replacing separated A and B atoms with a random mixture. Thus, if positive the mixing is endothermic and if negative the mixing is exothermic. The latter conclusions are only true at zero Kelvin because at elevated temperatures the entropy of mixing also enters the problem.

The primary term in the entropy of mixing is the configurational entropy, ΔS_{mix}. This is calculated by analysis of the number of combinations of N_A A and N_B B atoms which can be placed on the total number of lattice sites N_A+N_B in the solid. The number of configurations is:

$$\Omega = \frac{(N_A + N_B)!}{N_A! \ N_B!}, \qquad\qquad 4.26$$

from which the entropy is $\Delta S_{mix}=k\ln\Omega$. The logarithm of a factorial can be estimated with Stirling's approximation, $\ln(N!)\sim N\ln(N)$ from which:

$$\frac{\Delta S_{mix}}{k_B} \approx (N_A + N_B)\ln(N_A + N_B) - N_A \ln(N_A) - N_B \ln(N_B). \qquad 4.27$$

Taking $N=N_A+N_B$ and given that the difference of logarithms is the logarithm of the ratio of their arguments one may show that:

$$\frac{-\Delta S_{mix}}{Nk_B} = x\ln(x) - (1-x)\ln(1-x) \qquad 4.28$$

where we have also used the fact that $N_A/N=x$ and $N_B/N=(1-x)$. The total free energy change upon mixing per lattice site ($\Delta F/N$) can be shown to be the negative of the sum of Equation 4.25 and T multiplied by Equation 4.28. In other words:

$$\frac{\Delta F}{N} = k_B T[x\ln x + (1-x)\ln(1-x)] - \frac{x(1-x)\xi}{2}. \qquad 4.29$$

One can then plot F as a function of x for several values of T given a value for ξ. An example of such a calculation is given in Figure 4.18.

When the energy change per lattice site is positive the system is unstable. When negative the system will choose the lowest energy composition possible. When there is only one minimum in $\Delta F/N$ then there is an energy penalty to decomposing the mixture into separate phases. However, when two minima occur there is a reason to change the material into separate domains with each of the two low-energy compositions. In short, this favors the phase transformation. Note that the calculation provides a basis for understanding compositional instabilities and hence a process called spinodal decomposition.

Spinodal decomposition is phase separation resulting from fluctuations in composition that become thermodynamically stable up to some magnitude as the temperature decreases. The scale of stable fluctuations increases as the temperature decreases. This process does not produce a clear transition from a single phase to a two phase mixture in the microstructure because the first stable composition fluctuations have a very small magnitude. Thus, at least at first, the local differences in the separated phases are tiny. As temperature is lowered and the separation proceeds the separate phases become increasingly distinct. In many phase separation processes the initial small composition fluctuation regime is masked by the presence of a soluble liquid phase. Thus, for example, the spinodal decomposition marked by the solid line in Figure 4.18c is often cut off at the top as indicated by the dashed lines in the figure. Thus, by the time the phase separation begins the stable compositions are already far separated.

Full binary phase diagrams generally include other issues such as melting or boiling. Thus a complete treatment of the phase relationships in a mixture will involve additional calculations to estimate the energy of the other phases. The stable compositions and phases are those with the lowest overall energy. A consequence of

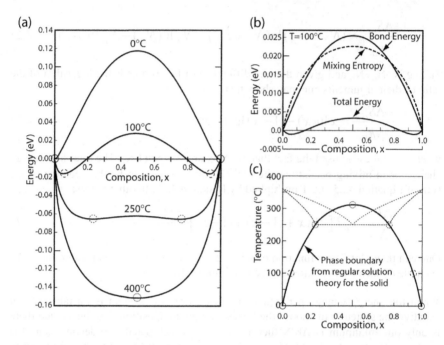

Figure 4.18: Free energy as calculated from Equations 4.25-4.29. (a) shows the free energy of mixing for several temperatures. Negative values indicate favorable mixing, while positive values indicate a preference for demixing. The more negative the energy the more favorable the mixing. (b) gives the two contributions to the energy, bond enthalpy and heat (temperature x entropy), whose difference gives the curves in (a). Notice in particular that the entropy dependence upon x is significantly more square than the energy dependence. (c) the locus of energy minima from (a) [circles] as a function of temperature determine the boundary of the stable phase regions and the compositions which are stable in the two-phase region. This locus determines the shape of the phase boundaries in the solid phase region of a binary phase diagram. To give an idea of how the rest of the phase diagram enters, the dashed curves are added. Calculation of these requires a separate determination of the energy of the various liquid phases.

including the liquid phase in the calculation would be the potential appearance of other phase boundaries such as are shown in the phase diagram in Figure 4.19.

4.6.2 Silicon crystal fabrication

Several of the concepts in this chapter are illustrated in the refining of Si from natural quartz or silica sand to electronic-grade Si and the subsequent growth of single crystals by the Czochralski process. Natural silica sand and quartz are nearly pure SiO_2. Unfortunately, getting rid of the oxygen in SiO_2 often degrades this initial purity dramatically. The reduction of SiO_2 is accomplished by mixing with a carbon source, normally coal, and other materials and heating the mixture to a high temperature (chosen to melt the component species, provide a rapid reaction, and favor the reduction of Si). The basic reduction process is shown in Figure 4.20.

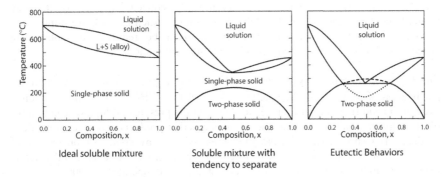

Figure 4.19: In a full calculation of a phase diagram one can include the liquid phase. An ideal solution produces a liquid-solid phase diagram as on the left side of the figure. As the bond energy difference from Equation 4.25 becomes more significant the solid tends to prefer to separate into two phases. This has consequences for both the low-temperature behavior described by Equation 4.29 and Figure 4.18 and for the liquid-solid relations. The result is the middle figure. A further increase in the tendency to phase separate leads to the behavior on the right and to a binary eutectic phase diagram as in Figure 4.8.

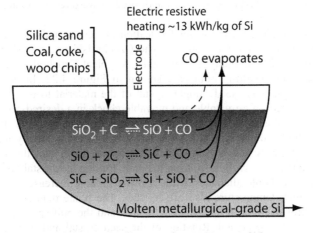

Figure 4.20: A schematic of the process for refining SiO_2 into metallurgical-grade Si using reduction by C. The product CO is removed by evaporation from the surface and is swept away in the gas. Because SiO has a moderate vapor pressure at these temperatures, some SiO also leaves the surface and condenses on cool surfaces nearby.

Oxygen reacts preferentially with C rather than Si making the reaction $SiO_2 + 2C \rightarrow Si + 2CO$ exothermic. [Note that CO is favored over CO_2 at high temperatures.] The removal of CO from the gas makes the process effectively irreversible (from Equations 4.14 and 4.16.) Thus, the Si is efficiently reduced, leaving only moderate O and C contamination. Unfortunately the C sources, such as coal, are impure and introduce contaminants, notably iron. These must be removed by subsequent processing. The result of the reduction is metallurgical grade Si.

The behavior described by Equation 4.14 is put to practical use as well in the purification of Si based on the reaction: $Si + 3HCl \rightarrow SiHCl_3 + H_2$. This reaction is exothermic and runs forward rapidly at 300°C in the presence of high concentrations of HCl gas and with removal of the gaseous reaction products. The tricholorosilane produced is easily separated from the unreacted HCl by condensation on a cool surface. It can then be fractionally distilled for purification. The reverse endothermic reaction forming Si and HCl will occur at higher temperature (~900°C) in the presence of hydrogen. This takes advantage of providing excess H_2 reactant and sweeping away product HCl in the unused H_2 to prevent the exothermic reaction. The trichlorosilane process was the standard method for purification of Si for many years. It is very straightforward but energy intensive and somewhat slow. A more efficient and speedy alternative has been developed based on silane (SiH_4) formation and decomposition. Both the Si reduction and purification processes are enhanced through the law of mass action. By removing reaction products the reverse reaction is slowed, thereby effectively increasing the desired forward reaction rate.

The final step in the production wafer fabrication process is making single-crystals from polycrystalline electronic-grade Si. The Czochralski process, shown schematically in Figure 4.21, normally achieves this. A typical Czochralski crystal growing system and the resulting boule are shown in Figure 4.22. In this method a small (4-6 mm diameter) "seed" single crystal of the material to be grown (Si, for example) is mounted on a rotatable and movable chuck in a desired orientation. As the crystal grows, this initial orientation determines the orientation of the resulting boule (bulk cylindrical crystal). The seed crystal is lowered gradually into a pot of liquid Si until it barely touches the surface. To grow the crystal the seed is then slowly (a few mm min^{-1}) withdrawn from the melt. Pulling the crystal out of the melt faster reduces the boule diameter while slower pulling increases the diameter. Changing the melt temperature also affects size. The boule diameter is measured dynamically through ports in the growth chamber and the pull speed is adjusted to achieve a given size of crystal. Rotation of the seed crystal and melt is needed to control the uniformity of the melt. This is critical to quality of the resulting crystal.

Inevitably a cold seed crystal put into contact with hot liquid experiences a strong thermal shock. This produces dislocations in the seed that must be prevented from propagating into the resulting boule. To accomplish this the growing crystal diameter is reduced to 3-4 mm and then enlarged again to the desired size (Figure 4.22). The entire weight (>100 kg) of the final boule is supported on this tiny neck!

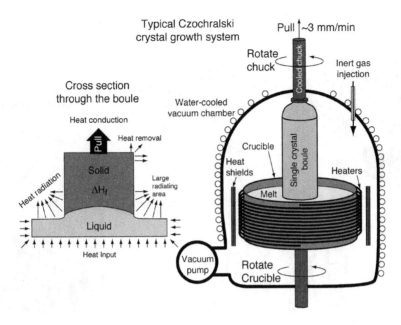

Figure 4.21: A schematic diagram of the Czochralski crystal growth process. A seed crystal mounted on a cooled chuck is lowered into a molten bath, which will solidify to start crystal growth. This seed is then withdrawn slowly and new crystal grows on the existing seed. If the seed is a single crystal then with proper care the boule will also be a single crystal with zero dislocations. The diameter of the boule is adjusted by control of the shape of the meniscus of the liquid and the shape of the solidification boundary.

One of the essential issues in growth of Si single crystals is temperature management. The seed crystal is cooled by radiation to the cold walls of the vacuum chamber in which the growth occurs. At the same time, heat is supplied to the crucible containing the melt. Finally, the latent heat of freezing, ΔH_f, provides an additional source of heat located, inconveniently, at the liquid-solid interface. A temperature gradient from the seed into the melt is established by these heat sources and sinks and the solid forms where the material passes through the melting temperature. If the crystal is growing uniformly the latent heat is generated equally across the solid-liquid interface. However, because the major heat sink is radiation from the surface of the melt and the boule, heat escapes most rapidly from the edges of the boule. In the center only conduction up the boule is available to remove the heat. Hence, this is the hottest region. A heat transfer model accounting for the various sources and sinks shows that the lower surface of the boule curves upward toward the center of the crystal, as shown in Figure 4.21.

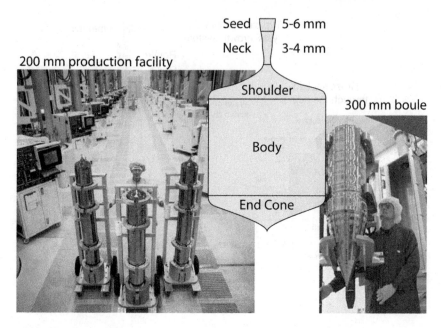

Figure 4.22: Photographs (left) of a 200-mm Si Czochralski growth facility, and a 300 mm Si boule (right) courtesy of MEMC Electronic Materials, Inc; St. Peters MO. The overlying schematic shows the regions of the boule.

The shape of the freezing surface is crucial to good crystal growth, as is control of convection in the melt. In this regard, the latent heat of solidification is particularly important as it can enhance convection. If freezing is nonuniform, local bursts of heat can be generated. When the boule surface is tilted, as is the case if it is significantly concave in the melt, a hot spot may be carried laterally across the boule surface and can cause melting of an adjacent area. These heating and cooling bursts produce locally high crystal growth rates, which can result in defects in the solid, as well as variations in doping. Convection currents in the melt are reduced in current Czochralski pullers by application of a large magnetic field (using superconducting magnets in some cases). The magnetic field interacts with moving atoms in the melt and acts as a brake, slowing convection significantly.

A modern Czochralski crystal growth system producing 300 mm-diameter 1200 mm long Si boules from a charge of Si in the crucible might typically stand 10 meters high, consume ~1 MW of electric power and over 760 L of cooling water and 400 L of Ar per minute. Roughly 2/3 of the electric power goes to energizing the magnetic convection-damping field (suggesting the motivation for a superconducting magnet).

The crucible generally holds ~250 kg of electronic-grade Si with both the crucible and boule rotated at up to 30 rpm. The crystal is grown at up to 8 mm min^{-1}. Thus, to grow a 1 m long boule would require ~1.5 hours to grow, consuming ~1500 kWh of power, 68000 L of water, and 36,000 L of Ar.

One of the observations about Czochralski growth is that the concentration of impurities in the solid is generally much different (and fortunately often much lower) than in the liquid. The ratio of the concentration in the solid to the concentration in the liquid is referred to as the segregation ratio. Typical values for segregation ratios of selected impurities are given in Table 4.1. Some of the more striking values are for O (which actually prefers to be in the solid phase) and Fe, which is in equilibrium when the liquid contains one hundred twenty five thousand times more Fe than does the solid. The latter is particularly fortunate as Fe is a major problem for semiconductor devices. Detectable degradations in performance may be found at the part-per-billion Fe level. Consequently, effective methods, such as those discussed above, to remove Fe are particularly important.

Table 4.1: **Segregation Ratios of Impurities in Silicon**

Impurity	C_{solid}/C_{liquid}	Impurity	C_{solid}/C_{liquid}
B	0.8	C	0.07
Al	0.002	Fe	8×10^{-6}
P	0.35	O	1.25
As	0.3		

The values in the table show why B would be better than Al for intentionally doping bulk Si grown by the Czochralski method. First, one would need 400 times more Al in the melt than B to achieve the same bulk doping level in the boule. Second, as the melt is used up in forming the solid the number of Al atoms in liquid is relatively unchanged while the Si is consumed. Thus the Al concentration in both the liquid and solid rises increasingly rapidly as the crystal grows. By contrast, because B is removed from the liquid at nearly the same rate as Si, the concentration in the melt and hence in the solid remains nearly constant.

To see why segregation might be expected, consider the composition and temperature of the liquid near the solid interface. This may be guessed based on the phase diagram of silicon and the impurity. Suppose the region of a generic binary eutectic phase diagram near the Si melting point is as shown in Figure 4.23. The "solidus" line marks the composition at a given temperature below which all of the material is solid, while above the "liquidus" line the system contains only liquid. Between the two lines the system is a mixture of solid and liquid phases.

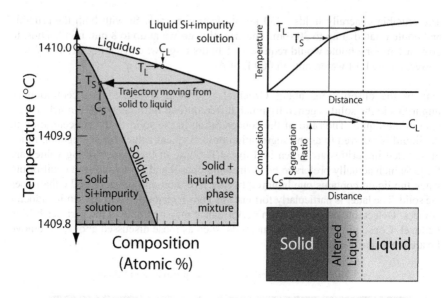

Figure 4.23: Schematic diagram of a typical binary eutectic phase diagram between Si and an impurity near the Si melting point. Insets at the right show the temperature and composition of the system across the solid/liquid interface, as well as a schematic of the interface. The binary phase diagram includes an indication of the compositions of the solid and liquid during growth and the change in both variables through the interface region. As shown, the composition of the liquid is enriched in impurity near the interface due to rejection of extra impurity from the solid during growth. This excess impurity must diffuse into the liquid and results in the composition change shown. Note that the observed segregation ratio will generally be lower during growth than the equilibrium value due to this diffusion rate limitation.

For any temperature, the segregation ratios given in Table 4.1 are the ratio of the composition of the solid to that of the liquid, given by the solidus and liquidus lines, respectively. However, near the solid-liquid interface there are two variables that could change across the interface, composition and temperature. Very close to the solid-liquid interface the temperature is nearly constant at the local freezing point for the local liquid composition. This corresponds to a horizontal line on the phase diagram, shown as the lower line on the trajectory through the interface on Figure 4.23. Because heat is generally transmitted more slowly in the solid, the temperature gradient there is likely to be larger. However, the impurity is also being rejected into the liquid as the solid grows due to segregation. Therefore, the liquid near the interface has an enhanced level of impurity determined by how quickly those atoms can diffuse away. At steady state a concentration distribution, as shown in Figure 4.23, will be established. A changing concentration of impurity in the liquid with position will also correspond to a change (generally a rise) in liquidus temperature. As a result, both temperature and composition will change across the solid-liquid interface region, as shown schematically by the trajectory in Figure 4.23. Therefore, the temperature gradient in the altered liquid layer is connected to the composition gradient there to some extent. For infinitely slow growth where diffusion in the liquid is very fast, the segregation ratio would be the ratio of solid to liquid composition at the solid-liquid interface. For faster growth the segregation ratio is effectively reduced due to the development of a concentration gradient in the liquid.

Much of this text is concerned with semiconductor alloys, particularly of the group III-V semiconductor compounds. These materials are not produced on Si substrates, but rather rely on GaAs, InP or other single crystals. Although boules of these materials are grown by the Czochralski process, there is a catch – the group V element in the case of III-V compounds is very volatile at the crystal growth temperature and would evaporate into the gas stream if something were not done to prevent its escape. One solution is the "liquid-encapsulated" Czochralski (LEC) method. In this process liquid boric oxide (B_2O_3) floats on top of the melt used to grow the boule (InP for example). Because it is insoluble in the boule material and vice versa the interface is stable. The B_2O_3 prevents melt material from dissolving in it and escaping to the surface. Because it is stable at the growth temperature the boric oxide itself will not decompose or evaporate. Finally, because it is stable with respect to GaAs, InP, or other materials being grown it will not dissolve or allow As or P to pass through it. Thus the boric oxide layer forms a barrier preventing escape of the volatile element in the compound semiconductor. This process has been used to produce compound semiconductor wafers up to 150 mm in diameter.

An alternate method is to allow the group V element to evaporate into the growth environment and to simply raise its pressure in the crystal puller until it is condensing at the same rate that it evaporates. The problem with this approach is that it requires a very high pressure of (somewhat dangerous) group V element in the puller and that the puller be designed to withstand this pressure safely. The major advantage of this method is that modest B and O contamination of the growing wafer does not occur. This permits lower ultimate doping levels in the wafers.

4.6.3. Rapid thermal processing

As a final example of materials science concepts in action, we briefly consider rapid thermal processing, used extensively in microelectronics manufacturing today. This technique is based on the fact that most processes from diffusion to reaction rates to nucleation in thin films have kinetic rate constants that depend exponentially on inverse temperature, as in Equation 4.15. Rapid thermal processing takes advantage of the fact that for two competing processes with different values of the activation energy E, the higher activation energy process is strongly favored at high temperature, while it is strongly suppressed relative to the other at low temperature. Thus, by adjusting the temperature one can adjust the relative rates of two processes.

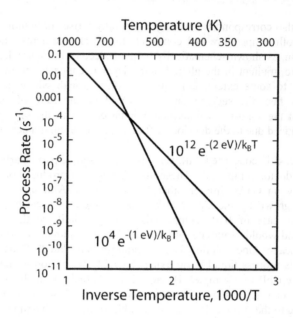

Figure 4.24: Rates for two hypothetical processes, one with an activation energy of 2 eV and a prefactor of 10^{12} sec^{-1} and a second with an activation energy of 1 eV and a prefactor of 10^4 sec^{-1}. The high activation energy process is favored by short times at high temperatures (left side of the graph) while the low activation energy process is dominant for long times at low temperatures (right).

It turns out that in many cases in microelectronics fabrication, the activation energy of a desired process is higher than for an undesired processes, as shown in Figure 4.24. Furthermore, in many processes, such as diffusion, we must only keep constant the product of diffusion rate and time (a product directly proportional to the square of the distance an atom can diffuse in that time). Therefore, by increasing the process temperature one can decrease the annealing time at that temperature. Hence the name, rapid thermal processing. While the desired process rate x time product is kept constant at the increased temperature, the undesired process rate has increased but to a lesser extent. Thus, when the time at elevated temperature is reduced, the undesired process has a relatively lower rate x time product and is suppressed relative to the desired process.

Rapid thermal processing can, in general, be used to suppress any low-activation energy process in favor of any high activation-energy process by raising the process temperature and decreasing the time. The reverse could be achieved with a low-temperature long-time heat treatment if a desired process had a very low activation energy.

4.7 SUMMARY POINTS

- All crystals form in one of 14 Bravais lattices.
- Planes and directions in these lattices are indexed with three-integers in cubic systems.
- Each lattice point on the Bravais lattice may have more than one atom associated with it.
- A "primitive cell" also may be defined containing exactly one formula unit of atoms of the compound.
- Hexagonal close packed and face-centered cubic crystals are both "close-packed" lattices having ABAB or ABCABC stacking sequences for atom planes.
- A reciprocal lattice may be defined based on Equation 4.2 and the basis vectors for the real-space lattice.
- Diffraction occurs when the change in wave vector in the diffraction event is a reciprocal lattice translation vector.
- A phase is a chemically and structurally homogeneous material.
- Thermodynamic stability requires that a phase or set of phases have a lower energy than any other set of phases that could be constructed from the same set of atoms under given conditions. This is equivalent to the set of atoms having their lowest average chemical potential.
- Entropy is the Boltzmann constant multiplied by the logarithm of the number of states accessible to the system. In reversible processes an isolated system exhibits no change of entropy while in irreversible processes the entropy must always increase. Entropy contributions to the energy of systems become increasingly important at higher temperatures.
- The number of degrees of freedom of a system is the number of constituents less the number of phases plus two. Single constituent systems have pressure-temperature phase diagrams such as that shown in Figure 4.7, binary phase diagrams are as in Figures 4.8 and 4.9, and ternary phase diagrams are as in Figures 4.10 and 4.11. Other phase diagrams also exist.
- The phases present in a multiphase region of a phase diagram are determined by the stable phases in surrounding regions of the diagram at a given temperature.
- Phase diagrams may be calculated from basic thermodynamics describing the interactions among atoms and the entropy associated with various arrangements of atoms.
- The less the atoms like to mix the more likely they are to form eutectic-type phase diagrams rather than showing large solubilities. This is determined largely by the relative energy of the bonds between different atoms compared to the bonds between like atoms.
- The law of mass action shows that the rate of a reaction depends upon a ratio of the reactants to the products multipled by a rate constant. This rate constant

depends exponentially on temperature and an activation energy and linearly on a prefactor.

- Multistep reactions may be limited by any of the rates of an intermediate reaction or by a combination of all rates.
- Any phase change creating a new interface is likely to require nucleation of the new phases. Nucleation is generally associated with an energy barrier and may be the limiting step in a reaction or phase transformation.
- Nucleation energy barriers depend upon the surface and volume energies associated with the reaction. A critical radius for the nucleus exists beyond which the new phase may grow spontaneously.
- Phase transformations often require transportation of atoms resulting from diffusion.
- The flux of transported species depends upon their concentration gradient and diffusivity.
- The diffusivity depends exponentially on temperature and an activation energy and linearly on a prefactor, usually related to a lattice vibrational frequency and the square of the jump distance over which the particle moves in a single average event.
- Organic molecules have covalently-bonded backbones with single (flexible), double or triple (rigid) bonds along their backbones.
- Interactions between chains are through weak Van der Waals or hydrogen bonds. The latter require charged subunits on the molecule.
- Properties of organic molecules are determined by the nature of functional groups, ligands, or moieties within the molecule, and by the arrangements thereof.
- Polymers are chains (thermoplastic) or networks (thermosetting) of small monomer molecules.
- Electronic grade Si is produced by reduction by C, followed by a purification step.

4.8 HOMEWORK

1. Derive Equation 4.21 for the critical nucleus of a cluster based on Equations 4.17 and 4.20.

2. Derive Equation 4.28 from Equation 4.27.

3. The basis vectors for a face-centered cubic real-space lattice, a_1, a_2, and a_3 are
$$\frac{a}{2}(1,1,0), \quad \frac{a}{2}(1,0,1), \quad \frac{a}{2}(0,1,1)$$
calculate the reciprocal lattice basis vectors b_1, b_2, and b_3.
Determine the lattice structure of the reciprocal lattice.

4. Light a match or candle and observe the resulting flame. Make a sketch of the flame showing the match stick or candle wick and the various visible parts of the flame. Indicate approximately on your drawing:

   a) the fuel rich region
   b) the region close to the ideal reactant mixture
   c) the fuel-poor region
   d) where the fuel is coming from
   e) where the oxygen is coming from.

5. Suppose we are attempting to nucleate a solid-phase reaction between a metal and a silicon substrate to produce a silicide intermetallic compound and carry the reaction to completion. Further suppose that we have a dopant in the solid which we do not wish to have diffusing as the silicide reaction takes place. Suppose the two processes both have rates defined by Equation 4.26. Given the kinetic parameters for the reactions as follows:

	k_0 prefactor	Activation Energy, E
Silicide reaction:	$1 \times 10^{12} \text{ cm}^2 \text{ s}^{-1}$	3.5 eV
Dopant diffusion:	$4.3 \times 10^3 \text{ cm}^2 \text{ s}^{-1}$	2.3 eV

The maximum temperature at which the silicide reaction can take place is 850°C and the minimum temperature is 450°C. If the product of rate and time for the desired process must be $2 \times 10^{-7} \text{ cm}^2$, calculate:

   a) the time necessary for the desired reaction process at 450°C
   b) the time necessary for the desired reaction process at 850°C
   c) the time-temperature products for the undesired diffusion at these temperatures.
   d) the ratios of the two process rates at the two temperatures

6. Consider the phase diagram of an impurity in Si near the Si melting point:

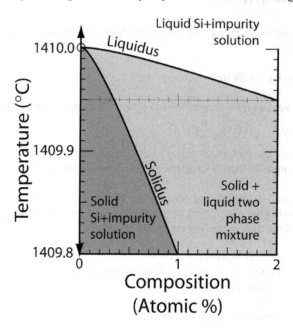

Suppose that you are growing a Si crystal by the Czochralski method from a liquid containing 2 at.% of the impurity with the corresponding phase diagram above.

    a) What temperature would you expect in the solid at the liquid interface in equilibrium? (You need not explain.)

    b) Calculate the segregation ratio for this impurity at this temperature.

    c) What is the effect on the temperature of the liquid near the interface of a chemically-altered liquid formed by rejection of the impurity into the liquid from the solidifying solid? (Explain in a sentence or two.)

    d) Briefly, how do you perform Czochralsky growth on a high vapor pressure material such as GaAs?

7. As an engineer at MegaJoule Industries your task is to grow bulk crystals by the Czochralsky process. One of the Czochralsky pullers has an air leak loud enough to hear during pump down. The worst part of the problem is that the leak cannot be fixed without replacing the entire unit. Because the whole growth facility is being replaced in a year it is not practical to fix this puller now. You find that the grown crystals contain both carbon and oxygen impurities as a result of the leak.

a) Could you remove the carbon that enters through the air leak by repeatedly melting and refreezing the boule? (This process is called zone refining and takes advantage of the tendency of some impurities to segregate from the solid to the liquid.) Explain briefly. (Segregation coefficient $C_s/C_1 = 0.07$)

b) Would you expect the concentration of oxygen to increase or decrease in the Boule during this process if a small amount of oxygen was present in the zone refiner as residual gas? (Segregation coefficient $C_s/C_1 = 1.25$)

c) In problem 6 above, there is a sketch of the phase diagram for Si and an impurity very close to the Si melting point and near pure Si. Make a similar schematic sketch for Si and O given the segregation coefficient in part (b) above.

4.9 SUGGESTED READINGS AND REFERENCES

Suggested Readings:

Cullity, B.D., *Elements of X-ray Diffraction*, 2nd edition. Reading: Addison-Wesley, 1978.

Shackelford, James F., *Introduction to Materials Science for Engineers*. New Jersey: Prentice Hall, 1996.

Shewmon, Paul G., *Diffusion in Solids* 2nd ed. Warrendale: TMS, 1989.

Shimura, Fumio, *Semiconductor Silicon Crystal Technology*. San Diego: Academic Press, 1989.

Van Vechten, J.A., "A Simple Man's View of the Thermochemistry of Semiconductors" in *Handbook on Semiconductors*, ed. T.S. Moss, Vol. 3, *Materials, Properties, and Preparation*, ed. S.P. Keller, North Holland, Amsterdam, 1980, Chapter 1.

Villars, P., Prince, A., and Okamoto, H., *Handbook of Ternary Alloy Phase Diagrams*, v. 7. Metals Park: ASM International, 1995.

References:

[1] Villars, P., Prince, A., and Okamoto, H., *Handbook of Ternary Alloy Phase Diagrams*, v. 7. Metals Park: ASM International, 1995.

[2] Gödecke, T., Haalboom, T., and Ernst, F., "Phase equilibria of Cu-In-Se. I. Stable states and nonequilibrium states of the In_2Se_3-Cu_2Se subsystem." *Zeitschrift Für Metallkunde*, 2000; 91: 622-34.

Chapter 5

ENGINEERING ELECTRONIC STRUCTURE

Advanced devices place strong demands on semiconductor properties. To obtain the highest performance it is necessary to engineer the properties of constituent materials. In some devices, this means designing the electronic energy band structures. In other cases, the natures of defects in the materials are most critical. In this and the following chapter we consider band engineering and leave defect design to Chapter 7.

In Chapter 2, we discussed the basic physics that determines the electronic structure of a periodic solid. Now we will put some of that understanding to work. In this chapter we will further develop the physics underlying trends in semiconductor bands as a function of the atomic electronic states from which they are constructed. From this we will see how controlled modification of the chemistry and structure of a semiconductor can be used to engineer its energy bands. We will also step back and see how trends in real energy bands can be found, understood, and exploited to control energy gaps, energy-momentum relationships, and band edge energies. Chapter 2 gives an idea of why bands of states form and the breadth of a band of states between energy gaps for given configurations of atoms. It does not provide any specifics as to why one material is different from another, nor does it suggest why bands of states exhibit behaviors such as indirect energy gaps. Here we will examine the chemical basis for the observed variability.

The objective of this chapter is to provide an insight into the results of quantum-mechanical band structure calculations without relying on quantum mechanics itself in detail. The following assumes that the majority of readers of this book will not have studied quantum mechanics beyond the brief review provided in Chapter 2. Details of the calculations of real band structures are therefore beyond the scope of this text. However, excellent, if more complex, explanations can be found in the classic texts *Solid State Physics*, by Ashcroft and Mermin, in more detail yet in *Electronic Structure* by Harrison, and in many other resources. Here we focus on the bigger picture of how solids organize their electrons. We will explore the functionalities of the bonding process and see how behaviors such as the common cation and common anion rules for heterojunctions arise, why energy band gaps decrease as the average size of a lattice increases, and many other typical behaviors.

5.1 LINKING ATOMIC ORBITALS TO BANDS

The quantum mechanics of bonding traces the evolution of the energies and distributions of electrons as atoms come together to form a solid. During this process, the states the electrons occupy develop from initial atomic orbitals to the energy bands of Chapter 2. The solutions of the Schrödinger Equation in the completed solid are constructed typically as a linear combination of the atomic orbitals (LCAO) with corrections as necessary. The LCAO method is considered in overview in Section 5.2. Before delving into LCAO, however, it may be helpful to review some of the results and to look at the bonding behaviors schematically.

In the discussion below we will need of the following terms:

Homopolar semiconductors: All atoms in the unit cell in these materials are the same so there is no charge transfer from one atom to another. Examples of such materials are Si and Ge. In these two examples, there are two atoms per primitive unit cell but both are the same kind of atom.

Heteropolar semiconductors: There are two or more different kinds of atoms organized regularly in the unit cell. These transfer electrons from one atom to the other and thus have different polarities. Examples of such materials are GaAs and AlN. In such materials the atoms are well organized into a compound (as distinct from a random alloy) with a specific lattice structure.

Note: A Si-Ge alloy has two kinds of atoms but they mix randomly and specific sites for one or the other cannot be distinguished. Therefore, such a mixture is an alloy of homopolar materials, not a heteropolar compound.

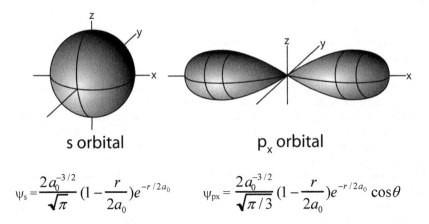

$$\psi_s = \frac{2a_0^{-3/2}}{\sqrt{\pi}}(1 - \frac{r}{2a_0})e^{-r/2a_0} \qquad \psi_{px} = \frac{2a_0^{-3/2}}{\sqrt{\pi/3}}(1 - \frac{r}{2a_0})e^{-r/2a_0}\cos\theta$$

Figure 5.1: Shows the shape of the s and p_x orbitals and the equations that describe them. a_0 is the atomic orbital size, r is the radius from the nucleus, and θ is the angle in the x,y plane.

5.1.1 Homopolar semiconductors

We begin the discussion by considering isolated atoms. Atoms have electron orbitals determined by solution of the Schrödinger equation for a positive core potential. The only case that can be solved explicitly is the hydrogen atom or a single electron orbiting a more highly charged nucleus. Atoms with multiple electrons are very hard to treat exactly, due to the difficulty in dealing with the correlations of motions of electrons with each other. Fairly good quantum chemical methods have been developed for approximating the correlated many-electron systems. While exact analytical solutions for the multi-electron case are not possible, the general phenomena found in single electron states holds true for multi-electron atoms, and correlation effects represent only perturbations on the single-electron behaviors.

Each atomic orbital is described by a series of quantum numbers (n, m, l, s), corresponding to the properties of total (n), orbital (m, with 0≤m≤n), and azimuthal (l, with |l|≤m) angular momenta, and spin (s=±1/2). For m=0 the orbitals are spherical and are termed "s" orbtials. For m=1 the orbitals have roughly figure-eight shaped probability distributions along one of the three coordinate axes and are termed "p" orbitals, one for each value of l. The m=2 and m=3 values give the "d" and "f" orbitals. Spin allows two electrons per orbital for each of the various s, p, d, and f states. The geometries and basic mathematical descriptions of the s and p atomic orbitals are given in Figure 5.1. Their derivations for the hydrogen atom may be found in most undergraduate quantum mechanics textbooks. Each orbital has a well-defined [binding] energy, which increases with increasing nuclear charge. The specific energies were observed over 120 years ago through their effect on the optical emissions of flames. The energies of all orbitals are commonly observable today by many techniques. Two of the more precise but accessible methods of direct observation of these states are x-ray photoelectron spectroscopy and Auger electron spectroscopy.

When two or more atoms bond together to form a molecule, the atomic orbitals may be considered to be mixed together to form molecular hybrid states. The hybrids are atom-like in that they are localized. However, they have the geometry of the molecule. Thus, they are generally termed molecular, rather than atomic, orbitals. For example, when a water molecule forms it has a bond angle between the two hydrogen atoms of 108°. This is because the three 2p orbitals and one 2s orbital of the oxygen have combined to form four $2sp^3$ hybrid orbitals pointing toward the vertices of a tetrahedron, and are therefore separated by 108°. Two of the orbitals are completely full of oxygen electrons while the other two have one opening each, which is filled by the hydrogen electrons. The shapes of sp^3 and similar sp^2 hybrid orbitals are shown in Figure 5.2. Both are common in semiconductor bonding. In the following discussion we will consider the consequences of sp^3 bonding on the energy of electrons in diamond or zincblende-structure semiconductors. A similar argument can be made for sp^2 bonding with respect to hexagonal wurtzite-structure materials. As we will see in Section 5.2, the hybrid orbital picture is equivalent to considering the interaction of each orbital with all others individually and adding the resulting interactions together. The advantage of the hybrid picture is that it shows why the resulting crystals have the symmetry and structure that they have.

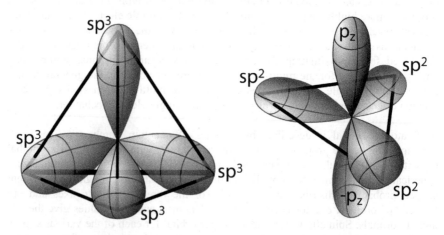

Figure 5.2: Shows the symmetry of the hybridized sp^3 and sp^2 molecular orbitals. The sp^2 orbitals lie in a plane perpendicular to the p_z orbitals and are equal lengths. The sp^3 orbitals are all equivalent to each other and stretch to corners of a tetrahedron.

Atoms from which diamond-structure and hcp semiconductors are constructed have three partially filled p valence-state (lowest binding energy) orbitals with energies close to the filled valence s-states. When the atoms come together to form the semiconductor, their one outer s state and three valence p states can be viewed as mixing to form sp^3 hybrid orbitals. This configuration maximizes the number of electron pairs, and makes each final bond as similar and as low energy as possible. The energy of the hybrid orbital is the linear average of the starting orbitals:

$$E_{sp^3} = \frac{1}{4}\left(E_s + 3E_p\right). \tag{5.1}$$

So far we have not changed the average energy of the orbitals, although the system began, for example in silicon, with two electrons in the 3s orbital and two in the three 3p orbitals. Therefore, the average energy of the electrons has increased as a result of the hybridization process, from the initial state $E_i = 2E_s + 2E_p$ to the final state $E_f = 4E_{sp3} = E_s + 3E_p$. This would be energetically unfavorable if it were not for the formation of bonds between these orbitals. Hybridization is always energetically unfavorable for individual atoms, which is why the orbitals for isolated atoms are as they are and do not occur as hybrids. To see why hybridization does occur in solids, we construct wave functions that are linear combinations of two sp^3 orbitals:

$$\begin{aligned}\psi_u &= \psi_1^{sp^3} - \psi_2^{sp^3} \\ \psi_g &= \psi_1^{sp^3} + \psi_2^{sp^3}\end{aligned}, \tag{5.2}$$

where ψ_g is the symmetric combination of the hybrid sp^3 wave functions of the individual atoms and ψ_u is the antisymmetric combination. (The g and u subscripts refer to the German words for symmetric and antisymmetric, gerade and ungerade. Some authors reverse the g and u labeling scheme.) These combinations are similar to those that appear in Chapter 2 for the nearly free electron model (Equation 2.5) and which gave rise to the higher or lower energy bands at the zone boundary (Equation 2.10). The symmetric combination lowers the energy of the electrons as the atoms approach each other and is referred to as a "bonding" orbital. The anti-symmetric combination raises the energy with respect to the starting states and is termed an "antibonding" orbital. When ψ_1 and ψ_2 are half-filled, the combination of the two contains enough electrons to exactly fill the bonding orbitals and leave the antibonding orbitals empty. The situation is shown schematically in Figures 5.3 and 5.4. Note that in this case we have gained as much energy from the bonding process as possible by placing all electrons in the lower energy states and no electrons in the higher energy states. This is why semiconductors have such strong bonds.

The nature of bonding can be seen immediately from a consideration of the effect on electron density of symmetric and antisymmetric linear combinations of the wave functions ψ_1 and ψ_2. Consider for simplicity the interaction of two s-orbitals. (The sp^3 hybrids behave essentially the same way but with more complex geometry.)

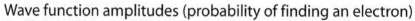

Wave function amplitudes (probability of finding an electron)

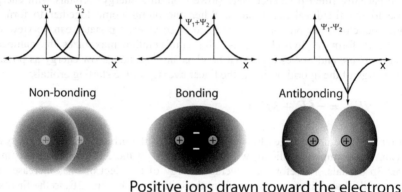

Figure 5.3: A schematic diagram illustrating the basis of cohesion in solids resulting from symmetric and antisymmetric combinations of atomic orbitals. The center of electron charge lies between the positive ions for a symmetric bonding orbital and outside of the positive ions for an antisymmetric orbital combination.

S-like atomic wave functions are peaked around the positively-charged nucleus of an atom and decay with distance from that atom (see Figures 5.1 and 5.3). A symmetric sum of the two, ψ_g, has a high intensity between the two atomic nuclei, twice the magnitude of a single wave function at half the interatomic distance. By contrast, the antisymmetric wave function, ψ_u, has a node ($\psi_u=0$) at the midpoint between the atoms. In the symmetric case there is a high density of negative charge between the atoms, which attracts the positive nuclei and holds the atoms together. This is the source of cohesion in the bonded pair. In the antisymmetric case, the wave function amplitude is depressed between the atoms and the positively-charged nuclei are relatively exposed to each other. This leads to repulsion. When only the symmetric state is filled with electrons, the maximum bonding occurs. When both states are completely filled the attractive and repulsive forces match and result in no net bonding.

The energy difference between the bonding or antibonding states and the hybrid molecular orbital energy is referred to as the homopolar energy, V_2 and the bonding-antibonding orbital energy difference is $2V_2$. V_2 can be shown [see Harrison, Ref. 1 for example] to depend approximately upon the inverse square of the interatomic distance, d, as:

$$V_2 \approx 4.4\frac{\hbar^2}{md^2} \text{ eV.}$$

5.3

Thus, as the atoms approach each other, the bonding strength increases rapidly (as $1/d^2$) until the repulsion of the positively-charged atomic cores begins to become significant. So far, however, we only have two atoms bonded into a dimer molecule. As more and more atoms are added, all of the atomic or molecular states of all of the atoms interact with each other. These interactions are between second, third, and increasingly higher-order neighbors and are progressively weaker as the solid becomes larger. It is these interactions which lead to the continuum of states we call a band. Their collective interaction is best described with waves as in Chapter 2.

A more complete analysis shows that the bands broaden following the same $1/d^2$ functionality as for the homopolar splitting V_2. However, because the broadening of the bands is relatively weak overall compared to the increase in their separation, the direct energy gap still increases with decreasing interatomic distance. In other words, the factor scaling $1/d^2$ in the band broadening term is smaller than the constant in Equation 5.3. The dependence of V_2 and energy gap on d is ultimately responsible for the pressure and temperature dependences of the energy gap in homopolar semiconductors (see Section 5.4).

The cohesive energy of the homopolar solid is just the amount by which the average electron's energy is reduced in going from the original atomic orbitals through the hybridization process to the formation of bonding and antibonding states and finally to the formation of bands. The greater the energy difference between the bonding and antibonding states ($2V_2$), the higher the cohesive energy of the material. For the simple bonding-antibonding states (not the bands) of Figure 5.4, the cohesive energy would be the average energy (increase) of an sp^3 hybrid state relative to the starting s and p states, $[(E_s+3E_p)-(2E_s+2E_p)]/4 = (E_p-E_s)/4$ per electron, less the energy gained by bonding, V_2 for each of the eight electrons in the unit cells (4 electrons per atom with a two atom basis in the diamond structure). Thus,

$$E_{cohesive} = 2\left(E_p - E_S\right) - 8V_2.$$

5.4

The true cohesive energy is this value modified by the average change in energy for electrons during formation of bands from individual bonded molecular orbitals. This difference corresponds to the average energy of the valence band relative to the bonding state. One can guess that this will be favorable for any material in which formation of the complete solid (consequently formation of bands) occurs. It could be unfavorable for cases such as diatomic gases although at low enough temperatures these materials do solidify. However, in all normal semiconductors the formation of bands is quite favorable and the average energy of the band is below that of the molecular orbital.

5.1.2 Heteropolar compounds

The picture in the previous section works well when there is only one type of atom in the material. When there are chemically different atoms, their electron densities, electron affinities, and nuclear charges differ. Because of this, their atomic orbitals

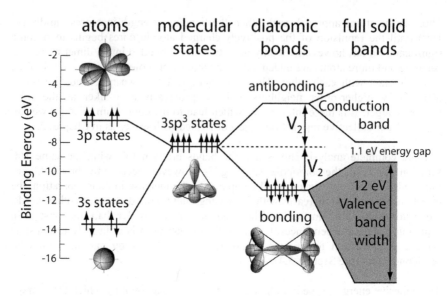

Figure 5.4: A schematic diagram of the evolution of bonding of Si atoms. The filled 3s and partially filled 3p atomic orbitals of two atoms combine to form half-filled sp³ hybrid molecular orbitals. These combine to form bonding and antibonding orbtials. As more atoms collect atoms collect to create a bulk solid, bands form.

have different energies. The easiest way to see this is to compare two atoms in the same row. For example, compare Ga with As. Both have filled 1s, 2s, 2p, 3s, 3p, and 4d core states and filled 4s valence states. In addition, each has a partially filled set of 4p valence orbitals. Ga has only one electron in its 4p orbitals, while As has three. The energy binding the 4p electrons is the same within a given atom, but the two additional positive charges in the nucleus of the As atom hold the three 4p electrons much more strongly than the weaker nuclear charge of the Ga nucleus holds its single 4p electron. This means that the 4p state in As is much lower in energy (stronger electron binding) than the 4p state in Ga. Likewise, the 4s orbital energies are affected by the additional nuclear charge. Consequently, the hybrid sp³ orbitals of the two atoms come out at much different energies. This energy difference is shown schematically in Figure 5.5 and is twice the quantity termed the chemical splitting, C. Mathematically,

$$2C = E_c^{sp^3} - E_a^{sp^3} \qquad\qquad 5.5$$

at some ideal interatomic distance, where E_c and E_a refer to the energies of the cation and anion molecular orbitals, respectively. The values of chemical splitting increase as the atoms move farther apart on the periodic table in a given row (larger difference in electronegativity). As the interatomic distance shrinks the chemical splittings increase, as discussed below.

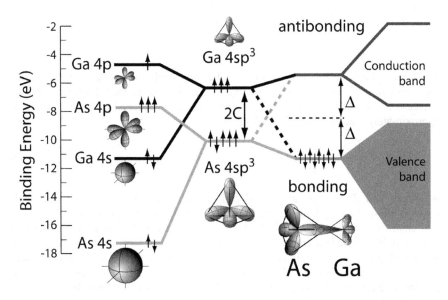

Figure 5.5: Shows the evolution of atomic orbital energies to form bonds and ultimately bands. The geometries of the atomic and hybrid orbitals are shown schematically as insets.

Heteropolar semiconductors can be thought to form sp^3 hybrid molecular orbitals exactly as do homopolar semiconductors. When we considered two atoms together in a homopolar semiconductor, bonding and antibonding states resulted from symmetric and antisymmetric mixtures of identical hybrid orbitals. The same combinations occur in a heteropolar semiconductor, but now the cation and anion hybrid orbitals sp^3_C and sp^3_A are more distinguishable and have different electron densities. Furthermore, the symmetric and antisymmetric states now have different contributions from the cation and anion molecular orbitals.

Thus, Equation 5.2 becomes:

$$\psi_u = u_2 \psi_a^{sp^3} - u_1 \psi_c^{sp^3}$$
$$\psi_g = u_1 \psi_a^{sp^3} + u_2 \psi_c^{sp^3},$$

5.6

where u_1 and u_2 are coefficients which describe the relative contribution of the anion and cation to the antibonding and bonding states. Both are related to the square root of the bond polarity, as discussed below. Furthermore $u_1^2 + u_2^2 = 1$. The separation of these states is now a combination of a homopolar splitting, V_2, type term and a chemical splitting, C term. The resulting separation, Δ, (see Figure 5.5) is given approximately (see Harrison, Ref 1, for discussion and derivation) as:

$$\Delta = \sqrt{C^2 + V_2^{\ 2}} . \qquad\qquad 5.7$$

The homopolar splitting component, V_2, of a heteropolar bond increases exactly as for a homopolar semiconductor according to Equation 5.3 as the interatomic spacing decreases. The chemical splitting also increases with decreasing bond distance d, approximately (see Ferry for discussion) as:

$$C = c\left(\frac{Z_A}{r_A} - \frac{Z_C}{r_C}\right)e^{-k_{TF}d} \qquad\qquad 5.8$$

where Z_A and Z_C are the anion and cation atomic numbers, r_A and r_C are their covalent radii, $d=r_A+r_C$ is the interatomic distance, k_{TF} is the Thomas-Fermi screening length (the distance over which electrons in a solid screen an extenally-applied electric field), and c is a constant. For typical semiconductors, k_{TF} is in the range of 0.1 nm. One should be cautious in applying the formulas 5.5 through 5.8 too closely as detailed band structure calculations yield somewhat different values (see discussion in Section 5.2) than one would infer from the simple formulas. However, these formulas provide illustrative examples of the general functional form of the variables.

Because C and V_2 increase rapidly with decreasing interatomic distance, the bonding and antibonding states move away from each other as interatomic distance decreases. This has important consequences. For example, as with the homopolar materials, this effect is directly responsible for the temperature and pressure dependence of the direct energy gap (see Section 5.4). Because the chemical splitting increases roughly exponentially with decreasing distance, while homopolar splitting increases only quadratically, the chemical splitting becomes increasingly dominant at small interatomic distances. This also leads to more polar bonding as bond length shrinks, as we will see below.

It is now time to return to the observation that the cation and anion molecular sp^3 orbitals do not contribute equally to the symmetric and antisymmetric molecular orbitals (Equation 5.6), i.e. $u_1 \neq u_2$. Because of the way in which orbitals combine, the molecular orbital that is closest in energy to the bonding or antibonding state contributes most to that state. Thus, the lower binding energy state associated with the cation contributes most to the antibonding orbitals. Likewise, the higher binding energy anion contributes most to the bonding state. This is indicated schematically in Figure 5.5 by the solid and dashed lines connecting the bonding and antibonding states to the cation and anion molecular orbitals. Solid lines show the primary contribution while dashed lines indicate a minority contribution. The mixing of the states will be discussed in more detail in Section 5.2, which provides a more specific justification for this difference.

If the antibonding states are empty and the bonding states are filled, the localization of the states results in a difference in the charge density around the cation and anion. In other words, as the bands become increasingly connected to given atoms, the bonding becomes more ionic. A quantitative relationship between chemical splitting and ionic character, α, of the bonds can be derived (see, for example Harrison) to be:

$$\alpha = \frac{C}{\Delta} = \sqrt{\frac{C^2}{V_2^2 + C^2}} \cdot \qquad 5.9$$

The increase in ionic bonding also applies as the atoms are forced closer together. Likewise ionic character will change with pressure and temperature. One can go further and determine that $u_1 = [(1+\alpha)/2]^{1/2}$ and $u_2 = [(1-\alpha)/2]^{1/2}$ (see Harrison, for example). One can then calculate u_1 and u_2 terms of C and V_2 using Equation 5.9.

$$u_1^2 = \frac{1 + \sqrt{C^2/(V_2^2 + C^2)}}{2} \text{ and } u_2^2 = \frac{1 - \sqrt{C^2/(V_2^2 + C^2)}}{2} \qquad 5.10$$

For the case of a completely ionic bond, $\alpha=1$, and the bonding and antibonding states are entirely composed of anion and cation states, respectively.

It may be helpful for the reader to consider some specific numerical values for some of these constants for common semiconductors. Atomic orbital energies for selected elements are given in Table 5.1. A full table of values may be found in Harrison. Table 5.2 gives data for homopolar and chemical splittings as well as minimum energy gaps for several semiconductors based on values from Ferry. The reader will find that the constants C and V_2 calculated using information in Table 5.1 based on the formulae above do not result in the numbers in Table 5.2. This is because of the various corrections to the results, which are necessary in an accurate calculation of a band structure and not included in the simplified approach resulting in the above. Some of these corrections have been included in the values in Table 5.2. Full band structures for several compounds are described in Section 5.3, and many more may be found in the literature. Once again, although *the formulas in this section are not sufficient to predict accurate band behaviors*, they do provide useful trends and explain the basis of many experimental results.

The observations about atomic contributions to bonding and antibonding states become more relevant when we realize that these states make up the valence and conduction bands. Therefore, a change in the cation, keeping the anion the same, will primarily affect the conduction band, while a change in the anion for a constant cation primarily affects the valence band. This leads to the common cation and common anion rules as well as to the relative magnitudes of the band offsets, which we encountered in Chapter 3. The ratio of the energy band offsets, $\Delta E_V / \Delta E_C$ directly reflects the relative magnitudes of the coefficients u_1 and u_2 in Equation 5.10 across a heterojunction.

Table 5.1: **Atomic and Molecular Orbital Energies**

Atom	E_s (eV)	E_p (eV)	E_{sp3} (eV)	Atom	E_s (eV)	E_p (eV)	E_{sp3} (eV)
C	17.52	8.97	11.11	Ge	14.38	6.36	8.37
Si	13.55	6.52	8.28	Sn	12.50	5.94	7.58
Al	10.11	4.86	6.17	P	17.10	8.33	10.52
Ga	11.37	4.90	6.52	As	17.33	7.91	10.27
In	10.12	4.69	6.05	Sb	14.80	7.24	9.13
Mg	6.86	2.99	3.96	S	20.80	10.27	12.90
Zn	8.40	3.38	4.64	Se	20.32	9.53	12.23
Cd	7.70	3.38	4.46	Te	17.11	8.59	10.72
Cu	6.92	1.83	3.10	Br	23.35	11.20	14.24
Ag	6.41	2.05	3.14	I	19.42	9.97	12.33
Values from Walter A. Harrison, Ref. 1.							

Table 5.2: **Bond Orbital Energies**

Molecule	Interatomic Distance (Å)	Homopolar Splitting $2V_2$ (eV)	Chemical Splitting, $2C$ (eV)	Bond Energy, 2Δ (eV)	Minimum 300 K Energy Gap (eV)
C	1.54	13.88		13.88	5.4
Si	2.35	5.96		5.96	1.107
AlP	2.36	5.92	3.48	6.87	2.5
Ge	2.44	5.52		5.52	0.67
GaAs	2.45	5.48	3.02	6.26	1.42
ZnSe	2.45	5.55	3.80	6.73	2.58
CuBr	2.49	5.37	5.59	7.75	2.94
Sn	2.80	4.20		4.20	0.08
InSb	2.81	4.16	2.56	4.88	0.165
CdTe	2.81	4.16	5.22	6.67	1.44
Values from D.K. Ferry, Ref. 2. Minimum energy gaps from Ref. 3.					

5.2 LCAO: FROM ATOMIC ORBITALS TO BANDS

Although the hybrid molecular orbital description of how bonds form is most convenient for a simple picture of the geometry of specific compounds, a more precise and general result can be obtained by keeping the original atomic orbitals in the scheme. This section considers how this may be accomplished and delves deeper into the methods for calculating of real E(k) diagrams. Even so, many details are glossed over. For a complete description of the methods the reader is referred to the suggested readings. For this discussion we will follow the linear combination of atomic orbitals (LCAO) approach. Note that in spite of the difference between considering atomic orbitals individually and using the hybrid orbitals, the results are identical as long as only the states that enter into the hybrid contribute to bonding.

The beginning of the LCAO approach is the valence atomic orbitals of the atoms forming bonds. In most semiconductors the important states are the s and p orbitals. In some cases other orbitals such as shallow-lying (outermost) d-states contribute significantly to the complete picture of bonding. For now we will ignore such complications. This is safe for the case of Si where no d-state electrons are present but not, for example, for GaAs. An analysis of orbital interactions requires selection of a coordinate system. It is conventional, for example, to orient the p-orbitals along the Cartesian coordinate axes. From this point it is easy to construct the sp^3 hybrid orbitals along bond directions in a diamond-structure material:

$$\psi_{[111]} = (\psi_s + \psi_{p_x} + \psi_{p_y} + \psi_{p_z})/2$$

$$\psi_{[\bar{1}\bar{1}1]} = (\psi_s - \psi_{p_x} - \psi_{p_y} + \psi_{p_z})/2$$

$$\psi_{[\bar{1}1\bar{1}]} = (\psi_s - \psi_{p_x} + \psi_{p_y} - \psi_{p_z})/2 \qquad 5.11$$

$$\psi_{[1\bar{1}\bar{1}]} = (\psi_s + \psi_{p_x} - \psi_{p_y} - \psi_{p_z})/2$$

Here the crystallographic indices in the subscripts refer to the hybrid molecular orbital directions. Because the Schrödinger Equation governing electron motion is linear, any combination of wave functions that solve it will also be a solution. In other words, choosing the hybrid orbitals or the atomic orbitals as a starting point for the calculation must yield identical results. The most flexible and general approach is not to be restricted to specific hybrid orbitals but rather to consider all possible orbital-by-orbital interactions of the fundamental atomic states. These states apply to a given atom in any environment. Thus, their use is valid for any material in which the atom occurs. As an example of a specific interaction, one can ask how does the p_x orbital on one atom interact with the p_z orbital on another atom.

To answer this, some additional terminology will be useful. Electron states having at least some component of their orbitals parallel to one another can interact in two ways. When the bonds (or projected components of the bonds) are parallel to the orbital axis, this results in a "σ" bond, while bonds perpendicular to the orbital axes are "π" bonds. Thus, s-orbitals always form σ bonds (they have no specific axis) while p-orbitals can have σ -like and π -like characters. When interacting orbitals do not lie directly along or orthogonal to a bond axis, one can decompose the resulting interactions into σ-like and π-like portions as well as a portion for which the orbitals are orthogonal to one another and therefore have no interaction. This leads to a series of geometric coefficients, which scale the interactions for each pair of states according to their orbital axes, relative positions, and symmetries. Sketches of the s and p orbitals in a diamond-lattice semiconductor along with some of their interactions are shown in Figure 5.6. Note how none of the p-orbitals point directly along bond directions for the conventional choice of axes. However, the orbitals can

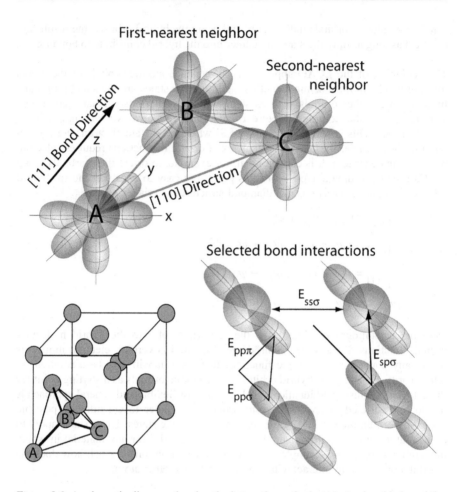

Figure 5.6: A schematic diagram showing the interactions of selected atomic orbitals and the geometry of these orbitals with respect to the crystal lattice in a zincblende or diamond structure material.

be decomposed into a component directed toward another orbital, a component parallel to another orbital (π-like bond), and a perpendicular component. If the axes were chosen such that a given pair of orbitals lay directly along that axis, the others would have reduced components along those directions and the result would be the same. Choice of coordinate axes is irrelevant to the result.

The coefficients of each interaction represent the relative strength and component of the interaction for a given bond. For the situation shown in Figure 5.6, the energies of the possible interactions may be shown to be [1]:

$$E_{ss} = V_{ss\sigma}$$

$$E_{sp} = -\frac{V_{sp\sigma}}{\sqrt{3}}$$

$$E_{xx} = \frac{V_{pp\sigma}}{3} + \frac{2V_{pp\pi}}{3} \quad , \qquad 5.12$$

$$E_{xy} = \frac{V_{pp\sigma}}{3} - \frac{V_{pp\pi}}{3}$$

where $V_{ss\sigma}$, $V_{sp\sigma}$, $V_{pp\sigma}$, and $V_{pp\pi}$ describe the strengths of the ss σ-bonds, the sp σ-bonds, the pp σ-bonds, and the pp π-bonds, respectively, for a given interatomic distance and type of atom. The interaction potentials, V, are then scaled by the projections of the p orbitals along the x,y, and z axes in the directions of the orbital with which they are interacting, leading to the final interaction energies of Equation 5.12. E_{xx} refers to the interaction of a p-orbital with another p-orbital on an adjacent atom along the same axis; for example, a p_x orbital interaction with another p_x. E_{xy} refers to the interaction of a p-orbital pointed along one axis with a p-orbital on an adjacent atom pointed along one of the other two orthogonal axes. (Although the axes are orthogonal, the orbitals are not, see Figure 5.6.) To repeat for emphasis: the V's are orientation and axis-independent interaction potentials while the E's are these interaction terms projected along specific axes. Harrison gives approximate formulas for atom-independent values of V from which values for the various E's can be estimated for a given lattice. According to Harrison, all V's, and hence all E's, scale as $\hbar^2/md^2$, from which Equation 5.3 results [$\hbar^2/m = 0.0762$ eV-nm^2]. In units of $\hbar^2/md^2$, the resulting values of the E's would be: E_{ss}=-1.4, E_{sp}=-1.06, E_{xx}=0.54, and E_{xy}=1.35 based on Equation 5.12. These coefficients, along with phase factors describing the phase of the Bloch waves in the solid figure into a matrix describing all of the possible pairwise bond interactions.

It is through the phase factors that a given electron momentum is defined. One might have expected this as, from the discussion of Chapter 2, band structures represent the interference of electron waves with the periodic potential of the lattice. For the wave functions in Equation 5.10, the corresponding phase factors are:

$$g_0(k) = e^{ik\cdot d_{111}} + e^{ik\cdot d_{\bar{1}\bar{1}1}} + e^{ik\cdot d_{1\bar{1}\bar{1}}} + e^{ik\cdot d_{\bar{1}1\bar{1}}}$$

$$g_1(k) = e^{ik\cdot d_{111}} - e^{ik\cdot d_{\bar{1}\bar{1}1}} + e^{ik\cdot d_{1\bar{1}\bar{1}}} - e^{ik\cdot d_{\bar{1}1\bar{1}}}$$

$$g_2(k) = e^{ik\cdot d_{111}} - e^{ik\cdot d_{\bar{1}\bar{1}1}} - e^{ik\cdot d_{1\bar{1}\bar{1}}} + e^{ik\cdot d_{\bar{1}1\bar{1}}} \qquad 5.13$$

$$g_3(k) = e^{ik\cdot d_{111}} + e^{ik\cdot d_{\bar{1}\bar{1}1}} - e^{ik\cdot d_{1\bar{1}\bar{1}}} - e^{ik\cdot d_{\bar{1}1\bar{1}}}$$

An examination of the terms for the g values will show that these simply represent the interference behavior of the electron waves with given wave vectors **k** interacting with atoms at positions defined by the real-space vectors **d** and at the origin. The g values include the free-electron-like behavior of Chapter 2. The calculation of the g factors becomes more complex when second-nearest neighbors and beyond are included, but the method is the same. The energy of an electron with wave vector **k** is the determinant of a matrix representing the energies of all possible orbital pairs. For example, for a zincblende semiconductor with no d-orbitals the LCAO matrix is [c.f. Ref. 4]:

	s_c	s_a	p_{xc}	p_{yc}	p_{zc}	p_{xa}	p_{ya}	p_{za}
s_c	E_{sc}	$E_{ss}g_0$	0	0	0	$E_{sp}g_1$	$E_{sp}g_2$	$E_{sp}g_3$
s_a	$E_{ss}g_0^*$	E_{sa}	$-E_{sp}g_1^*$	$-E_{sp}g_2^*$	$-E_{sp}g_3^*$	0	0	0
p_{xc}	0	$-E_{sp}g_1$	E_{pc}	0	0	$E_{xx}g_0$	$E_{xy}g_3$	$E_{xy}g_2$
p_{yc}	0	$-E_{sp}g_2$	0	E_{pc}	0	$E_{xy}g_3$	$E_{xx}g_0$	$E_{xy}g_1$
p_{zc}	0	$-E_{sp}g_3$	0	0	E_{pc}	$E_{xy}g_2$	$E_{xy}g_1$	$E_{xx}g_0$
p_{xa}	$E_{sp}g_1^*$	0	$E_{xx}g_0^*$	$E_{xy}g_3^*$	$E_{xy}g_2^*$	E_{pa}	0	0
p_{ya}	$E_{sp}g_2^*$	0	$E_{xy}g_3^*$	$E_{xx}g_0^*$	$E_{xy}g_1^*$	0	E_{pa}	0
p_{za}	$E_{sp}g_3^*$	0	$E_{xy}g_2^*$	$E_{xy}g_1^*$	$E_{xx}g_0^*$	0	0	E_{pa}

5.14

Subscript "a" designates atomic orbitals due to the anion while subscript "c" designates the cation orbitals. The energies on the diagonal of the matrix are the energies of the atomic orbitals for cations or anions. Values for such energies were given for selected elements in Table 5.1. The g values are complex numbers. To obtain proper behavior from the matrix, complex conjugates of g must be included such that elements across the diagonal are conjugates of one another. These conjugate values are indicated by an "*". The negative terms result because in some cases the negative lobe of the wave function is interacting with a positive lobe of another. Thus, the negative coefficient gives an attractive interaction.

Additional rows and columns should be added to the matrix for more complex compounds with more than two atoms and the formula for the g's becomes much more complex. The matrix also expands when shallow-lying d-orbitals must be taken into account. Simpler structures such as the diamond lattice have a smaller interaction matrix because there is no distinction between cation and anion sites. The matrix may also be modified by effects such as spin-orbit splitting. (Spin-orbit splitting is one of the corrections necessary to an accurate band calculation. It results from the interaction of the electron spin magnetic moment with the dot product of its velocity and the local electric field due to the positive atomic cores of the lattice.) Likewise, greater accuracy can be obtained if additional terms are included in the g values to account for second and higher neighbors.

One may wonder how to interpret this matrix in light of the nearly free electron model of Chapter 2 and the discussions at the beginning of this chapter. The Schrödinger equation enters into the matrix through the calculations of the individual interaction energies of various electron orbitals with one another; in other words, the $V_{ss\sigma}$, $V_{sp\sigma}$, $V_{pp\sigma}$, and $V_{pp\pi}$ terms that appear in Equation 5.12. These terms and the cation and anion orbital energies are also the only places that chemical differences appear in the problem. From the V values, the interaction energies for a given geometry of the lattice may be determined by projection of the orbitals onto the bond directions (Equation 5.13). As noted above, the matrix simply represents the set of equations for the linear combination of all atomic orbitals in pairs. Each possible pair corresponds to a specific element of the matrix. We encountered one such linear combination in the nearly free electron model in Equation 2.9, but for the combination of only two wave functions. This then led to the energies in Equation 2.10. Another example of a linear combination of two electron waves is shown in Figure 5.3. The individual elements of the matrix give the strength and phase of the individual interactions. Note that some of the matrix elements in Equation 5.14 are zero where the orbitals are orthogonal to one another. (For example, on the same atom orbitals must be orthogonal and cannot interact.)

The eigenvalues of Equation 5.14 describe the complete band structure for the solid when one considers all possible real-space translation vectors **d** of the lattice and all reciprocal lattice electron wave vectors **k**. In principle, calculating the band structure of a real solid, such as those shown in the next section, should be no harder than making the appropriate substitutions for **d** and **k** in Equation 5.13, calculating the phase factors (g's), substituting these into Equation 5.14, and calculating the eigenvalues. However, if one uses "standard" orbital energies (Table 5.1), estimates the interaction potentials V from simple approximate (but easy to solve) versions of the Schrödinger equation, and if one substitutes these into Equations 5.12-5.14 and calculates a band structure, one will not immediately arrive at the true structure. A high quality band calculation requires inclusion of numerous corrections to the potentials determining the V values, spin-orbit effects, higher neighbor interactions in the g values, and shallow-lying d-bands, when present, etc. Discussions of the details of these corrections may be found in the suggested readings.

Such detailed calculations are integrated into many computational models, so fortunately it is not necessary for the average semiconductor engineer to know how to perform the analysis. However, in spite of the convenience of computational methods, it is important to use them with caution. Even the most sophisticated methods generally perform much better for specific portions of the energy band structure and are relatively unreliable for other portions, especially in an *ab-initio* application (where one is not fitting experimental data). Much more reliable results are obtained by fitting experimental values to correct the calculation, making the band structure models interpretive rather than predictive. The most extreme version of the fitting approach is the empirical tight-binding method where one simply adjusts all of the matrix elements (with best guesses as to relative ratios) in order to match experimental data. One can then use the fit matrix elements to estimate energies of specific conformations of atoms via the g factors. The tight binding method makes little effort to actually determine the correct values of V from first-principles. In spite of the various concerns and caveats in the above discussion, band structure calculations are very useful in understanding trends in semiconductor behaviors and predicting optimal structures. Calculations are good for guessing the properties of a hypothetical semiconductor and therefore to semiconductor design.

We will consider some of the implications of the LCAO method next based on the above equations and ignoring detailed corrections. The simplest trend to understand is the behavior at Γ, where $\mathbf{k}=0$. In this case Equation 5.13 yields $g_0=4$ and g_1, g_2, and $g_3=0$. This makes the interaction matrix exceedingly simple and results in the following four energies [1]:

$$E\left(\Gamma_{ss}\right)=\frac{E_{cs}+E_{as}}{2}\pm\sqrt{\left(\frac{E_{cs}-E_{as}}{2}\right)^2+\left(4E_{ss}\right)^2}$$

$$E\left(\Gamma_{pp}\right)=\frac{E_{cp}+E_{ap}}{2}\pm\sqrt{\left(\frac{E_{cp}-E_{ap}}{2}\right)^2+\left(4E_{xx}\right)^2}$$

5.15

The $E(\Gamma_{ss})$ energy is for s-s bonds while the $E(\Gamma_{pp})$ energy results from the p-p bonds. Because at the Γ point $g_1=g_2=g_3=0$, there is no s-p bonding contribution. The "+" signs in Equation 5.15 represent antisymmetric antibonding states while the "-" signs correspond to the symmetric bonding states. Because the s orbitals have greater binding energies, the s-like states lie below the p-like states. Therefore, the top of the valence band and the top of the conduction band at Γ are composed of p-like states and the bottom of the valence and conduction bands are made up of s-like states. This means that **the band edges defining the energy gap of a direct-gap semiconductor of this type are p-like for the valence band and s-like for the conduction band.**

There are three p-p bonds with the same energy at Γ, one for each p-orbital axis. Thus, each of the two $E(\Gamma_{pp})$ energies refers to the energy of three separate branches of the E(k) diagram while only one branch occurs at each $E(\Gamma_{ss})$ energy. A more detailed analysis, including electron "spin-orbit" interactions in the calculation, [4] lowers the energy of one of the p bands. In addition, the curvatures of the two remaining bands are different and differ also from the split-off band. This leads to separate "light" (lower effective mass) and "heavy" hole behaviors. The two hole masses may be observed in some experiments sensitive to energetic holes.

Similar expressions to Equation 5.15 can be written for other parts of the band diagram. Relatively simple forms of the eigenvalues can be derived for the X and L points, [1] which describe, in part, the behavior of the bottom of the conduction bands in Si and Ge. The X point is primarily dominated by s-p and p-p bonding interactions while the L point behavior is the same as at the Γ point but with $(2E_{xx}+2E_{xy})$ replacing $4E_{xx}$ in the expression for the p-p energy in Equation 5.15. One can see from these results that to obtain an indirect-gap semiconductor it is necessary to have relatively large values of E_{xy} and relatively small values of bond length.

The formulas for the various symmetry points such as Equation 5.15 give a strong indication of the trends in the bands and their relationships to the fundamental chemistry of the material. The results are the basis of the formulas in Section 5.1. As an example of how such relationships can be derived, let us try to explain why the minimum energy gap in a semiconductor might increase with decreasing lattice constant even though the homopolar splitting and the bandwidth increase together and might be expected to cancel out the bond-length effect. For simplicity, consider a homopolar semiconductor. In this case $E_{cs}=E_{as}=E_s$ and $E_{cp}=E_{ap}=E_p$. Equation 5.15, then simplifies greatly and the band edges at the Γ point can be estimated (taking the appropriate signs) as:

$$\begin{aligned} E_c &= E_s + 4E_{ss} \\ E_v &= E_p - 4E_{xx} \end{aligned}$$

5.16

The direct energy gap in a homopolar semiconductor is then $E_{gap}=E_c-E_v$, or

$$E_{gap}(\Gamma) = E_s - E_p + 4(E_{ss} + E_{xx}).$$

5.17

Substituting from Equation 5.12, one may then obtain

$$E_{gap}(\Gamma) = E_s - E_p + 4[V_{ss\sigma} + (V_{pp\sigma} + 2V_{pp\pi})/3].$$

5.18

E_s and E_p are atomic orbital energies and are independent of bond formation and hence of bond length. If one examines Equation 5.18 one finds that the term in brackets resembles in some respects what one would get for a triple bond between two semiconductor atoms – contributions from one s-s σ-bond, one p-p σ-bond, and two p-p π-bonds. We can now answer the question at hand – how does $E_{gap}(\Gamma)$ change as the bond length changes. By examination of the values in Table 5.1, one finds that there is no consistent trend in E_s-E_p for the homopolar semiconductors. However, all of the wave function overlap energies V increase with decreasing interatomic distance. This shows why the net energy gap increases with decreasing atomic separation. A similar result is obtained in the heteropolar case. Therefore, we would expect a net change in energy gap, even though one might expect bandwidth and band-to-band spacing to offset one another.

One might also ask why it is necessary for an ordinary semiconductor engineer to worry about these details unless they were planning to pursue a graduate degree in semiconductor physics. The reason is that many aspects of semiconductor alloy and defect behaviors can be traced back to the phenomena discussed above. Furthermore, there are many exceptions to the simple rules of thumb. Even the common cation and common anion rules can appear to be violated in complex materials. Therefore, a more detailed understanding of the sources of bonding is necessary to have a good sense of how to engineer band structures and defects. To illustrate the value of the LCAO approach to understanding semiconductor behavior, consider the following case.

The individual bond-by-bond interactions contribute to different parts of the bands in materials. This may mean that while the valence band may be expected to result primarily from anion atomic orbitals, the states near the valence band edge may be dominated instead by particular atomic states from a cation. Such complications do not arise significantly in simple semiconductors with strongly covalent and strongly sp^3-like bonds, as in GaAs. However, in a complex ternary compound semiconductor such as $CuInSe_2$ this type of behavior has an important effect. In the case of this compound, the bottom of the conduction band is primarily derived from In atomic states, as one would expect based on the behavior illustrated in Figure 5.5 and represented by Equations 5.5-5.9. At the same time, the top of the valence band is primarily due to Cu-Se bonds. Therefore, replacing Se with S will primarily affect the valence band edge even though the cations are in common. Replacing In with Ga strongly modifies the conduction band edge and will have almost no effect on the valence band. This behavior illustrates the subtlety of bonding in complex compound semiconductors (and similarly complex insulators and metals) and shows how exceptions to common cation and common anion rules may occur or where alloying may have a surprisingly large effect on one or the other band edge. Likewise, breaking the symmetry of the system by strain may affect different parts of the energy gap differently (see band offset discussion for Si-Ge alloys in Chapter 6).

The bond-by-bond method is convenient for many purposes because we can consider any nth neighbor interaction between any orbital in the solid and any other as long as we can calculate appropriate g and E terms for insertion into the matrix of Equation 5.14. Standard methods for calculation of such parameters exist. In addition, we can easily add shallow-lying d-orbitals or other states that are not part of the normal sp^3 orbital geometry to the bonding. This is essential to an exact description of bonding in compound semiconductors involving elements below row three in the periodic table. When the d-orbital has a high binding energy, its wave function decays fast enough that it does not contribute significantly to bonding. However, the group IIb and IIIb metals such as Zn and Ga include d-orbitals shallow enough to be important. In some of the largest elements even f-states may contribute measurably (for example in HfN). Likewise, it is the second-nearest-neighbor atomic orbital interactions that distinguish between fcc and hcp crystal structures. Such interactions are small and change with the atomic number of the atoms involved. This explains why cubic and hexagonal forms may coexist in compounds such as ZnS or GaN and why one sees a transition from a stable cubic form to a stable hexagonal form with position of the constituent elements in the periodic table.

Table 5.3: **Energy Gaps and Lattice Parameters**

Semicon-ductor Class	Semiconductor	Lattice Parameter	Energy Gap, eV (at 20°C)	E_c (eV)	E_v (eV)
Cubic		nm			
IV	C (diamond)	0.35597	5.5		
IV	Si	0.54307	1.12 (indirect)	4.05	5.17
IV	Ge	0.56754	0.67 (indirect)	4.0	4.67
IV	a-Sn	0.64912	0.08		
III-V	GaP	0.54505	2.26 (indirect)	3.8	6.1
III-V	GaAs	0.56532	1.42	4.07	5.49
III-V	GaSb	0.609593	0.726	4.06	4.79
III-V	InP	0.58687	1.344	4.38	5.72
III-V	InAs	0.60583	0.354	4.9	5.25
III-V	InSb	0.6479	0.17	4.59	4.76
II-VI	ZnSe (cubic)	0.567	2.58	4.1	6.7
I-VII	CuBr	5.69	2.94	4.35	7.29
$I\text{-}III\text{-}VI_2$	$CuInSe_2$	0.578	0.98	4.0	5.0
$II\text{-}IV\text{-}V_2$	$ZnGeAs_2$	0.567	0.85		
Hexagonal					
III-V	AlN	0.3111 (a) 0.4978 (c)	5.9	0.6	6.5
III-V	GaN	0.3190 (a) 0.5189 (c)	3.45	4.0	7.4
III-V	InN	0.3533 (a) 0.5693 (c)	0.7 (note values vary greatly)		
II-VI	ZnS	0.3814 (a) 0.6258 (c)	3.911		
II-VI	CdSe	0.4299 (a) 0.7010 (c)	1.751 eV		

E_c (the electron affinity) and E_v (electron affinity + energy gap) measured with respect to the vacuum level. Lattice parameters in nm.

5.3 COMMON SEMICONDUCTOR ENERGY BANDS

Having armed ourselves with a more detailed idea of how energy bands develop in semiconductors, we now consider some specific examples. It is helpful to begin with an examination of some experimental data for some representative semiconductors. Table 5.3 lists lattice parameters and energy gaps for selected common semi-conductors. Several of the trends listed above are illustrated by these results.

For the group IV semiconductors the effect of the homopolar splitting on energy gap and its dependence on lattice constant are shown in Figure 5.7. Complete band

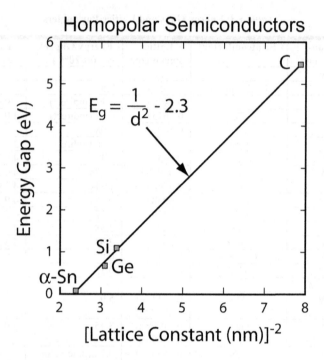

Figure 5.7: Shows the relationship of minimum energy gap to lattice constant for the common diamond-structure semiconductors.

structures will be discussed later in the section. The results range from diamond with the smallest lattice constant and largest gap to α-Sn with the largest lattice constant and smallest gap. The behavior fits well with a change in energy gap proportional to the inverse square of the lattice constant, as would be expected from Equation 5.3. The observation might be surprising as two of the materials have indirect gaps with minima at different symmetry points while the other two have direct gaps. Therefore, the detailed trends might not be expected to be as anticipated from Equation 5.3. Nonetheless, the general scaling behavior of the potentials is clear.

The situation is less surprising when one considers that the bonding-antibonding splitting has much more effect on the energy gap than do the details of the energy bands. As we found via Equation 5.18, bond-length-induced changes in the bonding-antibonding splitting have a much larger magnitude than do the changes in band-width. It is interesting to note that such an obvious trend in homopolar splitting is hard to observe in compound semiconductors because of the simultaneous change in chemical splitting. For example, the sequence BN, AlP, GaAs, InSb could, in principle, show a dominant effect of homopolar splitting. Certainly the lattice parameter changes are sufficient to expect a large change in homopolar splitting. However, the chemical changes in this sequence turn out to dominate the results, as might have been anticipated from Equation 5.8.

A comparison showing the effect of chemical splitting can be obtained most clearly based on the row-four semiconductors Ge, GaAs, ZnSe, and CuBr. The lattice constants in these materials are almost identical and all are cubic. Therefore, virtually all of the difference in their bands is the result of the chemical changes. The trend here is not as clear as in Figure 5.7 for the homopolar materials but for Ge, GaAs, and CuBr the results are still in good agreement with the relationship of chemical splitting to energy gap. A plot comparing the minimum energy gap in these materials with the bonding/antibonding splitting estimated using Equation 5.7 is given in Figure 5.8.

The general trends for chemical and homopolar splitting appear to be borne out by the behaviors of the minimum energy gaps in most cases. Let us now consider the complete band structures and look for additional trends there. The energy band structures for six semiconductors calculated by an enhanced LCAO method [5] are shown in Figures 5.9 (the series of group IV diamond structure materials Si, Ge, and α-Sn) and 5.10 (the fourth row series Ge, GaAs, and ZnSe).

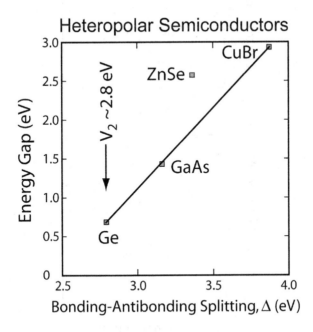

Figure 5.8: Changes in the energy gap of heteropolar semiconductors made up from elements in row 4 of the periodic table showing the effect of chemical splitting on the energy gap. Values used in generating this figure are from Table 5.2.

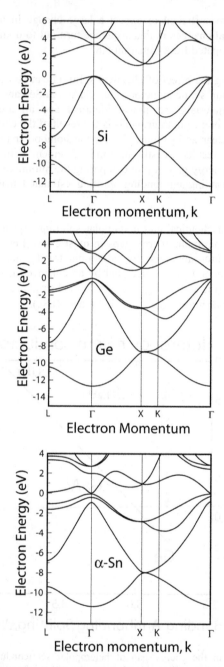

Figure 5.9: Energy band structures for the Group IV semiconductors Si, Ge, and Sn. Redrawn with permission from Chelikowski, J.R. and Cohen M.L. *Phys. Rev. B* **14**, 556-582 (1976). Copyright 1976, American Physical Society.

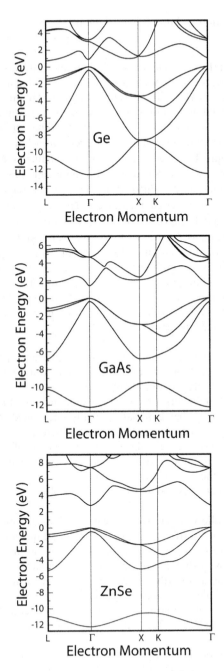

Figure 5.10: Energy band structures for the row 4 semiconductors Ge, GaAs, and ZnSe. Redrawn with permission from Chelikowski, J.R. and Cohen M.L. *Phys. Rev. B* **14**, 556-582 (1976). Copyright 1976, American Physical Society.

Details of the calculation and discussion of the band structures may be found in the original paper in which these results were presented by Chelikowski and Cohen [5].

There are several significant trends to notice about the band structures in Figures 5.9 and 5.10. For example, the homopolar group IV series, taking a vertical slice through the periodic table, shows very similar band structures. The most obvious change is that the conduction band minimum moves downward in energy relative to the top of the valence band, consistent with the reduction in bonding-antibonding splitting (V_2) as the bonds lengthen. More interesting are the relative behaviors at the Γ, X, and L points with bond length. The Γ point energy at the conduction band minimum shifts by more than 4 eV from Si to α-Sn, while the L point shifts by less than half of this energy and the X valley hardly shifts at all. Thus, some regions of the bands are much more sensitive to bond length than others. The large change at Γ and relatively small changes at X and L suggest that the sp σ-bonds and p-p π-interactions are becoming less significant as Z increases by comparison with the s-s and p-p σ-bonds. The p-p π-bonds turn out to primarily affect the width of the bands (or the dispersion with respect to **k**). The most significant change in the valence band as Z increases is the increased splitting among the p-states, primarily near the top of the band. This is due to increased spin-orbit interaction in higher Z elements. In addition, there is a modest change in valence bandwidth, resulting from the reduction in orbital overlap as the interatomic distance increases.

One can equally well consider a series of analogous heteropolar semiconductors moving down the periodic table; for example, by comparing AlP, GaAs, and InSb or ZnSe, and CdTe. Such comparisons show nearly the same trends with increasing Z as in homopolar semiconductors – the energy gap shrinks but the general features of the valence and conduction bands remain virtually unchanged. A notable difference is that by comparison to the homopolar materials there is even less change in the conduction bands. This is because the more ionic nature of heteropolar materials softens directionality and atomic orbital energies dominate over directional terms such as E_{xy}. Only in the smallest compounds such as AlP does one find indirect energy gaps in these materials and then only occasionally.

The situation as one follows a series of increasingly ionic compounds within a given row, Figure 5.10, is much different. As expected from Equation 5.7 and Figure 5.8, the energy gap increases with increasing chemical splitting. In addition, the valence bands change very obviously with the energies between the top and bottom of states due to a particular interaction (the band dispersion) decreasing as the atoms move apart in the periodic table. For example, while the p-like and s-like portions of the valence bands remain centered around roughly constant energies, the width of these parts of the valence band shrink. This causes a gap to open within the band.

The decreasing variation in energy with momentum across the diagram is not, perhaps, very surprising given the discussion in Section 5.1. As the chemical splitting increases, the bonds become increasingly ionic. **Ionic bonds are relatively**

non-directional and localized on a specific atom. A purely ionic bond results from electrostatic interactions, which are spherically symmetric (no dispersion with direction). Consequently, the bands become narrower as a function of **k** for increasing ionicity as direction becomes increasingly unimportant. The s-like portion of the valence band that merged with the p-like valence band in group-IV compounds (the bottom band in the -8 to -12 eV energy range) separates increasingly from the rest of the valence band as the polarity of the compound increases. Bands that crossed in the conduction band without interacting in the elemental semiconductor Ge react increasingly strongly in the less symmetric compound materials. As polarity of the materials increases the breadth of features in momentum space also changes with consequences for the effective mass of electrons and holes.

Unlike the homopolar materials that showed relatively dramatic changes in the conduction band as the atoms increase in Z, the heteropolar materials with fixed average Z show very similar conduction band behaviors. Note that the lower dispersion in ionic compounds results, among other effects, in these materials having direct energy gaps. It is nearly impossible for a highly ionic compound to have an indirect energy gap because the ionic bonds are much less directional than are covalent bonds. Consequently, the distinction between directions is smaller and it is less likely that a non-zero momentum wave vector will have a lower energy than for the zero-momentum Γ point. Indirect gaps are also absent from amorphous materials. In this case it is for the simple reason that specific directions do not exist.

An instructive alternative approach to visualizing the bonding and electron distribution in semiconductors is to construct a map of constant electron density surfaces in the material. An example of such a map for the electron density on a (110) plane of GaAs, along with the corresponding crystal lattice, is shown schematically in Figure 5.11. Similar maps can be constructed for slices through other atomic planes with different symmetries [see the original work by Chelikowski and Cohen, Ref. 5, for examples]. A sketch of the sp^3 hybrid orbitals for the anion (As in this case) is overlain on the diagram to show that the regions of high contour density correspond to the primary bond directions. Note that this is a plot of the electron density. Therefore, it corresponds to the amplitude of the wave functions of the valence band for various positions and hence reflects the symmetry of the valence band.

We can then compare such maps for different semiconductors. The electron density maps for the semiconductors for which the band diagrams are given in Figures 5.9 and 5.10 are shown in Figure 5.12. [5] One can see from a comparison of these diagrams that the electron density maps of the homopolar semiconductors are almost identical (the Si contour spacing is half that of the other diagrams). In heteropolar materials, by contrast, increasing the chemical splitting leads to increasing electron density around the anion and a corresponding decrease around the cation. Note that the diagrams in Figure 5.12 are for the valence band (filled states). Thus, the contours are enhanced around the anion because it makes a greater contribution to

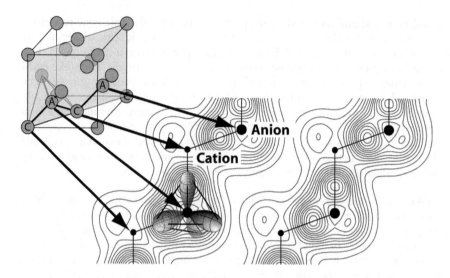

Figure 5.11: Shows a schematic diagram of the (110) plane in a zincblende-structure lattice with atoms marked C and A for cation and anion sites, respectively. The corresponding electron density map in this plane for GaAs is given as well based on the calculations of Chelikowski and Cohen. [5] The sp^3 hybrid orbitals of the anion are superimposed on a portion of the map to show the orientation of the orbitals. Two of the orbitals lie in the plane of the figure and two fall above or below the plane of the map. Note in particular the difference in electron density on the Ga compared to the As.

the valence band and acquires more electron density because of the ionic nature of the bond. A plot of the states in the conduction bands would be generally a complement of the valence bands and would be centered more strongly around the cation. Therefore, the symmetry of the states in the conduction band would vary more from material to material, especially for the homopolar materials. Concerning directionality of the bonding, note how much more spherically-symmetric the electron density becomes for ZnSe at the midpoint between the Zn and the Se atoms as compared to the same electron densities for GaAs and Ge. The much larger variation of electron density with angle for the smaller atoms is a reflection of the greater dispersion with angle in the energy band diagrams.

Summarizing the main trends:

- Materials with larger lattice constants have smaller energy gaps and less dispersion in their bands of states.
- Compounds involving atoms chosen from similar columns in the periodic table (group IV, III-V, II-VI etc…) have similar band structures.
- Compounds involving increasingly ionic bonds have less dispersion in their bands across momentum space (the E(k) diagrams).

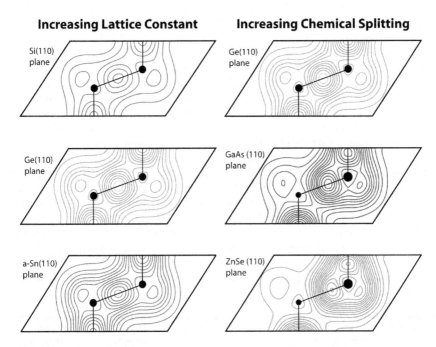

Figure 5.12: Electron density maps for a series of homopolar semiconductors of different lattice size showing the effect (or lack thereof) of lattice constant on electron density. Also shown is a corresponding series for increasingly ionic semiconductor compounds in row four of the periodic table. Contours for Si are at half the spacing of the other homopolar materials. Redrawn with permission from Chelikowski, J.R. and Cohen M.L. *Phys. Rev. B* **14**, 556-582 556-582 (1976). Copyright 1976, American Physical Society.

- Indirect gap semiconductors are made up of smaller, more covalent compounds or elements.

The detailed quantum mechanical basis for all of these changes is discussed in great detail in Harrison and many other sources but the reader is cautioned that a solid understanding of the notations of quantum mechanics is required to understand the discussion.

5.4 PRESSURE AND TEMPERATURE DEPENDENCE

We can combine many of the above points to understand the variation in the minimum direct and indirect energy gaps in semiconductors with hydrostatic pressure and temperature.

Hydrostatic pressure is uniform over the surface of a body and can be compressive or tensile. Compressive forces on an isotropic material push atoms slightly closer together while tensile forces pull them apart. Hydrostatic pressure, P, is a force per unit area (or stress). It induces a strain, ε, in a material which is linearly related to the pressure through the Young's modulus, Y: $\Delta P = Y \Delta \varepsilon$. The strain is defined as $\varepsilon = (l_f - l_i)/l_i$, where l_f is the final length of the material and l_i is its initial length. Bond length is directly related to strain as $\varepsilon = \Delta d/d$, where d is the interatomic distance. $\Delta \varepsilon$ is the change in this strain as the stress (pressure) changes. Thus, increasing the pressure decreases the interatomic distance. As this occurs, the bonding and antibonding states draw apart as is implied by Equations 5.3 and 5.6. The band widths also increase or change shape slightly as discussed in the preceding section resulting in changes in dispersion as well as changes in the minimum direct energy gap at the Γ point. For the direct gap, as it turns out, the increase in homopolar splitting and chemical splitting win out over the increase in band width as interatomic spacing decreases, leaving a roughly linear relationship between gap and pressure over the range of pressures normally studied. Because the direct gap increases with decreasing distance, the rate of change in energy gap with pressure (dE_{gap}/dP) is positive (increasing pressure, increasing gap). Some typical values are given in Table 5.4.

Table 5.4: **Pressure and Temperature Dependences of Selected Semiconductor Minimum Energy Gaps**

Material	E_g (300 K)	dE_g/dT		dE_g/dP (direct)	dE_g/dP (indirect)
	eV	A (eV/K) $\times 10^4$	B (K) for Eq. 5.11	$\times 10^{-6}$ eV·cm²/Kg	$\times 10^{-6}$ eV·cm²/Kg
Si	1.107	2.3	636	10.5	-1.09
Ge	0.67	3.7			+7.3
GaP	1.6	5.4			-1.7
GaAs	1.35	5.4	204	11	
InP	1.27	4.6		4.6	
InSb	0.165	2.8		15	
ZnSe	2.58	7.2		6	
CdTe	1.44	4.1		8	

Si and GaAs data from Sze (1981). [6] Remaining data from the CRC Handbook of Chemical Physics, 2001. [3] When no value is given it may be assumed that B<<300K.

The relationship between energy gap and pressure is different for indirect gap materials. In these there is, necessarily, a strong dispersion in the conduction band. This is essential to get a high-momentum part of the energy band below the zero-momentum minimum and explains why relatively ionic semiconductors have direct energy gaps. As Figure 5.7 shows, smaller interatomic distances lead to larger conduction band dispersions, strongly increasing the Γ-point energy and simultaneously reducing the energy of other points in the energy bands. This suggests that the behavior of the energy gap with pressure should be different and, indeed, opposite away from Γ; which is what is observed for most materials (Ge is an exception). Thus, in many indirect-gap semiconductors dE/dP < 0, indicating that as pressure increases, the gap decreases and becomes more indirect (see Table 5.4).

Theoretically, the opposite dependence of direct and indirect gaps on pressure could be used to convert indirect gap materials to direct gaps. However, a negative pressure (tensile stress) would have to be applied to achieve this conversion. Ceramics, including semiconductors, tend to be weaker in tension than in compression. Even by placing the indirect material in a strained-layer superlattice (see Chapter 7), which can achieve the highest tensile stress levels, it has been impossible to convert indirect semiconductors to direct gaps before the stress is relieved by formation of dislocations or by fracture.

The link between lattice constant and temperature also results in a change in energy gap. The quantitative relationship is less clear than in the case of pressure. The most straightforward connection is through the thermal expansion coefficient of the semiconductor, leading to an increase in interatomic spacing as the temperature rises. This causes a decrease in minimum direct energy gap with increasing temperature. In addition, interaction of electrons with phonon lattice vibrations changes as the phonon density changes with temperature and affects the band structure. The change in gap with temperature at constant volume is given roughly by:

$$\left(\frac{\partial E_{gap}}{\partial T} \right)_V = -3\alpha G \left(\frac{\partial E_{gap}}{\partial P} \right)_T \qquad 5.19$$

where α is the thermal expansion coefficient, G is the bulk modulus, and $\partial E_{gap}/dP$ is the pressure dependence of the energy gap, discussed above. More typically, an empirical relationship between temperature and lattice parameter is observed:

$$E_{gap}(T) = E_{gap}(T=0) - \frac{AT^2}{T+B}, \qquad 5.20$$

where A and B are constants. In other words, the gap depends quadratically on temperature at low temperatures and linearly on temperature at high temperature. Data for A (and B when available) in Equation 5.20 for several common semiconductors are given in Table 5.4. Values without a corresponding B show a linear change in gap with temperature. The linear behavior can be obtained directly from Equation 5.19, assuming a constant relationship of energy gap to pressure. In addition to broadening of the energy gap with decreasing T, structures in the density of states and features of the band structure also broaden.

5.5 APPLICATIONS

5.5.1 Experimental band structures

Direct application of the material in this chapter relates more to fundamental understanding of why semiconductors are the way they are than directly to making most semiconductor devices. Thus, one practical application for this chapter is in the interpretation of ultraviolet photoelectron spectroscopy (UPS) results. One can equally well view UPS as a way of measuring the band structures experimentally. It is UPS data which one generally fits to obtain, for example, an accurate E(k) diagram. We will begin this section by considering the UPS and inverse photoemission processes and how they can measure band structures directly. The experiments demonstrate that band structures are not hypothetical objects but real observables. The apparatus and electronic transitions in a solid associated with UPS photoemission are shown schematically in Figure 5.13.

To obtain valence band structure of a solid by normal photoemission, one arranges the experiment as follows. A linearly polarized photon strikes a smooth single crystal surface, ideally as free of defect states as possible, in a direction that is well defined with respect to the crystal lattice. This requires the ability to precisely tilt and rotate the crystal through a range of angles with respect to the incident photons. Both tilt with respect to the photon source and rotation in the surface plane are generally needed. It is helpful to be able to change the angle between the source and the detector. Because the photon is polarized and incident along a well controlled crystal direction, its wave vector is well defined. A similar ability to orient a detector relative to the crystal allows the momentum direction of emitted electrons to be determined. Finally, an energy analyzer is incorporated into the detector from which the photoelectron kinetic energy is determined. The energy of the state from which the electron originated relative to vacuum is then the difference between the energy of the photon and the vacuum level energy. For absorption of the photon and emission of the electron, the total momentum and energy are conserved. Consequently, the initial momentum and energy of the electron in the solid can be determined. Thus, measurements of photoelectron energies as a function of take-off angle give a measure of the band structure.

Inverse photoemission measures the conduction band states in a similar manner. In this case, an energetic electron beam strikes the solid from a fixed direction and with a fixed energy (thereby defining both the initial energy and momentum). The electron may be captured by the solid into a single unoccupied state with a consequent need to release the excess energy of the incident electron plus the binding energy of the state. Sometimes this energy is released by photon emission. Detection of the emitted photons in a given direction allows determination of their energy. From these pieces of information the energy and momentum of the electron in the bound state may be determined as for conventional photoemission.

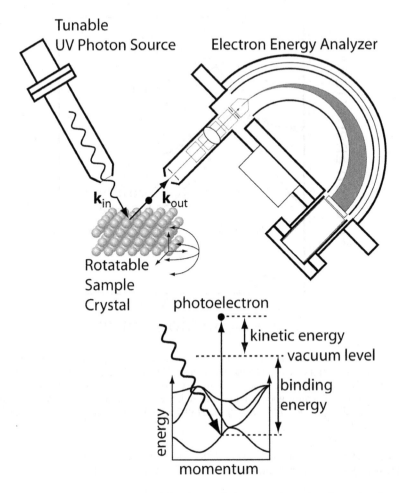

Figure 5.13: A schematic diagram of the photoemission process used to determine the structure of the valence band as a function of electron wavevector, **k**. The incident photon has a well-defined momentum $\mathbf{k}_{in}$ and the photoelectron has an outgoing wave vector $\mathbf{k}_{out}$. For an energy and momentum conserving process the initial energy and momentum of the electron can be determined, from which the band energy is known.

Typical experimental results for the valence and conduction bands of GaAs are shown in Figure 5.14 [7]. Not all values of incident and outgoing momentum can be probed effectively with a given apparatus. Furthermore, there may be portions of the energy-momentum space where it is too difficult to separate components of the band

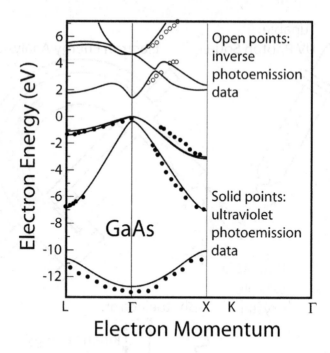

Figure 5.14: Experimental determination of a portion of the GaAs band structure as a function of electron momentum (points) by Ortega and Himpsel. [7] Also the calculation by Chelikowski and Cohen [5] (curves) for the band structure of GaAs. The agreement is excellent for the valence band and generally good for the conduction band. Adapted with permission from Ortega, J.E. and Himpsel, F.J. *Phys. Rev. B* **47**, 2130-7 (1993). Copyright 1993 by the American Physical Society.

structure in the output data. However, as the data in Figure 5.14 shows, it is possible to obtain very good experimental agreement with theory with some effort.

5.5.2 Gunn diodes

One can see a direct application of the concepts in this chapter in some more unusual electronic devices. An example is the Gunn effect, and resulting Gunn diodes, which have been used to produce high-frequency oscillators. In most semiconductors, as electrons accelerate they scatter increasingly strongly off the lattice, generating phonons and decelerating the electron. It is found that the carrier mobility becomes limited in proportion to the inverse of the electric field in the material [see, for example the discussion in Hess]. Consequently, a higher field produces a lower mobility. If we recall that the electron current density in the material is $J_e = q\mu_n n \mathcal{E}$ (Equation 3.6), then we see immediately that if μ_n is inversely proportional to $\mathcal{E}$ then J is independent of electric field.

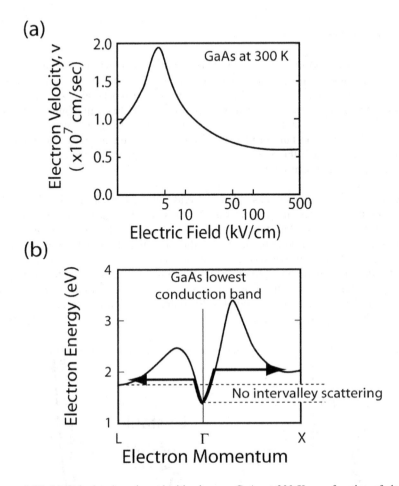

Figure 5.15: (a) Calculated carrier velocities in pure GaAs at 300 K as a function of electric field and (b) a schematic diagram of the lowest branch of the conduction band. Arrows indicate the acceleration and scattering of carriers to the X and L minima. At low energies no scattering is possible because states are only available around the Γ point. Once one reaches the energy of other band minima scattering begins and the carrier velocity starts to decrease. Part (a) redrawn with permission from Shichijo, H. and Hess, K. *Phys. Rev. B* **23**, 4197-4207 (1981). Copyright 1981 by the American Physical Society.

In other words, the electron velocity saturates. However, in semiconductors such as GaAs that have a low effective electron mass and modest phonon scattering rate, carriers may accelerate to relatively high velocities. The predominant scattering process becomes inter-valley scattering, rather than phonon emission within the same valley. In GaAs the scattering occurs to the X and L valleys as shown schematically in Figure 5.15. Electrons in these regions of the band structure have much higher effective mass (see Section 2.1.7 and Equation 2.18). Consequently, at the same energy they travel much more slowly.

The number of electrons in the high-mass valleys increases as the electric field increases because at higher fields electrons in the low-momentum valley near Γ accelerate very fast to a point where scattering to the other valleys is rapid. The consequence of this behavior is that the electron velocity has a maximum at a given field (see Figure 5.15) and decreases above and below this value. For a decrease in velocity to occur at constant (or increasing) electric field, the mobility must decrease, as one would expect if the effective mass increases. The resistance of the device is inversely related to the mobility.

When the resistance increases as voltage increases it produces a negative differential resistance in the device. In other words, the current voltage curve has a decreasing region at high voltage. Any time a device has a negative differential resistance an electronic circuit incorporating it is unstable and will switch between different operating points. The current/voltage curve for such a device is shown in Figure 5.16 with the unstable operating region marked. When the device is biased to a voltage in the unstable region it switches from one stable operating point to another at a given voltage or current. In short, it oscillates. Such an oscillation makes sense, perhaps, based on the following argument.

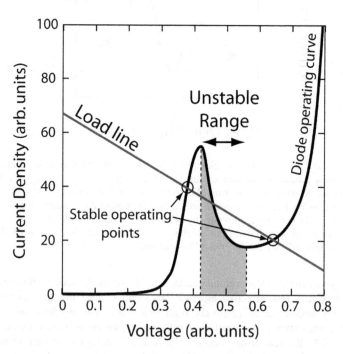

Figure 5.16: A schematic diagram of the current voltage curve for a Gunn diode. The device will not operate in a stable mode anywhere in the negative resistance region as marked. For the load line indicated the circuit has two stable operating points as shown.

At any point in the device where a local fluctuation in resistance raises the field in that region (higher resistance), scattering increases, mobility decreases, and resistance rises further in that area. This produces a local volume where a large fraction of the applied field is dropped. In this region, mobility decreases and velocity saturates. Carriers in low-resistance regions of the device continue to move well as the field is low. Although the carrier velocity is relatively low compared to the peak level in the high-field area, it is still of the order of the thermal velocity ($\sim 1 \times 10^7$ cm s^{-1}). Consequently the high field pocket can drift in the device in the direction of the positive contact. When it reaches the positive contact it disappears. As long as it is short enough, the device oscillates as high-resistance pockets form near the negative contact, drift to the positive contact, and are eliminated. The resulting circuit typically shows microwave oscillations.

So far, the description of the Gunn effect does not make use of designed materials. One could imagine that any direct-gap semiconductor with a sharp minimum at Γ and a broad minimum at some other momentum would suffice. However, we may wish to control the relative energies of the two minima such that a given amount of acceleration would occur before the onset of scattering. Such control can easily be obtained by either application of pressure, which allows adjustment of the relative energies of the direct and indirect gap minima, or, more practically, by alloying a direct gap material with an indirect gap material such that the appropriate energy difference between the direct and indirect gap minima would be obtained. Methods for band-gap engineering are discussed in detail in the next chapter where semiconductor alloys are described in detail. Note however that alloys typically have much lower carrier mobilities than pure compounds so the devices may function better with a strained layer than with an alloy.

5.6 SUMMARY POINTS

- Homopolar semiconductors consist of a single type of atoms.
- Heteropolar semiconductors consist of multiple atoms.
- Molecular orbitals are linear combinations of atomic orbitals.
- Symmetric combinations of atomic or molecular orbitals produce bonding states
- Antisymmetric combinations produce antibonding states.
- Homopolar splittings are the differences between bonding and antibonding state energies in the absence of chemical differences between atoms. The homopolar splitting varies inversely as the square of the interatomic distance. All other aspects of homopolar bonding, including the energy gap, tend to scale as $1/d^2$.
- Heteropolar compounds have both chemical and homopolar splittings contributing to their band structure and bond energy.
- Increasing chemical differences between atoms in heteropolar compounds lead to more ionic bonds, to an increasing connection of the valence band with the anion, and the conduction band with the cation. This is the basis of the common cation and common anion rules.
- Bonding can be treated as interactions of all individual atomic orbitals with all others as represented by the eigenvalues of the matrix of Equation 5.13. The matrix elements are affected by the electron wave vectors, the location of atoms contributing orbitals, and the energy of atomic orbitals and their interactions.
- The top of the valence band in common semiconductors is primarily derived from p-like states and has three branches resulting from the three p-orbitals. The bottom of the conduction band is s-like.
- The matrix elements related to orbital overlaps increase approximately as $1/d^2$.
- Materials with larger lattice constants have smaller energy gaps and less dispersion in their bands of states.
- Compounds involving atoms chosen from similar columns in the periodic table (group IV, III-V, II-VI etc…) have similar band structures.
- Compounds involving increasingly ionic bonds have less dispersion in their bands across momentum space (the E(k) diagrams).
- Indirect gap semiconductors are made up of smaller, more covalent compounds or elements.
- Increasing temperature causes thermal expansion of a material, increasing interatomic distances, and hence decreases energy gap.
- Increasing pressure decreases interatomic distance and hence increases energy gap.

5.7 HOMEWORK

1) Given the formulas for the s and p atomic orbitals in Figure 5.1, plot contours of constant wave function intensity in the r – theta plane.

2) Given the formulas for the p atomic orbital in Figure 5.1 and assuming that the p_y orbital is the same as the p_x but with $\sin\theta$ replacing $\cos\theta$, what is the sum of the squares of the p_x and p_y orbitals.

3) Suppose that a hypothetical homopolar semiconductor existed with a lattice constant of 0.4472 nm. Estimate its minimum energy gap. Explain briefly how you obtain this value.

4) In the energy-momentum diagram for ZnSe there is a narrow band of states between ~11 eV and 12 eV binding energies. For Ge, a similar band is found between ~ –9.5 and –13 eV. What is the atomic orbital most responsible for this band of states and why is the band narrower in ZnSe than in Ge?

5) What is the physical basis in bonding theory as described in this chapter for the common cation and common anion rules? Would you expect these rules to be more obvious when comparing GaP with GaAs or comparing CdS with CdTe? Explain briefly.

6) Equation 5.11 provides the combinations of atomic orbitals that make up the four sp^3 molecular orbitals. Write a similar set of equations for the sp^2 molecular hybrids shown in Figure 5.2.

7) Sketch a Harrison diagram similar to that of Figure 5.5, as close to correctly to scale as can reasonably be managed, for InSb using the data from the various tables in the chapter.

8) If one assumes that the contours in Figure 5.12 represent the same change in electron density except in the case of Si where the contour interval is halved, which homopolar and which heteropolar semiconductor has the stronger bonds? How do you know this to be the case and how would you connect the answer to the corresponding value of Δ for that semiconductor?

9) From the values listed in Table 5.4, calculate the hydrostatic pressure necessary to convert Si to a direct energy gap material (i.e. where the direct and indirect gaps are equal). Assume that the minimum indirect energy gap is 1.1 eV and the minimum direct gap is 3.3 eV at zero pressure.

10) Calculate and plot the values for bond polarization (α) for the series of semiconductors Ge, GaAs, ZnSe, CuBr. From these values, determine the values for u_1 and u_2 for each material.

11) Given that:

$$V_{ss\sigma} = -1.40 \; \hbar^2/md^2$$
$$V_{sp\sigma} = 1.84 \; \hbar^2/md^2$$
$$V_{pp\sigma} = 3.24 \; \hbar^2/md^2$$
$$V_{pp\pi} = -0.81 \; \hbar^2/md^2$$

Estimate the following values for GaAs using the formulas and data above and that $\hbar^2/m = 0.762$ eV-nm, and d $= a/\sqrt{3}$, where a is the lattice constant of the material:

a) The cohesive energy of GaAs
b) The four energies of the simple bands of GaAs at Γ
c) The minimum direct energy gap of GaAs
d) Compare the minimum direct gap estimated from these values to the correct value at zero Kelvin of 1.53 eV.
e) Compare the band energies calculated with those for the complete band structure for GaAs shown in Figure 5.10.

12) Could you use the Bloch wave sums in Equation 5.13 to simulate a hexagonal (wurtzite) semiconductor and distinguish the result from the corresponding cubic (zincblende) calculation? Explain why or why not.

13) Construct a version of Equation 5.13 that takes into account second-nearest neighbors.

14) Which semiconductor would make a better Gunn diode, ZnSe or GaAs? Explain briefly.

5.8 SUGGESTED READINGS & REFERENCES

Suggested Readings:

Ashcroft, Neil W. and N. Mermin, David *Solid State Physics.* Philadelphia PA: Saunders College, 1976.

David K. Ferry, *Semiconductors.* New York: Macmillan, 1991.

Harrison, Walter A. *Electronic Structure and the Properties of Solids: The Physics of the Chemical Bond.* San Francisco: Freeman, 1980.

Hess, Karl, *Advanced Theory of Semiconductor Devices*, Englewood Cliffs, NJ: Prentice Hall, 1988.

Van Vechten, J.A., "A Simple Man's View of the Thermochemistry of Semiconductors" in *Handbook on Semiconductors*, ed. T.S. Moss, Vol. 3, *Materials, Properties, and Preparation*, ed. S.P. Keller, North Holland, Amsterdam, 1980, Chapter 1.

References:

[1] Harrison, Walter A. *Electronic Structure and the Properties of Solids: The Physics of the Chemical Bond.* San Francisco: Freeman, 1980.

[2] Ferry, David K. *Semiconductors* New York: Macmillan, 1991.

[3] Lide, David R., editor, *CRC Handbook of Chemistry and Physics.* Boca Raton, FL: Chapman and Hall/CRC, 2001; also available on line at http://www.hbcpnetbase.com/.

[4] Chadi, D.J. and Cohen, M.L. "Tight binding calculations of the valence bands of diamond and zincblende crystals." *Physica Status Solidi B*, 1975; 68: 405-19.

[5] Chelikowsky, J.R. and Cohen, M.L. "Nonlocal pseudopotential calculations for the electronic structure of eleven diamond and zinc-blende semiconductors." *Phys. Rev. B,* 1976; 14: 556-582.

[6] Sze, S.M. *Physics of Semiconductor Devices.* New York: Wiley, 1981.

[7] Ortega, J.E., and Himpsel, F.J. "Inverse-photoemission study of Ge(100), Si(100), and GaAs(100): bulk bands and surface states." *Phys. Rev. B,* 1993; 47(4): 2130-7.

[8] Shichijo, H, and Hess, K. "Band structure-dependent transport and impact ionization in GaAs." *Phys. Rev. B,* 1981; 23: 4197-4207.

Chapter 6

SEMICONDUCTOR ALLOYS

As with all materials, engineering semiconductors primarily involves formation of alloys and control of defects. Defect engineering is discussed in detail in Chapter 7. This chapter considers the basics of alloying. The objective is usually to control the optoelectronic properties of the semiconductor, primarily through its energy band structure. Other properties change as well and some of these, such as lattice constant, are important to producing a high-quality material, as we shall see in Chapter 7.

We have briefly considered some of the applications of semiconductor alloys in Chapter 3. In current technology the use of alloys is seen in heterojunction bipolar transistors, laser diodes, and numerous other devices. Laser diodes used in fiber-optic communications, for example, are required to operate at a high modulation rate and at very specific wavelengths where the absorption coefficient in the fiber is unusually low. Thus, it is critical to these devices to know how to control the energy gap of the semiconductor effectively. Alloying is also necessary for improvement of laser efficiencies, which helps to allow the device to operate in a continuous mode.

6.1 ALLOY SELECTION

6.1.1 Overview

Unlike metals or ceramics where alloying is primarily aimed at engineering mechanical behaviors, forming semiconductor alloys concerns achieving specific optical or electronic properties. As we have seen in earlier chapters, optoelectronic properties are primarily determined by the semiconductor energy gap and band structure. An additional point of difference is that alloying in other materials aims at controlled production of second phases or microstructures, while semiconductor alloys must be single phase to be useful. Therefore, rather than being able to derive potential benefit from virtually any mixture of materials, in semiconductors we are restricted to those alloys which form single phases with nearly perfect mixing. In most cases we are further restricted to extremely high-quality dislocation-free single crystals as grain boundaries and dislocations can have a profound effect on the electronic behavior over a range of many microns (see Chapter 7). Such defects would normally be a problem for microelectronics where sub-micron sized devices are produced over many square centimeters of crystal surface.

Because of the need for single phases and single crystals, a major concern is the method of fabrication of the alloy. There are two common options for making high-quality single crystals – (1) to grow them as bulk materials (see Chapter 4) or (2) to form them as thin "epitaxial" single crystals on a large single-crystal substrate. (See Chapter 10. The term "epitaxial" refers to aligning the lattice of one crystal in a very specific way relative to another such that a thin crystal is grown on a thick substrate with a well-defined lattice orientation.)

It is difficult to grow bulk single crystals, and nearly impossible to grow bulk single crystal alloys with adequately-controlled chemistry, so for most alloys epitaxial growth is used. The substrate crystals are chosen from a limited, relatively available collection. Semiconductors for substrates include the common Si, Ge, GaAs, InP, Al_2O_3 (sapphire), SiC, the more rare GaSb, InAs, CdTe, and many more. To obtain a good epitaxial layer, the lattice constant of the alloy should be very close to that of the substrate (see Chapter 7). Given the limited number of bulk substrates, this is a significant constraint. Therefore, the alloy must be designed for both optoelectronic properties and lattice constant. As we will see, this can usually be accomplished but it adds significant complexity to the process.

Semiconductor alloys are categorized into one of the following groups:

- Binary alloys – mixtures of two elemental semiconductors (diamond, Si, Ge...)
- Pseudobinary [commonly "ternary"] alloys – a subset of the ternary alloys are mixtures of two compound semiconductors where only one component element is changed, keeping the remaining elements in common. A typical example is GaAs-AlAs, where both constituent compounds are arsenides.

Binary and pseudobinary/ternary alloys allow tailoring of minimum energy gap without the ability to affect lattice constant independently.

- Pseudoternary, quaternary or higher order alloys – mixtures of three, four, or more compound semiconductors.

  These alloys are used to engineer minimum energy gap with independent control of lattice constant.

  Pseudoternary alloys are mixtures of three compounds with one element in common, allowing the remaining constituents to vary in relation to one another. They are also a form of quaternary alloy. An example of a psuedoternary alloy is AlAs-GaAs-InAs. Quarternary alloys change four elements simultaneously. These are generally of the form $A_{1-x}B_xC_{1-y}D_y$. For example, $Ga_{1-x}In_xAs_{1-y}Sb_y$ is a quarternary alloy of GaAs, GaSb, InAs, and InSb where both the cation and anion are mixed. More complex alloys are also possible.

A second way of classifying alloys is to consider the valence and structure of the substituting species, thus an alternate set of groupings is:

- Isovalent (alloys where the valence of the atoms being exchanged is the same such as in GaAs-AlAs).
- Isostructural (both compounds have the same structure. For example, both compounds being zincblende structure.)

Conversely, one can have non-isostructural or non-isovalent alloys. Note, however, that because compounds with different structures or valences do not readily dissolve in each other, these latter categories are rare and must be produced by non-equilibrium processes such as sputter deposition (see Chapter 11). All commonly used alloys are both isostructural and isovalent, therefore this is not a very useful description. Henceforth, we will stick to the "pseudobinary, ..." identification. Examples of non-isovalent and non-isostructural alloys are considered in Section 6.5.

Alloying semiconductors, in general, affects the energy band structure and minimum energy gap, lattice parameter, mechanical constants (including elastic moduli, defect formation energies, etc.), optical and electronic conduction properties, and many other aspects of the resulting material. The majority of this chapter concerns the effect of alloying on the energy gap and energy band structure of semiconductors because this is the most critical issue in electronic applications. Furthermore, we can dispose of the alloy variation of many other properties relatively easily.

The lattice parameter usually can be calculated adequately using Vegard's Law, which states that the lattice constant of an alloy crystal, $a_0(A_{1-x}B_x)$, is an average of the lattice constants of the pure compounds, $a_0(A)$ and $a_0(B)$, weighted by the composition: $a_0(A_{1-x}B_x) = (1-x)a_0(A) + xa_0(B)$. In other words, the lattice constant changes linearly with composition of the alloy from that of one compound or elemental constituent to that of the other. Most mechanical properties also shift linearly with alloy composition. Because these properties are determined by substantial volumes of the semiconductor crystal, local composition fluctuations are averaged out and do not affect the property overall. Generalizing this observation, **any property of a material that is determined by an ensemble average over a significant volume of the material will generally scale linearly in the alloy with composition.** In spite of this observation, close inspection of most of these properties will reveal a small nonlinearity in the behaviors. An example may be found in Section 6.4 where the properties of Si-Ge alloys are discussed in detail.

Unfortunately, optical and electronic properties are usually determined by very small volumes and do not scale simply with composition, as we shall see. To make matters worse, one also needs to consider whether the energy gaps of the alloy constituents are direct or indirect. Optical devices virtually all work best with direct-gap materials while diodes and transistors may work with indirect-gap materials. In Chapter 5, we saw that smaller atoms and more strongly covalent bonds are more likely to result in indirect energy gaps. Thus, we are primarily concerned with direct-to-indirect gap transitions when we have elemental semiconductors or small atoms such as Al or P involved. We can expect, as will be seen below, that as we alloy one semiconductor with another the direct and indirect energy gap minima will shift at different rates as the alloy is changed. Not surprisingly, if one alloys a direct-gap semiconductor such as GaAs with an indirect gap semiconductor such as AlAs, there will be a transition from direct to indirect at some composition.

We saw in Chapter 2 that the mobility of a free carrier is determined by the rate of scattering of that carrier and the energy band structure (density of states or carrier effective mass) of the material. When we make an alloy we modify both the band structure and the scattering rate. Thus, we expect to find changes in the carrier mobility in alloys. Unfortunately, most of these changes are for the worse. The mobility is generally lower than the linear interpolation of mobilities of the end point compounds. Electrons moving in solids presume that the states they occupy will extend from one area to the next. However, in an alloy, what atom an electron will find in the next unit cell is not well determined (unless phases with superstructure ordering occur). Fluctuations in local chemistry lead to fluctuations in electronic states. These change the direction of carriers moving in the solid, an effect known as "alloy scattering", and reduce the mobility of carriers.

In addition to alloy scattering, scattering between states with similar energies but different momenta may occur. In particular, in mixtures of direct-gap semiconductors with indirect gap materials, there may be a range of compositions where the direct and indirect gap energy minima are nearly equal. In this region, the mobility of carriers may be anomalously low because the carriers will have a possibility of being scattered from one minimum to another. This process is shown schematically in Figure 6.1. An interesting result is that scattering of a carrier from one minimum to the other will result in sudden variations of effective mass. This is the basis of the Gunn effect, discussed in the applications section of Chapter 5.

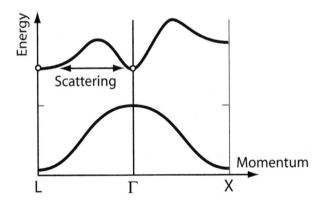

Figure 6.1: A schematic diagram illustrating scattering of electrons from one minimum in the conduction band to another. This occurs especially strongly in alloys near the direct-indirect transition.

6.1.2 Choosing alloy constituents

The choices of compounds from which to construct semiconductor alloys are limited. For example, the common elemental semiconductors are Si, and Ge. Other more rare examples include diamond, cubic Sn, and amorphous Se. The simple binary alloys therefore consist only of mixtures of these elements (excluding Se, which is not isovalent with the others) either in crystalline or amorphous form. All except Si-Ge alloys are extremely limited in usable compositions. Amorphous semiconductors have their own chapter (Chapter 8) and so will be ignored here. Carbon forms SiC rather than a Si-C alloy when mixed with Si. Sn has little or no solid solubility with the other materials. By contrast, Si and Ge are completely miscible. Si-Ge alloys are of sufficient importance that they are discussed in detail in Section 6.4. Until then, we will leave the elemental alloys in favor of the compound semiconductor alloys. These provide much more flexibility in the resulting properties but are also much more complex and difficult to work with.

The vast majority of semiconductors are compounds of two or more elements. Consequently, their alloys contain three or more elements. These are described in detail with a ternary phase diagram and are considered ternary alloys. We encountered a typical ternary phase diagram for Ga-Al-As in Chapter 4, which is reproduced in Figure 6.2 for convenience. There are no compounds that lie within the ternary triangle. This is not surprising in simple systems where two of the alloying elements are chemically very similar (as for Ga and Al). With no phases inside the triangle, compounds along the edges must be stable in each other's presence or must form an alloy. The tie lines that connect them (for example connecting GaAs with AlAs) define pseudobinary alloys of the compounds. The rest

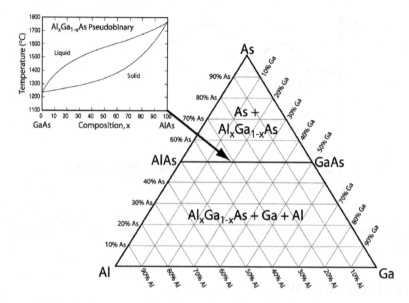

Figure 6.2: Shows the ternary phase diagram for Al-Ga-As. There is also significant solubility for Ga in Al (not indicated) and significant vapor pressure for As above a few hundred degrees C. The latter leads to As loss in open or vacuum environments. Inset based on data in Reference [1].

of the phase diagram is relatively uninteresting, consisting normally of two-phase regions.

Compound semiconductors are usually so particular about their compositions, especially in III-V materials, that they are restricted to lying along the tie lines. Hence, we can usually ignore the remainder of the ternary phase field. Thus, "pseudobinary alloys" are described with a simple binary phase diagram (see inset in Figure 6.2) where the end-points are compounds from the complete ternary diagram, and the remainder of the diagram is a slice of the phase space along the tie line. Pseudobinary alloys are often referred to by the more generic name "ternary alloys". We will generally use "pseudobinary" here, as it is more specific.

Because of the restrictions requiring the same structure and valence for high solubility, alloys are usually formed across common elemental groups; thus, III-V or II-VI alloys. Likewise alloys are generally formed from compounds with the same structure. Taking these factors together with the need for a good lattice constant match to the substrate, semiconductor alloys are generally represented graphically on energy-gap vs. lattice constant diagrams for given elemental groups. Figure 6.3 shows the energy-gap vs. lattice constant for the common III-V and Figure 6.4 the II-VI pseudobinary/ternary alloys along with selected common substrate materials.

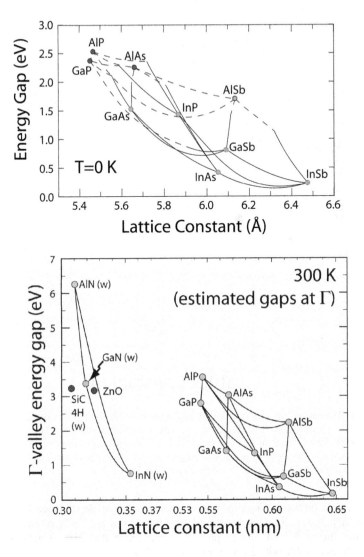

Figure 6.3: The energy gap of group III-V compounds as a function of lattice constants. The upper plot shows the minimum gaps for the alloys with dashed lines indicating indirect gaps and solid lines for direct gaps. The lower chart shows the minimum direct gaps. Based on data in References [2] and [3]. Note that the nitride values are for the hexagonal wurtzite phase (indicated with a "w"). The cubic nitride lattice constants are 0.438, 0.450, and 0.498 nm for AlN, GaN, and InN, respectively. [3] The cubic nitride energy gaps are similar to or smaller than the hexagonal phase values. Top figure redrawn with permission from I. Vurgaftman, J. R. Meyer, and L. R. Ram-Mohan, Journal of Applied Physics, 89, 5815 (2001). Copyright 2001, American Institute of Physics.

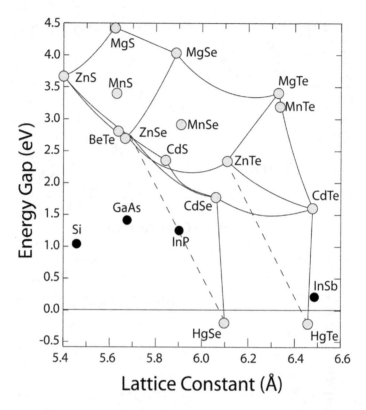

Figure 6.4: The energy gap as a function of lattice constant for II-VI compounds. All of these compounds are direct-gap. The dashed curves are linear interpolations. The real systems would exhibit lattice constant bowing. The black points indicate commonly available substrates on which the II-VI compounds can be grown. Note that HgSe and HgTe are metallic rather than semiconducting. Figure based on References [4] and [5].

Individual lines on these diagrams represent the energy gap dependence in individual pseudobinary alloys. Regions contained by four curves are for quaternary alloys, and define the region of phase space that should, if the alloy were completely miscible, be accessible with a given quaternary alloy.

Pseudoternary alloys would be defined by three lines connecting three specific compounds with a common cation or anion on this diagram. One of the most obvious and striking features of the energy-gap vs. lattice constant plots in these figures is that the end-point compounds are connected by lines bowed downward relative to a straight line. In other words, semiconductor alloys will have a lower energy gap than one might expect for a given composition. This phenomenon is known as bandgap bowing. Amazingly, the energy gaps of the alloys can actually lie below either of the endpoint compounds (see Figures 6.3 and 6.4). The reasons why bowing occurs and the magnitude of the bowing are described in Section 6.3.

The energy-gap vs. lattice constant diagrams of Figures 6.3 and 6.4 do not include any consideration of miscibility of one compound in the other. Indeed, it is very possible that the alloy will separate into the two compounds being mixed or some pair of partial alloys. The latter is known as spinodal decomposition and is a significant concern for semiconductor alloy designers. An alloy that undergoes spinodal decomposition is unstable against composition fluctuations. When a semiconductor alloy decomposes its resulting optoelectronic properties will typically be those of the lower energy gap compound of the resulting pair.

6.2 SEMICONDUCTOR ALLOY THERMODYNAMICS

6.2.1 Regular solution theory revisited

The issue of whether a semiconductor alloy is really soluble is essential to the success of its design. Fortunately, reasonable models of solubility exist and can be applied to these materials. The stability of semiconductor alloys can be treated at least approximately with regular solution theory (see Chapter 4), which works well enough to give an idea of the general behavior of alloys including ternary and quaternary systems.

As we saw in Chapter 4, the stability of a solution is determined by the change in Gibbs free energy, ΔG, with changes in composition:

$$\Delta G = \sum_i \mu_i \Delta N_i = \Delta E - T\Delta S + P\Delta V, \qquad 6.1$$

where μ_i is the chemical potential of species i and N_i is the number of atoms or molecules of that species. ΔE is primarily the change in bond energy of the system and is found by summing over all individual bond energies before and after mixing (Equation 4.25). The complication is that there are a very large number of bonds and, to be precise, second and higher neighboring effects should be included. ΔS is the change in entropy at temperature T upon alloy formation. Normally $\Delta V=0$. As with ΔE, ΔS consists of a number of terms but the dominant term for virtually any alloy system is the configurational entropy. In addition, there are entropy terms associated with changes in the vibrational modes of the system, which can be derived from the phonon density of states, and other terms. In most semiconductor alloys where the atoms are sufficiently similar for solubility, the vibrational mode changes result in only minor changes in total entropy.

When one configuration of a system has a distinctly lower free energy than another then that is the configuration the system chooses. In this section, we will consider an extension of the model in Chapter 4 to the calculation and minimization of this

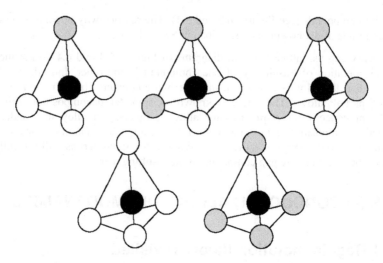

Figure 6.5: The five distinct alloy tetrahedra. White circles indicate A atoms, gray circles B atoms, and black circles C atoms in an $A_{1-x}B_xC$ type alloy. Note that each tetrahedron can be rotated to place any of the A or B atoms on a given vertex when forming the crystal. The energy of a tetrahedron is independent of its orientation. The model does not consider second-nearest neighbor interactions but can account for bond strains within each tetrahedron. A more accurate model must consider longer-range interactions.

energy for a pseudobinary/ternary solution of zincblende-structure semiconductors. Similar treatments can be created for diamond, wurtzite, and more complex forms. The discussion follows the approach of Ichimura and Sasaki [6], although a number of others have proposed similar models. The original description of the solution based on tetrahedral unit cells was by P.N. Keating [7].

There are five possible tetrahedra of first nearest A and B atom neighbors around a given C atom in an $A_{1-x}B_xC$ alloy, as shown in Figure 6.5. Each has a characteristic bond energy. In addition, each tetrahedron has a different size and shape, leading to strain energy contributions to its energy. If we know the net bond energies of each of these tetrahedra, we can estimate the total energy of the solid, assuming that this can be approximated as a linear sum over all tetrahedra present. The assumption that energies and other quantities can be estimated by averaging is known as the "virtual crystal" approximation. This is the same approximation that leads to the linear dependence of lattice parameter and other behaviors on alloy composition. Thus, we effectively have a virtual crystal bond but not a linear total-energy/composition dependence.

In conventional regular solution theory, the change of the bond energy for a simple binary AB solution upon mixing (as in Chapter 4) is given by

$$\Delta E = 1/2 \ (E_{AA} + E_{BB}) - E_{AB}, \qquad\qquad 6.2$$

where E_{AA}, E_{BB}, and E_{AB} are the energies of AA, BB, and AB bonds, respectively. This expression indicates whether mixing the atoms to maximize the AB bonds is preferable to phase separation to maximize the AA and BB bonds at T=0. As we noted in Chapter 4, this is a simple nearest-neighbor bonding result. When one has a pseudobinary rather than a simple binary solution one must turn to summing over tetrahedra because this accounts for second- (vertex-to-vertex) as well as first-nearest neighbor (vertex-to-center of tetrahedra) effects. The distinction is crucial to obtaining a reasonably accurate picture of the material behavior. This treatment provides a separate energy, E_i, for each of the five tetrahedra shown in Figure 6.5. The total bond energy of the system is then a sum over energies for all of the tetrahedra statistically expected to be present in the solid:

$$E = N\sum_j \alpha_j q_j E_j,$$
6.3

where N is the total number of tetrahedra, α_j is the multiplicity of a given tetrahedron (how many possible distinct orientations there are), q_j is the fraction of all tetrahedra which are of type j, and j indicates the number of B atoms in the tetrahedron. It is assumed here that the three orientations of the tetrahedra containing both A and B atoms have the same energy (i.e. third-nearest neighbor and longer range interactions involving atoms in other tetrahedra are ignored) and that all orientations are equally likely (giving well-defined values of the α_j).

The next step is to calculate the values of α_j. There are three rotations of the tetrahedra containing mixed A and B atoms that are, to some extent, distinct. The pure AC and pure BC tetrahedra cannot be distinguished upon rotation. Thus, for a diamond or zincblende structure, $\alpha_j=1$ for the two tetrahedra containing only A and C (j=0) or B and C (j=4) atoms, while tetrahedra including j=1 to 3 B atoms have $\alpha_j = C_j^4 = 4!/(j!(4-j)!)$ distinguishable rotations. Note that $\alpha_j q_j$ summed over all five values of j is unity, while the composition, x, of the alloy $A_{1-x}B_xC$ is $x = q_1+3q_2+3q_3+q_4$. One may further show that $q_j=x^j(1-x)^{4-j}$. [6] **Note: the difference between a random and a non-random alloy is in the q_j values.** The ignored third-nearest neighbor effects in the energies are generally much smaller than the second-nearest neighbor values and so this approximation is not normally a problem. Higher order interactions can be important in establishing, for example, the difference between cubic or hexagonal structures and the precise locations of phase boundaries.

The configurational entropy for a system with clustering described by the q_j distribution probabilities above has been estimated for a zincblende semiconductor alloy to be a modified regular solution entropy, [6]

$$S = k_B N\left(3[x\ln x + (1-x)\ln(1-x)] - \sum_j \alpha_j q_j \ln q_j \right),$$
6.4

where the familiar Sterling's approximation $x! \sim x \ln x$ has been used. Equation 6.4 reduces to Equation 4.28, $S=-k_BN[x \ln x + (1-x) \ln(1-x)]$, for a random alloy.

Combining Equations 6.1, 6.3 and 6.4 and assuming $\Delta V=0$ yields a composition-dependent free energy. The free energy can then be minimized using:

$$\sum_j \frac{\partial G}{\partial q_j} dq_j = 0. \qquad 6.5$$

Energy minimization also leads to: [6]

$$\sum_j \alpha_j \delta q_j = 0 \qquad 6.6$$

and

$$\delta x = \delta q_1 + 3\delta q_2 + 3\delta q_3 + \delta q_4 = 0, \qquad 6.7$$

where the δq_j are changes in the probabilities q_j upon a change in alloy composition (as in phase separation). Suppose that we can define an alloy order parameter:

$$\beta = 1 - \frac{P_{AB}}{x} \qquad 6.8$$

where P_{AB} is the probability that the second nearest neighbor of a given B atom is an A atom, and x is the average composition of the alloy. One can calculate the order parameter β from the q_j values:

$$\beta = 1 - \frac{q_1 + 2q_2 + q_3}{x(1-x)}. \qquad 6.9$$

The energy minimization criterion can be used to develop values of the various q_j's for given E_j values. These can, in turn, be used to calculate an order parameter. Consider for a moment the range of values β can take. For a completely phase separated alloy $P_{AB}=0$ and $\beta =1$. If a superstructure order develops in which $P_{AB}>x$ or $q_1>x(1-x)^3$, $q_2>x^2(1-x)^2$, and/or $q_3>x^3(1-x)$. This situation would yield $\beta<0$. **Hence positive β values indicate phase separation while negative β shows ordering.**

Note that because the energies for given tetrahedra can include strain energy and because the strain energy depends upon the average lattice constant of the structure and the average surrounding unit cell compositions, the strain energy, and consequently the E_j values are slightly composition dependent. Thus, when strain energy is a significant contributor to the energy of a unit cell, the q_j solutions need to be obtained separately for each composition value once the strain energy term is calculated for that x. The order parameter for the $In_{1-x}Ga_xAs$ pseudobinary alloy was determined by Ichimura and Sasaki [6] using the above approach. The results are shown in Figure 6.6. Note that this alloy is shown to exhibit superstructure ordering.

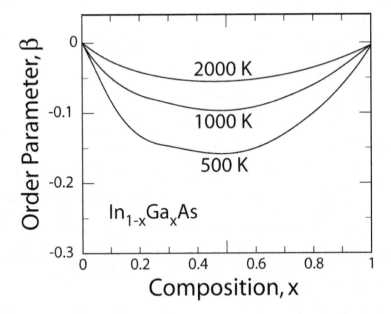

Figure 6.6: Shows the order parameter calculated for $In_{1-x}Ga_xAs$ as a function of composition at several temperatures. Redrawn with permission from Masaya Ichimura and Akio Sasaki, Journal of Applied Physics, 60, 3850 (1986). Copyright 1986, American Institute of Physics.

6.2.2 Ternary and quaternary solutions

Similar models can be developed for more complex systems such as ternary and quaternary alloys. As an example, we will consider a modified regular solution model for quaternary alloys proposed by K. Onabe [8]. This model does not include some of the detailed treatment in the Ichimura calculation [6] described in Section 6.2.1. Specifically, it does not include strain and assumes a random alloy entropy. However, the calculation is simple, illustrative of the behavior of multinary alloys, and can easily be extended to include the strain and entropy terms.

For a quaternary $A_{1-x}B_xC_yD_{1-y}$ alloy, the stability criterion is determined by the second partial derivative of the free energy change, ΔG, for mixing:

$$\frac{\partial^2 \Delta G}{\partial x^2} \frac{\partial^2 \Delta G}{\partial y^2} - \frac{\partial^2 \Delta G}{\partial x \partial y} = 0 . \qquad 6.10$$

The regular solution theory energy in Equation 6.2 can be generalized to multinary solutions. Suppose that w_{AB} is the bond energy of an AB compound, analogous to E_{AA} and E_{BB} in Equation 6.2. Further, suppose that $a_{AC-BC} = w_{AC}-w_{BC}$ being the energy of a given binary pair, is analogous to E_{AB}. The energies may be calculated based on a simple virtual crystal picture as in the definition of a_{AC-BC} just given or one can set up alloy tetrahedra as in Section 6.2.1. Onabe [8] used the virtual crystal approach, which yields the following result for the quaternary alloy:

$$\Delta E = w_{AC}(1-x)(1-y) + w_{BC}x(1-y) + w_{AD}(1-x)y + w_{BD}xy +$$
$$a_{AC-BC}(1-x)x(1-y) + a_{AD-BD}(1-x)xy + \qquad\qquad 6.11$$
$$a_{AC-AD}(1-x)(1-y)y + a_{BC-BD}x(1-y)y$$

The generalized regular solution entropy for a quaternary alloy is:

$$\Delta S = Nk\big[(1-x)\ln(1-x) + x\ln x + (1-y)\ln(1-y) + y\ln y\big] \quad 6.12$$

Substituting Equations 6.11 and 6.12 into 6.1 and the result into 6.10, a set of alloy stability curves in x and y are obtained for quaternary alloys. These curves are similar to points on the curve in Figure 4.18c at a given temperature and define a region of composition space in which the alloy is unstable and may decompose. In addition to direct calculation from known values, the bond energies in Equation 6.11 may be obtained by fitting observed spinodal decomposition data for pseudobinary alloys in a given quaternary system.

The resulting stability curves are ellipses or squared ellipses in the quaternary phase space. Each ellipse becomes smaller at higher temperature as the spinodal decomposition region shrinks (as is the case in Figure 4.18c), driven by the stronger entropy contribution. Onabe [8] carried out calculations for all of the common III-V semiconductor ternary and quaternary alloys. Selected results are shown in Figure 6.7.

The important points to note about the results in Figure 6.7, verifiable with more detailed modeling, are as follows:

- Systems with large lattice mismatch between the compounds on opposite corners of the diagram (for example, InSb and AlAs in the $Al_xIn_{1-x}As_ySb_{1-y}$ quaternary alloy) have large spinodal decomposition regions.
- Quaternary alloys may exhibit spinodal decomposition even though no decomposition will occur for any of the associated binary alloys at a given temperature.
- All of the quaternary systems except the $Ga_xAl_{1-x}P_yAs_{1-y}$ alloy are predicted to exhibit a spinodal decomposition in the simple equilibrium theory.
- The spinodal decomposition of an alloy within the decomposition instability will result in formation of alloy domains with compositions lying around the perimeter of the stability curve. There are no specific alloy compositions that must occur. Any fluctuation in x or y can take the alloy to a stable decomposed state. What actually occurs depends upon nucleation kinetics and strain energy (see Chapters 7 and 10).

- As with any spinodal decomposition, the curves shown are for an equilibrium condition. In a real crystal one almost never observes decomposition at the spinodal boundary, as the driving force for decomposition there is zero. Furthermore, there may be a substantial nucleation barrier preventing decomposition into small domains of different composition. (This reflects the contribution due to strain to some extent, which is ignored in the above treatment.)
- Growth techniques such as molecular beam epitaxy that are typically far from equilibrium may allow the alloy to be grown as a metastable solution without decomposition. However, subsequent heat treatments could lead to decomposition.
- The spinodal decomposition regions would, if the system was in equilibrium, eliminate certain portions of the quaternary phase space from accessibility, meaning that the full range of lattice constants and energy gaps one might expect to be able to obtain based on Figures 6.3 and 6.4 may not, in fact be possible.

Equilibrium Spinodal Decomposition Phase Boundaries

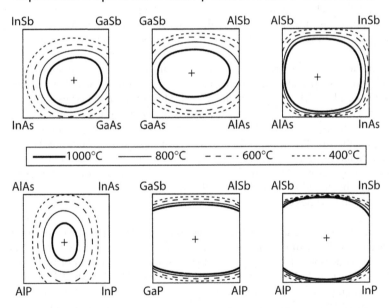

Figure 6.7: Spinodal decomposition curves for six quaternary semiconductor systems as calculated by Onabe in Reference [8]. Crosses indicate the midpoint of the spinodal decomposition regions. Ellipses are equivalent to slices through Figure 4.18c at various temperatures. Figure redrawn with permission from portions of Figure 1 in Kentaro Onabe, Japanese Journal of Applied Physics vol. 21 (1982), pages L323-5. Copyright 1982, Institute of Pure and Applied Physics.

Once again, it is important to note that the stability regions shown in Figure 6.7 are based on a simple regular solution model and do not include a detailed treatment of strain or a full treatment of configurational entropy. In spite of this, the strain effect is clearly visible in the model results (larger spinodal decomposition for larger differences in lattice size). This apparent contradiction is because, as we saw in Chapter 5, lattice size has a direct effect on bond energies and because the interaction energies are fit to pseudobinary behaviors. In spite of the limitations of the model, this approach and the results in Figure 6.7 should be qualitatively accurate.

6.2.3 More mechanisms for alloy ordering

Some semiconductor alloys develop a superlattice order rather than the random composition fluctuations described in the previous section. In some cases this is a thermodynamically-favored result (as in Figure 6.6 and Reference 5) while in other cases it is the result of the nature of the surface and the growth process. Thus, an $A_{1-x}B_xC$ alloy in which there is a preference for A atoms to have B atoms as second-nearest-neighbors and vice versa will favor structures in which such an organization is present. This leads to compound formation with a specific crystal lattice. In the chalcopyrite compounds such as $CuInS_2$ the ZnS zincblende lattice is replaced by an ordered organization with two Cu and two In per S. Although this compound is well organized, it is not surprising or unusual. It is simply a consequence of the attempt to balance the valence two S atom with two valence one and valence three metal atoms such that all atoms have their proper charge.

A less obvious result is ordering of a material in the absence of a strong chemical driving force. Well-known examples are the Si-Ge alloys, described in detail in Section 6.4 and the III-V pseudobinary alloys. Most of these materials show ordering as a result of details of their growth process and the surface plane on which the atoms form the crystal. Even though there may be no bulk equilibrium driving force for organization of the solid, ordering occurs in most cases. That this is usually a surface-related phenomenon may be deduced from the fact that such ordering is not found in alloys grown from the liquid phase. A detailed review of ordering III-V alloys may be found in Stringfellow [9] and other chapters in Mascarenhas (see recommended readings).

Either the "CuAu" structure in which ordering of the alloyed atoms occurs on (100)-type planes or the "CuPt" ordering on (111) type planes is observed in virtually all epitaxial pseudobinary III-V alloys (see Figure 6.8).[9] Occasional ordering on other planes is also found. The extent of ordering is generally described in terms of an order parameter, β (see Equation 6.8), which ranges from $\beta \rightarrow -1$ for a completely ordered phase [$P_{AB} \rightarrow 1$ and $x \rightarrow 0.5$ for Equation 6.8], to zero (a completely random alloy) to +1 (a perfectly phase separated structure with no AB units [$P_{AB} \ll x$]).

Ordering affects many properties of the material but primarily modifies the energy band structure. This has direct and measurable consequences for the optical properties of the final material, by which the order parameter is typically determined. In the most highly ordered alloys $-0.5 > \beta > -0.6$. The energy gap generally decreases quadratically with increasing $|\beta|$, in other words, $E_g = E_g(\beta=0) - \Delta E_g \beta^2$.

The $Ga_{1-x}In_xP$ pseudobinary alloy provides a good example of the behaviors in such ordered compounds. This alloy shows the CuPt ordering when grown by metal-organic vapor phase epitaxy (MOVPE). [See Chapter 12.] The extent of ordering is strongly affected by deposition temperature with a maximum order somewhat above 600°C, decreasing both above and below this value. The ordering decrease at lower temperatures indicates a thermodynamically-favored but kinetically-limited ability to organize the alloy. Above the optimal temperature disorder is increased by entropy due to increased diffusion among atomic layers in the bulk, leading to randomization of the buried structures. Likewise, enhancement in the group III flux relative to the group V flux supplying growth enhances ordering. Presumably this is because group III (cation) ordering originates on the surface and further that the surface phase responsible requires excess group III element to form.

The final proof that the process is the result of a surface phenomenon is that the presence of a surfactant reduces ordering dramatically. Surfactants are atoms that segregate strongly, collecting on the growth surface and causing up to a single atomic layer crust that floats on the surface as the solid grows. Surfactants lower the surface energy (which is why they surface segregate) and change the surface organization and growth mechanism completely, so it may not be surprising that ordering is eliminated.

Zincblende alloy with
CuAu ordering   CuPt ordering

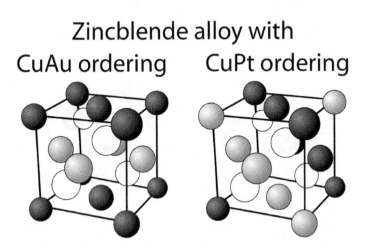

Figure 6.8: Schematics of a zincblende alloy of dark and light gray cations and white anions with CuAu and CuPt ordering on (100) and (111) planes, respectively. A common cation alloy has the same behavior.

One may see how CuPt or CuAu ordering occurs if one looks at the growth process in cross section. An example scenario is shown in Figure 6.9. The (100) surface of most zincblende semiconductors consists of long rows of pairs of atoms ("dimers") as shown in Figure 6.9 (a). If the surface of a III-V semiconductor is rich in group III element then the dimers usually are made of group III atoms while if rich in group V atoms, these define the surface dimers. In an alloy where one atom is larger than the other, it may be favorable to distort the dimers to provide extra space for the larger atom [as in Figure 6.9(b)]. This reduces the surface energy and should be favored at moderate temperatures where atomic mobility is sufficient to form the organized structure but not so high that entropy will destroy it. When the next layer of atoms is deposited on the previous layer, the buried atoms may not have the mobility to randomize themselves, trapping the atoms in an organized state. Furthermore, the strain due to the previously ordered large and small atoms transmits itself to the new surface layer, coupling the two layers together. Therefore, this strain field is preserved as growth proceeds. Thus, the arrangement of atoms in subsequent layers is connected to that of the underlying layers and an overall order of the lattice results [Figure 6.9 (c)].

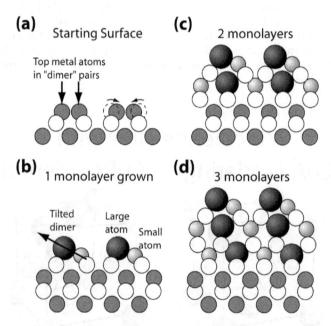

Figure 6.9: Schematic cross sections through a zincblende lattice along a (110) direction showing the dimerization of atoms on the surface of the material. The white circles represent common anions in an alloy and the large and small circles are the two cations. The medium gray circles represent the substrate cations. Strain resulting from placement of the large atoms in the second layer influences how dimers tilt on the surface. One side of these dimers favors the larger atom, which continues the underlying order.

It is important to emphasize again that this is not normally an equilibrium structure *in the bulk of the solid*. Calculations show that a random bulk alloy is typically favored over the CuPt or CuAu organization. Ordering of this type in alloys is a result of the growth process and is directly tied to the surface structure as growth occurs. This is why sometimes one order is found, sometimes another, and sometimes no order is present.

CuPt and CuAu ordering is not restricted to zincblende structure semiconductors. The chalcopyrites also exhibit both ordering structures. These materials include significant solubility for point defects on the metal sublattice (see Figure 4.11). The group III-rich phases are often found to include ordered point defects of one sort or another. $CuInS_2$ is particularly well known for forming ordered structures, even near stoichiometry. Unlike the III-V alloys, these structures are probably not the result of growth processes as they are found for a wide variety of methods and conditions, but rather are the result of deviations from stoichiometry and interaction of point defects. They are therefore thermodynamically stable even in the bulk. Unfortunately, these are generally detrimental to the operation of devices based on the materials.

In addition to organization of structures due to bond energy and surface growth effects, thermodynamically unstable bulk phases can result on the surface during certain processes for making a crystal. For example, the surface of a semiconductor has a different coordination and bonding than does the bulk and consequently there is a driving force for segregation of one alloy component to surfaces in some systems. This normally leads to a one or two monolayer thick enhancement of one alloy species relative to the other at a surface, although in some systems large second phase precipitates can appear on the surface or elsewhere.

6.3 BAND GAP BOWING

6.3.1 Binary and pseudobinary alloys

As discussed briefly in Section 6.1, observed semiconductor energy gaps lie below what one would expect for a linear interpolation between the end-point compounds. The direct energy gap can be written empirically as

$$E_\Gamma(x) = E_\Gamma^A + (E_\Gamma^B - E_\Gamma^A) x - c \, x(1-x). \qquad 6.13$$

The constant c is known as the bowing parameter and is four times larger than the negative deviation of the gap from the linear behavior at $x=0.5$, where the deviation from linearity is typically greatest. Similar expressions can be written and similar behaviors are observed for the gaps at other points in reciprocal space such as the indirect gap minima at X and L.

At first glance, the discussion of alloy miscibility in the previous section might suggest that bowing is simply a result of local composition fluctuations even in alloys that are miscible. This would lead to local domains with a low energy gap. Because most measurement techniques such as photoluminescence and most devices are sensitive only to the lowest gap in the material, composition fluctuations would lead to the appearance of bowing. While composition fluctuations certainly do occur, the explanation that we are just observing the normal band structure of local areas rich in one end-point semiconductor with a narrow gap is not sufficient to explain typical bowing behaviors. Furthermore, it cannot explain bowing that gives rise to gaps lower than the lowest end-point energy gap and it does not account for the fact that very small particles typically have larger rather than smaller gaps due to quantum confinement. We must look deeper into the effect of changes in atoms from one lattice site to the next on band structures to find the true causes in detail. Not surprisingly, the mischief can be traced to the Schrödinger equation.

Bowing can be explained from two fundamental principles, both based on an analysis of the effect of alloying on the electron potential term, U, in the Schrödinger equation. (1) "Symmetric" or averaged changes to the electron potential: If a random alloy can be treated with the virtual crystal approximation, then the atomic core potentials that appear in the Schrödinger equation and thus the matrix elements in the LCAO matrix (the $V_{ss\sigma}$, $V_{pp\sigma}$, $V_{pp\pi}$ values) are averaged between those representative of the end point compounds but the Bloch states of the electrons are relatively unaffected. (2) "Asymmetric" or localized changes in the core potential: In some cases interpolation of the core potentials while assuming a uniform periodic potential is not sufficient. One may have to account for more subtle effects such as those that entered into the determination of the bond energy in the alloy tetrahedra in Section 6.2. These changes include the effects of local strain, composition fluctuations, third- and higher-nearest-neighbor effects, and whether some or all atoms have shallow d-orbitals. As noted earlier, the latter is required for an accurate band structure calculation. Shallow d-orbitals tend to raise the level of the top of the valence band by repelling the p-orbitals that determine the valence band edge in diamond-structure semiconductors. As we will see, the asymmetric contributions are most important where the core potential differences are the largest.

The similar effects giving rise to both bowing and variations in the energy of the alloy tetrahedra in Figure 6.5 lead to an important conclusion. ***Bowing and solubility are directly linked. Alloys with large bowings should have large regions where spinodal decomposition would favor phase separation. Thus one may expect poor miscibility in alloys with large bowing.*** Examination of Equation 6.3 and the equations for the α_j and q_j terms also shows why bowing should have a maximum at $x{\sim}0.5$ and why it would depend roughly on the difference in bond energies between atoms – because this is where the amplitude of core potential fluctuations should be largest relative to the unperturbed lattice.

Let us explore the effect of core potential fluctuations in more detail. Referring to the discussion in Chapter 2, we begin by noting that we solved the Schrödinger equation using Bloch waves and taking a single Fourier coefficient of U in the analysis. More sophisticated calculations take more Fourier terms into account but the result is the same. The crystal potential can be decomposed into two components, one symmetric or random, U_s, representing the virtual crystal behavior, and an asymmetric U_a representing the local variations due to composition fluctuations. The symmetric term modifies the potential terms for the solid with the same Fourier coefficients as in the non-alloyed behavior. This means we do not need to modify the basic approach used to solve the Schrödinger equation because the symmetry of the lattice is the same and only the average potential changes. The asymmetric behavior represents potential terms with amplitudes on Fourier terms not normally associated with the crystal potential for a given symmetry – in other words they are local defects that clearly break the symmetry of the lattice overall.

For the virtual crystal of an AB semiconductor alloy the symmetric potential term is

$$U_s = \sum_{r'} [xU_A(r') + (1-x)U_B(r')].$$
6.14

The asymmetric part is best written as

$$U_a = \sum_{r'} [U_A(r'-\tau) - U_B(r'-\tau)]c_\tau.$$
6.15

The coefficients c_τ are pair-correlation function values. The pair correlation function describes the probability of finding an A atom (as opposed to a B atom) a given distance and direction from a particular lattice site. The pair correlation function can also be thought of as the real-space average lattice and gives us a global picture of the deviation of the solid from "average" as represented by the virtual crystal symmetric term. The pair correlation function is normally determined by Fourier-transforming a diffraction pattern or reciprocal space representation of the solid back to real space. If you want to see how to calculate a pair correlation function, try homework problem 4 at the end of the chapter. Averaging over all lattice sites, c_τ is $(1-x)$ for lattice sites containing an "A" atom and $-x$ when a "B" atom is present. A site containing a mixture of A and B atoms on average will have a value between these two limits. The value of τ is just the displacement of an atom from the lattice site r' where the potential is to be calculated. Such displacements can be strain driven or due to bond-length variations with local composition as occur in the tetrahedra in Figure 6.5. Thus, U_a terms corresponding to $\tau \neq 0$ are contributions due to strained structures, while the $\tau = 0$ terms are for composition fluctuations without distortions of the lattice.

Repeating for emphasis: Equation 6.14 describes that part of the alloy electron potential energy that is simply a perfectly uniform periodic potential as in Chapters 2 and 5 with a magnitude modified according to the average potential of the alloyed atoms. Equation 6.15 covers the non-uniform part of the potential having a spatial component given by the pair-correlation function and a magnitude equal to the difference in the alloyed atom atomic potentials. Thus, atoms with very similar potentials result in relatively small asymmetric potentials, regardless of local strain, and little additional bowing, while large differences result in large deviations from a virtual crystal behavior.

Solutions to the Schrödinger Equation can be found for $U \neq 0$ retaining higher order Fourier components of the potential. [Recall that in the nearly free electron model we took only the first component.] In a non-random alloy or in an alloy with large differences in atomic state energies, the local composition fluctuations can excite these higher order Fourier components exceptionally strongly, which modifies the band structure accordingly. Bowing above what would result from the virtual crystal approximation must be the result of the asymmetric part of the effective potential.

The virtual crystal model gives, for example, for Harrison's homopolar splitting:

$$V_2 [A_{1-x}B_xC \text{ alloy}] = V_2^{AC} + (V_2^{BC} - V_2^{AC})\, x. \qquad 6.16$$

Van Vechten [10,11] applied virtual crystal approaches such as interpolation of potentials to band gap calculations in some detail. Because the energy gap is not a linear function of V_2, especially when there is chemical splitting as well, the change in energy gap with composition is not a linear function of composition. Thus, even a calculation for a perfectly random (virtual crystal) alloy leads to band gap bowing, $c > 0$. *A full calculation of the energy gap based on the resulting parameters shows bowing for a virtual crystal scales with the difference in homopolar energies, $c \propto (V_2^{AC} - V_2^{BC})$.* A similar argument can be made based on the potentials in the LCAO matrix. In short, even for a virtual crystal the larger the difference in atomic state energies, the larger the bowing, as expected from the above arguments. The virtual crystal approximation also leads to parabolic decreases in the density of states around the band edges as in the nearly free electron model.

Examples of alloys that follow virtual crystal behavior (random component of bowing at least twice the non-random contribution) are InAs-InP, and the direct gap in GaAs-GaP, suggesting that these behave as nearly random alloys. [2] Note that this random alloy bowing can lead to decreases in energy gap below either end point compound gap. The entire band structure is affected and band widths as well as density of states center of mass can change with composition.

There are practical problems in achieving a true virtual crystal. For example, to avoid some amount of disorder in the lattice, the average number of A atoms which were second-nearest-neighbors to B atoms in an $A_{1-x}B_xC$ alloy would have to be constant. At a composition of $x = 0.5$, this would mean that every B atom would have two A atoms and two B atoms for second-nearest-neighbors. This arrangement exists in some materials and leads to structures such as chalcopyrite. Such a well-organized structure is no longer a random alloy. Rather, it is a new compound with unique symmetry and, consequently, a different band structure. Thus, even in the perfectly distributed alloy case, one can expect to have deviations from a virtual crystal behavior because a perfect distribution is not random. The reason a virtual crystal

Table 6.1: **Semiconductor Alloy Bowing Parameters**

Alloy	Bowing Parameter (eV)			Alloy	Bowing Parameter (eV)		
	Γ	X	L		Γ	X	L
AlGaAs	-0.13	0.55	0	AlAsP	0.22	0.22	0.22
AlInSb	0.43			AlAsSb	0.8	0.28	0.28
AlInP	-0.48	0.38		AlPSb	2.7	2.7	2.7
GaInAs	0.48	1.4	0.33	GaAsP	0.19	0.24	0.16
GaInSb	0.41	0.33	0.4	GaAsSb	1.43	1.2	1.2
GaAlN	1.0			InAsSb	0.67	0.6	0.6
GaInN	2.4			GaAsN	120		

All values are for the zincblende structure of the alloy. Most values and especially GaAsN, and AlGaAs and AlInP Γ-point values are composition dependent. Data are from Reference [2].

behavior can exist at all is that **sufficiently short-ranged potential fluctuations can be effectively screened by electrons in surrounding states and composition fluctuations are thereby averaged locally** and reduced in their effects.

NOTE: while simple bowing parameters represent convenient ways to discuss and estimate composition dependences of energy gaps, they are often significant oversimplifications of the actual situation.

Energy band structures for all of the common III-V semiconductors have recently been reviewed by Vurgaftman et al. [2] and values for bowing parameters for a wide range of binary alloys are given. Table 6.1 summarizes results for selected alloys. Clear trends in the data are limited but some observations can be made. First, in general, the greater the difference in atomic state energies the greater the bowing parameter, as predicted by Van Vechten. Materials with higher chemical splitting differences and larger miscibility gaps tend to have larger bowing parameters, as anticipated from the asymmetric potential equation 6.15. Second, common anion alloys often have lower bowing coefficients as compared to common cation alloys. This is because the shape of the conduction band changes much more than does the valence band shape as alloying occurs, partially obscuring the bowing effect. Finally, the nitride alloys tend to have very large bowing parameters. The most extreme case is for the common cation nitride/arsenide alloys where bowing parameters are many electron volts. As one might expect, there is very limited solubility for the common cation nitride alloys.

One of the most striking features of the bowing parameter values in Table 6.1 is the negative value for the direct gap in AlGaAs and AlInP. This is particularly

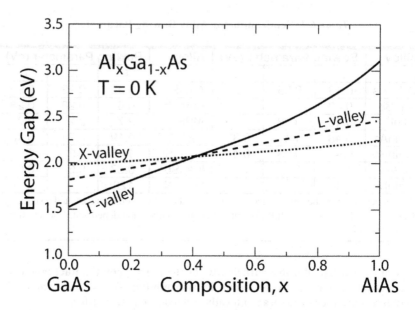

Figure 6.10: Minimum transition energy values for three primary gaps in AlGaAs alloys as a function of alloy composition. Note the direct-indirect transition at approximately 43% Al. Redrawn with permission from I. Vurgaftman, J. R. Meyer, and L. R. Ram-Mohan, Journal of Applied Physics, 89, 5815 (2001). Copyright 2001, American Institute of Physics.

important, as AlGaAs is one of the most heavily used of the pseudobinary alloys. The various energy gaps in this alloy have been measured by a number of groups but inconsistent values have been obtained. Based on an extensive review Vurgaftman et al. [2] propose the band gap behaviors for AlGaAs shown in Figure 6.10. The point is that the bowing is not even close to the idealized quadratic behavior described above. Indeed, several observations suggest no bowing behavior. It is easy to see how the complex curvature in Figure 6.10 could be measured as straight or with a variety of curvatures, especially in a noisy experiment or in measurements at different temperatures. The complexity of the behavior is further not surprising given the nature of the asymmetric contribution to the atomic potential (Equation 6.15) and all of the various strain and bonding terms that enter into determining it.

6.3.2 Bowing in quaternary alloys

Band gap bowing occurs in ternary and quaternary alloys as it does in binary and pseudobinary alloys. The behavior of quaternary alloys and can be treated in the same way. For a typical quaternary of the form $A_{1-x}B_xC_yD_{1-y}$, Equation 6.13 for the energy of a specific part of the band structure (for example, the direct gap) can be generalized to [2]:

$$E_{ABCD}(x,y) = \frac{x(1-x)[(1-y)E'_{ABD}(x) + yE'_{ABC}(x)]}{x(1-x) + y(1-y)}$$
$$+ \frac{y(1-y)[xE'_{ACD}(y) + (1-x)E'_{BCD}(y)]}{x(1-x) + y(1-y)}.$$

     6.17

Here E_{ABCD} is the energy of a given band in the quaternary and E_{ABD}, E_{ABC}, E_{ACD}, and E_{BCD} are the energies of the same band that would have been determined from Equation 6.13 for each individual pseudobinary alloy. Thus, Equation 6.17 is a linear interpolation of the pseudobinary alloy values. This approach produces a reasonable estimate of the gap values. [2] There are other methods for estimating the alloy energy gap that increase the effective bowing but the performances of these methods are not particularly improved.

An obvious difference between Equations 6.17 and 6.13 is the absence of a specific (new) bowing parameter for the quaternary. This is perhaps not so surprising as the variety of combinations of elements that can figure into the tetrahedra of Figure 6.5 is similar to what one would expect from the various binaries. Thus, a given tetrahedron can have any combination of cations around a given anion (as in a common anion pseudobinary). Likewise one can have any collection of anions around a cation. To mix cations and anions simultaneously requires a larger volume than a single tetrahedron and, consequently, a larger interaction range (beyond second nearest neighbor). Therefore the magnitude of additional bowing one might expect in a quaternary alloy is much smaller than the pseudobinary bowing and can be neglected. A similar result is obtained for pseudoternary alloys of the type $A_xB_yC_{1-x-y}D$ (where A,B, and C are all either cations or anions and D is an anion or cation, respectively). For a detailed discussion and references to other methods of estimating bowing parameters in quaternary alloys as well as a development of many of the relevant parameters for specific compounds, see Vurgaftman et al.. [2]

6.4 SILICON-GERMANIUM ALLOYS

Silicon is the most commonly applied semiconductor and is the basis for the vast majority of microelectronic devices. It has proven durable in the face of intense competition from other materials for many reasons. The more significant of these are its mechanical strength, high performance oxide dielectric, and reproducible contacts. (Interestingly, none of these are related to its properties as a semi-conductor.) Because most manufacturing processes are adapted to silicon and because many of these advantages survive alloying, at least to some extent, silicon-germanium alloys have become technologically important. Silicon-germanium alloys are one the more heavily studied of the semiconductor combinations and a large amount of data on the system exists. Interpretation of this data is also relatively straightforward, as modeling binary alloys is much simpler than for pseudobinary and more complex systems. The Si-Ge alloy exhibits a number of rather unique and interesting features that make it worthy of discussion in any case. This section reviews some of these points. A more detailed review of this topic may be found in Kasper and Lyutovich (see recommended readings).

6.4.1 Structure and solubility

Silicon-germanium alloys have an fcc Bravais lattice with the diamond structure. The alloy exhibits complete solid solubility across the entire phase space. Below, ~170 K, theory suggests there could be a spinodal decomposition regime. However, the atomic mobilities at such a temperature would prevent decomposition of an existing alloy on any time scale. The Si-Ge phase diagram is shown in Figure 6.11.

The lattice parameter of this alloy has been carefully examined, as it is important to determining how the material responds to strain due to a lattice mismatch with a substrate (see Chapter 7). A small deviation from Vegard's law has been found. The lattice parameters for Si and Ge are, respectively, 0.54310 and 0.56575 nm. The difference is 0.02265 nm so Vegard's law would predict an alloy lattice constant $a(Si_{1-x}Ge_x) = 0.5431 + 0.02265x$ (nm). Experimentally, however, the lattice parameter has been found to be a roughly quadratic function of composition with corrections decreasing as the order of the term increases [12]:

$$a(Si_{1-x}Ge_x) = 0.54310 + 0.01992\,x + 0.002733\,x^2 \text{ (nm)} \qquad 6.18$$

A more accurate fit to the experimental data can be obtained if one uses a third-order polynomial. One should be cautioned that these values are temperature dependent and the results at this level of accuracy are for 300 K. Values at other temperatures will be significantly modified, at least at the scale of Equation 6.18.

Mechanical properties of the alloys have also been examined, although the literature on this topic is smaller. The elastic moduli have been measured and simulated theoretically. Some studies show elastic constants greater than what one would obtain from a linear average,[14] while a more recent work suggests that the linear average provides a good approximation.[15] Additional studies also exist. The differences are most likely the result of a variety of methods of sample preparation and measurement. However, overall a linear average seems a workable approximation.

Although $Si_{1-x}Ge_x$ alloys exhibit full solubility in their equilibrium binary phase diagram, thin films grown by molecular beam epitaxy [see Chapter 11] show evidence of bulk superstructure ordering, and anomalous phase separations are observed in some cases. This result has been obtained by numerous laboratories and results in a CuPt-like structure in which alternating (111) planes have an excess of Si and then Ge. This phenomenon occurs in films grown on (100)-oriented substrates, thus the ordering planes are inclined with respect to the substrate surface as was shown in Figure 6.9. It is generally thought to be a consequence of the (2x1) surface reconstruction [formation of pairs of atoms on the (100) surface in elongated strings] resulting in a surface strain field. The strain field causes ordering in the atoms on the

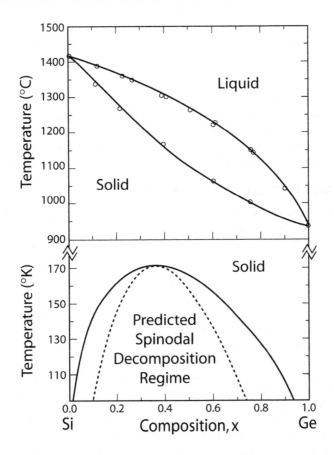

Figure 6.11: The phase diagram for Si-Ge binary alloys. High temperature phase diagram after References 12 and 13. Low temperature data based on Reference 13. Figures redrawn with permission from Herzog, H.J.; "Crystal structure, lattice parameters and liquidus-solidus curve of the SiGe system," page 45 and Jäger, W.; "Ordering in SiGe alloys," p. 50, both in *Properties of Silicon Germanium and SiGe:Carbon.* Erich Kasper and Klara Lyutovich, eds., London, INSPEC, 2000. Copyright 2000 The Institution of Engineering and Technology. The equilibrium spinodal decomposition boundary in the lower figure is the solid curve, while the dashed curve shows the expected region of strong instability. The low-temperature spinodal is expected based on the parameters found for fitting the melting behavior.

surface as described in Section 6.2.3. Because of their limited mobility, they remain on the same lattice sites and the surface-induced order is preserved in the bulk, resulting in organization of atoms along (111) planes as growth proceeds.

The interaction of strain fields can also be used to drive three-dimensional ordering of Ge or SiGe alloys grown as strained layers on Si substrates. Finally, strain may drive Ge to segregate to the surface of alloys during growth such that the surface Ge concentration exceeds the bulk concentration. This is a commonly observed phenomenon during epitaxial growth of single crystals in which one constituent is significantly larger than the other. That component of the alloy is usually enriched on the growth surface.

It is important to remember that everything about phase separation in these alloys refers to non-equilibrium behaviors resulting from the nature of vapor-phase crystal growth. One may often take advantage of these non-equilibrium processes to produce materials that are not possible based on the equilibrium phase diagram.

6.4.2 Band gap engineering

$Si_{1-x}Ge_x$ alloys are primarily intended for band gap engineering. Addition of Ge to Si gradually reduces the minimum energy gap of the alloy until x~0.85. At this point the energy band minimum at X has the same energy at the minimum at L. For higher Ge fractions the L point minimum decreases rapidly to the standard value for Ge. The energy gap changes as a function of alloy composition are shown in Figure 6.12.

All of the alloy energy gaps increase sublinearly as temperature decreases below 300 K in a manner roughly similar to that of pure Si. The minimum energy gap in SiGe alloys is strongly affected by their state of strain (Figure 6.12b). Virtually all applications of SiGe alloys involve growth of the alloy on a Si substrate. Because the alloys all have larger lattice constants than Si, they will be under compressive strain when sufficiently thin. At a critical film thickness this strain will begin to be relieved, thus it is possible to grow essentially unstrained material under the right conditions. If a thick strain-free alloy layer is grown with a high Ge content, then subsequent growth of a low-Ge alloy will result in that layer being under tensile strain. The details of strain and strain relief are described in Chapter 7. For now, the point is that films under either compressive or tensile strain in the plane of the film can be produced. Based on the discussion of Section 5.4, one might presume that a film under tension would have a higher energy gap, while a film under compression would have a lower gap. However, experimental observations (c.f. Figure 6.12 (b)) show that all films under strain have lower gaps, regardless of the sign of the strain. This is because a film under compression in the plane of the film attempts to maintain a constant film volume by a tensile expansion normal to the surface (i.e.: the Poisson effect). Because measured properties are generally sensitive to the minimum gap, only the lowest value is detected. Normally, the minimum gap includes equivalent contributions from all three symmetry axis directions for the cubic semiconductors. Strain breaks the symmetry and results in both raised and lowered energy gaps and a corresponding decrease in the band edge density of states as the band edge is broadened. The strain-based reduction in energy gap is frequently used in design of devices.

Because many devices use SiGe alloys to accelerate or trap charges in either the conduction band edge or the valence band edge, it is important to know the band offsets in addition to the bandwidths. Various measurements of band offsets have been reviewed recently along with a detailed theoretical model of band edge behaviors in strained and unstrained films. [17] The results indicate a valence band offset of ~0.22 eV for Si on Ge and 0.74 eV for Ge on Si. One might expect that these values should be identical. For a truly unstrained system the band offset in one

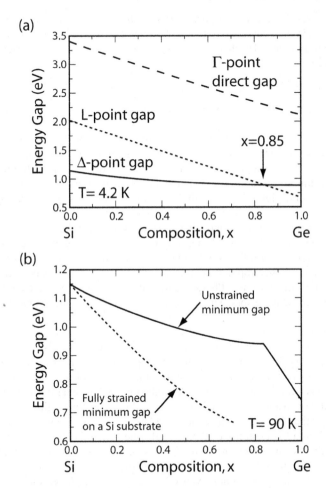

Figure 6.12: Energy gaps (a) for three important minima as a function of alloy composition, x, in SiGe alloy and (b) comparing strained and unstrained minimum energy gaps. Redrawn with permission from Penn, C.; Fromherz, T; and Bauer, G.; "Energy gaps and band structure of SiGe and their temperature dependence," in Kasper, Erich, and Lyutovich, Klara, editors, *Properties of Silicon Germanium and SiGe:Carbon.* London: INSPEC, 2000, p.125. Copyright 2000, the Institution of Engineering and Technology.

direction should be identical to the offset in the other direction. Consequently, there must be more to the experiments than meets the eye. There are many ways in which band offsets can be affected during a measurement and the results are notoriously unreliable. However, an averaged value of 0.58 eV was proposed by Van der Walle. [17] In a more recent work a fit to a range of compositions was proposed for $Si_{1-x}Ge_x$ alloys relative to Si as [18]:

$$\Delta E_V = 0.6x - 0.07x^2. \tag{6.19}$$

This formula results in a valence band offset of 0.53 eV for pure Ge on Si, which is equal to the 0.58 eV average of Van der Walle, within the reliability of the formulas. The same group concluded that the corresponding conduction band offset should be approximately

$$\Delta E_C = 0.44x - 0.09x^2. \tag{6.20}$$

Both Equations 6.19 and 6.20 apply to strain-free structures.

The strained cases are most clearly dealt with in the review by Van der Walle, [17] although many other treatments are available in the literature. The result, however, is relatively simple for strained alloy layers on Si substrates – **virtually all the band offset is accommodated in the valence band**. This turns out to be particularly advantageous for design of certain bipolar junction transistors, as we shall see later.

The reader is cautioned that band offsets have traditionally been hotly debated and are highly sensitive to individual measurements and materials preparation methods. The most reliable method for establishing the offsets has traditionally been to model the performance of heterojunction devices, which have performances that depend strongly on the offset values. Photoelectron spectroscopy is easier to use but more subject to experimental errors. The behavior that you observe may vary.

SiGe layers allow engineering of the energy gap of the resulting alloy but not the lattice constant, as was the case for pseudobinary alloys. As discussed above, the level of strain has a dramatic effect on band gap and band offsets. Therefore, it is of considerable interest to be able to adjust the strain independent of energy gap. To do this, one needs a more complex alloy. A moderate level of separate control of energy gap and lattice constant can be achieved by adding C to SiGe alloys. Because C is much smaller than Si or SiGe, it reduces the lattice constant, improving the resulting lattice match to a Si substrate. Carbon can also be added directly to Si to form Si-C alloys. In this case, however, the solubility of C in Si is very limited and heating to a sufficient temperature for a sufficient time will almost inevitably lead to SiC precipitates. One can sometimes even observe such precipitates on the surface of "pure" electronic grade Si wafers which have been heated to ~1200°C for a few minutes due to residual C in the nominally carbon-free wafers. Clearly, Si-C alloys are not in thermodynamic equilibrium. Nonetheless, they are often useful if high temperatures can be avoided. Fortunately, C solubility improves with the addition of Ge to the system to form a ternary alloy.

Because of the large size differences, bonds between either Si or Ge and C are highly strained in a SiGe matrix. The large strain results in a substantial amount of band gap bowing in $Si_{1-x}C_x$ alloys and reduces their energy gap below that of Si (by about 27 meV at 1 atomic % C), even though diamond has a much higher gap. Likewise, C reduces the energy gap in SiGe alloys, allowing desired values to be reached with less Ge than would be needed in a simple Si-Ge alloy and with compensation of the lattice size change. Furthermore, addition of C has the effect of reducing dislocation mobilities by accumulating in the dislocation cores. This inhibits strain relief by formation of misfit dislocations (see Chapter 7). By adjusting the C and Ge contents independently, it is possible to obtain alloys lattice matched to Si substrates. The energy gap of these alloys decrease by ~0.01 eV per atomic percent of Ge, which is about half the rate of decrease of pure Si-Ge alloys as described by Figure 6.10 but is still sufficient to produce a useful change in the band edges. To obtain such strain compensation, one adds approximately one C atom per eight Ge atoms in the alloy. Unfortunately, the large difference in atomic size and chemistry of C relative to Si makes C relatively insoluble and limits its addition to one or two atomic percent. The alloys are stable to 800°C. Higher temperatures provide sufficient atomic mobilities that SiC precipitates form rapidly.

6.4.3 Alloying and carrier mobility

Alloying disturbs the regularity of the crystal. This inevitably leads to scattering of moving carriers and a reduction in carrier mobility. Both electrons and holes are affected by such scattering. Results for $Si_{1-x}Ge_x$ alloys have been reviewed recently by F. Schaffler [19] and typical behaviors are shown in Figure 6.13. Both electron and hole mobilities fall well below a linear interpolation between end point compounds and both have minimum values between 30 and 50 atomic % Ge. There is no obvious loss of mobility resulting from scattering from one band minimum to the other around the transition in minimum energy gap behavior. The L minimum yields an electron mobility roughly twice that in the Δ minimum (near the X point) due to the difference in electron effective mass.

Several processes contribute to the mobility behavior in the alloy. The mobility is reduced by alloy scattering due to fluctuations in the atomic core potentials. Also, in biaxially strained thin films the cubic symmetry of the solid is eliminated by the strain field, and individual branches of the band structure are split to different energies. This reduces scattering from one band to another near the band minimum and consequently increases mobility. Alloy scattering dominates the behavior overall.

Although the carrier mobilities are reduced in thin film and bulk alloys relative to the pure semiconductors, mobilities within very thin quantum well structures can be significantly enhanced by reduction in possible scattering directions. Thus, artificially structuring thin films of alloys can recover and even improve the performance of the material.

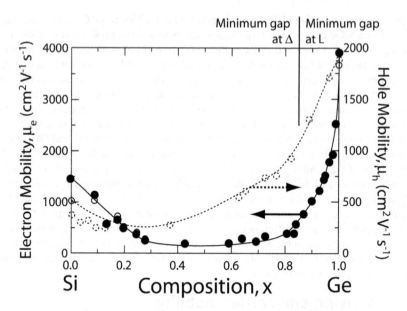

Figure 6.13: Electron and hole mobilities in $Si_{1-x}Ge_x$ alloys. Solid lines and figures refer to electron properties while dashed lines and points represent hole behaviors. Figure redrawn with permission from Van der Walle, C.B.; "SiGe heterojunctions and band offsets," in Kasper, Erich, and Lyutovich, Klara, editors, *Properties of Silicon Germanium and SiGe:Carbon.* London: INSPEC, 2000, p. 149. Copyright 2000, The Institution of Engineering Technology.

6.5 METASTABLE SEMICONDUCTOR ALLOYS

Although phase diagrams may be calculated for semiconductor alloys, many crystal growth methods result in non-equilibrium materials. Because of their strong bonds and refractory natures, semiconductors usually have low atomic diffusion rates. Consequently, if an alloy can be induced to form during growth of the material, it is generally found to be sufficiently stable to be used, if not heated too much. Such an alloy, which is thermodynamically unstable but kinetically stable, is referred to as "metastable". At room temperature these materials are often forecast based on extrapolations from high temperature data to be stable over effectively infinite times. Metastable alloys of metals, semiconductors and insulators in amorphous, poly-crystalline, and single-crystalline forms have all been grown successfully. We encountered metastable crystal structures earlier in Section 6.2.3. Now we will examine metastable alloys in more detail.

Metastable alloys have been deposited as epitaxial thin films on a variety of substrates by a variety of techniques. Deposition methods (see Chapters 10-12) have included sputtering, metal-organic vapor-phase epitaxy (MOVPE), molecular beam epitaxy (MBE), and many more. A description of these techniques may also be found in the recommended readings. A full review and list of everyone who has produced metastable alloys is not provided but some examples may be found as well in the references and recommended readings.

The general approach to obtaining a metastable alloy is to deposit the material one atomic layer at a time under conditions in which atoms land in random locations (producing an alloy) but where atomic mobility is sufficiently limited or nucleation of second phases is so suppressed that phase separation on the surface is not possible. Once the surface atoms are buried in a well-mixed state under other atoms, their mobility is greatly reduced. Hence phase separation can be avoided if it can be prevented on the surface.

A classic example of the growth of a metastable semiconductor alloy was the growth of $(GaSb)_{1-x}Ge_{2x}$ by Cadien et al.. [20] and by Shah [21] using sputtering onto (100)-oriented GaAs or glass substrates. The resulting metastable phase diagram is shown in Figure 6.14. The investigation also included study of the effect of energetic particles on the formation of the metastable alloy. This dependence is given in Figure 6.15. The results are similar to those found for other materials and are illustrative of the general aspects of metastable alloy behavior. The results in Figures 6.14 and 6.15 may be understood on the following basis. It may also be helpful to refer to Chapter 10 for a description of thin film nucleation and growth.

Because the $(GaSb)_{1-x}Ge_{2x}$ was deposited by a vapor-phase growth technique where atoms strike the surface in random distributions, the material is created as a well-mixed alloy. To initiate phase separation, separate phases nucleate. In other words, sufficiently large composition fluctuations must develop to be stable.

As there is no significant equilibrium solubility between Ge and GaSb, this means compositions completely excluding one species or the other from given regions must form. Because the two constituents of the alloy (GaSb and Ge) are very different chemically, such a separation would necessarily involve formation of grain boundaries between nearly pure GaSb and Ge. This arrangement will have a high interface energy and consequently a high nucleation barrier (see Chapter 4). Furthermore, because GaSb is a polar compound, there is a driving force to require that equal numbers of Ga and Sb atoms fall into locally segregated regions of GaSb. All together it appears difficult to nucleate the separate phases, as is, in fact, observed.

There is clearly an increased tendency to phase separate as one approaches the x=0.5 alloy as can be seen from Figure 6.14. The minimum in metastable solubility occurs because the mixture must contain the largest number of bonds between atoms that intrinsically do not like to be mixed. The behavior is similar to what one might expect for a non-ideal liquid solution as in the middle diagram of Figure 4.19.

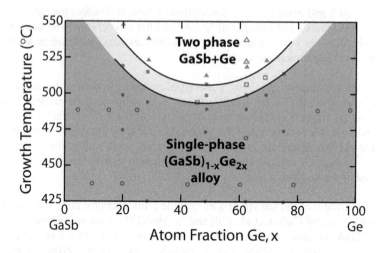

Figure 6.14: A phase diagram for the metastable $(GaSb)_{1-x}Ge_{2x}$ alloy showing the phases present as a function of deposition temperature and composition based on data in Cadien et al. [20]. The darker gray area is the single-phase alloy region. The light gray region is a mixture of alloy and two-phase regions, and at sufficiently high temperature only the two equilibrium phases are found. Open data points are for films on GaAs (100) single-crystal substrates while the closed points are for films on glass.

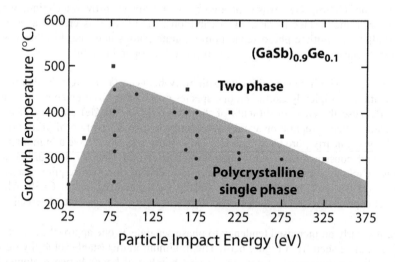

Figure 6.15: A phase diagram for the metastable $(GaSb)_{0.9}Ge_{0.1}$ alloy showing the phases present as a function of deposition temperature and fast particle bombardment energy. Figure is based on data in Shah [21]. Circles indicate single-phase metastable alloys formed while square points show two-phase samples.

Furthermore, as the alloy contains a greater concentration of the minority material (as x approaches 0.5), the distance minority atoms must diffuse to find other similar atoms decreases. At the same time, the magnitude of composition fluctuations occurring to initiate nucleation increases, thus enhancing the probability of a nucleation event. Even so, although there is a strong driving force for nucleation of phase-separated regions, the nucleation barrier is apparently sufficient to prevent formation of the separated phases even at relatively high temperatures.

Diffusion kinetic limitations must certainly also be important as evidenced by the presence of the mixed region including domains of both the metastable alloy and the two equilibrium phases. If nucleation were the only problem, then this region would not exist because the equilibrium phases provide nuclei for further phase separation. The only explanation is that even when nucleation has occurred in some locations, diffusion prevents atoms in the alloy from reaching their preferred phases in adjacent grains. Furthermore, it is clear that even under these conditions nucleation is sufficiently rare that it does not occur everywhere.

One common observation is that it is far simpler to form metastable phases under a wider range of conditions in the presence of a high flux of energetic particles bombarding the growing film surface. This is why sputtering frequently results in metastable alloys even when unintended, while evaporation and chemical vapor deposition rarely produce these materials. To understand this, let us turn to Figure 6.15 to see the effect of fast particle bombardment on the materials. As the energy of some particles striking the surface during growth increases above a few electron volts, the single-phase metastable region expands to cover a wider temperature range. This is because energetic particles striking small nuclei will tend to break them up. Ion modification of growth has been studied theoretically by a number of groups. The results suggest that even particles with 25-50 eV of energy can probably completely disrupt clusters as large as seven to ten atoms. Higher energies can damage or destroy even larger clusters. Consequently, as a nucleus forms it must become large enough not only to be energetically stable (often requiring only a few atoms), it must also remain stable following the impact of an energetic particle that can disrupt it. The higher the energy of the bombarding particles, the larger the nucleus must be to withstand the impact. Nucleation under these conditions becomes far more difficult. To produce the larger nucleus requires faster and longer-range diffusion of atoms and thus higher temperatures.

A peak occurs in the single phase region as a function of ion bombardment energy beyond which further increases in particle energies cause a decrease in the metastable temperature region. This is because the energy deposited in the surface nuclei saturates and further increases in energy go into underlying layers. This energy provides no help in preventing nucleation but can increase atomic mobilities both in the bulk and on the surface. It can also cause preferential sites on the surface for atoms to bind and nucleate new phases.

Typically in sputtering and related processes in which energetic particles are common the particles striking the surface with high kinetic energies are a few percent of all incident atoms. If more of the particles can be accelerated then the energy needed in each particle can be reduced. This is the basis of techniques such as ionized physical vapor deposition where one intentionally ionizes as many of the gas phase species as possible but in which these are accelerated only to modest energies.

Overall, the optimal energy for particle bombardment to enhance the formation of single-phase metastable alloys is just sufficient to break up nuclei and raise the critical nucleus size.

Metastable alloys produced by methods such as metal organic vapor phase epitaxy (MOVPE) that do not include fast particle bombardment of the growing surface rely on restricted atomic mobilities and low equilibrium-phase nucleation rates to avoid phase separation. For example, Shin et al.. [22] obtained a quaternary GaInAsSb alloy with a composition well within the spinodal decomposition range (see Figure 6.7) by MOVPE. In this case, the process required group-V-element-poor growth conditions and low temperatures. Under optimal conditions, single-crystal epitaxial layers were obtained in this work. Achieving epitaxy is difficult and indicates that surface atomic diffusion must have been significant – sufficient to allow the atoms to settle into desired lattice sites. It was observed that the surface roughness increased as the alloy moved farther into the spinodal region. This shows that atoms were attempting to find low-energy positions by moving laterally on the surface to more desirable locations. Apparently this was insufficient to allow phase separation. In many similar processes surface roughening is a first step to loss of epitaxy. It is likely that if growth had continued longer either the epitaxy would have been lost or phase separation would have occurred or both.

6.6 APPLICATIONS

Semiconductor alloys are essential to the design and engineering of many devices taking advantage of heterojunctions. Indeed, it is rare to be able to achieve the ideal heterojunction performance with pure semiconductors. In transistors, alloy hetero-junctions and graded alloy layers may be used to control electron injection and base transit times. In many optical devices such as laser diodes (see Chapter 3) the energy gap of the material must be tailored to produce optical emission at a specific wavelength. This is difficult in a pure semiconductor. Likewise, the dielectric properties are a function of alloy composition and can be tailored to suit the needs of a specific device. Two examples of devices employing semiconductor alloys are given here.

6.6.1 Heterojunction bipolar transistors

The two primary problems limiting the performance of bipolar transistors are (1) injection of current from the base to the emitter in addition to the desired emission from emitter to base, and (2) low minority carrier velocities and high recombination rates in the base. These problems can both be reduced by using semiconductor alloys for the base and taking advantage of heterojunction band offsets at the junctions. Heterojunction bipolar transistors (HBT's) are modified bipolar junction transistors. They have the same basic circuit design and operating principles as conventional bipolar transistors, and are optimized primarily for switching speed. Because the need is for speed, most HBT's are npn type devices as the mobility of electrons is significantly higher than for holes. Commercially, most HBT's are produced from SiGe or SiGeC alloys because of the general preference in manufacturing for Si-based processing. Further improvement in performance can be obtained using III-V semiconductor compounds and alloys, although these are much more difficult to manufacture.

To improve emitter injection performance, the base is formed from a semiconductor alloy such that the barrier for emission of carriers from emitter to base is significantly lower than from base to emitter. This means using a lower gap material for the base. SiGe alloys are just such a material. The band edge diagram of a typical HBT is shown schematically in Figure 6.16. The asymmetric emitter barrier leads to a much higher injection current from emitter to base than from base to emitter. The device shown incorporates a graded gap alloy base that also accelerates minority electrons toward the collector due to the slope of the conduction band edge.

To futher optimize the efficiency of electron injection into the base relative to hole injection base to emitter, it is advantageous to maximize the fraction of the band offset that occurs in the valence band as compared to that in the conduction band (for the common straddling-gap heterojunctions found in most heterojunctions). Strained (pseudomorphic) SiGe on Si produces such a heterojunction with a lower conduction band offset than in the unstrained material. Even better is to use strain-compensated $Si_{1-x-y}Ge_xC_y$ alloys that both reduce or eliminate strain in the base layer (and consequently reduce the chances of dislocation formation) and also transfer the band offset primarily to the valence band even without strain. Thus carbon doped SiGe alloys are used for HBT base layers in current devices.

Higher current gain resulting from the improved emitter injection efficiency is not a very large issue for most high-speed circuits. However, the higher current gain permits more heavily doped base regions because reverse current in the emitter junction increases with base doping. Therefore, the improved injection efficiency can offset effects due to stronger base doping. Higher base doping, in turn, decreases the series resistance of the base and allows a thinner base region at the maximum allowable resistance. The primary benefits of improved emitter injection efficiency are thus in formation of a thinner or lower resistance base. Both increase the device speed.

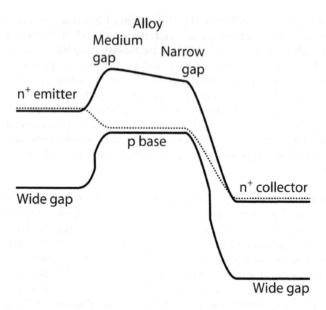

Figure 6.16: A schematic of a band-edge diagram for a HBT biased for on-state operation. The base alloy has a narrower energy gap than the surrounding emitter and collectors. In addition, when possible the alloy is graded to decrease the gap across the base, resulting in acceleration of injected electrons and reducing recombination. This may be impractical in some cases where a higher alloy may lead to defects so most current devices do not bother with base grading.

One of the primary factors controlling the switching speed of a bipolar transistor is the time it takes for carriers to transit the base. After all, it is hard to turn the transistor off until the previously injected carriers cross the base and are swept into the collector. Consequently, HBT speed is also a matter of base transit time. A particularly clever way to enhance this speed is to use a graded alloy composition in the base. Heavy p-type doping forces the valence-band edge to remain near the Fermi energy. Thus, a decrease in energy gap results in a decrease in the conduction band edge energy across the base. An electron injected into the base from the emitter experiences acceleration toward the collector, while a hole does not. The situation is shown schematically in Figure 6.16. In addition to accelerating carriers, grading the base reduces recombination and consequently decreases base current and increases transistor gain. However, because of the high quality and very narrow bases now used, most carriers transit the base rapidly and recombination is a relatively small contribution to base current by comparison with injection current into the emitter. Hence the motivation to use a graded base has been reduced in recent years. Furthermore, the graded base generally requires a greater degree of alloying, which can result in either spinodal decomposition or strain relief by dislocation formation.

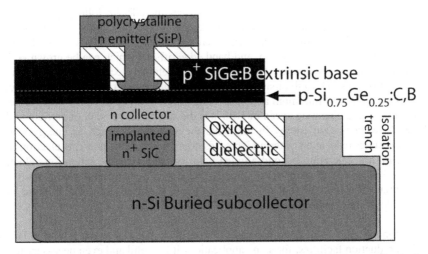

Figure 6.17: A device structure used in a recent high-performance HBT device. The device structures and performances were originally reported in Rieh [22] and Jagannathan [23].

Both are detrimental to device performances. These factors also motivate against using graded bases. Still, in some devices it can present an advantage. A significant feature of the graded base is that it cannot be formed from single semiconductors. The energy gap grading can only be achieved through alloying.

Devices similar to the structures shown in Figure 6.16 have been produced using Si-Ge alloys. A recent report by Rieh et al. [23] described a device achieving a maximum operating frequency of 350 GHz. The structure is shown schematically in Figure 6.17. It makes use of a number of geometric design features to enhance performance through minimization of capacitance and base resistance. For example, the base is divided into two sections, one thin $Si_{0.75}Ge_{0.25}$: C,B doped to ~1% B for high p-type conductivity and with C to reduce stress. A second extrinsic base region is doped to saturation with B outside of the active base area for maximum hole-concentration, reducing resistance and improving contact to the base. The emitter is polycrystalline P-doped Si. In this device the dimensions are so small and the base so thin that a graded Ge layer was not used. The buried subcollector and implanted SiC collector under the base reduce collector series resistance and improve reverse breakdown characteristics of the collector junction. The primary advances in this device relative to prior devices are due to a reduction in scale. Larger but otherwise similar devices give operating frequencies of ~100 GHz.

The above devices use Si and SiGe alloy bases. Similar performances are found in HBT's based on III-V materials and alloys. These devices are produced with arsenides, phosphides, and nitrides in various alloy combinations. The primary advantages of these over group IV based devices include the following. 1) Alloys with higher electron and/or hole mobilities can be used. The higher mobility allows higher frequency operation in otherwise identical devices. 2) Alloys exist with larger ranges of miscibility, energy gap, and band edge offsets, which can be grown lattice matched to specific substrates. This permits more detailed electrical engineering with fewer constraints due to the materials used. 3) Larger differences in energy between the conduction band minimum and other band minima can be obtained. A significant limitation to the operating speed of HBT's is the saturation velocity of carriers in the base. As carriers gain energy in the band, they eventually reach an energy where multiple states exist and scattering increases dramatically. This prevents electrons (or holes) from accelerating beyond a given velocity, known as the saturation velocity. In III-V alloys, the saturation velocity achieved is increased significantly in appropriate alloy combinations. The current record for fastest transistor is based on a III-V alloy single-heterojunction HBT by Hafez et al.. [25]

Even though III-V alloys and materials are far more difficult to work with in large-scale production facilities, for some applications it is useful and even necessary to go to the trouble to produce devices from these materials.

6.6.2 Solar cells

Solar cells (properly termed "photovoltaics") produce electric power from light. The basic process is shown schematically in Figure 6.18. In their simplest form, solar cells are simple homojunction diodes in which electron hole pairs are created by light absorption and are separated by the built-in voltage of the junction. In many respects they are similar to light-emitting diodes but where light is absorbed and current produced rather than the reverse. Indeed, many solar cells can be operated as light-emitting diodes for diagnostic purposes. For optimized operation, most of the highest-performance solar cells depend upon careful application of semiconductor alloys. Sometimes this is achieved through separate single-composition alloys in each layer of a multilayer device, while other devices use alloys graded through the thickness of the structure to control the band-edge energy of both the conduction and valence bands.

Solar cells have an optimum efficiency when the energy gap of the device is adjusted based on the spectrum of light being absorbed. The solar energy spectrum is very close to a black-body spectrum for an ~5800 K material, modified by absorption. (This is the temperature of the gas in the sun at the photosphere surface where the gas becomes effectively transparent to visible light.) The peak intensity of sunlight is in the yellow-green region of the visible spectrum, which is why our eyes detect light around this wavelength so well. For our sun a simple single-junction solar cell diode should have an energy gap of ~1.4 eV for highest efficiency. This is very close to the energy gap of GaAs, so in principle an alloy semiconductor should not be required. However, one loses all of the energy in the solar spectrum below 1.4 eV for such a device. (It is transparent to this low-energy light.) A more efficient structure uses multiple junctions, each absorbing a progressively lower-energy portion of the solar spectrum, and transmitting the remaining light to the device below. The record performance for a solar cell is 36.9% in a three-junction device using an InGaP alloy for the wide-gap absorber, GaAs for the intermediate-gap layer, and Ge for the lower-gap absorber. [27]

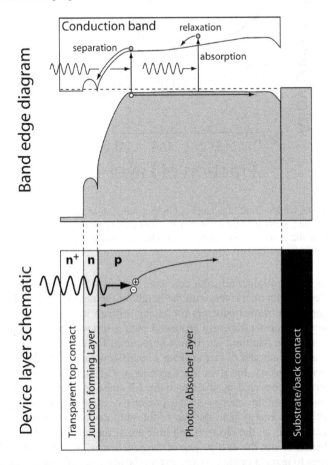

Figure 6.18: Schematic diagrams of the band-edges for a graded-gap alloy solar cell similar to a pseudobinary $CuInSe_2$-$CuGaSe_2$ alloy device. [25] This is a heterojunction cell in which the p and n-type materials are different with layers as shown in the lower portion of the diagram. Light passes through the transparent conducting top contact, through the heterojunction-forming compound, and into the absorber layer where the electron-hole pairs are produced. The wider gap at the back of the device prevents conduction band electrons from reaching the back contact (Schottky barrier) and recombining. The increased gap at the front of the structure drives holes away from the junction and can be used to tailor the inversion of carrier type at the junction (to n-type, as shown).

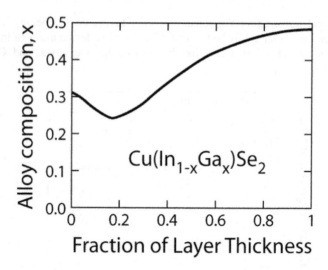

Figure 6.19: The composition of a $Cu(In_{1-x}Ga_x)Se_2$ alloy solar cell as a function of distance from the collecting heterojunction through the absorber layer. Note that both the front and back regions of the absorber layer contain Ga. The resulting presumed band-edge diagram is shown schematically in Figure 6.18.

To picture how a multijunction solar cell works, think of each junction as analogous to a hydroelectric dam on a river, with the height of the dam related to the difference in energy gap of the semiconductors producing a given junction relative to the one above it. Power is most efficiently produced with a series of moderate-height dams rather than one dam. Multiple junctions are required for the same reason – one can not produce a large amount of power from a single low dam at the bottom of a river running down a canyon – the river loses its energy as it runs down the canyon. That energy cannot be recovered by the low dam. Likewise, a single junction solar cell cannot recover all of the energy of the photons it captures. If those photons are absorbed with more energy than the gap of the semiconductor the remaining energy is lost to heat until both the electron and hole are at the respective band edges. This happens whether or not the energy gap of the junction is graded. Nonetheless, there are advantages to using a graded energy gap in the absorber of a single-junction solar cell.

To illustrate these advantages we consider the highest performance solar cells based on polycrystals, which use a graded $Cu(In_{1-x}Ga_x)Se_2$ [CIGS] alloy absorber layer. [26] The structure of this device is shown schematically in Figure 6.18 and composition of the graded alloy is given in Figure 6.19. In addition to the CIGS light absorber there is a thin CdS layer which forms the heterojunction and induces an n-type region in the surface of the CIGS, a transparent top contact [usually ZnO:In or $(In_2O_3)_{1-x}(SnO_2)_x$], and a Mo back contact. The majority of the absorber layer is p-type with $p \sim 10^{16}$ cm^3, thus the Fermi energy is within $\sim$70 meV of the valence band edge.

The alloy is designed as follows. The Ga content is increased toward the back of the layer to widen the energy gap. A wider gap produces a change in electron chemical potential that aids in collection of electrons generated in the back of the structure. This is due to the chemical gradient built into the alloy and results from the changing electronic structure of the atoms in the lattice. The electric field of the junction only penetrates $\sim$0.5 μm into the absorber CIGS. If it were not for this field, carriers generated closer to the back contact than to the depletion region would most likely be lost to the back contact. In thick devices the limiting factor may be the minority carrier diffusion length rather than back contact recombination. The graded alloy should also improve this situation as in the graded-base HBT. The presence of a weak Schottky diode at the back contact (see Figure 6.18) further enhances the need for the electron accelerating region as electrons would be trapped by the field of the barrier and forced into the metal contact. In the real devices, the need for grading is reduced by the very high absorption coefficient of CIGS, which results in most carriers being generated near the front of the device. However, in very thin structures (under $\sim$1 μm) some clear improvement due to the grading may be observed.

As noted above, the CdS heterojunction-forming layer induces a thin n-type region at the surface of the CIGS absorber layer. This causes a strong band bending and type inversion near the junction as shown in Figure 6.18. In spite of the electric field of the depletion region, holes generated near the junction may reach it and recombine through defect states. It was found that increasing the Ga content in the alloy within the depletion region is beneficial to the device. Note, however that if the Ga were to penetrate too far (beyond the depletion region) there would be a hill in the conduction band that would reduce the chances of electrons generated behind the hill reaching the depletion region and being collected. In some devices of this type the surface is treated to replace some of the Se with S. The S alloy should primarily reduce the valence band energy, leading to an improved band offset at the collecting heterojunction and stronger type conversion. Because this replacement is only made in the depletion region, no barrier to electron collection results. Unfortunately, replacement of Se with S also leads to increased CuPt or CuAu ordering which has been shown to reduce device efficiency.

Alloys are also employed in multijunction amorphous (Si,Ge,C)-based solar cells (see Chapter 8) and other devices. The most popular solar cells, however, are based on pure elemental Si crystals. These are not the highest performance devices but are very easy to manufacture and produce the largest amount of power per unit cost. As with many technologies, it is hard to beat Si.

6.7 SUMMARY POINTS

- Semiconductor alloys are designed to optimize optical and electronic properties rather than mechanical properties as in most engineering alloys.
- Unlike other engineering alloys where defects and second phases may be used to improve performance, semiconductor alloys must be single phase, free of harmful defects, and usually must be single crystals.
- Alloys include binary (two elements), pseudobinary/ternary (two compounds), or multinary (more than two elements or compounds). Most alloys are isostructural and isovalent.
- Because of the difficulty of growing bulk single crystal alloys, most alloys are produced as epitaxial single crystals on available substrates.
- Binary and pseudobinary alloys allow tailoring of the energy gap *or* lattice parameter. Ternary and higher alloys may permit independent control of these.
- Properties of semiconductors determined by relatively large volumes of the material (hundreds of atoms) generally change linearly with alloy composition. Examples include mechanical properties and lattice constant.
- Properties determined by first- and second-nearest-neighbor bond distances show considerable non-linearity. Examples include energy gap, optoelectronic and acoustic properties.
- Solubility of semiconductor alloys is determined primarily by first- and second-nearest neighbor bonds. In pseudobinary alloys the alloyed elements are second-nearest neighbors. Regular solution theory using energies of tetrahedral clusters of atoms is effective in modeling solubility and may be applied to binary, pseudobinary, and higher-order alloys.
- Alloys between materials with large chemical or size differences in the alloyed atoms have the lowest miscibilities (mutual solubilities).
- Spinodal decomposition in quaternary alloys does not result in well-defined compositions but rather in composition fluctuations to points on the boundary of the spinodal region. Which points are chosen will be determined by minimization of strain or other interface energies or by nucleation or diffusion kinetics.
- Surface phenomena related to specific crystal growth conditions may lead to surface atomic ordering in alloys. This order may be preserved in the bulk of the alloy resulting in thermodynamically metastable superstructures, most commonly the CuPt and CuAu structures.
- The same phenomena that lead to limited solubility (chemical and size differences between atoms) are responsible for band gap bowing. Thus, large bowing and limited solubility typically go together.
- Bowing results from symmetric or asymmetric changes to the effective atomic potentials that enter into the Schrödinger equation. These changes are reflected in the chemical and homopolar splittings or equivalently in the LCAO matrix elements. These, in turn, result in the observed changes in minimum energy gap.

- Semiconductor alloys generally have lower carrier mobilities due to alloy scattering (except where other scattering mechanisms dominate).
- SiGe alloys have complete solid solubility at or above room temperature.
- Lattice strain of either sign reduces observed energy gap in SiGe alloys and enhances the fraction of band offset in the valence band relative to Si.
- Adding C to SiGe reduces lattice parameter and energy gap and enhances the fraction of the band offset in the valence band relative to Si.
- Some deposition processes may permit metastable alloy formation. This is generally the result of kinetically-limited nucleation of the equilibrium phases combined with slow bulk and surface diffusion kinetics.
- Fast particle bombardment of surfaces can effectively increase the critical nucleus size for formation of equilibrium phases (or any other nucleation process for that matter) and thus can enhance formation of metastable phases.
- HBT's utilize alloy base layers to improve emitter injection efficiency. Recombination in the base and base transit time may also be reduced. Consequently, base doping may be increased which is primarily responsible for faster device performances.
- Solar cells using alloys either in single or multijunction devices may show improved performances relative to non-alloyed devices.

6.8 HOMEWORK

1. What are the primary properties of a semiconductor that we seek to improve using alloying to improve device performance?

2. Given the availability of only GaAs, InP compound semiconductor substrates of high quality, recommend semiconductor alloys to use to obtain the following results:
    i.   A direct energy gap of 1.7 eV.
    ii.   A direct energy gap of 0.9 eV.
    iii. A direct energy gap of 3.0 eV (assuming you can grow a II-VI compound on the III-V substrate of adequate quality).

3. $Hg_{1-x}Cd_xTe$ has long been used as an infrared sensing material. What are the range of minimum photon energies to which this alloy can be made sensitive (i.e.: what energy gaps can be obtained)?

4. Suppose that we have a square-planar lattice of A and B atoms distributed as follows:

A	B	B	A	B	B
B	B	A	B	B	A
A	A	B	B	A	A
A	A	B	B	A	B
B	B	A	B	B	A
B	B	A	A	B	B

    Assuming periodic boundary conditions (ie: if you go off the left side of the above lattice you come back in on the right, likewise for top and bottom edges), calculate and sketch the pair correlation function for the structure by giving the average probability of finding a B atom in a given direction and distance from the origin. Your plot should be the same size as the lattice above.

    [To calculate this set up a matrix the same size as the lattice above. Begin with any lattice point with an A atom on it and stepping horizontally and vertically across the lattice until you find a B atom. Add one to any cell in the matrix that is displaced from the origin by the same horizontal and vertical distance as you have moved from your origin. Repeat this procedure starting from all other sites containing A atoms in the lattice shown. Note that while the origin from which you search for B atoms moves, the origin from which you determine the position in the matrix does not. The probability of finding a B atom at site 1,1 where you begin should be zero because you always begin on an A atom. Divide the final number in each matrix element by 14 (the number of A atoms) and multiply by 100% to get the probability of finding a B atom on a given site. Since there are 14 A atoms and 21 B atoms the average chance should be 60%.]

5. What do the dashed lines on Figure 6.3 indicate and why do the lines have sudden changes of slope where the line converts from dashed to solid?

6. Given the equations for q_j and α_j: $q_j = x^j(1-x)^{4-j}$ and $\alpha_j = C_j^4 = 4!/(j!(4-j)!)$,
   i. Calculate and plot the q_j values as a function of x.
   ii. Show that these values result in the sum $\Sigma(q_j \alpha_j)$ over j=0 to 4 yields unity for x=0.5 using your calculated numerical values of q for part (i).
   iii. Suppose that $E_j = 2(1+0.25j)$. Calculate the total energy of the solid for N=100 as a function of x.
   iv. If you have a completely random alloy with a random distribution of alloy tetrahedra (calculated with the formulas given at the start of this question), what does the answer to part (iii) tell you about the energy of the system?
   v. Using your calculated values for the q_j above and the corresponding α_j, show that for this random alloy the entropy calculated from Equation 6.4 yields the same values as does the regular solution theory entropy: $S = -k_B N[x \ln x + (1-x) \ln(1-x)]$ from Chapter 4.

7. Consider the two pseudobinary alloys: $(Ga_{1-x}In_x)As$ and $(Al_{1-x}In_x)As$ where we will assume that the latter is restricted to the direct-gap portion of the composition range. Which of these would you expect (without looking at the tables) to have:
   i. The larger bowing coefficient?
   ii. The greater tendency to spinodal decomposition?
   Briefly explain the basis for your answers using the information provided in the text. In the case of bowing, specifically consider the implication of Equation 6.16.

8. Pseudobinary alloys exhibit large bowing coefficients in some cases, well away from the average energy gap of the starting compounds. The average energy gap of a quaternary alloy is essentially the average of the constituent binary alloys. Why is there no additional bowing in the quaternary alloy?

9. Discuss the effect of lattice strain on the energy gap of Si-Ge alloys.

10. Suppose that a metastable binary alloy is formed and that the time necessary for phase separation of half of the alloy into stable compounds follows the expression:

$$t_{1/2} = 1.6x10^{-19} \sec \quad e^{3.5eV/k_BT}$$

In other words, a process with a rate-limiting step having an activation energy of 3.5 eV.

    i.   Calculate the time required for decomposition of half of the alloy at 27°C and 627°C. (For times above 10^4 s, convert the answer to hours. For times above 10^7 s, convert the answer to years.)

    ii.   If the material must function for 10 years with less than 1% of the $t_{1/2}$ time reached, what is the maximum operating temperature for the material?

11. Calculate and plot the energy gap of the $(Ga_{1-x}In_x)As$ alloy as a function of composition of the alloy.

12. Discuss the benefit(s) of using a strained Si-Ge alloy for the base in a hetero-junction bipolar transistor rather than using an unstrained layer based on topics in this chapter.

6.9 SUGGESTED READINGS & REFERENCES

Suggested Readings:

Greene, J.E, Epitaxial crystal growth by sputter deposition: applications to semiconductors II. *CRC Critical Reviews in Solid State and Materials Sciences*, 1984; 11:189-227.

Green, Martin A.; Emery, Keith; King, David L.; Igari, Sanekazu; and Warta, Wilhelm, Solar Cell Efficiency Tables. *Prog. Photovolt: Res. Appl.* 2004; 12:55-62.

Kasper, Erich, and Lyutovich, Klara, editors, *Properties of Silicon Germanium and SiGe: Carbon*. London: INSPEC, 2000.

Kikuchi, Ryoichi, A Theory of Cooperative Phenomena. *Phys. Rev.*, 1951; 81: 988-1003.

Kroemer, Herbert, "Nobel lecture: quasielectric fields and band offsets: teaching electrons new tricks". Rev. Modern Phys., 2001; 73:783-793.

Mascarenhas, Angelo, editor, *Spontaneous Ordering in Semiconductor Alloys*. New York: Kluwer, 2002.

Van Vechten, J.A., "A Simple Man's View of the Thermochemistry of Semiconductors" in *Handbook on Semiconductors*, ed. T.S. Moss, Vol. 3, *Materials, Properties, and Preparation*, ed. S.P. Keller, North Holland, Amsterdam, 1980, Chapter 1.

References:

[1] Villars, P., Prince, A., and Okamoto, H., *Handbook of Ternary Alloy Phase Diagrams*, v. 7. Metals Park: ASM International, 1995.

[2] Vurgaftman, I., Meyer, J.R., and Ram-Mohan, L.R., "Band parameters for III-V compound semiconductors and their alloys." *J. Appl. Phys.*, 2001; 89:5815-75.

[3] Schubert, E.F., *Light Emitting Diodes*, 2nd edition. Cambridge: Cambridge University Press, 2006, Chapter 13.

[4] Wei, S-H., and Zunger, A., "Band offsets and optical bowings of chalcopyrites and Zn-based II-VI alloys." *J. Appl. Phys.*, 1995; 78: 3846-56.

[5] Faschinger, W., "The energy gap E_g of $Zn_{1-x}Mg_xS_ySe_{1-y}$ epitaxial layers as a function of composition and temperature." *Semicond. Sci. Technol.*, 1997; 12: 970-3.

[6] Ichimura, M., and Sasaki, A., "Short-range order in III-V ternary alloy semiconductors." *J. Appl. Phys.*, 1986; 60: 3850-5.

[7] Keating, P.N., "Effect of invariance requirements on the elastic strain energy of crystals with application to the diamond structure." *Phys. Rev.*, 1966; 145: 637-45.

[8] Onabe, K., "Unstable regions in III-V quaternary solid solutions composition plane calculated with strictly regular solution approximation." *Jpn. J. Appl. Phys.*, 1982; 21: L323-5.

[9] Stringfellow, G.B., in Mascarenhas, A., editor, *Spontaneous Ordering in Semiconductor Alloys*. New York: Kluwer, 2002, pp. 99-117.

[10] Van Vechten, J.A., "Quantum dielectric theory of electronegativity in covalent systems. II. Ionization potentials and interband transition energies." *Phys. Rev.*, 1969; 187:
1007-20.

[11] Van Vechten, J.A., and Bergstresser, T.K., "Electronic structures of semiconductor
alloys." *Phys. Rev. B*, 1970; 1: 3351-8.

[12] See Herzog, H.J.; "Crystal structure, lattice parameters and liquidus-solidus curve of the
SiGe system," in *Properties of Silicon Germanium and SiGe:Carbon*. Erich Kasper and
Klara Lyutovich, eds., London, INSPEC, 2000, p. 45.

[13] Jäger, W.; "Ordering in SiGe alloys," in *Properties of Silicon Germanium and
SiGe:Carbon*. Erich Kasper and Klara Lyutovich, eds., London, INSPEC, 2000, p. 50.

[14] Bublik, V.T., Gorelik, S.S., Zaitsev, A.A., and Polyakov, A.Y., "Diffuse X-ray
determination of the energy of mixing and elastic constants of Ge-Si solid solutions."
Phys Status Solidi B, 1974; 66: 427-32.

[15] Floro, J.A., Chason, E., Lee, S.R., Petersen, G.A., "Biaxial moduli of coherent $Si_{1-x}Ge_x$
films on Si(001)." *Appl. Phys. Lett.*, 1997; 71: 1694-6.

[16] Penn, C.; Fromherz, T; and Bauer, G.; "Energy gaps and band structure of SiGe and their
temperature dependence," in Kasper, Erich, and Lyutovich, Klara, editors, *Properties of
Silicon Germanium and SiGe:Carbon*. London: INSPEC, 2000, p. 125.

[17] Van der Walle, C.B.; "SiGe heterojunctions and band offsets," in Kasper, Erich, and
Lyutovich, Klara, editors, *Properties of Silicon Germanium and SiGe:Carbon*. London:
INSPEC, 2000, p. 149.

[18] Galdin, S.; Dollfus, P.; Aubry-Fortuna, V.; Hesto, P.; and Osten, H.J., "Band offset
predictions for strained group IV alloys: $Si_{1-x-y}Ge_xC_y$ on Si(001) and $Si_{1-x}Ge_x$ on $Si_{1-z}Ge_z$
(001)." *Semicond. Sci. Technol.*, 2000; 15: 565-572.

[19] Schaffler, F.; "Electron and hole mobilities in Si/SiGe heterostructures," in Kasper, Erich,
and Lyutovich, Klara, editors, *Properties of Silicon Germanium and SiGe:Carbon*.
London: INSPEC, 2000, p. 196.

[20] Cadien, K.C., Elthouky, A.H., and Greene, J.E., "Growth of single-crystal metastable
semiconducting $(GaSb)_{1-x}Ge_x$ films." *Appl. Phys. Lett.*, 1981; 38: 773-5.

[21] Shah, S.I.U., Ph.D. Thesis, *Crystal growth, atomic ordering, phase transitions in pseudobinary constituents of the metastable quaternary $(GaSb)_{1-x}(Ge_{2(1-y)}Sn_{2y})_x$*. University of
Illinois, 1986.

[22] Shin, J, Hsu, T.C., Hsu, Y., and Stringfellow, G.B., "OMVPE growth of metastable
GaAsSb and GaInAsSb alloys using TBAs and TBDMSb." *Journal of Crystal Growth*,
1997; 179:1-9.

[23] Rieh, J.S., Jagannathan, B., Chen, H., Schonenberg, K.T., Angell, D., Chinthakindi, A.,
Florkey, J., Golan, F., Greenberg, D., Jeng, S.-J., Khater, M., Pagette, F., Schnabel, C.,
Smith, P., Stricker, A., Vaed, K., Volant, R., Ahlgren, D., Freeman, G., Stein, K., and
Subbanna, S., "SiGe HBT's with cut-off frequency of 350 GHz." *IEDM*, 2002: 771-4.

[24] Jagannathan, B., Khater, M., Pagette, F., Rieh, J.-S., Angell, D., Chen, H., Florkey, J.,
Golan, F., Greenberg, D.R., Groves, R., Jeng, S.J., Johnson, J., Mengistu, E.,
Schonenberg, K.T., Schnabel, C.M., Smith, P., Stricker, A., Ahlgren, D., Freeman, G.,

Stein, K., and Subbanna, S., "Self-aligned SiGe npn transistors with 285 GHz f_{MAX} and 207 GHz f_T in a manufacturable technology." *IEEE Electron Device Letters*, 2002; 23: 258-60.

[25] Hafez, Walid; Lai, Jie-Wei; and Feng, Milton; "Vertical scaling of 0.25-μm emitter InP/InGaAs single heterojunction bipolar transistors with f_T of 452 GHz." *IEEE Electron Device Letters*, 2003; 24: 436-8.

[26] Ramanathan, K.; Contreras, M.A.; Perkins, C.L.; Asher, S.; Hasoon, F.S.; Keane, J.; Young, D.; Romero, M.; Metzger, W.; Noufi, R.; Ward, J.; Duda, A.; "Properties of 19.2% efficiency ZnO/CdS/CuInGaSe$_2$ thin-film solar cells." *Prog. in Photovoltaics: Research and Applications*, 2003; 11: 225-30.

[27] King, R.R.; Fetzer, C.M.; Colter, P.C.; Edmondson, K.M.; Ermer, J.H.; Cotal, H.L.; Hojun Yoon; Stavrides, A.P.; Kinsey, G.; Krut, D.D.; Karam, N.H., "High-efficiency space and terrestrial multijunction solar cells through bandgap control in cell structures." *Proc. 29th IEEE Photovoltaic Specialists Conference*, IEEE, 2002: 776-81.

Chapter 7

DEFECTS IN SEMICONDUCTORS

As with all other classes of materials, one of the primary keys (if not THE key) to engineering a semiconductor is control of defects in its structure. Defects can be divided into classes according to their dimensionality. Thus, zero (point), one (line), two (plane) and three (volume) dimensional defects occur in semiconductors and each is significant is considered in turn, although two and three-dimensional defects will be lumped together as they behave similarly. Furthermore, the behaviors of two and three-dimensional defects can be considered to be extensions of zero and one-dimensional behaviors. Therefore, we will spend more time on the latter two. In this chapter we will consider only defects in crystalline materials. Amorphous semiconductors, the ultimate in defective materials, are considered in the following chapter.

7.1 POINT DEFECTS

Zero dimensional defects, generally known as point defects, represent the substitution of one atom for another on a given lattice site in the structure, the absence of an atom, or atoms lying between sites in a crystalline material.

Missing atoms are known as vacancies. In an elemental semiconductor there is only one type of vacancy, although it may have multiple allowed charge states. Compounds consist of ordered arrangements of chemically different atoms. Vacancies can occur on any of these positions. Thus, in GaAs, one can have a Ga vacancy (V_{Ga}) or an As vacancy (V_{As}). More complex compounds can have more types of vacancies, one for each distinct lattice site. Vacancies always occur in crystalline solids, as we shall show below, and are therefore intrinsic to materials.

Substitutional defects are either impurities or antisites. "Impurities", as one would expect, refers to atoms which are not constituents of the semiconductor host. They are not intrinsic to the nature of a solid, but result from its incomplete purification or intentional contamination. Thus, they are referred to as extrinsic defects. Antisites, as with vacancies, are intrinsic to compounds. Antisite defects are found only in crystals with more than one sublattice and having different atoms on each. Si has two sublattices, both fcc Bravais lattices, translated by {1/4, 1/4, 1/4} with respect to each other. However, because the atoms on both are the same, moving a Si atom from one sublattice to the other has no effect. By contrast, GaAs has two sublattices, one on which only Ga atoms reside, the other contains only As. Thus, there are two antisite defects, a Ga on an As site (Ga_{As}) or an As on a Ga site (As_{Ga}). Antisite defects may, and often do, have multiple charge states in the energy gap.

Interstitials are atoms lying between lattice sites. They can be impurities or matrix constituent atoms. Thus, both intrinsic and extrinsic interstitials are found in semiconductors. The density of interstitials can be very small or very large and they can change the properties of that material entirely. A familiar example of the effect of interstitial impurities is the change in properties going from pure iron to low alloy steel. Such steels consist of carbon atoms dissolved on interstitial sites within the body-centered cubic iron lattice. The carbon changes the properties of the iron matrix enormously, increasing hardness and strength. In general there are a variety of interstitial sites in a given lattice. The face centered cubic lattice of common semiconductors has two interstitial sites, known as the tetrahedral and octahedral sites. These are illustrated schematically in Figure 7.1. Note that the diamond lattice may be viewed as being constructed by filling half of the tetrahedral interstitial sites in an fcc lattice with atoms.

Frenkel defects are vacancy-interstitial pairs formed by an atom leaving a lattice site vacant and transferring to an interstitial position. While such defects can be created by diffusion processes, they more commonly occur as a result of an energetic particle striking an atom, displacing it into an interstitial position, and leaving a vacancy behind. Creating defects through fast particle collisions with the lattice generally is not as strongly affected by the formation energy and entropy of the defect because the impacting particles typically have an energy far above the minimum necessary to create the Frenkel defect.

Point defects often result in states in the energy gap of a semiconductor and therefore can cause doping, formation of tails of states around the band edges, minority carrier traps, and favored centers for recombination of minority with majority carriers. Therefore, they represent a critical factor determining the behavior of semiconductors. Understanding the formation, charge-state, and interaction of free carriers with point defects is therefore essential to controlled engineering of semiconductors and semiconductor alloys.

7.1.1 Electronic states due to point defects

Any point defect modifies the crystal potential of the structure [i.e. U(**r**) in the Schrödinger Equation] locally and therefore perturbs the bands near it. For defects with atomic orbitals very similar to those of the matrix atoms this generally leads to a

(a)

(b)

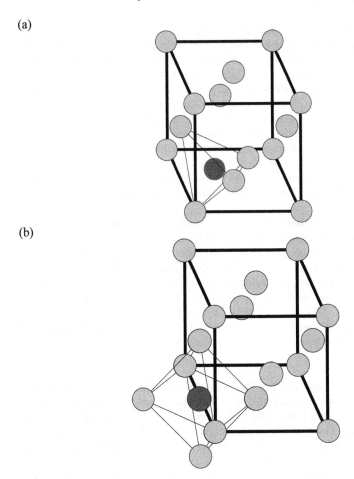

Figure 7.1: (a) Tetrahedral and (b) octahedral interstitial sites in the fcc lattice. Note that the tetrahedral interstitial site is the location of the second atoms in the diamond structure. Thus, the diamond structure is constructed from the fcc lattice by placing a second atom on half of the tetrahedral interstitial sites. The octahedral interstitial sites are the edge and body centers of the cube.

small perturbation of one of the states already present, producing a level close to the band edge. This is the case for all normal dopant atoms. An impurity atom with a larger difference in chemistry usually leads to a major change in bonding or to an essentially non-bonding state. Such states are often far from any band edge and cause major problems in the performance of devices. In analogy, a rat among mice might behave in very similar ways and essentially looks like a large mouse. An iguana among mice looks and behaves in a completely different manner and could never be mistaken for a large mouse.

It is common to divide impurity states according to the energy difference between that state and the appropriate band edge. This energy is measured from the conduction band for a donor and from the valence band for an acceptor because in an otherwise pure material these are the bands to which or from which electrons go/come to charge the defect. When the energy difference is on the order of a few thermal energy units ($k_BT = 25.8$ meV at 300 K), the state is generally referred to as a "shallow" level. When the energy difference is much larger than k_BT, then the state is called a "deep" level. Under this conventional temperature-based definition, levels can change from deep to shallow as the temperature changes; thus, the boundary between the two is not clearly defined. A distinction in the fundamental physics of the two classes of states, deep and shallow, is needed, which the temperature-based description does not provide. Consequently, different definitions are used here:

Shallow levels are states derived from the *matrix* of the semiconductor and modified slightly by the impurity atom. These levels are distributed over a large volume of real space encompassing many matrix atoms in addition to the impurity. Because the shallow state is distributed over much of the surrounding matrix, it must be matrix-like (i.e.: band-like). This is important because when such a state occurs in a semiconductor alloy, as the matrix is changed the state changes too, following the band edge as it moves. Thus, it remains shallow in all alloys of the semiconductor. These states are delocalized in real space and consequently are localized in reciprocal space near the band minimum/maximum. Shallow levels are discussed in detail in the next section.

Deep levels are states derived from the point defect atomic orbitals, regardless of the energy difference between the state and the appropriate band edge. An impurity atom that does not participate in the normal bonding of the solid will have an impurity-like state. For example, a transition or rare-earth element having d or f shell valence orbitals will not naturally hybridize to form a tetrahedral geometry as needed by the lattice. *The state being defect-like rather than matrix-like means that in an alloy the state remains at roughly the same energy while alloying moves the band edge toward or away from it.* Deep defects are generally strongly localized in real space around the impurity atom and consequently are distributed in reciprocal space. The distribution in reciprocal space is important in that it allows a deep level to interact with carriers of both types across a wide range of electron momenta. Furthermore, the chance of an electron interacting with a state in the energy gap increases as the state moves away from the band edge. Thus, unfortunately, deep levels are particularly effective at interacting with free carriers.

Deep levels can be divided into states that primarily trap minority carriers and release them at a later time, and those that primarily mediate recombination of minority and majority carriers. These states are, consequently, referred to as traps and recombination centers. The two behaviors are illustrated schematically in Figure 7.2. Although traps and recombination centers can each act as the other under some circumstances, there are general "rules of thumb" that help to distinguish them. These are listed in Table 7.1. An initially doubly-charged point defect will be strongly repulsive to majority carriers, as they carry the same charge, and strongly

Table 7.1: **Ideal Properties for Traps and Recombination Centers**

Defect Property	Traps	Recombination Centers
Charge state before capture of a minority carrier:	Doubly charged, opposite to minority carriers	Singly charged, opposite to minority carriers
Cross section for minority carriers before capture:	10^{-9} to 10^{-12} cm^2 (strongly attractive)	10^{-12} to 10^{-15} cm^2 (moderately attractive)
Cross section for majority carriers after capture:	10^{-18}-10^{-20} cm^2 (moderately repulsive)	10^{-15} to 10^{-17} cm^2 (neutral)
Depth (energy from minority carrier band edge):	Relatively small	Relatively large

attractive to minority carriers. After capture of the minority carrier, because the sign of the defect charge remains the same as for the majority carriers, the trap remains relatively repulsive to them. If it is close enough to the minority-carrier band edge, the trap typically releases the minority carrier back into the band before it captures a majority carrier. The opposite occurs for a recombination center where the energy to reach the band edge and release the minority carrier is large enough that the state captures a majority carrier before releasing the minority carrier. This is assisted if the state was initially singly charged, in which case the majority carrier is not repelled by the defect after capture of a minority carrier.

Traps and recombination centers can have dramatic effects on the performance of devices. Normally these defects are a problem. In a defect free diode or bipolar junction transistor the speed of the device can be limited by the time to remove charge from the base. Because recombination centers reduce this time, they can, in theory, make a transistor switch faster. However, the other losses of performance due to recombination generally outweigh the gain in recombination speed. There are also other ways of accelerating devices that do not involve degrading them (see Chapter 6). Traps slow device switching and can lead to transient capacitance effects because they collect charge and release it slowly when the applied voltage changes. Recombination centers cause lower gain in transistors, reduced power production in solar cells, lower output in light emitting devices, and more reverse leakage current in junctions of all types. Thus, both traps and recombination centers are almost universally problems to be avoided. These defects are one of the primary reasons why semiconductors need to be purified to such a great extent.

Before moving on, let us consider briefly the cross sections for capture of a charged carrier by one of the defects listed in Table 7.1. The covalent radius of an atom is of the order of 0.1-0.2 nm. Thus, the area which an atom occupies is roughly this radius

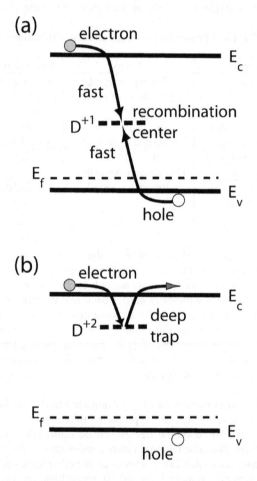

Figure 7.2: Schematic diagrams of (a) recombination centers and (b) trap states relative to their band edges. Figures are sketched for a typical situation of a p-type material.

squared and multiplied by π, or 3×10^{-16} cm^2 to $\sim10^{-15}$ cm^2. The chances of an electron interacting with a defect for which it has neither a strong affinity nor a strong repulsion will be roughly the chance that it passes through the area which that defect occupies multiplied by the probability that the carrier can be captured by that defect with proper conservation of energy and momentum. The interaction and conservation chances are high, so the probability for interaction is roughly the cross-sectional area of the atom. Consistent with this picture, the interaction probability is termed a cross-section and is in area units. Cross-sections which are larger than $\sim10^{-15}$ cm^2 suggest that the carrier is interacting attractively with the defect. The atom appears larger because it tends to reach out and pull the carrier in. The larger the area over which the atom can pull the carrier in, the larger its cross-section will be. Conversely, when the charge on the carrier is the same as that on the defect, an interaction effectively requires the carrier to hit the defect very close to straight on. Slightly off-center trajectories lead to the carrier being deflected rather than being captured. The net result is that defects with the same charge compared to the carrier have smaller cross-sections than the size of the atom. A quantitative model of electrostatic charge scattering for oppositely charged carriers and defects suggests that the effective radius of the defect is (c.f.: Bube) $r_d = zq^2/4\varepsilon k_B T$, where z is the defect charge. This produces a cross section of $\sim7\times10^{-12}$ cm^2 for a singly charged defect in Si at room temperature, and is much larger than an atom, as might be expected. A doubly charged defect cross section in this model would be four times greater, although real defects tend to be larger than this.

A more extensive discussion of the behavior of traps and recombination centers can be easily constructed by writing rate equations for trapping of the various carriers in various charge states of the defect and their subsequent release. Such a detailed description, along with all of the associated mathematics, may be found in the recommended readings, for example see Bube or Hess. The discussion is not reproduced here, as such a level of detail is not necessary to most readers.

7.1.2 Shallow levels

Shallow level defects can be understood, at least approximately, based on the "hydrogenic" model. This approximates the defect as a H atom in the dielectric medium of the semiconductor. An impurity that is chemically similar to the matrix atom it replaces but has one more or one fewer electron and proton (a "monovalent" substitution) often results in this type of state. Examples of such impurities are P (considered in detail below) or Al in Si. The hydrogenic model can be applied to any monovalent substitutional point defect in either an elemental or a compound semiconductor and to both n and p type dopants.

A P atom can be viewed as a Si atom plus a H atom – one extra electron and proton. If one ignores the proton for the moment, an electron added to a Si solid would necessarily wind up in the conduction band, as the valence band is full. Of course, one cannot ignore the extra proton, to which the electron will be attracted. Thus, the electron is not completely free to move about the solid, as it would normally be in

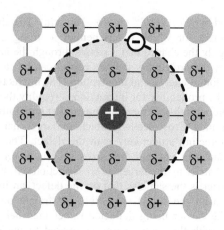

Figure 7.3: A schematic of the dielectric response of a semiconductor to a charged impurity atom. The illustration is for a donor state. An acceptor is similar but with opposite charges. The δ^+ and δ^- indicate positive and negative charges induced in the silicon by the presence of the proton and electron of the impurity.

the conduction band, but is bound to the proton by electrostatic energy. If the binding were as strong as in a simple H atom, this energy would be 13.6 eV. In a dielectric medium with the dielectric properties of the matrix (e.g.: Si), as the electron begins to leave the proton, it notices the presence of the positive charge left behind far less than for a H atom in free-space. The remaining electrons in the solid screen the positive charge from the view of the added electron by moving in toward the proton. This screening reduces the attraction of the electron for the proton dramatically, as shown schematically in Figure 7.3, leaving it only loosely connected. Because the resulting attraction is weak, the conduction band state, which held the electron is only slightly reduced in energy – it remains a nearly normal conduction band state. The energy difference between the lowered-energy defect level and the rest of the band edge is the binding energy of the electron in the hydrogenic state.

A complete analysis of the H atom in a dielectric medium with a relative dielectric constant ε_r and effective mass m^*/m, shows that the atom would have a larger atomic radius, r, and smaller ionization energy E than for H as:

$$r = \frac{m}{m^*}\varepsilon_r a_0,$$

7.1

and

$$E = \frac{m^*}{m}\frac{13.6 eV}{\varepsilon_r^2}$$

7.2

where a_0 is the normal atomic radius for H (0.037 nm) and 13.6 eV is its normal ionzation energy. Given the relative dielectric constant for Si: $\varepsilon_r=11.9$, and effective mass ($m^*=1.1m$), one finds that a monovalent hydrogenic state in Si should have a radius of ~12 times that of H, and an ionization energy of ~32 meV. A similar analysis can be carried out for holes, resulting in an acceptor hydrogenic state with an ionization energy of 53 meV. These energies are very close to the energies obtained for monovalent (groups III and V) substitutional elements close in size to Si, suggesting that the hydrogenic state is a useful model. Note that as the impurity becomes chemically very different from the element for which it substitutes, the observed behavior begins to deviate strongly from the hydrogenic value. This is because the atomic potential U differs significantly from that of the host plus or minus one proton. In addition, lattice strain effects may be significant.

The same analysis can be conducted for divalent substititional elements (groups II and VI in Si), equivalent to placing a He atom in a dielectric medium. The results are the same as in Equations 7.1 and 7.2 except that the ionization energy is four times larger for the doubly-charged state (scales as the charge squared) and the doubly-charged radius is half that for the monovalent substitution. The observed behavior for divalent substitutional elements does not fit experiments as well as for monovalent substitutions because interactions of the two electrons with each other and with the matrix are not accounted for. Furthermore, the chemical nature of the divalent substitutional element is significantly different from the host element and the smaller radius suggests that the ability of the matrix to screen the nuclear charge is less. Consequently, the assumption that the atomic potential, U is approximately that of the matrix plus two additional screened charges is no longer valid and the screening dielectric is no longer fully effective.

The physical size of a defect may also affect its ionization energy. A plot of the depth of most of the levels due to monovalent substitutional elements in Si as a function of the atom size is shown in Figure 7.4. One can see immediately that donor atoms, even though some are much larger than the Si matrix are almost hydrogenic in their ionization energy. However, acceptor defects become rapidly higher in binding energy compared to the hydrogenic state. The electronic structure of the two groups of atoms is nearly the same for atoms in the same row of the periodic table. Thus, perturbations due to the altered core states of the atoms are unlikely culprits in the increasing depth of the acceptors. Two possible reasons could explain the difference. First, the electronic structure of the acceptors acting on the valence band could have a much greater effect than the donor atoms acting on the conduction band. This is possible but seems unlikely in a simple semiconductor such as Si with no chemical splitting. A second possibility is that lattice strain is having an effect. The donor, giving up an electron, presumably would become smaller and thus should fit the lattice better. By contrast, an oversized acceptor atom would become even larger by taking on an electron. Presumably this is unfavorable and might require an

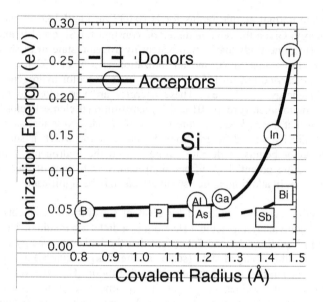

Figure 7.4: Dopant ionization energies as a function of dopant covalent radius relative to Si. Note that the acceptors show much greater variation with size than do donors.

energy penalty to put the extra electron on the impurity atom. The magnitude of the size change may also be related to the amount of localization of the state. Presumably a smaller diameter more localized state (less screening) would show a greater size change. This would amplify the strain energy effect in larger atoms and greatly increase the ionization energy of the larger acceptors. An analysis of the strain energy of hydrogenic defects in Si suggests that the size-effect provides a good explanation of the behavior of donors as compared to acceptors. [1]

Strain is also well known to govern the incorporation of dopant atoms into solids. **An oversized atom tends to be rejected to the surface of a crystal as it grows,** (it "surface segregates") because the strain energy near the surface is far lower than the strain energy when the atom is inside the bulk crystal. This means that it is very hard by many deposition techniques to insert large atoms such as Sb into smaller matrices such as Si. This, in turn, causes a much higher concentration of dopant on the surface than in the bulk of the material. Sometimes the surface concentration can be over a million times the bulk concentration. This is highly useful for purifying a semiconductor, but a nuisance when trying to dope it. Consequently, most common dopant atoms are nearly the same size, or smaller, than the host semiconductor. *A dopant that segregates to the surface can significantly lower the surface energy.* Such elements act as surfactants and can improve the smoothness of epitaxial layers.

7.1.3 Depth of intrinsic defects

An examination of the experimentally determined depth of various impurity states in semiconductors shows that the more covalent a semiconductor (the smaller the chemical splitting) the higher the defect formation energy and the farther states associated with that defect are from the appropriate band edge. This can be understood based on the approach outlined in Section 5.1.

In a purely covalent elemental semiconductor such as Si, a vacancy represents the lack of atoms on a given lattice site with which other atoms can bond. Consequently, the hybrid molecular orbitals (e.g.: sp^3) do not have anything to bond with and might be expected to end up in the energy gap at roughly the molecular orbital energy, as shown in Figure 7.5. Such a state could be exactly in the center of the energy gap and could have either a positive or negative charge. However, the atoms surrounding the vacancy can reorganize their dangling bonds somewhat (see Chapters 8 and 10) so vacancies are not perfect non-bonding molecular orbitals.

In Si, the vacancies have three possible charge states, $+1$, -1, and -2, in the energy gap. The cost of placing an electron in the vacant site is roughly half of the energy gap, as would be expected for a dangling molecular orbital. However, the price of removing an electron to produce a positive charge on the vacancy is significantly higher (nearly the gap energy), as is the price of adding a second electron to an already negative vacancy. This is because an electron can be added to the vacancy without severely disrupting the reorganization of the dangling bonds, while removing an electron undoes some of this reorganization, and therefore costs more energy. It is unlikely that ionizing a non-bonding state will ever cost much less than the energy of the parent molecular orbitals relative to the surrounding bands. Similar arguments can be made for an interstitial, which does not participate in bonding. Thus for an elemental semiconductor the intrinsic states are all expected to be deep, and are generally found to be so.

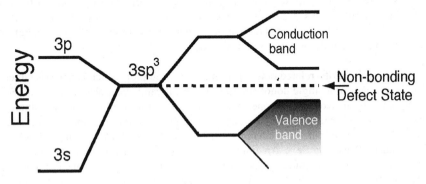

Figure 7.5: Shows how a non-bonding molecular hybrid orbital resulting from a vacancy point defect causes deep levels near midgap in Si.

A compound semiconductor has molecular orbitals with energies separated by the chemical splitting. Consequently, these orbitals are, in the absence of reorganizations, closer to the band edges than for an elemental semiconductor. Thus, the conduction band is primarily composed of the molecular orbitals closer to it (from the cation) while the valence band is composed mostly of anion orbitals. The cation vacancy is an acceptor, while the anion vacancy is a donor. The dangling non-bonding orbitals are the opposite type – an anion vacancy has cation dangling bonds and vice versa. The acceptor states of the cation vacancy, being anion dangling bonds, are closer to the valence band than for an elemental covalent semiconductor, lowering the ionization energy of the acceptor defect in the more ionic material. In general, vacancies tend to be shallower as the chemical splitting increases. Reorganizations of the dangling bonds enter into the picture and can make the associated defects deeper than one might expect based on the simple non-bonding state arguments. However, reorganization does not overcome the basic trend. Furthermore, as the chemical splitting increases, the opportunity for energy gain by bond reorganization decreases. Similar arguments for interstitials do not work well but antisite defects follow the same trends in general.

Repeating: the greater the chemical splitting in a compound semiconductor the shallower the defects are likely to lie. Elemental Group IV semiconductor vacancies are close to the center of the band gap. III-V semiconductors have shallower vacancy and antisite defect states. II-VI semiconductors have even shallower defects as do other compound semiconductors with high ionic component to their bonding or large chemical splittings. This is also a significant part of the reason why semiconductors with higher chemical splittings are less sensitive to the presence of dislocations and grain boundaries.

7.1.4 Ionization of defects

A point of common confusion with point defects is their charge as the Fermi level moves in the energy gap and especially where the charges go to or come from in order to charge or discharge the defect. Furthermore, the way in which defect ionization energies are indicated on band-edge diagrams is often confusing. Therefore, a pause to consider these points seems in order.

Consider an impurity-related defect [Sn in Si, for example], which can act as either a single acceptor or a single donor (in other words, it can have a charge state Sn^+, Sn^0, or Sn^-). Suppose also that a second impurity [Se in Si, for example] can be a single or double donor (charge states Se^0, Se^+, or Se^{+2}). Suppose further that these defect states have the following ionization energies: 0.27 eV for Sn^-, 0.25 eV for Sn^+, 0.25 eV for Se^+, and 0.40 eV for Se^{+2}. An immediate question is, why is there no ionization energy for the neutral states? This involves two concepts. First, ionization energies are always the energy cost for converting a species from one charge state to another. For +1 charge states this means the energy cost to remove an electron from a neutral atom, for −1 charge it is the cost to add an electron to the neutral state, for −2 charge it is the cost to add a second electron to a defect that is already in the -1 charge state, etc. Therefore, the charge zero state is the reference state for the first ionization step. Second, solids are charge neutral under normal circumstances and atoms are always introduced into a solid without a charge or with a compensating charged species. The $+1$ charge on an atom therefore means that an electron has gone somewhere else within the solid. Usually this means, for example, to the conduction band but it can also mean to a compensating defect, as we will see below. The neutral state is the natural condition for an atom. As a consequence, there is no energy needed to create a neutral defect once the impurity atom is introduced into the material.

The defect ionization energies are marked with respect to the band to which or from which a carrier must come if the defect exists in an otherwise pure material. As long as clusters of defects do not form, this energy remains valid if one introduces a variety of impurities at once. Thus, when one says that the ionization energy of the process $Se^0 \rightarrow Se^+$ is 0.25 eV, this must be with respect to the conduction band because that is where the electron must have gone when it left a neutral Se atom to leave behind the Se^+ ion. Likewise, the Sn^- ion must have accepted its electron from the valence band. Hence the 0.27 eV ionization energy is measured with respect to the valence band. In summary, all negatively charged defects have their ionization energy determined with respect to the valence band edge, while all positively charged defects have their energy measured relative to the conduction band edge. We typically plot these states in the energy gap according to their ionization energies relative to the appropriate band edges. The defects described above would be marked as shown in Figure 7.6.

The next question that arises is what is the charge state of these defects for a given Fermi level. This is easily determined, as occupation is set as usual by the Fermi

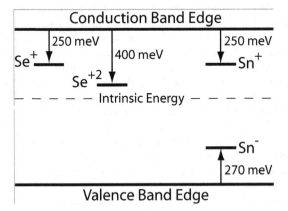

Figure 7.6: A schematic showing the location of example impurity states in the energy gap of Si. In this case the states for Se and Sn are represented with the depths of each level marked relative to the appropriate band edge. A positive state may lie in the lower portion of the gap (below the intrinsic level if its ionization energy is sufficiently high).

function at the indicated defect level relative to the Fermi level at a given temperature. For the situation in Figure 7.6, if the material is strongly n-type (Fermi level near the conduction band edge) at T=0 K, the Se atoms are neutral (both +1 and +2 charge states lie below the Fermi energy and are thus filled with electrons and the Sn atoms are in the −1 charge state. These conditions are achieved by the n-type dopant atoms (not shown in Figure 7.6) donating electrons to the Sn impurity atoms. Likewise, for a strongly p-type material the Se atoms are in the +2 charge state and the Sn is in the +1 charge state. Clearly the number of Se and Sn atoms must be significantly below the number of dopants for this to be the case. To figure the position of the Fermi energy in a less heavily-doped material it is necessary to include all impurity states in the calculation. However, as long as one dopant/ impurity dominates all others sufficiently, it is likely to dominate in determining the Fermi energy. A more complex possibility is if Si is doped with a relatively large amount of Se and a modest amount of Sn and no other dopants. In this case the Sn could donate to the Se^{+2} making it Se^+ but it could also accept an electron from the Se to become Sn^-. The energy gained in the latter is so much greater than the gain in the former that the Se donates to the Sn. The resulting charge states would be primarily Se^+ or Se^0 depending upon the Sn concentration and Sn^-. If there is a modest amount of Sn then more of the Se atoms will remain neutral than will become ionized since ionization is endothermic for any Se not donating to the Sn atoms.

7.1.5 Point defect densities

Vacancies are important to the behavior of many semiconductors and, along with antisites, are typically the primary source of at least one type of doping in relatively ionic semiconductors. Indeed, vacancies or antisite defects can be so prevalent that it is nearly impossible to dope some semiconductors one or the other type. In these materials, for every dopant atom added, an intrinsic point defect of the opposite type is also added, causing complete compensation ("autocompensation") and no net change in conductivity. To understand this, we will now examine what determines the density of point defects in a solid. The following discussion refers primarily to vacancies but can relate equally well to any intrinsic point defect. Many aspects of this discussion are derived from an analysis of the diffusivity of dopants in Si, although the results are quite general. For a detailed development of the models and description of how the results relate to experiments, see, for example, F.F.Y. Wang (in the suggested readings) and in particular Chapter 7 by R.B. Fair [2].

In the following discussion we will begin with a conceptually easy argument for vacancy formation that does not work in some applications and then consider a less obvious approach that gives a more accurate picture of defect formation.

The concentration of vacancies in a solid (including in semiconductors) is determined by minimization of the Gibbs free energy change due to addition of the defect, ΔG_{Vac}. As in earlier chapters, to find ΔG_{Vac} we need the energy of formation of the defect, ΔE_{Vac} per defect and the corresponding change in entropy, ΔS_{Vac} at constant volume. The energy of formation of an uncharged vacancy depends upon the bonding structure and bond energies of the solid, the presence of compensating charges and the presence of impurities. In an elemental undoped semiconductor, ΔE_{Vac} is roughly the heat of sublimation. For compound semiconductors, the heat of sublimation varies for the different constituent species but also gives an indication of ΔE_{Vac} for each constituent. Thus, the formation enthalpy of the defects on different sublattices is generally very different for compound materials.

The entropy of formation of a vacancy is primarily the configurational entropy with a small contribution from changes in the vibrational entropy. The configurational entropy for a small (n<<N) number of vacancies, n, in the lattice of N sites is:

$$S = k_B \ln \Omega = k_B \ln \left[\frac{N!}{(N-n)!n!} \right] \approx k_B [N \ln N - (N-n)\ln(N-n) - n\ln n]. \quad 7.3$$

The third form of the equation was obtained via Stirling's approximation, as usual. The equilibrium condition is:

$$\frac{\partial \Delta G_{Vac}}{\partial n} = 0, \text{ or} \qquad\qquad 7.4$$

$$\frac{\partial \Delta G_{Vac}}{\partial n} = \Delta E_{Vac} - T\Delta S_{vib} - T\frac{\partial S}{\partial n} = \Delta E_{Vac} - T\Delta S_{vib} - k_B T \ln \frac{N-n}{n}. \quad 7.5$$

Setting Equation 7.5 equal to zero and solving for n,

$$[V_0] = n = N \left(e^{\Delta S_{vib}/k_B} \right) e^{-\Delta E_{Vac}/k_B T}, \qquad\qquad 7.6$$

where $[V_0]$ is the concentration of neutral vacancies. The term in parenthesis in Equation 7.6 is a correction due to vibrational entropy and is generally of the order of unity. For typical semiconductors $N \sim 5 \times 10^{22}$ cm^{-3}.

Equation 7.6 shows that the concentration of neutral vacancies increases exponentially with increasing temperature. For typical vacancy formation energies the concentration of the defects is very small, even at high temperatures. This is why the diffusivities of atoms in Si and other highly covalent semiconductors are low even at very high fractions of their melting points.

The situation changes dramatically when the semiconductor is doped. Equation 7.6 is the behavior for an uncharged vacancy. As long as the vacancy formation energy does not change with doping, this equation holds. However, as doping changes the Fermi energy, some of the vacancies become charged. This does not, however, necessarily invalidate Equation 7.6. Indeed, experimental data shows that the concentration of vacancies increases rapidly with doping in many cases. If the neutral vacancy population were to remain fixed and if the charged defect concentration is related to the neutral concentration by a Fermi function, as one would expect, one can write the number of charged vacancies with ionization energy $E_{\pm1}$ of a given charge state ±1, $[V_{\pm1}]$ relative to $[V_0]$ as:

$$\frac{[V_{\pm1}]}{[V_0]} = e^{\pm(E_f - E_{\pm1})/k_B T}. \qquad\qquad 7.7$$

A more highly-charged defect such as +2 would be determined by the same formula but the ratio would be with respect to the concentration of the less charged (e.g. +1) defect rather than the neutral vacancy. The total number of vacancies in the material is then the sum over all charge states.

To illustrate the effect of Equations 7.6 and 7.7 on the vacancy concentration, let us consider the case of undoped and doped Si. Here vacancies can occur in three charge states, +1, −1, and −2, with energies 1.0 eV below the conduction band, and 0.6 and 1.05 eV above the valence band at 300 K, respectively. The overall energy gap of Si is 1.12 eV. The energy of formation of a neutral vacancy is 2.4 eV, the vibrational entropy change per added vacancy is 1.1 k_B, and the site density in the Si lattice is 5×10^{22} cm^{-3}.[3]

First, let us consider the concentration of neutral vacancies. At 650°C (923 K), Equation 7.6. gives:

$$[V_0] = 5x10^{22}\, cm^{-3} \cdot e^{1.1} \cdot e^{-2.4/(8.617x10^{-5} \cdot 923)} = 1.2x10^{10}\, cm^{-3}. \qquad 7.8$$

Thus, even at 650°C, the equilibrium neutral vacancy population is only a fraction of a part per trillion of sites in the material. At 1420°C (the Si melting point) the equilibrium vacancy population is only 10^{16} cm^{-3} or roughly 200 ppb.

We should pause for a moment here to consider whether Equation 7.8 truly applies to a piece of Si. In other words, is the solid in equilibrium? Formation of vacancies can be thought of as diffusion from an infinite source (the vapor phase) into the solid. Rapid diffusion is needed to bring the concentration to equilibrium throughout the solid. In silicon, temperatures over ~900°C are necessary to provide enough diffusion for vacancies to move through a normal wafer thicknesses in a reasonable time. As the temperature is lowered even moderately slowly the vacancy population "freezes" (it is no longer able to adjust continuously during cooling) and is therefore no longer characteristic of the lower temperature as cooling continues. The vacancy population in a crystal grown from a melt (such as a Si wafer) will thus be characteristic of some temperature closer to the melting temperature than to room temperature even after cooling and will vary depending upon the distance from a vacancy source or sink. Even so the concentration of neutral vacancies will be low on a percentage basis. Repeating: *calculations of equilibrium vacancy concentrations here assume vacancies can move extremely rapidly. This is not generally the case and defect populations may not be in equilibrium throughout the material.*

With the caveat that equilibrium may not be achieved, it is still useful to know what the equilibrium situation is, as the system always moves in this direction. Therefore, let us return to the concentration of charged vacancies in doped Si that would be in equilibrium with the neutral vacancies at a given temperature. Suppose that the Fermi energy is at the center of the ~1.12 eV energy gap or at ~0.6 eV from the valence band edge. For the negative one charge state the denominator in Equation 7.7 is two (at any temperature), and so half of the total number of vacancies will have this charge. The two remaining states (–2 and +1) lie 0.45 and 0.4 eV from the Fermi energy. The denominators of Equation 7.7 for these two values are 3.6×10^7 and 5.3×10^6 at 300 K, respectively. Less than one part per million of the total vacancy population will be in these charge states at room temperature. Overall, in undoped Si (at any temperature), there should be twice as many total vacancies as neutral vacancies and the number of negative and neutral vacancies will be roughly equal.

Now, let us consider the number of defects in heavily n-type Si where the Fermi energy is, say, 0.05 eV below the conduction band edge (the same position as the double negatively-charged vacancy). Not surprisingly, there will be very few positively-charged vacancies in n-type Si. However, the concentration of single negatively-charged vacancies will be $e^{0.45/k_BT}$ or 36 million times the number of neutral vacancies at 300 K. The double negatively charged vacancies will be equally common, as $E_{-2}-E_f = 0$.

The number of charged defects in Si in equilibrium with one neutral vacancy at 300 K is shown in Figure 7.7. Clearly, whenever the Fermi level is very high or low in the energy gap, virtually all of the vacancies in the material are charged. More importantly, the total number of vacancies must be much larger than in the undoped material. Such a behavior, while not necessarily intuitive, is observed in many strongly semiconductors.

There is a problem with Equations 7.6 and 7.7 in one critical point. The Fermi energy in an uncompensated material depends logarithmically on the number of electrically active doping impurities added, while Equation 7.7 implies that the number of vacancies or other defects should rise exponentially with Fermi level. This should mean that *for every impurity added a vacancy would also be added*. Because the charge of the vacancy is the same as that of the free carrier the dopant generates, the vacancy should exactly compensate the impurity and *no net doping would result*. Clearly this cannot be the case in Si or GaAs where very large doping levels can be achieved. Dopability suggests that the proportionality between the number of vacancies and the number of impurities must be small over a wide range of doping levels in these materials. However, in ionic compounds autocompensation is very much the case and adding impurities often results in no doping. Vacancy compensation is most obvious in II-VI semiconductors, in which only one type of extrinsic doping usually works. To understand this it is better to think in terms of a

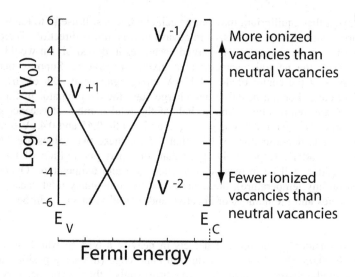

Figure 7.7: Shows the relative concentration of vacancies in silicon at equilibrium for various Fermi level positions at 300 K. Redrawn with permission from Fair, Richard B, "Concentration profiles of diffused dopants in Silicon," in Wang, F.F.Y., ed., *Impurity Doping Processes in Silicon* (North Holland, Amsterdam, 1981), chapter 7. Copyright Elsevier, 1981.

change in vacancy formation energy with doping level rather than simply insisting that Equation 7.6 must hold for all doping levels based on a single formation energy.

The contradiction between Equation 7.7 and the observation that most semi-conductors can be doped requires a better explanation of the behavior. The better approach is as follows. A direct connection between Fermi energy and vacancy formation energy has been established based on density functional methods by Baraff and Schlüter [4] and by experiments. [5] These studies led to the proposal of pinning levels (the Fermi energy is forced to remain at a given energy) in semiconductors by S.B. Zhang et al., [6] which provides a relatively simple picture for the resulting behaviors. The formation energy ΔE_d for a given defect is found to be a function of the change in Fermi energy ΔE_f (from an intrinsic reference energy) and its charge state, q: [4]

$$\Delta E_d(q, E_f) = \Delta E_0(q) - q\Delta E_f, \qquad 7.9$$

as shown by Baraff and Schlüter for GaAs, where ΔE_0 is the energy to form an uncharged defect. Note that the number of defects in a solid according to Equation 7.6 depends exponentially on ΔE_d. Also note that the rate of change of the defect formation energy with Fermi level depends upon the defect charge. If Equation 7.9 is applied in Equation 7.6, the resulting vacancy concentration is the total rather than the neutral concentration. Because ΔE_f changes logarithmically with the number of dopant atoms, the number of defects can vary linearly with the number of dopant atoms. This is exactly the conclusion from Equation 7.7 under the assumption of contant ΔE_d in Equation 7.6. Thus, we can approach the problem of defect density by either conceptual model (constant ΔE_d in Equation 7.6 plus the use of Equation 7.7 or by use of Equation 7.9 in Equation 7.6 directly) and get the same answer as long as $\Delta E_0 < q\Delta E_f$. Using Equation 7.9 in Equation 7.6 works in all cases.

Based on Equation 7.9: when doping works it is simply a case that ΔE_0 is much larger than ΔE_f. A failure of doping implies $\Delta E_0 << \Delta E_f$. A consequence of Equation 7.9 is that depending upon the charge of the defect and ΔE_0, the energy to form that defect may become zero or even negative (exothermic). This means that defects will form spontaneously when the Fermi energy moves beyond a given value. Once again, this leads to compensation of dopants by point defects and to effective "pinning" of the Fermi level at a given energy (it can not be moved beyond that energy due to spontaneous compensating defect formation).

Returning to the question of whether the system will be in equilibrium as the Fermi level is changed by doping, note that insertion of vacancies will occur at the same time as insertion of dopants by any technique. Therefore the compensation process usually is not limited by diffusion rates, as vacancies will move at least as fast as the impurities causing doping. The question is what is the temperature establishing the equilibrium? It would normally be at or near the doping temperature.

In most III-V compound and group IV elemental semiconductors the formation energy for all intrinsic defects is high enough that the Fermi energy can never overcome it. Thus, the Fermi energy is not generally pinned by defects and high doping levels of both types are possible. Group III nitrides are an exception to this behavior as it is reasonably easy to form nitrogen vacancies. Furthermore the large energy gap in the material results in large possible values of ΔE_f. This means that V_N donor defects will compensate acceptors, making p-type doping highly difficult. There is no easily formed intrinsic acceptor in GaN so n-type doping is usually possible (by nitrogen vacancies if by no other means). Relatively ionic compound semiconductors, including many of the II-VI compounds and analogous I-III-VI$_2$ semiconductors have very low defect formation energies. Furthermore, the electrostatic energy of charge-unbalanced defects lowers the formation energy of compensating defects significantly. Therefore, these materials often are nearly impossible to dope one or the other type (and sometimes both).

For other intrinsic defects (antisites, interstitials,...), their concentration depends upon their formation energy and entropy exactly as for vacancies and the formation energy will change with the Fermi energy or in the presence of other defects as well.

Summarizing: the population of intrinsic point defects changes radically with Fermi level and frequently prevents doping of a given type above a given level or can prevent doping entirely. The higher the defect formation energy, the less likely is autocompensation while the wider the energy gap the more likely autocompensation will occur. The predominant charge of the point defect will be negative if the material is n-type and positive if p-type (presuming that the Fermi energy lies closer to the appropriate band edge than the defect level).

7.1.6 Vacancies and dopant diffusivity

The change in vacancy concentration as a function of dopant density has a direct effect on the diffusivity of that dopant. [2] For low dopant concentrations the diffusivity of the dopant increases as the vacancy concentration increases if the dopant diffuses via a vacancy mechanism, as most do. This effect may be detectable even if the vacancy concentration is too small to have a measurable impact on doping. The charge on a vacancy can also have a significant effect. If the vacancy concentration gets too high, then defect clusters form and the diffusivity drops abruptly. To see how this can work in detail, we will consider the cases of As and P diffusivities in Si.

The diffusivity of the dopant can be different through vacancies of various charge states. The total diffusivity of a dopant might be, for example: [2]

$$D_{impurity} = D_{q=0} + \sum_{q \le -1} D_{-q} \left(\frac{n}{n_i} \right)^{-q} + \sum_{q \ge 1} D_{+q} \left(\frac{p}{n_i} \right)^{q} \qquad 7.10$$

where D_q is the diffusivity of the impurity through a vacancy of charge q, and n, p, and n_i are the electron, hole, and intrinsic carrier concentrations, respectively. As an example, for diffusion in Si at 1127°C (1400 K), the diffusivity for As is: [2]

$$D_{As} = 0.066 e^{-3.44/k_BT} + 12 e^{-4.05/k_BT} \left(\frac{n}{n_i} \right). \qquad 7.11$$

The first term is diffusion through neutral vacancies. The second term is for charged vacancies. The ratio of n/n_i is due to the connection between dopant and vacancy concentrations. For $n=10^{20}$ As cm^{-3} (all As ionized) and $n_i=1.5 \times 10^{19}$ cm^{-3} at 1400 K,

$$D_{As} = 2.3 \times 10^{-14} + 1.7 \times 10^{-13} cm^2 / \sec \qquad 7.12$$

In this situation, typical of As movement, diffusion through charged defects dominates the result. It is clear that a constant (concentration independent) diffusivity cannot be expected for As in Si. As the As concentration falls within the sample, the diffusivity also falls because of the change in n. Consequently, the concentration gradient for As diffusing into a Si wafer is greater than one might expect for a constant diffusivity.

When the vacancy concentration rises sufficiently, the diffusivity of an impurity may drop dramatically if the impurity and the vacancy form clusters. This is the case for both As and P. For As, when the dopant concentration exceeds ~10^{20} cm^{-3}, As atoms cluster with each other, resulting in electrically inactive groups. The processes can be expressed as follows: [2]

$$3As^+ + 3e^- \underset{<900^\circ C}{\overset{>1000^\circ C}{\rightleftharpoons}} As_3^{+2} + 2e^- \xrightarrow[25^\circ C]{} As_3 \qquad 7.13$$

This occurs only at high temperatures where the As atoms are all ionized. The resulting cluster becomes neutral when the temperature is reduced. In theory the cluster can decompose by a reverse reaction below 900°C as indicated to the right in Equation 7.13. However, if the sample is cooled moderately rapidly cluster decomposition may not have sufficient time to occur, in which case As_3 results.

The above is an example of clustering of a dopant with itself. Clusters can also occur with compensating sites, especially vacancies. In any compensated material, a donor transfers an electron to an acceptor, becoming positively-charged and resulting in a negative acceptor. These are attracted to each other forming a multisite defect often with no net charge. The cluster does not scatter moving carriers as strongly as separated charged defects might. As in the case of As above, formation of clusters requires a sufficient number of defects that they stick together faster than entropy drives them apart. Clusters typically form at a critical defect concentration.

Compensation is not a sure sign of cluster formation because the required concentrations are not always present. However, strong compensation with high quantities of oppositely-charged defects generally leads to clustering. Indeed, it almost has to at some point because it will become difficult for defects of one charge to get out of the range of defects of the opposite type. Clusters are also common in more complex semiconductors and insulating materials as these often dissolve higher concentrations of intrinsic defects.

As an example of clustering effects with vacancies we examine P in Si. This system combines virtually all of the phenomena discussed above: dopant enhanced diffusivity below the clustering concentration, decreased diffusivity above it, and limited electrical activity. A typical concentration depth profile behavior for P diffused into Si is shown in Figure 7.8. P-vacancy clusters are found to form and decompose at a temperature-dependent and doping-dependent concentration. Higher temperatures require higher doping levels for clustering, as does the presence of a significant number of compensating acceptor dopants. Below this concentration the clusters decompose, P moves more rapidly, and is more electrically active.

As long as we are on the subject of phenomena that affect diffusivities, let us briefly consider the fact that a charged defect should move differently in an electric field, E_{el}. Doping changes the Fermi level in a material and induces strong electric fields. Thus, there should be a correction to diffusivity resulting from the electric field as well as from vacancy and clustering effects. From Fick's first law, one can determine the flux of impurity atoms diffusing in a solid:

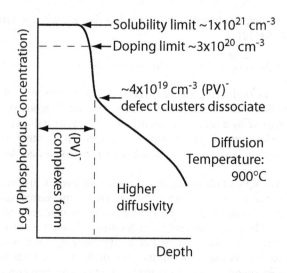

Figure 7.8: Phosphorous concentration resulting from diffusion into silicon. Formation of dopant clusters at high concentrations limits the diffusivity. Autocompensation limits the electron concentration to a value (the doping limit) below the solubility limit. Redrawn with permission from Fair, Richard B, "Concentration profiles of diffused dopants in Silicon," in Wang, F.F.Y., ed., *Impurity Doping Processes in Silicon* (North Holland, Amsterdam, 1981), chapter 7. Copyright Elsevier, 1981.

$$F = -D\frac{d[I]}{dx} - q\frac{D}{k_B T}[I]E_{el}, \qquad 7.14$$

where [I] is the impurity concentration, D is its diffusivity, and q is the dopant charge. The first term describes diffusion based on concentration gradients while the second describes the electric field dependence. $D/k_B T$ is the dopant mobility. This equation can be simplified by noting that the electric field can be written:

$$qE_{el} = \frac{d}{dx}\left(E_f - E_i\right). \qquad 7.15$$

In other words, the electric field is directly related to slope of the band edges in a semiconductor. However, for an uncompensated semiconductor, $E_f - E_i$ is directly related to the dopant concentration,

$$E_f - E_i = k_B T \ln\left([I]/n_i\right) \qquad 7.16$$

from which,

$$qE_{el} = k_B T \frac{d}{dx}\left[\ln\left([I]/n_i\right)\right]. \qquad 7.17$$

Substituting Equation 7.17 into Equation 7.14 gives

$$F = -D\frac{d[I]}{dx} - D[I]\frac{d}{dx}\left[\ln\left([I]/n_i\right)\right] = -hD\frac{d[I]}{dx} \qquad 7.18$$

where

$$h = 1 + [I]\frac{d}{d[I]}\ln\left([I]/n_i\right) \qquad 7.19$$

or, doing the derivative,

$$h = 1 + \frac{[I]}{n_i}\left[\left(\frac{[I]}{2n_i}\right)^2 + 1\right]^{-1/2} \qquad 7.20$$

Note that the smallest h can be is one for [I]~0 and the largest is two for [I]>>n_i. Thus the maximum effect of an electric field on diffusivity is a factor of two. This stands in stark contrast to the effect of Fermi energy on vacancy formation energy and clustering, which can both change diffusivities by orders of magnitude.

In summary, the largest effect on substitutional diffusivity of an impurity is the enhancement of vacancy concentration resulting from the change in Fermi level due to doping. This can change diffusion rates by orders of magnitude. Note however that one must calculate the vacancy concentration at the diffusion temperature. This can affect the conclusions greatly. At very large concentrations clusters are likely to form. These reduce diffusivity abruptly to a much lower value. Finally, the formation of electric fields by doping can additionally change diffusivity by up to a factor of two (much smaller than the effect due to vacancy concentration).

7.2 LINE DEFECTS

Line defects are one-dimensional and are commonly known as dislocations. These involve the displacement of atoms in the solid, leading to a strong distortion of the lattice along a line of sites. Dislocations have two basic forms, edge dislocations and screw dislocations. Edge dislocations, see Figure 7.9, can be visualized as being created by slicing partially through a crystal and inserting an extra partial plane of atoms into the structure. The line of the dislocation is the edge of the partial plane where it terminates within the solid (Fig 7.9.a.). If one begins on a given lattice site and traces a loop around a dislocation core (the line of the dislocation) moving a fixed number of positions right, down, left, and up, one will not return to the starting position (see Fig 7.9.b.). This procedure is known as following a Burger's circuit and the displacement between the beginning and ending point is connected by the Burger's vector, **b**. For a "perfect" dislocation the Burger's vector stretches from one lattice site to another in the same Bravais Lattice. For a pure edge dislocation, **b** is perpendicular to the dislocation line, *l*, so in vector dot product notation, **b**•*l* = 0.

A screw dislocation can be visualized as taking a cylinder of material, slicing it from the edge to the center and displacing the outside of the cylinder down one atomic spacing on one side of the cut, as shown in Figure 7.10. The dislocation line runs down the center of the cylinder and the **b** is parallel to l, so **b**•l=bl. For a screw dislocation (unlike an edge dislocation) bonds are bent but not broken.

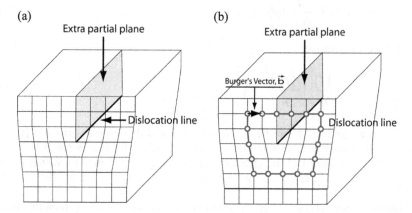

Figure 7.9: (a) A schematic diagram of an edge dislocation. (b) Shows the method for determining the Burger's vector of the dislocation by tracing a similar number of atom sites in a loop around the dislocation line. The Burger's vector connects the beginning point of the loop with the ending point.

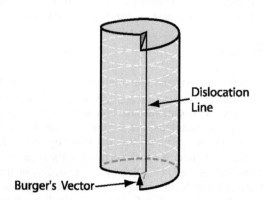

Figure 7.10: A schematic diagram showing the lattice distortion associated with a screw dislocation. The Burger's vector is parallel to the dislocation line direction. Note that traveling once around a screw dislocation core generally brings one a single atomic height step distance up or downward. Crystals often grow by adding atoms to the step. By this means the step progresses around the core, adding layers to the crystal without adding step area to the surface.

Things of particular note about dislocations are as follows. First, in the case of both the screw and edge dislocations, virtually all of the distortion of the crystal due to the defect is accommodated within the first few atom spacings around the dislocation line (see Figs 7.9 and 7.10). The magnitude of the distortion decreases as 1/r, where r is the distance from the dislocation line.

Second, a dislocation line can never end within a solid. It can close on itself to form a loop, but it cannot end. To see how this is the case, imagine an extra partial plane of atoms forming an edge dislocation within the solid. If the plane ends it produces a dislocation. To say an edge dislocation could end within a solid would be equivalent to saying a plane can end without having an edge. Clearly this is impossible in a crystal. The only place an extra plane can end without producing a dislocation is at the surface of the solid. A dislocation loop can result from a small piece of lattice plane stuck into the solid and surrounded by dislocation line or a small piece of vacant space (a missing piece of plane). Again, the vacant region is surrounded by a dislocation loop. These cases are shown in Figure 7.11.

Third, dislocations may be of mixed character – partially screw-like and partially edge-like. In this case the Burger's vector is neither perpendicular to nor parallel to the dislocation line. One can see how this can happen as follows. Slicing the solid and moving an irregularly-shaped plane of atoms to form a dislocation has the same displacement, and hence the same Burger's vector across its entire surface. However, the line of the dislocation is forced to follow the edges of the plane, wherever they

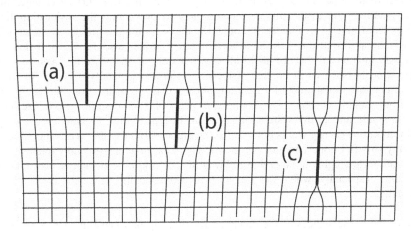

Figure 7.11: A schematic diagram of various edge dislocation configurations. (a) a classic edge dislocation with an extra half plane inserted into the material. This type of dislocation forms a loop which closes at the crystal surface. Thus, threading segments will occur. (b) and (c) are loops resulting from coalescence of interstitials and vacancies, respectively to form small planes. When these result in stacking faults, the faults are termed extrinsic and intrinsic for (b) and (c), respectively and have dislocations at their edges.

go. Consequently, some edges of the dislocation line surrounding a plane may be edge like, some screw like, and some mixed.

The energy to form a dislocation is significant in most materials but is particularly large in strongly bound materials. As a consequence, these materials are brittle. The energy per unit length of a screw dislocation can be shown to be:

$$\Gamma_{screw} = \frac{G}{4\pi}\left[b^2 \ln\left(\frac{r}{b}\right)\right] + E_{core}.$$ 7.21

G is the bulk modulus of the matrix material. A similar formula to Equation 7.21 can be developed for the energy of an edge dislocation. Equation 7.21 is determined by integrating the strain field of the dislocation from the dislocation core to the edges of the crystal in which it is located. However, because the strain diverges as 1/r, the integral would diverge if integrated from r=0. To deal with this an inner radius $r = b$ is defined within which the dislocation is assumed to have a core energy E_{core} per unit length. This inner limit is the basis of the "b" in the denominator of the logarithm of Equation 7.21. In many cases E_{core} is approximated by a scale factor on the inner cutoff radius that becomes a fitting parameter. Thus by adjusting the inner cutoff of the integral one can include E_{core} in the logarithm. The result is the same.

The energy of the dislocation diverges logarithmically as the system size goes to infinity. Although real crystals have very large r values, they are not infinite and so the real dislocation line energy is finite. In addition, the free surface changes the strain field of the dislocation and contributes to preventing the dislocation line energy from diverging. Finally, although the dislocation gains energy as the crystal grows, the ability of the dislocation to move to a surface also decreases, making it less than guaranteed that the dislocation will escape from the solid. Even in very large crystals (such as Si boules) dislocations can and generally do exist. In large single crystals Equation 7.21 essentially means that the dislocation line energy is large but roughly constant. For example, a factor of 3 increase in system size only adds a single unit to the system energy term. In typical bulk Si samples, the edge dislocation line energy works out to be of the order of 100 eV nm^{-1} (corresponding to roughly two full bond energies per atom). This is a huge energy and explains why dislocations are highly unfavorable and why Si is so brittle. Nonetheless, dislocations do form, especially when the material is mechanically damaged as in wafer polishing.

One of the consequences of the quadratic relationship between Burger's vector and dislocation line energy is that the energy of two dislocations with short Burger's vectors can be lower than the energy for one dislocation with a longer vector (e.g. $2^2+2^2<3^2$). This can drive decomposition of a perfect dislocation with a long Burger's vector into two partial dislocations. Such a case is shown in Figure 7.12. A perfect dislocation represents a displacement of an atom from one lattice site to another. For example, the displacement of an atom from a cube corner to a cube face center in the diamond or zincblende structures is the displacement by a perfect dislocation with a Burger's vector of a/2 [110]. A displacement from the site at (0,0,0) to (1/4, 1/4, 1/4) does not take an atom from one equivalent site to another, as these two points are not on the same Bravais lattice. Therefore, such a displacement creates a stacking fault (see Section 7.4.1) in the solid.

The dislocation associated with such a displacement is called a partial dislocation. A second partial dislocation that displaces atoms from (1/4, 1/4, 1/4) to (1/2, 1/2, 0) brings the system back to the original lattice and terminates the stacking fault. The two partial dislocations with Burger's vectors shown in Figure 7.12 are a/6 [111] type and have a lower total energy for the than the one perfect a/2 [110] type dislocation. (The a/2 and a/6 refer to the Burger's vector length, while [111] and [110] give the vector direction.) Thus, the perfect dislocation is driven to decompose. However, because separating the partial dislocations creates an area of stacking fault, and because there is an energy penalty associated with this, there is an equilibrium (and usually small) separation of the partial dislocations. Partial dislocations in compound semiconductors generally also have a large amount of charge associated with them as partial dislocations generally move one type of atom onto the other atom's sublattice and so create a plane of antisite defects. Such defects are highly energetic. Thus, in partially ionic materials separated partial dislocations generally do not occur.

One of the more important aspects of edge and screw dislocations is their strain field. This is crucial to accommodating differences in lattice constants in heterostructures. Strain is the distortion of the lattice caused by stretching, bending, or shearing it. An edge dislocation includes a "hydrostatic component" of strain, which means that the lattice is to some extent uniformly expanded or compressed by the dislocation. The stress, σ, on a material is the force per unit area and is linearly related to the strain, ε.

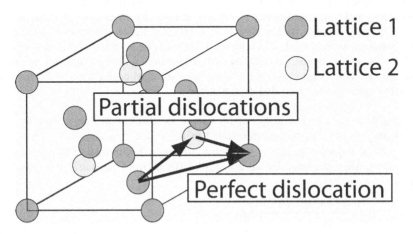

Figure 7.12: A schematic diagram of the Burger's vectors associated with decomposition of a perfect dislocation into two partial dislocations.

Thus, σ=Yε, where Y is the elastic or Young's modulus. Looking at the drawing of the edge dislocation, one can see that the lattice is compressed relative to its normal spacing on the side of the line where the partial plane ends. On the other side of the dislocation core the lattice is expanded (it is in tension). In the plane perpendicular to the extra half plane running through the core the strain is zero.

Mathematically, the radial component of the dislocation stress may be written as:

$$\sigma_{rr} = -\frac{Gb}{2\pi(1-v)r^2 l}\vec{r}\cdot\vec{\xi}.$$

$$\text{7.22}$$

where $\vec{\xi}=\vec{b}\times\vec{l}$, G is the bulk modulus (similar to the elastic modulus), and v is the Poisson's ratio. The latter is the amount by which a material contracts along a direction perpendicular to a direction it is being stretched, or vice versa. One can understand the origin of the Poisson's ratio by noting that the volume of a solid tends to remain constant upon distortion. Therefore, stretching a solid in one direction leads to a corresponding contraction in all other directions. The behavior of σ_{rr} is shown schematically in Figure 7.13 for an edge dislocation. The angle θ and radial distance r are defined in the figure. Note that for a screw dislocation, the dot product is zero so there is no radial stress term. The radial stress in Equation 7.22 gives a corresponding total hydrostatic stress (a pressure), P_{edge} for an edge dislocation of:

$$P_{edge} = -\frac{(1+v)Gb\sin\theta}{3\pi(1-v)r},$$

$$\text{7.23}$$

with the same behavior as shown in Figure 7.13. The hydrostatic stress of a screw dislocation is zero and a mixed character dislocation has an intermediate stress.

One consequence of Equation 7.23 is that edge dislocations attract impurity atoms. Impurities (and intrinsic defects) are nearly always the wrong size for the site on which they lie. An oversized atom will be attracted to the tensile portion of the edge dislocation stress field while an undersized atom will be attracted to the compressive region. In other words, essentially all impurities will be attracted to edge dislocations. This is good when the impurities are detrimental but can be harmful when the impurities are intended dopant atoms.

Because there are a large number of sites along the dislocation line, a large number of impurity atoms can be accommodated in the dislocation core. The interaction energy of an impurity located at position r and angle θ with respect to a pure edge dislocation, where the impurity has a size ΔR different from the matrix atom radius R can be shown to be:

$$E = \frac{4}{3}\frac{(1+v)}{(1-v)}\frac{R\Delta R}{r}\sin\theta,$$

$$\text{7.24}$$

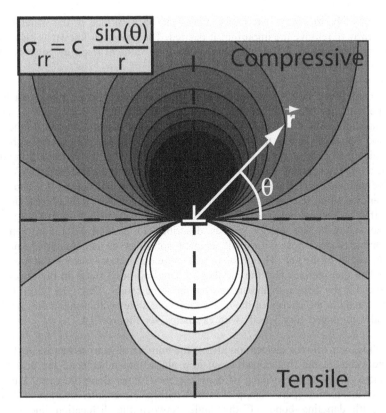

Figure 7.13: The strain field distribution around an edge dislocation line. Lighter colors represent tensile strains while darker colors represent compressive regions. The constant c is $c = -Gb/2\pi(1-\nu)$ from Equation 7.22. The inverted "T" at the center of the figure is the conventional symbol for the core of a dislocation.

Note that this interaction energy scales with the magnitude of the misfit between the impurity and the matrix, is strongest directly above or below the core of the dislocation, and increases in strength as $1/r$ as the atom approaches the dislocation line. Also note that if a dislocation core collects a large number of impurity atoms by this mechanism, its future movement is inhibited as it must break free from these impurities (and overcome the energy attracting them to the dislocation core).

One can take advantage of the attraction of impurities to dislocation cores to reduce undesired species by intentionally introducing dislocation loops on the back surface of a wafer. This is done by mechanically damaging the back side after the polishing and etching of the front surface is complete. Most detrimental impurities in Si diffuse much faster than dopant atoms. Many are interstitial atoms and move exceptionally fast. Therefore, annealing a wafer with back surface damage collects point defects there, out of harm's way. After this anneal the back surface of the wafer can be etched away to remove the impurities if desired. The process is known as dislocation gettering of impurities. Screw dislocations have no value as impurity getters as they have no hydrostatic stress term.

Another important fact, which applies only to dislocations having edge character, is that they include dangling bonds. One can see this because a plane of atoms terminates along the dislocation line. Consequently, the atoms along this plane edge have too few other atoms to bond with directly. The situation is modified by any reorganization of the atoms in the dislocation core. However, as a general rule, there will be a large number of dangling bonds (of the order of one per atom) in the core of an edge dislocation. The consequence is that the material behaves as if it had a line of vacancies along the dislocation. (The defects are not exactly like vacancies and have different reconstructions and defect state energies. Nonetheless, dangling bonds produce states in the energy gap as described in Section 7.1.) For elemental semiconductors and those with small chemical splittings the dangling bond states are near the center of the gap. The problem with edge dislocations becomes immediately clear when one realizes that the number of dangling bond states in the dislocation core exceeds the doping concentration within some radius. Thus, the Fermi energy will be forced to the energy of the dislocation core states within that radius as if the material were doped very heavily with vacancies, see Figure 7.14.

The radius over which a dislocation affects the number of free carriers is roughly the average radius to the first dopant atom in a plane perpendicular to the dislocation core and as thick as the spacing of dangling bond states along the core. A mixed-character dislocation (part edge, part screw) will have a smaller fraction of its core atoms with dangling bonds. If the angle between the dislocation line and the Burger's vector is α, then the average number of lattice spacings, a, between dangling bonds will typically be $a/\sin(\alpha)$, on average. Each dangling bond may or may not collect an electron or hole depending on the Fermi function, the difference in energy between the Fermi energy and the dangling bond state energy and the temperature. Assuming the opposite charge on the dopant as on the dangling bond (they compensate one another), the average distance, R, around a dislocation core where the material is virtually denuded of free carriers is found from:

$$\pi R^2 N_d = \frac{f \sin \alpha}{a},$$ 7.25

where f is the fractional filling of dislocation dangling bonds with carriers (determined by the Fermi function, the doping or Fermi energy, and the temperature). The left side of this equation is just the area of semiconductor perpendicular to the dislocation line necessary to encounter one dopant atom on average. The size of R determines the size of the dead region around a disclocation core. Consider the case of Si with

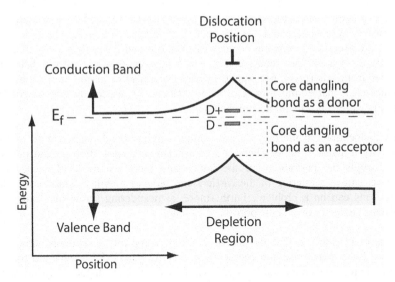

Figure 7.14: Shows the effect of a dislocation on local Fermi level in a semiconductor. The dangling bonds associated with the dislocation give rise to states in the energy gap. When the dangling bond density exceeds the local doping density (which is virtually assured within some distance) the Fermi energy is pinned at or among the energies of the dangling bond states.

6×10^{15} cm^{-3} dopant atoms containing a single edge dislocation. If one assumes a 50% filling of the defect states ($f = 0.5$) and $a \sim 0.25$ nm, one obtains:

$$R = \sqrt{\frac{f}{a \pi N_d}} = \sqrt{\frac{0.5}{2.5 \times 10^{-8} cm \bullet 3.142 \bullet 6 \times 10^{15} cm^{-3}}} = 3.3 \times 10^{-5} cm \qquad 7.26$$

In other words, 1/3 of a micron around the core has effectively no doping. This is enough to prevent any moderately small device through which an edge dislocation passes from working. Even if the device is much larger than this the change in conductivity and carrier recombination can ruin it.

Even when the change in Fermi level is not, in itself, disastrous, the core of the dislocation tends to be conductive along its length because of carriers hopping from one dangling bond to the next. This can occur in either p or n-type material in most semiconductors. Such conduction provides a shunt path in junctions and causes increased leakage currents. At high doping levels where one might expect the Fermi-level change to be less important, edge dislocation cores can scatter carriers off of the electric field of their depletion region, thereby reducing mobilities. In addition, they can act as efficient trapping or recombination centers.

These problems, taken together, are why one works so hard to eliminate edge dislocations from semiconductors. Typical single crystal Si boules, when sliced into wafers, are specified to contain *no* edge dislocations and fewer than a given number of screw dislocations per square centimeter. When one thinks about the dislocation densities in metals (given per square micron generally) or in common ceramics, it becomes clear that elimination of *all* edge dislocations, especially after sawing and polishing wafers, is a major accomplishment.

Screw dislocations, while they do not exhibit dangling bond states, are not without the ability to cause trouble. There is a significant shear distortion around the core. This disrupts the periodicity of the lattice and modifies the local energy band structure, causes scattering of the carriers and reduces mobility. Thus, even screw dislocations can be a problem. Furthermore, a meandering screw dislocation will have edge character over some portion of its length.

An important point is that the above discussion applies to semiconductors with dangling bond states near the center of the energy gap. II-VI semiconductors and other ionic materials have shallow-lying vacancy (and hence dislocation) states. This means that these materials are far less susceptible to edge dislocations. Shunting is still a problem but the dislocation core is more likely to act as a heavily doped region than as an intrinsic region. Furthermore, the charge compensation mechanism in polar materials tends to produce defects that compensate the charge on a dislocation core. Behaviors due to grain boundaries are similar to the effect of dislocations which is why the II-VI and other complex compound semiconductors can often be used as polycrystalline materials rather than requiring the extreme crystalline quality of the group IV and III-V semiconductors. The down side of II-VI materials being less sensitive to dislocations is that the dislocation formation energy is much lower. Consequently, the materials are soft, easily damaged, and do contain a large density of dislocations no matter how careful one is during crystal growth and wafer preparation. The inability to eliminate dislocations was a major reason for the failure of II-VI semiconductors to produce efficient laser diodes.

7.3 STRAIN RELIEF IN HETEROSTRUCTURES

In creating heterojunctions of semiconductors, it is common to find that the lattice constant of one semiconductor does not match exactly to that of the second semiconductor (see Chapters 6 and 10). This situation is shown schematically in Figure 7.15. Typically in heterojunctions, thin films of one semiconductor are deposited on much thicker substrates of another. As with a rubber bands stretched between pins on a surface, the film may be stretched (or compressed) to match the substrate lattice spacing. However, for every atomic layer added to the film, another layer's worth of strain energy is added to the system (as if another rubber band is stretched between the pins). Thus, the strain energy in the entire system increases linearly with the thickness of the film. Eventually, in the analogy, one would expect the strength of the pins holding the rubber bands to be overwhelmed and the pins would break. This is, indeed, what happens in strained thin films. Above a critical thickness the bonds at the interface between film and substrate break and the film relaxes, at least partially, with the resulting mismatch in the lattices accommodated at the heterojunction by formation of dislocations (Figure 7.15). The two states of mismatch between the film and substrate are given the following names:

Pseudomorphic or commensurate: means the film is stretched/compressed such that its lattice constant matches that of the substrate and bonding is continuous.

Incommensurate: means that the film has relaxed at least partially such that the lattice constants of film and substrate do not match. Edge dislocations accommodate the mismatch due to relaxation.

Although in principle the substrate could also distort, the vast majority of the strain of a thin layer on a thick substrate is in the film. For materials with roughly equal modulus, the strain is distributed according to the ratio of the film and substrate thicknesses. Thus, a 1 micron film on a 100 micron thick substrate will have 100 times more strain in the film than in the substrate. Because the strain energy depends upon the strain squared, the strain energy *distribution* goes as the ratio of the squares of the thicknesses. Thus the vast majority of the strain energy is in the film.

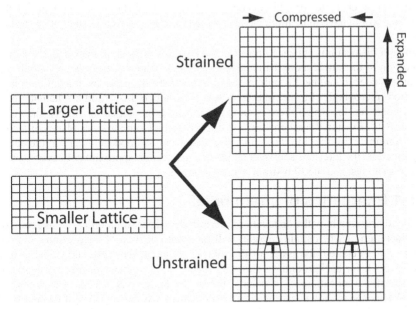

Figure 7.15: A schematic diagram of the results of joining two layers of different lattice constant. In this case it is assumed that a thin film with a larger lattice constant is growing on a thick substrate with a smaller lattice. The films can be joined by stretching the film lattice to match the substrate (commensurate) [top right] or by forming dislocations at the interface (incommensurate) [bottom right].

While the strain energy in a thin film is increasing linearly with film thickness, the energy that a dislocation would cost is increasing only logarithmically (Equation 7.21). Thus, however small the strain in the film, sooner or later as the thickness increases the strain energy will exceed the energy associated with strain-relieving dislocations. Above this "critical thickness" dislocations tend to be introduced into the film to relieve the misfit stress, hence they are called "misfit" dislocations.

Misfit dislocations must have at least some edge character with the Burger's vector having a component in the interface between the two films. This can be understood in either of two ways. Edge dislocations have extra partial planes of atoms associated with them. If this extra partial plane enters the material with a smaller bulk lattice constant and ends at the heterojunction between the two materials, then the extra partial plane will accommodate some of the misfit size difference. See, for example, Figure 7.15. A second way to understand the situation is to note that any dislocation with at least partial edge character has a hydrostatic strain field associated with it (Figure 7.13). If this strain field is opposite to (or at least partially opposed to) the strain due to misfit, then the sum of the two strains will be smaller. The result is that for every percent of misfit between two lattices, a perfect edge dislocation would have to be introduced every 100 lattice spacings in the interface to completely eliminate strain from the overlying layer. A two-dimensional interface requires a two-dimensional dislocation network. Dislocations with partial edge character and partial screw character or edge dislocations having their Burger's vector out of the interface would have to be more closely spaced to relieve the same amount of strain.

Although dislocations appear to result in very large strains at an interface, the actual distortion of the lattice more than one or two atomic distances away is very small. As a consequence, it can be difficult to pick out the misfit dislocations in a lattice-image electron micrograph of a real interface. A typical lattice image and a schematic of a heterojunction, for which there is an exceptionally large lattice misfit, are shown in Figure 7.16. Even in the schematic, it is difficult to pick out the extra planes of atoms that terminate in the interface and the distortion is modest even in the core. Nonetheless, the interface shown contains one dislocation approximately every seven lattice spacings, and the film strain is fully relieved.

7.3.1 Energetics of strain relief

Before discussing how the dislocations are introduced for strain relief, let us return to consideration of when strain energies will favor introduction of dislocations in more detail. For a detailed discussion of the following model, see Hirth and Lothe or any similar book for background on elasticity theory and Matthews or other sources for the specific application of elasticity theory to lattice-mismatched structures. A complete analysis shows that the elastic strain energy, E_{el}, in a film of thickness h on a much thicker substrate is given by:

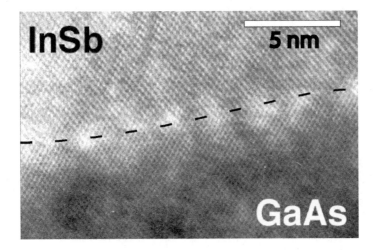

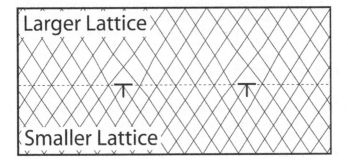

Figure 7.16: A cross-sectional transmission electron micrograph of a GaAs/InSb heterojunction. The mismatch in this system is approximately 14.6%, giving rise to a misfit dislocation roughly every seven lattice spacings. In this image, the core regions of the edge dislocations appear as brighter regions. The schematic below shows a sketch of the dislocation structure in the image above. Each marked dislocation terminates two of the (111)-type lattice fringes as shown and is located in the whiter regions along the marked interface.

$$E_{el} = \frac{2G(1+v)}{1-v} f^2 h ,$$　　　　　　　　　7.27

where v is the Poisson's ratio of the film, f is the fractional lattice mismatch between the two bulk lattices, and G is the bulk modulus of the film. For lattices without strain in either material,

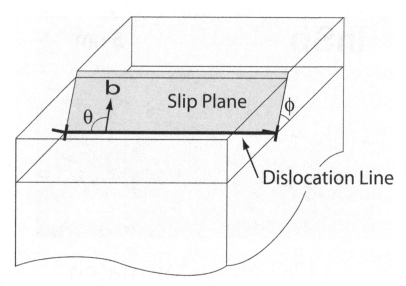

Figure 7.17: A schematic diagram of a misfit dislocation lying in an inclined slip plane in a thin film. Note that introduction of this dislocation leaves a slip step on the film surface. For a single perfect dislocation, the step would be one atomic spacing high.

$$f_\infty = \frac{\Delta a}{a} = \frac{a_f - a_s}{a_s} \qquad\qquad 7.28$$

with a_s being the bulk lattice constant for the substrate and a_f being the value for the film. The dislocation line energy for a single dislocation in a film is likewise:

$$E_{dis} = \frac{Gb^2}{4\pi} \frac{1 - v\cos^2(\theta)}{1 - v} \ln\left(\frac{\alpha h}{b}\right) \qquad\qquad 7.29$$

where α is a constant to correct for the dislocation core energy and θ is the angle between the Burger's vector and the perpendicular to the dislocation line, see Figure 7.17. For a pure edge dislocation, $\theta=0$, while for a pure screw dislocation, $\theta=90°$. The first dislocation will be introduced, at equilibrium, just beyond the point where the dislocation energy and the strain energy are equal. Note that for full strain relief we need one edge dislocation every $1/f$ lattice planes, so the larger the misfit the more dislocations we need. In other words, Equation 7.29 must be multiplied by f

when balancing it against Equation 7.27. Solving $E_{el} = f\,E_{dis}$ for h gives the critical thickness, h_c, above which there should be dislocation introduction:

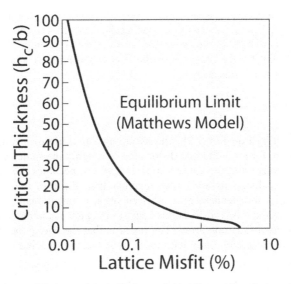

Figure 7.18: The equilibrium critical thickness for a film with a Poisson's ratio of 0.22 assuming strain relief by 60° dislocations (θ, ϕ =60°). The curve was calculated using Equation 7.29, with α=1.

$$\frac{h_c/b}{\ln(\alpha h_c/b)} = \frac{1 - v\cos^2\theta}{8\pi f(1+v)\cos\phi}. \qquad\qquad 7.30$$

where ϕ, defined in Figure 7.17, is the angle between the glide plane (see Section 7.3.2) and the misfit interface. This equation is not a simple function in the form h_c = some function of f. However, because the logarithmic dependence in the denominator of the left hand term varies much more slowly than the numerator, it is usually taken as roughly constant and h_c is, at least locally, inversely proportional to the misfit, f. In Equation 7.30, h_c appears only as a ratio with respect to the strain relieving dislocation Burger's vector. Consequently, the ratio is general for any vector orientation. The behavior of h_c/b for a typical thin film heterostructure is shown in Figure 7.18.

Typical analyses end at this point. We have a relationship between the equilibrium thickness at which strain relief begins and the misfit. One can quickly see, however, that introduction of even a single dislocation into the interface decreases the effective misfit of the film and substrate, at least locally. Therefore, after the first dislocation is added at h_c, the thickness should have to increase before the second dislocation is introduced. The result is that it is better to discuss the equilibrium spacing between misfit dislocations for a given misfit and film thickness, rather than discussing the expected onset of strain relief. This equilibrium spacing can be estimated as follows. [7] Suppose that a uniform array of misfit dislocations exists at a semiconductor heterointerface with a spacing D. The elastic energy in the film (Equation 7.27) is still valid. However, the misfit strain (Equation 7.28) needs to be modified to account for the presence of the dislocation array. The average residual misfit in the presence of this array is:

$$f(D) = f_\infty - \frac{b\cos\phi}{D}$$ 7.31

where f_∞ is given by Equation 7.28, and assuming a uniform dislocation array. The dislocation energy for an individual dislocation is roughly the same as that given by Equation 7.29, except that the value h in the logarithm becomes the smaller of h or D. Thus, if the dislocations are closer together than the film thickness, then the logarithm becomes independent of h. This equation is insufficient, however, when the strain fields of the misfit dislocations begin to interact, as interaction modifies the total energy of the dislocation array. The result is that the energy per unit area of a two-dimensional dislocation array with uniform dislocation spacing D must be used. This energy is:

$$E_{array} = \frac{Gb^2(1 - \nu\cos^2\theta)}{8\pi(1 - \nu)D}\ln\left(\frac{\alpha h}{b}\right),$$ 7.32

for h<<D. Equating the energy of the dislocation array (7.32) to the residual strain energy in the film leads to:

$$D(h) = \frac{b\cos\phi}{\dfrac{\Delta a}{a} - \dfrac{b(1 - \nu\cos^2\theta)}{8\pi(1 - \nu)\cos\phi}\dfrac{\ln(\alpha h/b)}{h}},$$ 7.33

which can be rewritten more compactly as:

$$D(h) = \frac{b\cos\phi}{\dfrac{\Delta a}{a} - C\dfrac{\ln(\alpha h/b)}{h}},$$ 7.34

where C is a constant for a given type of dislocation. The critical thickness, Equation 7.30, is simply Equation 7.34 for D = ∞. (In other words, when the denomenator is zero.) The analysis of the dislocation network energy can be helpful in interpreting the local deviation from equilibrium of a misfit dislocation array that has a non-ideal spacing.

A complication in dealing with dislocation spacings is that the dislocation network *is* usually far from its equilibrium, especially locally. An example of a misfit dislocation array is shown in Figure 7.19. One can see that the array is non-uniform in both sets of dislocations, and further that one set has more dislocations per unit length than the other. Both of these are typical phenomena in misfit dislocation arrays. Because the strain from a dislocation decreases rapidly with distance, the effect of a dislocation on the local misfit in a solid more than about one film thickness from the dislocation core is relatively modest. Consequently, the actual dislocation network in the film does not tend to follow either the critical thickness analysis or the analysis of equilibrium dislocation spacing. One can, however, use the formulas to determine how far the film is from equilibrium locally and determine the driving force for addition of more dislocations in a given area.

The process for introducing dislocations into thin films inevitably adds disorder to the dislocation network because, as we will see, dislocations can not be added at any point (you require a source) and dislocations can move sideways in the interface to organize themselves only with some difficulty in typical heterojunctions. It is therefore useful to examine the real dislocation distribution and the mechanism for introducing misfit dislocations to understand the behavior of real systems.

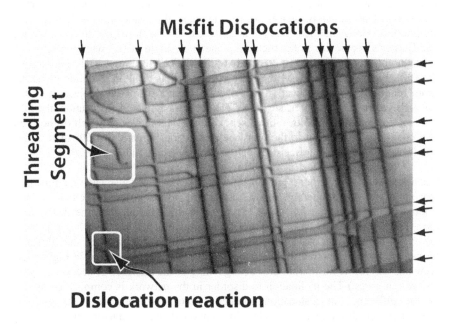

Figure 7.19: A transmission electron micrograph looking down upon the interface between a GaAs substrate and an InAs epitaxial layer. Both threading segments stretching from the heterojunction to the surface and reactions between dislocation segments are visible in the micrograph. The lines in the image are the misfit dislocations in the interface. [8]

7.3.2 Misfit dislocations

Before describing how dislocations are inserted into a thin film to relieve strain, a brief description of the dislocations themselves seems in order. The ideal misfit dislocation is a pure edge dislocation with its Burger's vector lying in the interface plane between the two semiconductors. This dislocation relieves the maximum misfit strain for the minimum cost in dislocation line energy. Consider a heterojunction between two diamond-structure semiconductors with the boundary between being a cube face. In principle, two perfectly-ordered perpendicular arrays of edge dislocations could form in the heterojunction, stretching from side to side of the film and leaving no dislocation segments trailing up through the layer to disrupt electron motions. This situation is, unfortunately, seldom observed. The primary reason is that dislocations have to be inserted by sliding lattice planes past one another. This is done most easily on specific planes, called glide planes.

The glide planes for fcc lattices are the close-packed {111}-type planes. However, semiconductor films generally do not grow well on these planes. Most semiconductor heterojunctions have cube faces [{001} planes] as their interface because this surface grows well. Unfortunately, as it is not a glide plane, the dislocations have to glide to the interface along inclined {111} planes, tilted ~55° with respect to the interface. The Burger's vector lies in the inclined plane at an angle of 60° with respect to the dislocation line. Hence, they are termed "60°" misfit dislocations. Their geometry is shown in Figure 7.20. Because the Burger's vector has to lie in the glide plane, the strain relief of such a dislocation is the edge component of the projection of this Burger's vector onto the interface plane. The result is that a 60° dislocation, the most common form of misfit dislocation in semiconductors, relieves only half of the strain of a pure edge dislocation in the interface plane. The formation of 60° dislocations also leaves single layer surface steps on the film, as shown in Figure 7.20.

Because the interface is not a glide plane, misfit dislocations also cannot adjust their spacing easily. Consequently, misfit dislocation arrays generally contain a large amount of disorder (Fig. 7.19). An exception to this case is when the dislocations form in the very early stages of epitaxial growth when the overlying layer is extremely thin. In this case, the network of dislocations grows into the film as it forms, and is nearly perfect from the beginning. (The heterojunction in Figure 7.16 grew this way. Note how strain drove roughening of the interface as decorated by the dislocation cores.) The inclination to disorder in the network is complicated as well by the difference in dislocation energetics and strain necessary for glide of dislocations on different glide planes. In compound semiconductors, this generally leads to one of the two perpendicular dislocation arrays having a significantly smaller dislocation density and more disorder in the dislocation spacing (Fig 7.19). A discussion of this phenomenon may be found in Beanland.

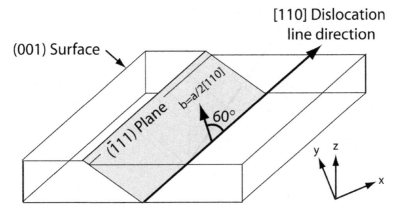

Figure 7.20: A schematic diagram showing the geometry of a 60° dislocation on an inclined (111) type plane. Note the surface step resulting from slip on the glide plane.

7.3.3 Dislocation dynamics

Early studies showed that many films grown at sufficiently high temperatures exhibited evidence of strain relief beginning around the critical thickness estimated from Equation 7.30. Because a significant driving force is required to move dislocations and because the resolution of the analysis techniques does not generally allow observation of the first motion of any dislocation in the material, the observed critical thickness was always greater than the value given by Equation 7.30. However, the qualitative trends are correct. A detailed review of the experimental background and a discussion of the various mechanisms by which dislocations may be introduced into thin films is available in the recommended readings. This section provides only a summary of some of the major mechanisms of strain relief.

All substrates contain some density of dislocations, although one hopes in semiconductor substrates that their density is low. In addition, faults occur at the onset of film growth, usually due to surface contamination, creating additional "threading" dislocation segments. Threading dislocations pass through the thickness of the film to the surface and extend as it grows. This situation is shown schematically in Figure 7.21.a. Forces on the threading segment can cause it to bow out on a glide plane, see Figure 7.21.b and 7.19. As this occurs a segment trails behind in the interface between the film and substrate (Figure 7.21c).

If this segment has an edge component of the right sign, it may provide strain relief. When the critical thickness is exceeded, the force resulting from strain relief exceeds the force necessary for dislocation extension (the "Peach Koehler force" [9]) and a misfit relief segment is created.

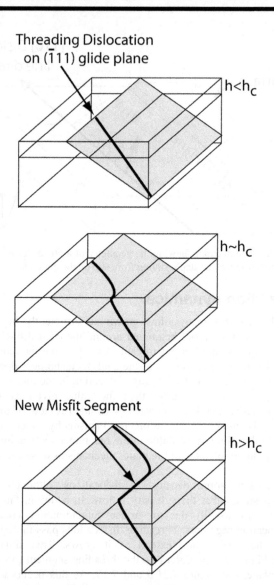

Figure 7.21: Shows the mechanism for introduction of misfit dislocations by extension of an existing threading dislocation. (a) below the critical thickness the threading segment passes through the film in a relatively straight segment. (b) Around the critical thickness the dislocation begins to bow outward as a driving force for strain relief begins to act upon it. (c) Above the critical thickness the threading segment extends rapidly, leaving a misfit segment behind. Effectively half of (c) may be observed in Figure 7.19.

This mechanism occurs very readily because the threading segments are pre-existing in the film, thus there is no nucleation barrier. When the film is of sufficiently high quality, threading segments can glide a long distance before encountering an obstacle. Unfortunately, when the film is of such quality, or if the film is etched into mesas that are small enough to contain, on average, no threading segments this mechanism produces little or no overall relief of strain. Therefore, such films can often be grown well in excess of the critical thickness with no strain relief. Even when there are numerous misfit dislocations, the level of strain relief observed exceeds that which could be obtained from the initial threading dislocation density. Clearly some form of dislocation multiplication must be possible.

A second mechanism for introduction of misfit dislocations is homogeneous nucleation of dislocation loops at a surface and their expansion into the film. Here, a small volume of film slips, beginning at a surface step for example, creating a loop of dislocation on an inclined plane. In common semiconductors with fcc Bravais lattices and (001) type surface planes, such dislocations would be the 60° type described above. This situation is shown schematically in Figure 7.22. Such a mechanism eventually creates two threading segments and a misfit dislocation.

Dislocation segments interact with each other through their strain field, and segments of opposite line direction and the same Burger's vector are attracted. This is the case for segments on opposite sides of a dislocation loop. Thus, one would expect that it would require significant energy to force the segments apart in the process of forming a dislocation loop. Indeed, it is this attraction that drives dislocation loops to collapse, and is a mechanism for elimination of (energetically unfavorable) dislocations from solids. Therefore, one would think that it would be difficult to form misfit dislocations by loop nucleation and expansion. Furthermore, the initial dislocation loop provides only a very small amount of strain relief as it affects only a very limited volume of material. A detailed analysis [c.f. Beanland] shows that the energy gain due to strain relief by the loop increases as the radius squared, while its line energy increases as its radius. Thus, for small enough loops the energy cost of the loop is much greater than the energy gained by its formation. This argument suggests that a significant nucleation energy may occur for formation of dislocation loops.

The dislocation loop nucleation energy increases as the dislocation energy increases. Thus, relatively soft materials such as II-VI semiconductors would be more likely to nucleate misfit dislocations by loop formation than would harder materials such as Si. Materials with small misfit strains are less likely to form dislocations by loop expansion, because the loop would have to be much larger before the energy payback from strain relief would compensate for the cost of the loop. Stress concentrators can reduce the loop nucleation barrier, although in many materials such concentrators are rare. If the effective additional stress associated with the concentration point adds to the misfit stress, the local energy barrier to loop nucleation can be reduced or eliminated.

Examples of stress concentrators include edges of the strained layer on patterned substrates, surface or growth defects (second phases, and other imperfections), or local misorientations of the substrate-film interface relative to a low-index plane.

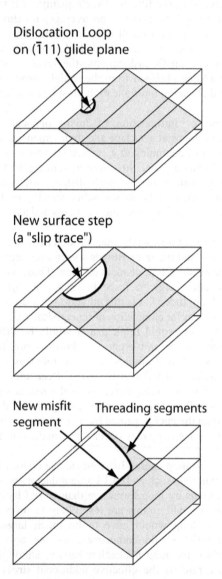

Figure 7.22: Nucleation of a dislocation loop at a surface resulting in creation of a misfit dislocation segment and two threading segments. The loop expansion leaves a new segment of step edge on the surface and relaxes the strain in the film.

Finally, dislocations with shorter Burger's vectors such as partial dislocations are more likely to form by loop nucleation because the dislocation line energy scales as the square of the Burger's vector. However, partial dislocations leave stacking faults behind which are easily detectable in these materials and have high energies. No stacking faults are observed in typical compound semiconductor strained layers, showing that partial dislocation loops are not occurring.

In the absence of stress concentrators, it has been estimated that misfit dislocation formation by half-loop nucleation and growth will only occur in highly-mismatched semiconductors which have a very small critical thickness for strain relief. This is because for such a situation the loops will nucleate before the film gets very thick. Consequently, the resulting threading segments are short relative to the strain energy relieved and the critical radius for a stable loop will be small enough to allow a reasonable nucleation rate. If a loop nucleation mechanism can occur it makes strain relief more likely. For common III-V semiconductors and typical growth temperatures the misfit strain required is of the order of 6% to allow loop nucleation.

Films of high misfit such as might lead to dislocation loop formation do not grow as uniform thin layers, but rather form islands on the substrates. The critical thickness analysis does not hold well for such a system, as strain relief can occur both by dislocation formation and by relaxation of the island without dislocations. Relaxation without dislocation formation has been extensively modeled using linear elasticity theory by Johnson and Freund [10], as in Figure 7.23. The model shows that for moderate misfits and moderate film thicknesses, strain can be maintained in small islands well beyond the critical thickness calculated with Equation 7.30.

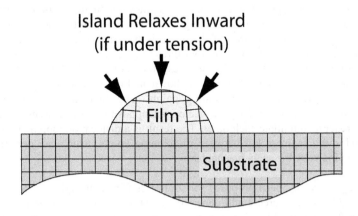

Figure 7.23: A schematic diagram showing the relaxation of an island on a substrate in which the island was stretched (tensile stress) to fit the substrate lattice. To accommodate part of the lattice mismatch it relaxes inward toward the upper part of the island.

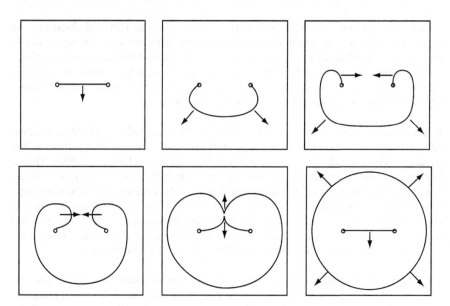

Figure 7.24: A schematic diagram of the operation of a Frank Read source generating dislocation loops. The diagram views the source from the top. The fixed position circles in each diagram show pinning points. A segment connecting the pinning points bows outward forming a loop. When the loop segments meet they react to annihilate each other. (Bottom left and center figures.) This leaves a segment between the pinning points and a loop of dislocation. Thus, the source generates loops as long as the stress continues.

The most likely mechanism for formation of new misfit dislocation segments is through dislocation interactions leading to dislocation multiplication. A classic example of a dislocation source of this type is the Frank-Read source in which a misfit dislocation segment connects two threading segments, which act to pin the moving dislocation. Under sufficient stress the mobile misfit dislocation bows outward between the pinning points as shown in Figure 7.24. The bowed region eventually reaches around the pinning points. The two sides of this loop of dislocation have the same Burger's vector but opposite line directions. Therefore, they can react to annihilate each other. The result is a dislocation loop that continues to expand and a segment left between the pinning points, which can generate new dislocation loops by the same mechanism.

A similar process has been proposed for strain relief in semiconductors as shown in Figure 7.25. Such a source could generate almost any number of misfit dislocation segments. The only problem is that it would generate misfit dislocations either at the same or closely spaced locations. It would then be necessary for these dislocations to move laterally in the interface to space themselves appropriately for strain relief.

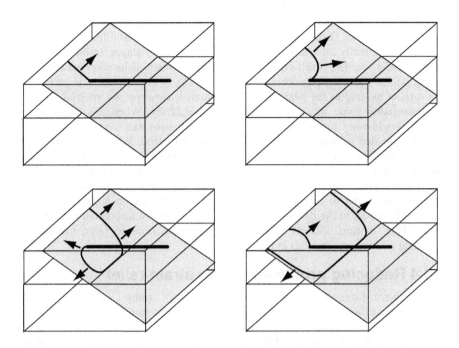

Figure 7.25: Shows the operation of a dislocation loop-type source for misfit segments. A pinning point is assumed such as the horizontal dislocation segment shown in the drawing. The segment in the glide plane (gray) bows under the misfit stress until it reaches the substrate interface. It then extends forming two threading segments and a misfit segment and leaving behind a new segment to continue operation of the source. [12]

Misfit segments have been observed in the transmission electron microscope (TEM) moving into strained layers and forcing adjacent segments to move aside. [11] Such a situation must necessarily involve not only sufficient strain to drive misfit segment extension (the Matthews critical thickness criterion of Equation 7.30) but also the additional strain necessary to drive the lateral movement and ordering of the dislocation network. The latter strain may be especially large as the interface plane is not generally a glide plane. The resulting dislocation array should be under slight compression to account for the residual force driving ordering. Therefore, a source in which dislocations must push each other around to achieve the proper spacing will operate at a thickness above the critical thickness. Finally, it is likely that the source will shift along the pinning object with each loop. This could permit a well-ordered dislocation network to form from a source without lateral dislocation motion. However, such a source would rapidly reach the end of the pinning segment and cease to function. The Frank-Read-like mechanisms for dislocation introduction are most relevant to films with low misfit and relatively perfect substrates.

It is possible to grow films with very high misfits where the critical thickness is less than one monolayer. Dislocations are grown into the interface as it forms in this case. Such films generally form as isolated islands in their early stages. In this case, each island rapidly acquires a self-contained equilibrium misfit dislocation network with highly uniform spacing (see for example, Fig 7.16). Such a situation is perfect for strain relief as long as the islands remain separated. However, as the islands grow, they inevitably come into contact with their neighbors. When this happens, the islands should want to merge, but their dislocation networks, formed in isolation, must be adjusted to match each other in the merged island. When the islands meet side to side, there is a strong force perpendicular to the dislocation lines driving the reorganization of the network. This generally allows the islands to merge, resulting in a larger island with a relatively perfect dislocation network. By contrast, when islands meet corner to corner, the driving force for dislocation alignment is at an angle with respect to the dislocation lines themselves, and a disordered network can result. This, in turn, inhibits island coalescence and leads to large numbers of threading dislocation segments in the resulting film.

7.3.4 Reducing problems due to strain relief

As the above discussion suggests, relieving stress in strained layers is not only necessary at sufficiently great thicknesses, but is also not easily achieved. The result is that strain relief is non-uniform and dislocation sources leave a large number of threading dislocation segments behind in the film. Such segments are inevitably detrimental to the performance of devices in the grown films. Several methods for reducing such problems have been explored and we will consider some of the more significant ones here.

The first approach is to discourage strain relief as much as possible by keeping the strained layer thinner than the point at which dislocations begin to form. As noted above, this may be well above the critical thickness of Equation 7.30 if sources of misfit dislocations can be eliminated. The most straightforward method to reduce misfit dislocation nucleation is to pattern the substrate into mesas in hopes that the average mesa will not contain any previously existing threading segments. This has been shown to be effective in increasing the critical thickness for strain relief substantially.

A second method to inhibit dislocation formation is to grade the interface through a range of alloy compositions to increase the strain gradually in the layer. Several factors contribute to this inhibition of dislocations. First, the initial strain in the grown layer is low, increasing the chance of forming a smooth film. A smoother film forms a more uniform dislocation network more easily. In addition, grading the junction means that misfit dislocations do not have a specific interface at which to occur, reducing the chances of dislocation tangles and potentially enhancing dislocation mobility. Such grading may also inhibit the operation of some dislocation sources.

Although it is possible by controlling alloy compositions and choosing appropriate substrates to reduce misfit strains, it is not always possible to obtain the desired film composition without strain relief. In practical terms, strain relief always leads to formation of threading dislocation segments in the film. Although one cannot usually prevent their formation, it is possible to reduce the number of these segments that propagate into the film as growth occurs. The most common method is to produce a superlattice of layers having alternating strains (compressive and tensile). The strain field of a threading dislocation interacts with the strain dipole at each interface. One sign of the interfacial dipole strain is likely to react favorably with the dislocation. Consequently, there is a driving force to bend the threading segment into an interface of the superlattice, where it is buried by subsequent film growth. This works best when the thickness of the strained layers is between one and two times the critical thickness for strain relief. Strained-layer superlattices are well known to reduce threading dislocation densities significantly and can reduce the overall amount of strain relief that takes place for a given thickness of film.

Unfortunately, overall it has generally proven impossible to eliminate threading dislocations to the point that a device in the film operates at its peak performance. Consequently, most film growth processes and alloy compositions are designed to achieve as close as possible to a lattice match between the film and the substrate, such that dislocations will not form and few threading segments will necessarily occur, or to keep the film thickness below the critical thickness.

7.4 PLANAR AND VOLUME DEFECTS

Planar defects can be divided into two categories, those that result from a rotation or displacement of the lattice with no change in long-range chemistry, and surfaces or interfaces between unlike materials or between like materials misoriented by a large angle. The former category includes twins and stacking faults, while the latter is primarily surfaces, interfaces between unlike materials, and grain boundaries in polycrystals of a given material.

7.4.1 Twins and stacking faults

Twins are defects that involve a change in the lattice orientation across a plane, known as the twin plane. Twin defects are formed by a rotation of the lattice by 60° or 180° about an axis perpendicular to the twin plane (axial type I twins), by shear (type II twin), or by reflection about a mirror (twin) plane (growth twins). The latter occur only during thin film growth, while axial and shear twins can occur during growth or as a result of deformation of the material. The rotation axis is usually along a bond direction. Figure 7.26 shows a transmission electron micrograph lattice image of a twinned region in an InSb epitaxial layer on GaAs. Note that in such a twinned region, the lattice is reflected about the twin plane. The reflection is often reversed at a second twin nearby as in the figure.

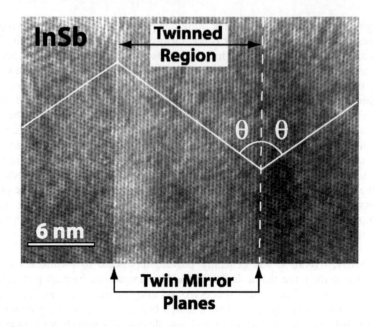

Figure 7.26: A transmission electron microscopy lattice image of a growth microtwin in an InSb epitaxial layer. The mirror (twin) planes are marked and one set of lattice fringes are highlighted to emphasize the mirror reflection.

Because twins represent rotations or reflections of the lattice, there are no broken bonds associated with them. Indeed, twins are only significant at the level of second-nearest neighbors in the lattice. Consequently they have relatively little effect on the electronic properties other than reducing carrier mobility through increased scattering.

Stacking faults, as their name implies, are due to errors in the stacking of atoms to form a crystal lattice. To produce such a fault from a perfect lattice it is only necessary to pass a partial dislocation through the material as partial dislocations fail to carry atoms on one lattice site onto an equivalent site. Thus, it is inevitable that a partial dislocation will have a stacking fault associated with it. One of the simplest forms of stacking fault occurs in fcc and hcp crystals, converting one to the other.

As we saw in Chapter 4, fcc crystals have a stacking sequence denoted "ABCABC", in which the letters represent the positions of planes of atoms with respect to each other. This situation is shown schematically in Figure 7.27. Likewise, the hcp structure has an "ABABAB" stacking sequence. Therefore, an error in stacking results in a sequence such as "ABCABCBACBAC" in a fcc structure. Here, the left

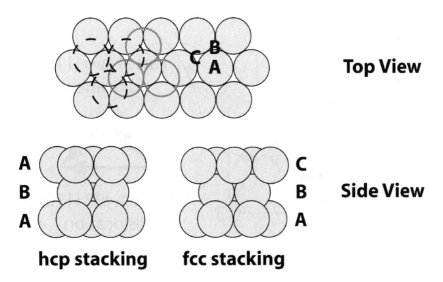

Top View

Side View

hcp stacking   fcc stacking

Figure 7.27: A schematic diagram comparing the stacking of a fcc and a hcp lattice. In the top view, the open gray circles indicate the position of atoms on a "B" site, while the dashed circles indicate the location of atoms on a "C" site. The C sites are rotated by 60° relative to the B sites.

side of the sequence and the right side are related by a reflection in the ordering. The junction between the two is the fault "BCB" region. This region appears to be a small domain of hcp stacking. Likewise, a similar stacking fault in an hcp material appears to be a small domain of fcc stacking. A second fault will return the overall stacking sequence to its correct order. The faults in these cases appear to be 60° rotations of the crystal lattice around a bond direction. Such faults are nearly indistinguishable from twins in the electron microscope, even at the atomic level. Stacking faults of this type can be relatively benign, as are twins, carrying with them no dangling bonds and no charged defects (in homopolar materials for example). Their primary consequences for the structure are in the strain fields associated with the partial dislocations that bound them. This type of stacking fault usually occurs during crystal growth, especially at lower temperatures and is an indicator of insufficient growth temperature or excessive growth rate. In some materials such as SiC, stacking faults are extremely common under all conditions and limit the performance of the material.

Not all stacking faults are so inconsequential. Consider, for example, a stacking fault in GaAs that switches the cation and anion sublattices. A perfect a/2{110} dislocation in GaAs can, for example, decompose into two a/6{111} partial dislocations as shown schematically in Figure 7.28. This decomposition results in a plane of Ga-Ga bonds at one side of the fault and As-As bonds at the other side. This is highly undesirable and causes dramatic electronic effects. Consequently, this type

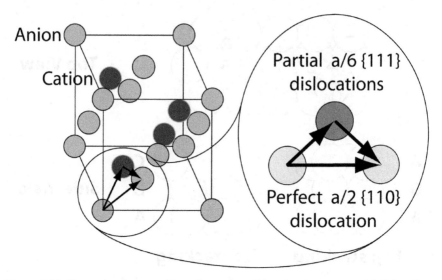

Figure 7.28: Shows the decomposition of a perfect dislocation into two partial dislocations giving rise to cation-anion sublattice disordering.

of fault is uncommon in compounds and has a high energy. The farther the two partial dislocations move away from each other, the higher the energy of the system, resulting in a strong attractive force drawing the partial dislocations together. In more complex compound semiconductors there may be additional ordering on either the cation or anion sublattices, as in the chalcopyrites. In these materials, the energy of a stacking fault that disorders the stacking of that sublattice is much lower than the cation-anion disordering faults. Consequently, such faults are much more common. Furthermore, point defects may occur which can compensate the charge on the fault boundaries and allow the faults to separate over larger distances.

When a stacking fault occurs in an ordered material it results in an ordering fault in the crystal structure known as an *antiphase boundary*. Thus, for example, a stacking fault in an fcc stacking sequence of ABCABC might result in a stacking of ABCCABC. It may also change a stacking such as GaAsGaAs to GaAsAsGaAs. In other words, it results in a change in the *phase* of the stacking without changing the crystallographic directions or axes.

7.4.2 Surfaces, interfaces, grain boundaries

This section gives a brief summary of the most critical aspects of surfaces, interfaces and grain boundaries. Numerous books have been written on this subject. For a single source on the physics of surfaces the reader is referred to Desjonquères or Zangwill in the recommended readings.

Anyone who has tried to make an ohmic contact to compound semiconductors or who has studied the variability of threshold voltages in metal-oxide-semiconductor field-effect transistors has probably experienced the effects of surfaces and interfaces on electronic properties first hand. Similarly, anyone who tries to make a device on polycrystalline group IV or III-V compound semiconductors has discovered the problems associated with grain boundaries. All of these structures generally have a significant number of dangling bonds associated with them. The few exceptions include a perfect epitaxial heterojunction or the interface between amorphous SiO_2 and a Si surface where defect densities may be less than ten parts per million. This is a primary reason for the choice of SiO_2 as a gate dielectric in field-effect transistors. The inability to achieve such low interface state densities between GaAs and an overlying dielectric layer is one of the primary reasons why devices based on GaAs have rarely been successful on a large scale. Dielectrics on GaAs with acceptably low interface state densities are usually lattice-matched wide-gap epitaxial semiconductor layers.

Even in a heterojunction, as we saw above, lattice mismatch can result in the formation of dislocations with edge character at the junction and have a significant number of dangling bonds. Consider, for example, a heterojunction between two semiconductors with a 1% lattice mismatch. If this mismatch is perfectly relieved by dislocations, then roughly 1% of the interface atoms will lie in the core of an edge dislocation and have a dangling bond as a result. This gives an effective concentration of defects of $5x10^{20}$ cm^{-3} if converted to a volume quantity. To effectively dope the semiconductor in this region, it would be necessary to have more dopant atoms than this. Such dopant levels are nearly impossible to achieve in most materials. Consequently, the electronic behavior of interfaces between semiconductors and other materials is usually dominated by the density of interfacial dangling bonds, not by doping. Therefore, even a strain-relieved heterojunction containing misfit dislocations is generally not an acceptable electrical interface. An exception to this description is in polar compounds where charge compensation may drive reorganization of the interface and reduce the effective dangling bond density.

The effect of large densities of defect states at an interface was alluded to in Chapter 3 in the discussion of real Schottky barriers. Here even the presence of a metal in contact with the semiconductor does not supply enough electrons to fill all of the defects and still deplete the semiconductor, as would be expected for an ideal Schottky diode.

Surfaces are generally similar to interfaces but even more extreme in their effects. Surface state behaviors are similar to what one would expect for a sheet of vacancies. So many are present and in such close proximity that the surface includes a band of defects. This may have significant dispersion as a function of electron wave vector as in other band structures. Examples of surface states measured and calculated on the Si(001) 2x1 surface are shown in Figure 7.29. A band of surface states provides an excellent site for recombination of minority and majority carriers. This effect of

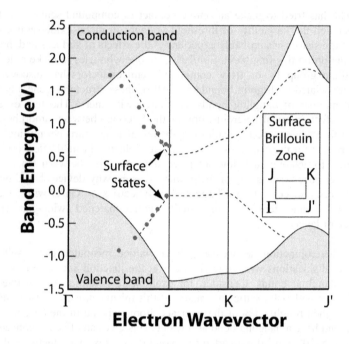

Figure 7.29: The surface states for Si(100) in the 2x1 reconstruction. Circles indicate experimental data and dashed curves show the calculated surface states. The bands indicated represent a range of energies covered by the surface component of the bulk band structure. Bulk band details are conventionally omitted in such diagrams to emphasize the surface states. Because the surface is reconstructed the surface Brillouin zone is rectangular. The results demonstrate why a maximum in the surface-state density of states occurs at an energy of ~0.5 eV. After *Surface Science* 299/300, Himpsel F.J., "Electronic structure of semiconductor surfaces and interfaces.", 525-540 (1994) with permission, copyright Elsevier 1994.

surface states showed up in the discussion of laser diodes in Chapter 3 where the recombination current for minority carriers at surfaces is so great that it generally limits the maximum power output of the diode. Furthermore, surface states usually pin the Fermi energy at the defect band making doping impossible near there.

The effect of surface defects can sometimes be reduced by surface treatments that "passivate" dangling bonds. This is usually done by covering the surface with a material that eliminates the dangling bonds or reduces their electrical activity. The passivating coating also reduces the surface energy of the sample. Hence, these passivating materials are surfactants. Usually they can also be thought of as surface dopants, providing a high level of doping and filling or emptying the dangling bonds completely. Because surfactants reduce the energy of a surface, these materials generally segregate strongly to that surface, especially if the surfactant is significantly larger than the atoms in the bulk. Being oversized simplifies the use of the surfactant, as it will not readily dissolve in the solid.

Typical surfactants have a difference in valence of one electron relative to the atom they replace. Thus, As atoms passivate Si surfaces, while S atoms can be used to passivate GaAs. H treatment of a Si surface reduces its reactivity significantly and passivates surface states. However, H is a small atom and relatively weakly bound to the surface. Furthermore, it reacts readily with other atoms that may adsorb. Hence, H passivation is delicate and of limited long-term value. Of more significance is that H can fit into many vacancies in solids, will passivate their dangling bonds, and so reduces the number of electrically-active defects. This is one reason why many chemical vapor deposition crystal growth processes include copious amounts of H as a dilutant gas. It not only contributes to control of the deposition reaction process, but it helps reduce the effects of growth defects.

Grain boundaries usually also produce bands of states in the energy gap. Sometimes these can be strongly compensated but usually their density is so high as to prevent doping near the boundary. For most group IV and III-V semiconductors this makes the optoelectronic application of polycrystalline materials nearly impossible (unless the grains are huge on the scale of the electron mean free path and diffusion lengths). II-VI and other ionic compounds are more effective as polycrystals, not because states do not occur, but rather because they are shallow and furthermore because the two types occur in such nearly equal quantities as to almost perfectly compensate one another. Major efforts have been made to find surfactants that could be intro-duced into grain boundaries in III-V and group IV materials to eliminate the boundary states but so far none are sufficiently effective to allow these materials to be used as small-grained polycrystals.

At this point, we will not belabor the issue of surfaces and interfaces further. To sum-marize the situation, their primary behavior results from large densities of dangling bonds. The effects of these dangling bonds are similar to the effects of vacancies and their influence may be reduced by the presence of surfactants if appropriate ones can be found.

7.4.3 Volume defects

The last class of defects considered here are volume defects. These are due to precipitates and domains of materials different from the matrix in which they lie. There is little new to add concerning these as the majority of the problems associated with them are due to interface states at the boundary between one material and another. An interface between different materials, even if perfect structurally, will generally have a contact potential that will produce an electric field and trap one type of carrier. If it has a lower energy gap it may trap both types of carriers. Such a second-phase region is used to advantage in a laser diode, in which the active quantum well traps both types of carriers (see Chapter 3).

Inclusions of second phase material may be harder or softer than the surrounding matrix. If harder, they are likely to cause plastic deformation (dislocation formation) in the softer material. If a two-phase material is heated, a difference in thermal expansion coefficient may cause deformation of the surrounding matrix as well.

When we considered the miscibility of semiconductor alloys, we noted that spinodal decomposition is likely to occur in many alloys having a significant lattice mismatch or difference in bonding chemistry. Such decomposition results in domains of one composition of material lying within or adjacent to another. If the compositions of the two phase-separated materials are not too different, then the decomposition results in coherent domains bound together without dangling bonds. It is less likely that a spinodal decomposition will proceed beyond the point where coherent domains can be formed as the additional cost of dislocations at interfaces would motivate against further phase separation. When modeling such a decomposition, the strain as a function of phase composition present must be included in a description of the free energy of the system and tends to stabilize the alloy.

7.5 SiC: a case study in stacking faults

Silicon carbide is an important wide-energy-gap semiconductor used for high power switching devices and as a substrate for some optical devices. It is stable, easily created and the elements from which it is formed are inexpensive and common.

The basic structure of SiC is the hexagonal wurtzite structure and based on the hexagonal close packed structure. The wurtzite unit cell is shown in Figure 7.30. What is most interesting for our purposes is that the stacking fault defect that determines how the planes stack has a very low energy. Furthermore, regular stacking fault structures occur spontaneously and alter the band structure. An ideal hexagonal close packed structure such as shown in Figure 7.30 stacks planes in alternating "ABABAB" type stackings (see Figures 4.4 and 7.27), while a face-centered cubic stacking is "ABCABC". The fcc stacking for GaAs is shown in Figure 4.3, while Figure 4.4 shows how the hexagonal planes are projected in the cubic symmetry.

SiC occurs in both the ideal fcc and hcp stacking sequences. Which of these occurs depends upon the conditions under which the structure is formed. The fcc sequence is known as "3C" SiC because three hexagonal planes are required to make the basic stacking creating the cubic structure. The basic hexagonal close packed stacking is known as the "2H" SiC structure. As with other fcc and hcp structures, the plane stacking sequences are ABCABC and ABABAB, respectively (Figure 7.30). Other materials including AlN, GaN, InN, CdSe and CdTe all can be created relatively easily in both the cubic and hexagonal structures. As with SiC, their band structures and electronic properties are modified by the change in stacking.

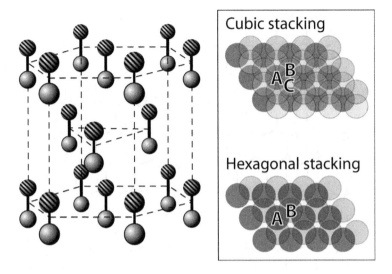

Figure 7.30: The basic lattice structure of SiC in the hexagonal structure (left) and a top view of the stacking sequences of the cubic and hexagonal structures (right).

What is unique about SiC is that it also occurs in many other stacking sequences, the most common of which are the 4H and 6H stackings. The 4H stacking sequence is "ABACABACA". In other words stacking faults occur regularly in the sequence. Once could view the 4H structure of SiC as fcc with regular stacking faults or as hcp with regular stacking faults. It is half fcc and half hcp in the stacking structure. The 6H stacking is "ABCACBABCACB". This is 2/3 cubic and 1/3 hexagonal.

Note that for the SiC to know that it is in a 6H crystal structure, interactions must occur that allow electrons to "know" where atoms are over 6 hexagonal plane distances. Each plane has two atoms (Figure 7.30) so this is equivalent to a *12*th-*nearest neighbor interaction distance*. If this is not remarkable enough, there is a recognizable rhombohedral structure known as 15R SiC which is 2/5 hexagonal and 3/5 cubic and has a 15 plane or 30 atom repeat distance. Viewing these structures along a <1120> direction (in the hexagonal indexing scheme) the structures can be shown schematically as in Figure 7.31. A transmission electron micrograph of a 4H SiC crystal is shown in Figure 7.32.

The various structures are stable enough to be possible to create individually but are not separate phases. Therefore, a separate name was needed to describe them. **The different faulted structures are known as polytypes**. The stacking fault energy in SiC is low, which allows the mixture of stacking sequences present in the various

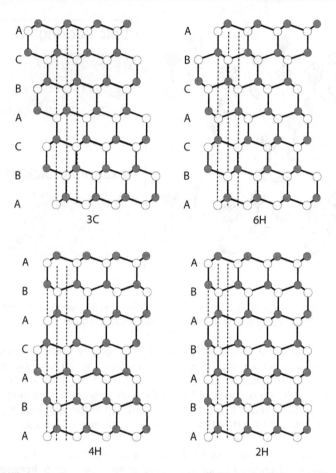

Figure 7.31: Shows the four most common polytypes of SiC viewed from a <1120> direction. The positions of atoms in the A, B, and C stacking positions are indicated by the dashed lines. Note that where the stacking faults are located depends upon whether you consider the basic structure to be cubic or hexagonal.

polytypes. However, this also means that it is difficult to create a pure single polytype structure and faults in the faulted stacking sequences of the polytypes are common.

The differences in lattice plane stacking have a direct effect on the electronic properties of SiC. For example, the energy gaps for the various polytypes are 3.33, 3.26, 3.02, and 2.39 eV for the 2H, 4H, 6H, and 3C; respectively. [14] All four polytypes are indirect gap materials with a corresponding direct gap in excess of 4 eV.

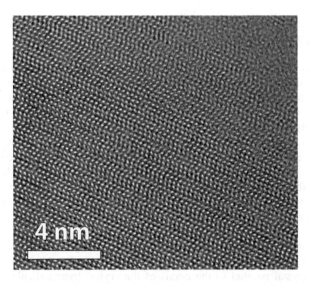

Figure 7.32: A lattice image high resolution transmission electron micrograph for a 4H SiC single crystal. Notice the appearance of regular twins in the structure (compare the diagonal twin planes in this image with the vertical twin planes shown in Figure 7.26). (Image courtesy P.O.Å. Persson and L. Hultman, Linköping University.)

There is limited data on the electron affinities of these phases. However, data for Schottky barrier heights on the 4H and 6H polytypes suggest that the electron affinity of the 6H polytype is significantly greater than that of the 4H polytype. [15] The same data suggests that the conduction band edge of the C-terminated polar (0001) face lies above the conduction band edge for the Si-terminated face. This is what one would expect based on the charge transfer between C and Si. Values quoted in Reference [15] are 4.1 and 4.8 eV for the two phases, respectively. However, significant discrepancies exist and the Schottky barrier heights are very sensitive to processing conditions. Data in Reference [14] suggests that the work functions of the 6H and 3C polytypes are similar but no carrier concentrations were given. Therefore it is not possible to estimate a work function based on the data provided. However, the results are suggestive that the electron affinities of these phases are similar. A lower work function for the 3C polytype was also reported in the same work.

The changes in electronic structure of the various SiC polytypes have a corresponding effect on the electron and hole effective masses. Typical values are, for example, 1.1 m_0 in the 3C polytype and 1.75 m_0 and 0.66 m_0 in the 4H structure. The two latter values are for conduction in and perpendicular to the basal plane of the hexagonal structure, respectively. [Egilsson et al. in Reference 15] The values for the 6H structure are similar to those of the 4H structure but somewhat lower. Note that the effective mass in the 3C structure is roughly intermediate between the values for the hexagonal structures. The effective mass of electrons is typically 0.25-0.7 depending upon the structures.

In addition to changes in carrier effective mass, the presence of stacking faults also has a direct affect on the carrier mobilities of the materials, with the 4H material exhibiting a uniformly greater mobility for both electrons and holes (approximately 400 and 20 cm^2 V^{-1} s^{-1} at 10^{18} carriers cm^{-3}) compared to the 6H polytype (approximately 200 and 18 cm^2 V^{-1} s^{-1} at 10^{18} carriers cm^{-3}). Electron mobilities decrease by roughly a factor of two as the electron concentration increases from 10^{14} to 10^{18} cm^{-3}. Likewise, the hole mobilities decrease with increasing carrier concentration. [Kimoto and Matsunami in Reference 15] Higher mobilities are also reported in other works. For example, Casady et al. report 800 and 140 cm^2 V^{-1} s^{-1} for electrons and holes, respectively but carrier concentrations were not given. [16] These values are consistent with undoped material.

Finally, one of the typically specified properties of SiC crystals is micropipe density. When SiC single crystals are grown it is common to find small voids propagating through the crystal parallel to the c-axis of the crystal (perpendicular to the basal plane of the hexagonal structure). A detailed analysis of these defects shows that they are actually screw dislocations with very large Burger's vectors. Therefore micropipes may also be described as large-core screw dislocations. Currently available SiC wafers are specified to have fewer than 30 micropipes cm^{-2} with the highest quality commercial wafers having fewer than 5 micropipes cm^{-2}. [17] Experimental reports of as low as 1.1 micropipe cm^{-2} have been published. [15] Micropipes are critical defects because they are large enough to allow metal atoms to diffuse down into their cores. This can result in a conducting path through the device which is a typical cause of reverse breakdown failure in diodes. Because the devices are commonly used for high voltage high power switching devices, micropipes are a major concern in device yield and reliability.

7.6 SUMMARY POINTS

- Point defects are vacancies, interstitials, antisites and impurities. The first three of these are intrinsic defects while impurities are extrinsic. Antisite defects only occur in a lattice with a superstructure order.
- Point defects modify the crystal potentially locally. This causes in a change in the bonding and the distribution of electronic states as a function of energy and momentum.
- "Shallow" electronic states are derived from the matrix of the semiconductor and modified slightly by the impurity atom. The result is a hydrogenic state that can be modeled as one charge of carrier orbiting a screened nuclear charge of the opposite sign. Such states maintain a roughly constant energy with respect to the band edge as long as the dielectric constant and effective mass are constant.
- "Deep" levels are derived from the point defect atomic orbitals, regardless of the energy difference between the state and the appropriate band edge. These states move relatively little if the band structure is modified by alloying.
- Trap states capture and release minority carriers faster than they cause recombination. This is most likely in doubly-charged defects.
- Recombination centers capture a minority carrier and then a majority carrier before the minority carrier is released into the band. This is most likely for initially singly-charged defects.
- The depth of a vacancy or antisite defect is generally reduced in more ionic compounds relative to more covalent materials.
- The ionization energy of a donor state is measured with respect to the conduction band edge while that of an acceptor is determined relative to the valence band, regardless of how deep the state is or what other defects are present. Charge state is determined by the Fermi energy relative to these levels.
- The equilibrium density of intrinsic defects is determined by a Boltzmann factor including the energy of formation of the defect. This energy is a linear function of the Fermi energy.
- When the energy gap is large, the energy of formation of a defect is small, or a material is very ionic it may be difficult or impossible to dope the semiconductor one type or more than to a given level. Occasionally both types of doping may be impossible. This is because the formation energy for the compensating defect becomes small.
- The diffusivity of atoms moving by a vacancy mechanism is enhanced by doping due to the formation of more vacancies.
- The diffusivity is reduced when clusters of the moving species form or when these atoms form clusters with other defects such as vacancies. (Clusters are less mobile.)
- Electric fields can enhance diffusivity by at most a factor of two.
- Edge dislocations have a hydrostatic component of their strain field and dangling bonds in the dislocation core while screw dislocations do not.

- The hydrostatic component of edge dislocations may be used to getter (trap) impurities in the dislocation core.
- Dangling bonds in dislocation cores may affect doping and free carrier concentration to a large radius around the core and may act as shunt paths in diode junctions.
- Pseudomorphic or commensurate structures involve a thin film grown on a substrate such that its lattice constant is the same as that of the substrate. When a lattice mismatch is present this is accommodated by strain in the film.
- Incommensurate structures consist of an unstrained lattice-mismatched film on a substrate in which the mismatch is accommodated in full or in part by dislocations.
- Dislocations with edge-character can reduce lattice-misfit strain in hetero-junctions.
- Strain energy in commensurate layers increases linearly with film thickness and as the square of the lattice mismatch.
- A critical thickness exists for thin films above which it is favorable to transform the film from commensurate to incommensurate by dislocation formation. This critical thickness scales roughly inversely with lattice misfit.
- Real misfit dislocation networks contain significant disorder due to the nature of the sources of misfit dislocations.
- In general the thickness of the layer required for strain relief to occur by dislocation formation must exceed the critical thickness sufficiently to provide a driving force for organization of the dislocation network as it forms and to operate sources.
- Small lattice-mismatched islands may relieve strain by distortion of the entire island.
- Misfit dislocations in (100)-oriented interfaces between diamond- or zincblende-structure semiconductors are generally "60°-type" (Figure 7.20). Such dislocations are half as effective in relieving strain as a pure edge dislocation with its Burger's vector in the heterojunction would be.
- Misfit dislocations may be formed by glide of existing threading segments, by expansion of dislocation loops from the surface or by operation of sources resulting from pinned dislocation segments.
- Threading segment sources are limited by the density of threading dislocations.
- Homogeneous loop nucleation (not requiring a pinning point or defect) operates only in highly-mismatched systems.
- Strained-layer superlattices may be used to reduce propagation of threading dislocations resulting from strain relief in underlying layers.
- Twins and some stacking faults produce bond distortions but no broken bonds and are usually of limited consequence. Stacking faults producing antiphase boundaries in ordered materials are high energy structures and result in significant numbers of charged defects.
- Surfaces, interfaces, grain boundaries, and second phase precipitates may accumulate charge due to local variations in work function as in any heterojunction.

Defect states associated with them form defect bands and may pin the Fermi level even in Schottky barriers.

- Surfactants may be used to reduce the effects of surface states and to minimize surface energies. These are typically impurities with compatible molecular orbital structures to the atom they replace but with a single extra or fewer electron. An oversized impurity will segregate onto surfaces for size reasons as well.

7.7 HOMEWORK

Consider a III-V semiconductor "CB" with energy gap 1.3 eV, electron affinity 4.4 eV, and lattice parameter 5.7 Å. Assume that the following data for atoms C, B, and an impurity G are available:

Element	Group	Electro- negativity	Radius nm
C	III	1.6	0.13
B	V	2.0	0.12
F	IV	1.8	0.15
G	I	1.9	0.12

States in the energy gap:

Semiconductor: CB

Conduction Band

$-V_B^+-$ $-G^{-2}-$ $-F^+-$ $\equiv H^+\equiv$

$-V_C^--$ $-G^+-$

$-G^--$ $-F^--$

Valence Band

State	Ionization Energy (eV)	State	Ionization Energy (eV)
V_B^+	0.45	V_C^-	0.45
H^+	0.04		
F^+	0.10	F^-	0.10
G^+	0.75	G^-	0.15
		G^{-2}	1.10

Note: Atom G in semiconductor CB does not donate an electron to itself; hence it will not occur simultaneously in charge state +1 and −1. The vacancies for atoms B and C are designated V_C and V_B, and have normal charge states +1 and −1, respectively when not neutral.

1) When semiconductor CB is doped with 1×10^{18} cm^{-3} G atoms *and no other impurities,* what are the charge states of

     i)   the majority of the G atoms?
     ii)   the two types of vacancies V_B and V_C?

2) Which of the two vacancies is most common when the material is doped with this number of G atoms? (Explain briefly) – Assume that the formation energies for the two neutral defects are the same.

3) A p-n homojunction is formed in semiconductor CB by doping with atom G on the p-type side and with another atom (H) on the n-type side. H is a simple hydrogenic donor. However, there are some G atoms (0.01 x the number of H atoms) on the n-type side. Would you expect the residual G atoms on the side doped heavily with H atoms to be traps or recombination centers (explain briefly)?

4) Atom F is normally found to be an acceptor (charge state –1) in compound CB even though it is a group IV element.
    a. On which sublattice would you expect to find F atoms normally?
    b. Would you expect to find F to have a greater or lesser tendency to segregate to the surface compared to atom G?
    c. Which atom, F or G would you expect to be more likely to work as a surfactant on semiconductor CB? Explain briefly.

5) The Fermi energy in the material is given by:
$$E_F - E_i = k_B T \ln\left[(N_d - N_a)/n_i\right]$$
where E_i=0.65 eV above the valence band edge, n_i=2x10^7 cm^{-3}, and T=300K. Suppose that a single crystal of BC contains 2.0x10^{17} cm^{-3} H atoms, 9.8x10^{16} G atoms, and no F atoms.
    a. Estimate the Fermi level position in the solid if there were no G atoms present. Use this Fermi level for the two following parts.
    b. Calculate the ratio of G atoms in the –1 charge state relative to the neutral charge state based on:
$$[G^-] = [G_0] e^{(E_f - E_{G^-})/k_B T}$$
    where [G$_0$] is the concentration of neutral G atoms.
    c. Calculate the ratio of G atoms in the –2 charge state relative to the –1 charge state based on:
$$[G^{-2}] = [G^-] e^{(E_f - E_{G-2})/k_B T}$$
    d. Estimate the concentration of G atoms in each of these three charge states based on the answers given in parts b and c.
    e. Calculate the approximate error in the Fermi energy if you had included the G atoms in the estimate in part (a).

6) Doping semiconductor BC with H atoms affects the vacancy population in the semiconductor. Assuming the formation energies for the two vacancies, V$_B$ and V$_C$ are 2.0 and 3.0 eV, respectively, and the density of sites in the lattice is 5x10^{22} cm^{-3}, estimate the vacancy concentration in the lattice for a Fermi level position of 1.25 eV above the valence band edge. Assume equilibrium at 900 K.

Assume that you want to grow a $Ga_{1-x}In_xAs$ layer for use as a photodetector. The mole fraction of the film has been chosen to be x = 0.813 such that the detector will have a 1.0 eV energy gap. Assuming a linear relation between lattice constant and alloy composition (Vegard's Law) you estimate the alloy lattice parameter to be 0.572 nm for this alloy.

7) You have the choice of GaAs or InP substrates. Given that the lattice parameters of GaAs and InP are 0.565 and 0.587 nm, respectively. Calculate the misfit f between the film and the two substrates.

8) In units of the Burger's vector, **b**, estimate the critical thickness for the $Ga_{0.187}In_{0.813}As$ alloy grown on GaAs if strain relief is accomplished by 60° type misfit dislocations. (Assume that the core correction factor $\alpha=1$ and that the Poisson ratio is 0.29). Hint: begin by determining angles θ and ϕ for these dislocations.

9) What is the most likely way in which dislocations would be introduced into the film when the thickness is much greater than the critical thickness? Briefly describe how this process works.

10) Suppose we have three dislocations as possible candidates for misfit strain relief: (1) a perfect edge dislocation with its Burger's vector lying in the plane of the heterointerface but not on a glide plane, (2) a 60° partial dislocation with its Burger's vector lying in a glide plane inclined with respect to the interface, and (3) a perfect screw dislocation with its Burger's vector lying in the plane of the interface and in a glide plane.

    a. Which dislocation gives the maximum strain relief (assuming you can chose the Burger's vector orientation correctly for each given case).
    b. Which dislocation is most likely to form in response to misfit strain relief?

11) If the alloy is doped n-type by a hydrogenic donor with $N_d=1.1 \times 10^{15}$ cm^{-3}, estimate the average radius over which a single pure edge dislocation threading through the epitaxial layer would deplete the carriers. Assume that the dislocation produces states at mid-gap, the average dangling bond spacing is 0.572 nm, and that dangling bonds are 100% filled.

7.8 SUGGESTED READINGS & REFERENCES

Suggested Readings:

Beanland, R.; Kiely, C.J.; and Pond, R.C.; "Dislocations in heteroepitaxial films", in *Handbook on Semiconductors*, ed. T.S. Moss, Vol. 3A, *Materials, Properties, and Preparation*, ed. S. Mahajan, North Holland, Amsterdam, 1994, Chapter 15.

Bube, Richard H, *Electrons in Solids : an Introductory Survey*. Boston: Academic Press, 1992.

Feng, Zhe Chuan and Zhao, Jian H., *Silicon Carbide: Materials, Processing and Devices*, v. 20 in the series *Optoelectronic Properties of Semiconductors and Superlattice*, ed. by Manasreh, M.O., New York: Taylor and Francis, 2004.

Hess, Karl., *Advanced Theory of Semiconductor Devices*, Englewood Cliffs, NJ: Prentice Hall, 1988.

Hirth, John Price and Lothe, Jens, *Theory of Dislocations*. New York: John Wiley, 1982.

Kröger, F.A., *The Chemistry of Imperfect Crystals* (North Holland, Amsterdam, 1964).

Matthews, J.W., *Epitaxial Growth*, parts A and B. New York: Academic Press, 1975.

Toshio Mura, *Micromechanics of Defects in Solids* (Martinus Nijhoff, The Hague, 1982).

Van Vechten, J.A., "A Simple Man's View of the Thermochemistry of Semiconductors" in *Handbook on Semiconductors*, ed. T.S. Moss, Vol. 3, *Materials, Properties, and Preparation*, ed. S.P. Keller, North Holland, Amsterdam, 1980, Chapter 1.

Wang, F.F.Y., editor, *Impurity Doping Processes in Silicon* (North Holland, Amsterdam, 1981)

References:

[1] Rockett A., Johnson D.D., Khare S.V., Tuttle B.R., "Prediction of dopant ionization energies in silicon: the importance of strain." *Phys. Rev. B*, 2003; 68: 233208-1-4.

[2] Fair, Richard B, "Concentration profiles of diffused dopants in Silicon," in Wang, F.F.Y., ed., *Impurity Doping Processes in Silicon* (North Holland, Amsterdam, 1981), chapter 7.

[3] Van Vechten, J.A., "A simple man's view of the thermochemistry of semiconductors" in *Handbook on Semiconductors*, ed. by T.S. Moss, v. 3, *Materials, Properties, and Preparation* ed. by S.P. Keller, North Holland, Amsterdam, 1980 (Chapter 1).

[4] Baraff, G.A. and Schluter M., "Electronic structure, total energies, and abundances of the elementary point defects in GaAs." *Phys. Rev. Lett.*, 1985; 55: 1327-30.

[5] For one of the earliest observations, see, for example, Schockley W., and Moll J.L., *Phys. Rev.*, 1960; 119: 1480.

[6] Zhang, S.B.; Wei, Su-Huai; and Zunger, A.; "A phenomenological model for systematization and prediction of doping limits in II-VI and I-III-VI$_2$ compounds." *J. Appl. Phys.*, 1998; 83: 3192-6.

[7] Rockett A., Kiely C.J., "Energetics of misfit- and threading-dislocation arrays in heteroepitaxial films." *Phys. Rev. B*, 1991; 44: 1154-62.

[8] Kiely C.J., Rockett, A, Hsieh K.C., York P.K., Kalem S., Coleman J.J. and Morkoc H. "TEM Studies of InAs/GaAs heterostructures grown by MOCVD and MBE techniques." Unpublished.

[9] Peach M., Koehler J.S., "The forces exerted on dislocations and the stress fields produced by them." *Phys. Rev.* 1950; 80: 436-9.

[10] Johnson H.T. and Freund L.B., "Mechanics of coherent and dislocated island morphologies in strained epitaxial material systems." *J. Appl. Phys.* 1997; 81: 6081-90.

[11] R. Hull, private communication. The reader is referred to the many excellent works by Hull and collaborators that describe the results of these observations. As an example of such work, see Hull R., and Bean J.C., "New insights into the microscopic motion of dislocations in covalently bonded semiconductors by in-situ transmission electron microscope observations of misfit dislocations in strained epitaxial layers." *Phys. Statu. Solidi A*, 1993; 138: 533-46.

[12] Beanland, R.; Kiely, C.J.; and Pond, R.C.; "Dislocations in heteroepitaxial films", in *Handbook on Semiconductors*, ed. T.S. Moss, Vol. 3A, *Materials, Properties, and Preparation*, ed. S. Mahajan, North Holland, Amsterdam, 1994, Chapter 15.

[13] Himpsel F.J., "Electronic structure of semiconductor surfaces and interfaces." *Surface Science*, 1994; 299/300: 525-540, and references therein.

[14] Harris, Gary L., *Properties of Silicon Carbide*, London: INSPEC, Institution of Electrical Engineers, 1995.

[15] Feng, Zhe Chuan and Zhao, Jian H., *Silicon Carbide: Materials, Processing and Devices*, v. 20 in the series *Optoelectronic Properties of Semiconductors and Superlattice*, ed. by Manasreh, M.O., New York: Taylor and Francis, 2004, various sections.

[16] Casady, J.B.; Agarwal, A.K.; Seshadri, S.; Siergiej, R.R.; Rowland, L.B.; Macmillan, M.F.; Sheridan, D.C.; Sanger, P.A.; and Brandt, C.D.; "4H-SiC power devices for use in power electronic motor control", *Solid-state Electronics*, 1998; 42: 2165-76.

[17] Cree materials catalog, retrieved from www.cree.com, February 2007.

[18] See, for example, Persson, P.O.A.; Skorupa, W.; Panknin, D.; Kuznetsov, A.; Hallen, A.; Hultman, L., "Structural defects in ion implanted 4H-SiC epilayers", *Materials Research Society Symposium – Proceedings*: 2001; 640: H6.2.1-H6.2.5.

Chapter 8

AMORPHOUS SEMICONDUCTORS

Amorphous semiconductors are somewhat of a niche area of electronic materials. However, they are critical to a number of important applications. It would be worth spending some time studying them based on these applications alone. More significantly, these materials are very distinct in their optical and electronic nature. Understanding their properties is highly instructive in a general sense.

Most of the discussion in this chapter focuses on amorphous group IV semiconductors (Si and Ge) which are by far the most commonly applied and technologically important. Amorphous Si (stabilized by hydrogen) is widely used as a photosensitive and photoconductive material in copying machines, for thin film transistors in devices such as flat-panel displays, and for energy conversion in photovoltaic solar cells. Other amorphous semiconductors are primarily chalcogenide glasses. As_2S_3, As_2Se_3, and Te_4Ge are all examples. Some materials occur in both crystalline and amorphous phases, and can be converted from one to the other by proper heat treatment. Such phase transformations have been the basis for data storage devices. Other common amorphous materials include silica (SiO_2) [and related alloys such as common window glass], silicon nitride (Si_3N_4), and most organic compounds. The silicon oxides and nitrides have very large energy gaps and are generally considered insulators, rather than semiconductors. Likewise, organic semiconductors are very different from other amorphous materials and are described in the next chapter.

8.1 STRUCTURE AND BONDING

Amorphous semiconductors can be divided into two general classes, those with tetrahedral bonding and four-fold coordination such as the group IV elemental materials, and those with two-fold coordination as in the chalcogenides (containing group VI elements) and pnictides (containing group V elements).

The structure of the two-fold amorphous semiconductors roughly resembles that of cross-linked organic molecules without the hydrogen. This is not surprising as Se and other group VI atoms are similar to group IV C atoms but with two extra electrons. A carbon chain with two hydrogens per carbon, as in alkane molecules, gets its two extra electrons from the hydrogens, while a Se chain does not need the hydrogens to achieve the same effect. Numerous defects can occur in the chain structures. These are similar to the carbocation and carbanion charged defects in organic chains (see Chapter 9) and have a net charge. The structure of a typical amorphous chalcogenide and some of the common defects are shown schematically in Figure 8.1. The bonding of amorphous chalcogenides is similar to bonding in C chains and results in the same types of molecular orbital structures described in Chapters 4 and 9 for organic molecules. There are also features common to the tetrahedrally bonded amorphous materials described below. We will not consider the chalcogenide materials further as their nature is similar to other more common materials discussed in detail. The interested reader may find additional information in the readings, especially in Brodsky.

The structure of four-fold-coordinated amorphous semiconductors consists of randomly arranged tetrahedra of atoms as shown in Figure 8.2. The bonding is generally covalent, thus this structure is referred to as a random covalent network. Bond distances are close to those of the corresponding crystalline materials (generally $< \pm 1\%$ deviation). The major change is in the bond angle (distortions typically $< \pm 10°$ relative to the tetrahedral 109° 28'). In a diamond-structure crystalline semiconductor, the smallest circuit through individual bonds and back to the starting atom is via six-membered rings. In a random covalent network, the distortion of the bonds leads to anywhere from four to eight membered rings (and occasionally to even more heavily-distorted structures), although the most probable structure is still the six-membered ring. None of the more modest distortions have severe detrimental effects on the optical and electronic properties. The problem comes in the most highly-distorted bonds, and, even more significantly, in dangling bonds resulting from occasional tetrahedra which are left with one corner having no other tetrahedron to bond to. These are rare, but even a few percent of such defects leads to large concentrations of states in the energy gap.

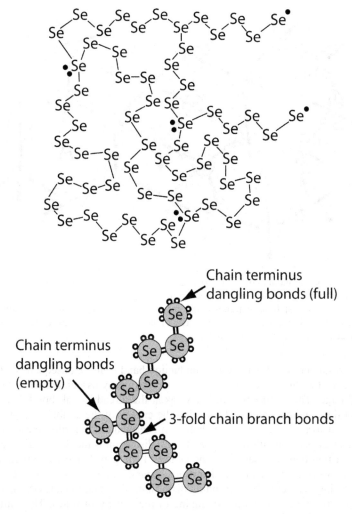

Figure 8.1: Schematic diagrams of the bonding in amorphous Se. Each Se atom is normally linked to two others. Defects occur in which one or three bonds to a Se occur. These junctions have extra electrons or partially-filled states and act as donors as acceptors as a consequence.

If we consider the bonding structure described by the Harrison diagram for Si in Figure 5.4, we find that the weakest bonds are those at the upper edge of the valence band and the lower edge of the conduction band. This is most easily seen by noting that the electrons in the top of the valence-band have the highest energy of those in any filled state. Therefore, one suffers the smallest energy penalty by distorting them.

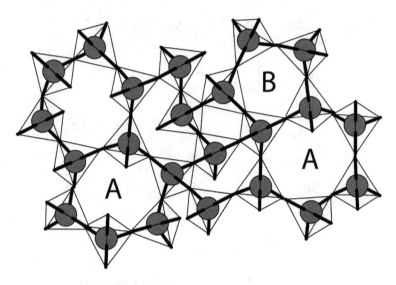

Figure 8.2: Shows a random covalent network of tetrahedrally-bonded atoms such as Si or Ge. Note that the coordination of atoms changes. Some regions show the usual six-membered ring structure found in crystalline materials (rings marked "A") . Other areas contain smaller (for example the five-membered ring, marked "B", or larger rings. It is rare to find rings larger than eight atoms or smaller than four atoms.

A similar argument can be made for conduction band states. Consequently, the bond distortions of the random covalent network are primarily associated with states near the band edge. The distortions further weaken the already weak bonds, raising the energy of the bonding states and lowering the energy of the antibonding states. The higher the distortion, the larger the change in energy and the farther the state moves into the energy gap. The ultimate case of bond distortion is the dangling bond, which results in a non-bonding state near the center of the gap. For the complete random covalent network a continuum of distortions occurs, which gives rise to a continuum of states filling the energy gap. Bonding distortions also modify states throughout the band, resulting in effective local averaging of the density of states. This situation is shown schematically in Figure 8.3.

Several models for the density of states have been proposed and measurements have been made to understand the nature of states in the energy gap of amorphous tetrahedral semiconductors. These generally include the following components: (1) delocalized band-like states, (2) localized band-like states leading to band tails, (3) highly-distorted states deep in the gap, and (4) non-bonding states in the gap. Each of these components is, in some sense, distinct in its behavior, contribution to the density of states, and contribution to conduction behavior, although a sharp boundary may not exist between the groups. Nonetheless, it is useful to make the distinction for reasons we will see below.

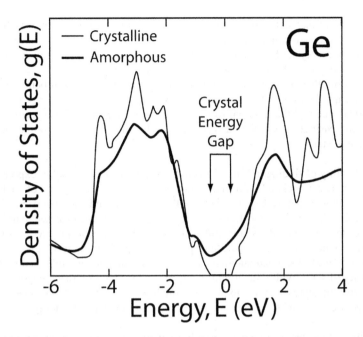

Figure 8.3: The density of states for amorphous and crystalline Ge. Note that the density of states for the amorphous material does not go to zero (there is no true energy gap). Amorphous Si has a similar behavior. Redrawn with permission based on Kramer B and Weaire D., "Theory of electronic states in amorphous semiconductors," Figure 2.8 in Brodsky M.H., *Amorphous Semiconductors* in *Topics in Applied Physics*, v. 36. Berlin: Springer-Verlag, 1979. Copyright Springer, 1979.

Delocalized band-like states are the result of relatively unperturbed bonding-antibonding structures (it does not take much regularity to form them). These account for the bulk of the density of states diagram in the energy range covered by the middle of the crystalline valence and conduction bands. With reference to the LCAO theory of Chapter 5, the relatively undisturbed bond length means the effective matrix elements for any local structure are essentially undisturbed. It is even possible to treat the bulk of the band structure of amorphous semiconductors with a generalized Bloch wave method as in other band structures, although no spherical asymmetry can exist when averaging over the entire material. Electrons and holes in these distributed states move freely at any temperature as they would in a crystalline energy band.

Note that an indirect-gap semiconductor is impossible in an amorphous material as crystalline directions for indirect minima are undefined. Therefore, all amorphous semiconductors have direct gaps.

Localized band-like states are distorted to the point that they lie well away from the energies of the delocalized band-like states. Because localized states are widely scattered in energy and position, they are too far apart to overlap much with each other (true even at third or fourth neighbor distances). Indeed, to be physically isolated from other states of like energy is a *necessary component* of being localized. Therefore, an electron or hole in one such state must change its energy in order to move. If one cooled the material to near 0 K the movement of these carriers would become increasingly sluggish and would finally be due only to long-range tunneling from state to state.

The probability of finding a state of given distortion decreases roughly exponentially with the energy associated with that distortion. Therefore, the density of such states decreases approximately exponentially with energy around the "band edge". This band edge is defined as the transition from localized to delocalized bonding. In principle, this is a nebulous transition. However, because the interaction of states changes exponentially with separation, practically speaking there is a rather abrupt transition between localized and delocalized states. The resulting exponential tails in the density of states within the mobility gap are referred to as Urbach edges. Because the density of states can be expressed roughly as $e^{-(E-E[edge])/U}$, the energy U is characteristic of the width of the tail, and is known as the Urbach energy.

The ability of carriers to move is strongly reduced near the transition from localized to delocalized states, therefore their mobility drops rapidly there. A "mobility gap" can be defined based on this transition for a given amorphous semiconductor. This mobility gap takes the place of the energy gap in crystalline materials. While states exist continuously within the mobility gap, the optical absorption properties and the behavior of diodes based on these materials are governed by the mobility gap in much the same way that the energy gap governs the behavior of crystalline materials. The mobility gap changes with both pressure and temperature, as does the energy gap of a crystalline semiconductor but with a different functional form. The effect is most obvious in the optical gap of the material as discussed in Section 8.4.

The third type of state, deep states in the energy gap are distinct from the band-like localized structures in that their probability ceases to decay rapidly with energy as one draws away from the mobility edge. They represent a "noise background" of states present across the mobility gap at a roughly fixed concentration determined essentially entropically. They are distinct from the fourth type of state as they include a distinct, if very heavily distorted, bond-like character. Such states could include various reconstructions of the interior of a vacancy or multisite vacancy cluster, which would allow dangling bonds to interact at least weakly to form some sort of bond. This situation is shown schematically in Figure 8.4.

Finally, the fourth type of state is the non-bonding dangling bond. Dangling bonds have energies near the middle of the gap where non-bonding states would be

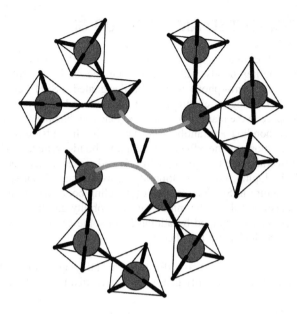

Figure 8.4: Structure of a vacancy in an amorphous semiconductor showing bonding reconstruction.

expected in covalent crystalline materials, as discussed in Chapter 7. Dangling bonds have several important effects on the material. First, they are more common than other deep states in the energy gap as they result from relatively well-defined defects rather than specific rare types of distortions in the structure. Second as non-bonding states they are half filled with electrons. Third, because they are so numerous, they tend to pin the Fermi energy at the center of their energy range. That is to say, it is difficult to move the Fermi energy away from the energy of these defects without either reducing their density or doping the material very heavily. Fourth, because they are non-bonding states, it is possible to insert a small atom (e.g. H) in the structure, which can seek them out and bond to them without otherwise changing the amorphous structure. This is the basis of the stability of hydrogenated amorphous silicon, discussed below.

It is possible to produce some amorphous compound semiconductors such as amorphous GaAs. These materials have a distinction between dangling bonds associated with the cation and the anion. Therefore, non-bonding states of type (4) would not necessarily lie at or near the middle of the mobility gap. In principle, H passivation of such states is also possible. These materials are generally less stable and less homogeneous in amorphous form than other amorphous materials and are not widely used in technology applications.

8.2 HYDROGENATED AMORPHOUS Si

As noted above, adding hydrogen to amorphous Si (or Ge) provides an opportunity for bonding to dangling bonds without affecting the basic amorphous covalent network. Hydrogenated amorphous Si will be designated a-Si:H henceforward. A hydrogen atom, discounting its electron is just a proton and is very small. It can fit into virtually any space in the material including on lone dangling bonds. The hydrogen-matrix bonds have bonding and antibonding levels that lie outside of the mobility gap. This removes most of the effects of dangling bonds, reduces the density of states in the gap overall, (see Figures 8.3 and 8.5) and makes it much easier to move the Fermi level via doping. Hydrogenation also provides significant stabilization of the amorphous structure in Si. The removal of dangling bonds has a direct effect on the density of states not only in the middle of the mobility gap but also at the band edges. Similar results are obtained by adding F. In some cases, optimal results are obtained by adding both H and F together.

One can also consider hydrogen as effectively an alloying element. It has been found, for example, that the mobility gap is directly related to the H_2 pressure in the

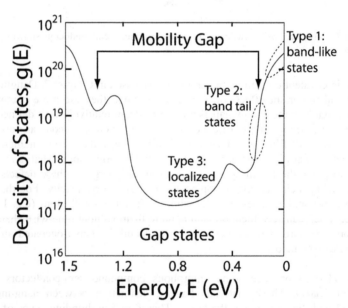

Figure 8.5: Density of states in the mobility gap region of amorphous silicon. Type 4 dangling bond states do not appear in this material because hydrogenation has reduced their density below the highly distorted type 3 localized states. Figure redrawn with permission based on Lecomber, P.G. and Spear, W.E., "Doped Amorphous Semiconductors," Figure 9.1, in Brodsky M.H., *Amorphous Semiconductors* in *Topics in Applied Physics*, v. 36. Berlin: Springer-Verlag, 1979. Copyright Springer, 1979.

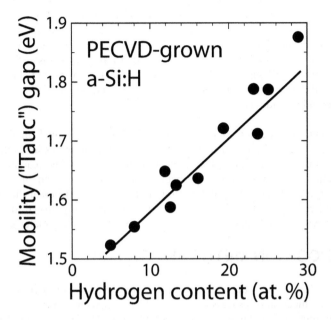

Figure 8.6: The change in mobility gap in a-Si grown by plasma-enhanced chemical vapor deposition as a function of H content. Data courtesy C. Wronski, described in detail in Reference [3].

deposition reactor (Figure 8.6) in a-Si grown by plasma-enhanced chemical vapor deposition. [3] This behavior is also observed in a-(Si,Ge) and a-(Si,C) alloys. Increasing the activity of H_2 in the process increases the probability of hydrogen breaking a highly distorted bond and incorporating into the two broken bonds. This has the net effect of both decreasing the electrical activity of distorted bonds and simultaneously increasing the number of Si-H bonds. Because H is monovalent, its addition fundamentally changes the nature of the a-Si bonding, reducing the need for tetrahedral hybridization and reducing the energy of certain distorted bonding structures. This contributes to increasing the mobility gap.

Unfortunately, Si-H bonds are relatively easily broken (only one single bond needs to be severed) and H is a relatively mobile species. For example, introducing an electron into any Si-H antibonding state eliminates the bond. Once the bond is broken, the H atom may escape into the matrix, leaving an unpassivated dangling bond behind. This results in an increase in electrically active states in the mobility gap under light or forward bias conditions and is known as the Stabler-Wronski effect.[4] In this effect, the density of states in the mobility gap increases upon injection of minority carriers into the material either by photogeneration (as in a solar cell) or by direct injection as in a forward-biased diode. Both of these are detrimental to resulting devices. The effect can be mostly reversed by annealing at moderate temperatures. This allows repassivation of dangling bonds and a return to better performance. Unfortunately, most devices using a-Si:H involve minority carrier injection and do not operate at sufficiently elevated temperatures to anneal during operation. Thus the Stabler-Wronski effect leads to an overall loss of performance and has been a significant barrier to broader application of a-Si:H. For a detailed description of defect states in the energy gap of a-Si, and a-SiGe alloys, see Cohen in the recommended readings.

Finally, intentional addition of H_2 during a semiconductor growth method such as chemical vapor deposition can increase the density of vacancies and other growth defects because the H_2 can lower their formation energy by reacting with the resulting dangling bonds. This is probably responsible for much of the behavior shown in Figure 8.6. Generally the minimum amount of H_2 necessary to get the job done should be used. On the brighter side, H covers up many growth mistakes that would otherwise degrade device performances even in crystalline materials.

8.3 DEPOSITION METHODS FOR a-Si

Much of the discussion in this chapter refers to changes in the properties of deposited materials as a function of process conditions. Hence, it seems appropriate to mention the relevant methods used in producing a-Si:H briefly. These can be divided into two approaches, chemical vapor deposition (CVD) [see Chapter 12] and sputtering [see Chapter 11], with the CVD contributing the vast majority of the processes. CVD methods, in turn, can be divided into conventional CVD, plasma-enhanced CVD (PECVD) operated either with a dc plasma or an rf plasma, and hot-wire CVD. All of these involve supplying Si (or the atoms for whatever semiconductor is to be deposited) in the form of a gaseous reactant. Typical reactants for a-Si:H are silane (SiH_4), disilane (Si_2H_6) and hydrogen (H_2). In the reactor the reactant, say SiH_4, diluted with H_2, decomposes as in $SiH_4 \rightarrow Si + 2H_2$. The Si then deposits on the substrate as a-Si with incorporation of H atoms depending upon the partial pressure of the H_2 gas in the reactor. This partial pressure also controls the forward reaction kinetics of the source gas decomposition according to the law of mass action [see Chapter 4]. The kinetics and reaction products can be dramatically altered by adding energy to the reactant gases prior to their striking the growth surface. This reduces the need for thermal energy driving the decomposition to be supplied by the substrate surface; which, in turn, reduces the required deposition temperature. Low temperatures are essential to obtaining good amorphous material.

The most straightforward method for supplying energy to the reactants is by passing the gas through a partially- or strongly-ionized plasma. Such plasmas may be generated by dc or rf electric fields. Either may be used in PECVD. When the reactants pass through the plasma, the molecules are ionized, partially decomposed or simply vibrationally excited. Any of these result in a dramatic increase in the reactivity of the molecule. Alternately, hot-wire CVD increases the reactivity of the gas by passing it by a heated filament. This also partially decomposes the reactants.

By choice of reactant, reactant partial pressure, and reaction enhancement conditions, large changes in the details of the reaction process can be achieved. For example, variation of the frequency of excitation of the gas in rf-PECVD changes the types of species present in the gas, which alters the average reactivity. The primary experimental variables in the CVD deposition processes that control aspects such as crystallinity or amorphous nature and defect density in the resulting film are the gas partial pressures, temperature, and the method and conditions of stimulating the reaction.

Sputtering methods use a partially-ionized inert gas such as Ar to bombard a target such as Si. Atoms are knocked off of the target by the collisions and thus vaporized. The resulting Si atoms deposit on a substrate. For low-temperature processes, this results in amorphous material. Hydrogen is supplied by addition of H_2 to the sputtering gas. The sputtering gas pressure determines the number of collisions a molecule undergoes between the target and the substrate. Particles that travel directly from target to substrate without collisions may have significant energies. Collisions generally slow the species, often to thermal energies. This means that the gas pressure influences the growth process dramatically through control of the energy distribution of species striking the growing film surface.

8.4 ELECTRONIC PROPERTIES

8.4.1 Carrier transport and mobility

Free carrier movement in a-Si:H, and to an even greater extent in other amorphous semiconductors, is difficult because there are a large number of defect states in the material into which a free carrier might fall. Once trapped in a defect state, the carrier has several possibilities for further motion, illustrated in Figure 8.7. First, it can gain energy from the heat in the system sufficient to carry it into the delocalized states, in which it is free to move as a carrier in a crystalline material would,

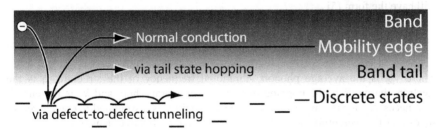

Figure 8.7: Conduction mechanisms illustrated schematically for an electron. Similar processes exist for holes. Normal band mobility requires thermal excitation from a trap state into the band.

although with much higher scattering rate. While it lasts, the mobility of such a carrier is not dramatically different from that of a normal crystalline semiconductor. However, after a short time this carrier is trapped again and becomes effectively immobile. Carriers are excited to the mobility edge mostly from states at or near the Fermi energy. The number of carriers, n for electrons, excited above the mobility edge is therefore given approximately as: [5]

$$n = kTN(E_C)\left(e^{-(E_C - E_f)/kT}\right)$$ 8.1

where E_C is the "conduction band" mobility edge energy, $N(E_c)$ is the effective density of states at the mobility edge, and E_f is the Fermi energy. Mott [6] derived a relationship for mobility corresponding to this carrier behavior and having the functional form:

$$\mu = \frac{C}{T},$$ 8.2

where C is a constant. Different treatments have derived different forms for the constant but all agree upon the inverse temperature dependence. The models yield typical room temperature mobilities for a-Si:H of the order of 1-10 cm^2 V^{-1}s^{-1}, well above that actually observed. Taken together with Equation 8.1 this model suggests an essentially simple exponential temperature dependence of the conductivity by this mechanism. It also shows that this type of conduction freezes out rapidly as the temperature decreases, especially when, as is typical, the Fermi energy is far from a mobility edge. Band-like conduction is generally only significant at high temperatures or when the material contains a relatively low density of states in the mobility gap.

Conduction can also operate in the band tails below the mobility edge when such tails are wide and have a sufficiently high density of states to allow rapid hopping of a carrier from one site to another via thermally-activated processes. In this case the mobility of a carrier depends exponentially upon the amount of energy it must gain to leave a given state and hop to the next. Thus, it was proposed that the mobility will have the form [5]

$$\mu_{hop} = \frac{qv_{ph}R^2}{6kT}e^{-W/kT}$$ 8.3

where v_{ph} is the average phonon vibrational frequency of the structure, q is the electron charge, R is the hopping distance in an average hop, and W is the energy gain necessary for the average hop. The net mobility is a weighted average over all states and the weighting factor is determined by the effective density of states at a given energy. The density of states typically falls rapidly (exponentially) below the mobility edge, while the probability for the energy gain decreases exponentially as the barrier for the hop becomes larger. The net result is equivalent to what one would obtain directly from Equation 8.3, using W=kT, leaving the exponential term as 1/e. Note that once again, as in Equation 8.2, the mobility decreases roughly inversely with temperature but with a different constant. The mobility obtained is of the order of 0.01 cm^2 V^{-1}s^{-1} for typical values of the remaining constants, or about two orders of magnitude smaller than for the conduction above the mobility edge. The band tail mobility is closer to typical values observed for mobilities.

In the final conduction pathway is phonon assisted tunneling. This requires states sufficiently close in energy and momentum and physically close enough in the solid for tunneling from one to the next to occur. The farther the carrier energy is from the mobility edge the slower this process, in general, due to the lower density of states and the higher the effective tunneling barrier. However, in some cases tunneling may be faster than for the mechanisms discussed above. In particular, if there is a relatively high density of states at the Fermi energy then this mechanism can operate relatively easily. Such a situation is typical when the Fermi energy lies at the energy of dangling bond states and when there are many dangling bonds present in the material. As with hopping in band tails, the phonon-assisted tunneling mechanism conduction has a highly path-dependent effective mobility. Carriers that tunnel between states that are very close to one another in the solid move rapidly and have a higher effective mobility. More widely spaced states result in extremely small effective mobilities. The jumping rate, r, is given by [7]

$$r = v_{ph} e^{-W/kT} e^{-2\alpha R},$$ 8.4

where tunneling yields the exponential dependence of rate on jump length R, α is the characteristic decay length of the integrated overlap of the wave functions of the two states, and W represents an effective energy barrier. In this case, W is assumed to be modest. From Equation 8.4 an effective mobility can be estimated as [7]

$$\mu = \frac{qR^2}{6kT} r = \frac{q v_{ph} R^2}{6kT} e^{-2\alpha R} e^{-W/kT},$$ 8.5

which is obtained by noting that the diffusivity is D=rR^2/6, and using the Einstein relation, μ=qD/kT. Equation 8.5 is nearly identical to Equation 8.3, with the exception of the addition of the exponential tunneling term and includes the same inverse temperature dependence as in the previous formulae. This would be convenient if it were not for a much more complex behavior in the average jump distance R. Furthermore, the necessity to weight the probability of each jump for the likelihood of occurrence of a given configuration of states of a given type causes a very different net temperature dependence. A complete derivation is beyond the scope of this book but may be found in Mott [7]. A properly averaged jumping rate gives a term of the form:

$$r = v_{ph} e^{-A/T^{1/4}}.$$ 8.6

A temperature dependence of the conductivity scaling as an exponential of T$^{-1/4}$ is a standard signature of variable-range phonon-assisted tunneling conduction. A similar behavior is found for conduction in bands of defect states in crystalline semiconductors.

8.4.2 Mobility measurements

A standard method to determine the mobility of carriers in semiconductors is the Hall effect. In this method, a current passes through a sample, which is subjected to a magnetic field normal to the conducting plane. The Lorentz force deflects the moving carriers in the plane, resulting in a voltage build-up perpendicular to the moving current. The sign of this voltage gives the type of carrier moving, while the magnitude gives the mobility. For variable-range hopping, this behavior is very unpredictable, as either electrons or holes may dominate the conduction process. This is because the carriers move through states near the Fermi energy. Consequently, it is nearly as easy for electrons to move as for holes. Whichever carrier has the higher probability of hopping from state to state will dominate. This shifts as temperature changes, leading to instabilities in the measurements.

An alternative method of measuring mobility is to determine how far photogenerated charge in a sample drifts in the presence of an electric field as a function of time. The experimental method for such measurements is shown schematically in Figure 8.8. An electric field is applied along the length of a bar of sample. A flash of illumination creates electrons and holes. These drift in opposite directions according to their mobilities, the applied electric field, and time, resulting in a current at the contacts as carriers arrive. This current may then be modeled to extract the mobility. Because the carriers are charged, a voltage difference is associated with the separated charges. This voltage may also be sensed at contacts along the test material. In a crystalline semiconductor with a well-defined carrier mobility, the carriers drift together down

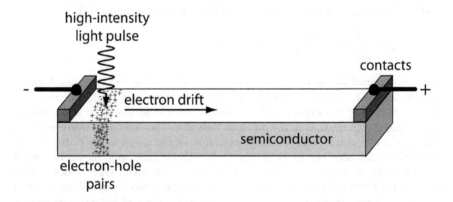

Figure 8.8: A schematic diagram of the method for making drift mobility measurements. Current is measured to determine the number of photogenerated carriers reaching the contacts per unit time.

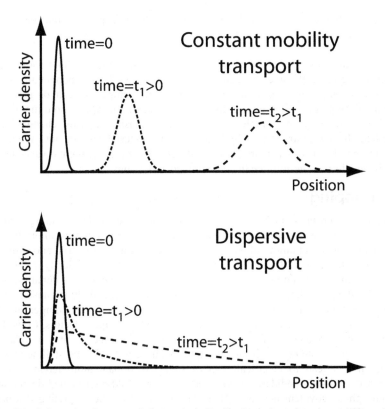

Figure 8.9: Schematic diagram of the typical distribution of carriers in a drift mobility measurement of an a-Si:H sample for several times. [For details see Zallen, Ref. 8.]

the sample in a bunch. The bunch spreads out due to diffusion while recombination eliminates minority carriers gradually. The net result is the carrier density as a function of time shown in the upper part of Figure 8.9 for normal conduction.

In an amorphous semiconductor the carrier mobility varies dramatically from one carrier to another. When these carriers drift in an electric field the result is very different from in a crystalline material and the distribution for this "dispersive transport" is shown in the lower half of Figure 8.9. The most common situation is that a carrier has no significant mobility (most carriers are trapped), thus the peak in the concentration of carriers is fixed at its initial location. A fraction of the carriers become mobile and drift in the field according to the mechanisms outlined in Figure 8.7. Even though the most common location of carriers is where they started, the average position moves. The behavior is referred to as dispersive transport because there is a dispersion in the mobility values.

Experimental measurements of carrier mobilities in a-Si$_{1-x}$Ge$_x$:H alloys show that the mobility decreases [9] and the defect density increases [10] exponentially with x. Deposition temperature, post deposition annealing temperature, hydrogen partial pressure during deposition, photon flux, and other variables affect both mobility and defect state densities in complex ways. Most of density of states changes can be traced to the concentration of hydrogen in the material and its interaction with dangling bond states, which are strongly affected by the experimental variables. For a more detailed discussion, see the recommended readings.

8.4.3 Doping

A significant problem in application of amorphous materials including a-Si:H is doping. Doping, as discussed in previous chapters, involves adding impurity atoms that change the total number of electrons (and protons) in the material and shift states near the band edges. These are the same states that are typically affected by distortions of the bonding so one might expect that the behavior of a dopant might be altered by the amorphous nature of the structure. Even more likely to cause problems is the very high density of defects. Each defect can potentially compensate for a dopant atom, as was the case for vacancies in crystalline semiconductors. Considering the intrinsically defective structure of amorphous semiconductors, it is not clear that addition of monovalent substitutional impurities used in doping crystalline materials should in fact lead to electrically-active states. A dopant atom having an extra electron is an ideal candidate for location on a three-fold-coordinated defect site in the amorphous covalent network. Therefore, it would not be surprising if a dopant atom added during the deposition process would simply induce formation of a defect site, as was the case for doping partially ionic semiconductors. In spite of these concerns, a significant fraction of dopant atoms have been observed to be electrically active. This is presumably because either most of the dopants are in tetrahedrally coordinated sites, or adding dopants adds fewer additional coordination defects to the material than the number of dopant atoms. The complexity of the materials makes prediction of doping effectiveness from theory nearly impossible so it is useful to look at the experimental data.

A-Si:H and related materials are typically doped with B, As and P, as is crystalline Si. These dopants have a very high solubility in both the amorphous and crystalline phases and can be electrically active to concentrations exceeding 10^{20} cm^{-3}. Typical defect densities in the mobility gap range from 10^{17} - 10^{19} cm^{-3} eV^{-1} and vary with deposition technique. Therefore, the possibility of doping depends somewhat upon the details of the material processing, making process control critical to the yield of resulting devices. Nonetheless, if the electrically active dopant concentration were 10^{20} cm^{-3}, then one might expect reasonable doping. In plasma-enhanced chemical vapor deposition and similar vapor phase processes, roughly 1/3 of the dopant atoms incorporated into the amorphous structure are electrically active. In PECVD, doping is generally accomplished by adding gases such as diborane (B_2H_5), arsine (AsH_3) or phosphine (PH_3) to the deposition environment. One can also ion-implant the dopant into a-Si:H, but the efficiency of the doping that results is low, even when the implant is carried out at elevated temperatures. Doping by addition of phosphorous or boron to a-Si:H by MOCVD generally allows changes in conductivity in excess of five orders of magnitude, although the Fermi energy rarely is less than 0.2 eV from the mobility edges. Nonetheless, the built-in voltage in an a-Si junction may exceed 1 V. Even with an optimal process, it is difficult to obtain a sufficiently high electrically active dopant concentration to move the Fermi energy all the way to the mobility gap edges which would yield the full 1.5eV built in voltage one might expect based on the mobility gap value.

A consequence of the difficulty of doping is that most a-Si:H devices are based on "p-i-n" structures. In these, one dopes the ends of the device as strongly as possible, leaving the central region undoped (intrinsic). The highly doped regions are used to generate depletion, built-in electric fields, and for effective contacts to the material. The central intrinsic region is generally the "active" part of the device as in a solar cell or field-effect transistor. The reasons for the p-i-n structure include the fact that optimization of the material for high doping may not lead to the best properties for stable operation of the device. With the p-i-n structure, the intrinsic region can be produced under conditions favoring higher or more stable device performance.

8.4.4 Short-range order

The most straightforward method to obtain high doping levels is to use polycrystalline rather than amorphous material for the heavily doped regions. The difficulty with polycrystalline Si is that, when lightly doped or undoped, the properties and performance are heavily influenced by grain boundaries. The density of such boundaries is determined by the deposition technique and conditions used, which renders lightly-doped or intrinsic polycrystalline materials unpredictable or unreliable for devices. The solution to both the doping problems of a-Si:H and the unreliability of lightly-doped or intrinsic polycrystalline Si is to use polycrystals for heavily doped regions and in conjunction with the contacts, and amorphous material for the lightly-doped or intrinsic areas.

One does not need to obtain large grain sizes in order to produce a much more dopeable material. Thus, the standard contact regions in current a-Si:H devices use a heavily-doped micrograined or nanograined polycrystalline material. Recent experiments by Feng et al. have further demonstrated that such nanograined material can be controllably obtained by simple changes in the deposition conditions for sputter-deposited a-Si:H. [11] The process uses dc magnetron sputtering [see Chapter 11] with a Si target with a mixture of Ar and H_2 gas. Crystallinity increases as hydrogen partial pressure increases above a critical pressure, which depends upon the details of the processing equipment, over a range of temperatures. It is also straightforward to produce amorphous C-Si-Ge alloys this way by mixing CH_4 with the sputtering gas

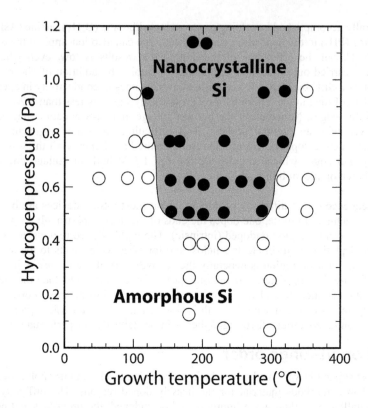

Figure 8.10: Experimental results showing how H_2 partial pressure and substrate temperature affect crystallinity in sputter deposition of a-Si:H in Ar-H_2 gas mixtures. Filled circles represent nanocrystalline Si while open circles correspond to amorphous Si. Figure courtesy J. Abelson, for discussion see Reference 11.

and by cosputtering Si and Ge targets simultaneously. The experimental results are shown in Figure 8.10. The process is the most easily controlled but not the only method for producing crystalline/amorphous transitions within a structure layer. When one introduces nanocrystalline regions in a device, the result is a doping profile with large variations in electrically active dopant concentration between the amorphous and nanocrystalline regions.

8.5 OPTICAL PROPERTIES

Optical absorption in tetrahedral crystalline semiconductors such as Si and GaAs shows variations with energy corresponding to variations in the bonds and the details of the density of states obtained from the energy-momentum relationships [see Section 2.1]. The lowest energy absorption derives primarily from transitions from p to s orbitals on adjacent atoms in common semiconductors [Chapter 5]. When the same material is produced as a glass with tetrahedral subunits, the first and second nearest neighbor bonds are largely unchanged. This means that the basic absorption behavior is primarily unchanged, although distortions in bond angles modify the energies of absorption edges and maxima somewhat. Loss of organization at longer ranges blurs peaks in the absorption spectrum. As bond distortions cause states to move from the band edges into the gap, the effective gap for optical absorption generally increases. Finally, because amorphous materials cannot have indirect gaps, a transition from indirect to direct upon amorphization also involves a loss of dispersion with momentum direction and a change in density of states. For an indirect gap material such as crystalline Si, a conversion to amorphous would be accompanied by a loss of the low energy minimum, which can also lead to an expansion of the energy gap.

Absorption coefficient data and various contributing mechanisms of a typical amorphous semiconductor are shown in Figure 8.11. [12] There are three primary regions. The high absorption region I corresponds to absorption due to electron transitions across the mobility gap from delocalized (band like) states to other delocalized states. Typically, for band-like transitions as the photon energy approaches the mobility gap energy, the absorption coefficient exhibits a power-law decrease. Region II is absorption from band like states to band tail states or band tail to band tail transitions where the density of states is still rather high. In this region the absorption coefficient is typically falling exponentially. Region III is due to transitions to, from, or among localized states in the mobility gap and absorption is roughly constant.

For a free-electron system the band density of states decreases parabolically, as discussed in Chapter 2. The same parabolic decrease is also generally found in amorphous semiconductors. We know that a power-law expansion of any function around an extremum has quadratic dependence when sufficiently near the maximum or minimum point. Thus, it should not be surprising that the density of states follows a similar behavior in crystalline and amorphous materials. This quadratic behavior is exhibited in region I of Figure 8.11. An absorption edge can be defined based on fitting this portion of the absorption spectrum based on the function

$$\alpha(v) = b\left(hv - E_0\right)^2 \qquad\qquad 8.7$$

where E_0 is the absorption edge energy, hv is the photon energy, and b is a constant. Equation 8.7 assumes free-electron band edges. The true bands in amorphous semiconductors are not exactly parabolic and may distort this behavior to the point that a quadratic relationship is no longer observable.

In all semiconductors and particularly in amorphous materials there are states in the energy gap that decrease in density roughly exponentially as one moves into the gap. These are the band tails described earlier. The resulting absorption coefficient

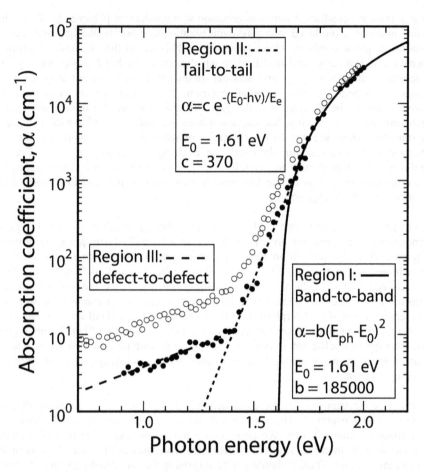

Figure 8.11: Optical absorption data for two a-Si:H films deposited by rf-PECVD using SiH₄ source gas. [12] Solid circles indicate as-deposited film results while open circles show the effect of annealing at 500°C for 30 min. Data was fit using Equations 8.7 (solid curve) and 8.8 (small dashed curve) for regions I and II. Region III also shows a simple exponential fit (larger dashes). The indicates shows that annealing increases the defect density in the film.

decreasing exponentially below the energy E_0 of Equation 8.7 can be written approximately as

$$\alpha(h\nu) = const. \; e^{-(E_{0'}-h\nu)/E_e} \qquad\qquad 8.8$$

where $h\nu$ is the photon energy, $E_{0'}$ is the same as the parabolic energy edge and E_e is the characteristic width of the dominant band edge, typically <0.1 eV. The wider band edge will set the apparent band-tail width and determine the absorption coefficient behavior. Which edge this is depends upon the details of the semiconductor involved. In a homopolar semiconductor such as amorphous silicon, the band tails will be relatively symmetrical with differences primarily resulting from the effective masses of the bands. In more complex ternary or multinary compound semiconductors, the sublattice on which the most substantial disorder occurs may determine the greatest band edge width. Various models have been proposed to explain the source of Equation 8.8 and can be found in Connell. [13] It is sufficient here to assume that the behavior is due to distorted localized bonds in the structure.

The optical gap (related to the mobility gap) in amorphous Si and other amorphous semiconductors exhibits temperature-dependent behaviors that are qualitatively similar to the temperature-dependences of crystalline materials. For example, a-Si:H shows a decrease in optical gap as temperature increases, as one would expect. Unlike the case for crystalline materials, the temperature dependence of a-Si:H is primarily related to changes in phonon-electron interactions. Thus, rather than a linear-quadratic relationship to temperature, one finds a change in optical gap of

$$E_{gap}(T) = E_{gap}(0) - \frac{2a_B}{e^{\Theta/T} - 1} \qquad 8.9$$

where a_B is the electron-phonon interaction strength and Θ is the effective temperature of phonons in the solid.

The effect of pressure on the optical gap in amorphous semiconductors also differs from the behavior of crystalline materials. In indirect-gap semiconductors we found a negative change in gap with increasing pressure in Si and some other materials due to increasing angular dependence of bonds at short distances. Direct gaps were all found to increase with increasing pressure due to greater orbital overlaps at shorter bond lengths. In amorphous semiconductors one finds an increasing gap with increasing pressure as expected for direct-gap materials. However, there can also be structural changes induced by increasing pressure that can lower the gap. The latter is observed in a-Si:H, while normal, positive changes in gap occur in a-Ge:H.

8.6 AMORPHOUS SEMICONDUCTOR ALLOYS

Amorphous semiconductors can be alloyed in exactly the same way as can crystalline materials, although not with as good results. Currently, the most common inorganic amorphous semiconductor alloys are a-(Si,Ge,C):H. These have been investigated in some detail and some are in use in select applications. The carbon alloys have proven less effective due to difficulty in reducing defect densities in the mobility gap, as discussed below. The mobility gap can also be adjusted by controlling the hydrogen content, as described above and shown in Figure 8.6, which has proven more effective for increasing the gap than the use of C to the maximum gap obtainable with hydrogen addition.

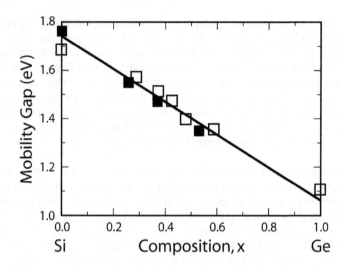

Figure 8.12: Shows typical values for mobility gap in a-Si$_{1-x}$Ge$_x$ alloys as a function of alloy composition at ~300K. Open points after Unold et al. [15]. Closed points after MacKenzie and Paul.[16] Numerous other sources of similar data also exist. Values vary somewhat with deposition conditions as might be expected from Figure 8.6.

From a consideration of the structure and bonding of amorphous materials, one can imagine how difficulties can arise. Consider the local bonding variations in crystalline alloys that normally cause bandgap bowing (see Chapter 6). Such variations will also occur in amorphous alloys and will change the energy of states in the material. Distortions in the amorphous network structure augment these effects. The alloy will therefore spread out the energies of distorted bonds more than in a non-alloyed material. Both broadening of the band edges and normal band-gap bowing can be expected in amorphous alloys. These two effects act in opposite ways, with an enhancement of states in the gap causing an increase in its width, while bowing reduces the gap. The counteraction of effects leads to less bowing in the amorphous alloys than in crystalline mixtures. The opposing effects also lead to broader Urbach edges and more states in the mobility gap. Such states can be the downfall of amorphous alloys in defect-sensitive devices.

As one would expect, Ge can be used to lower the mobility gap from 1.5-1.9 eV for a-Si to 0.9-1.0 eV for a-Ge in a continuous and relatively linear fashion. Figure 8.12 shows typical results for optical gap as a function of alloy composition. [15,16] Because, as discussed above, the mobility gap depends upon the structural distortions of the amorphous material as well as its composition, the mobility gap is also dependent upon the processing of the materials. Deposition processes using disilane

(Si_2H_6) as a Si source gas in a high H_2 partial pressure with careful control of ion bombardment of the growing film have been shown to produce relatively high-quality materials.

Amorphous Si/a-$Si_{1-x}Ge_x$:H alloy heterojunctions have band edge offsets as do crystalline materials. Roughly 80% of the change in energy gap is accommodated in the conduction band for this alloy.[17]

Unfortunately, deposition of a-$Si_{1-x}Ge_x$:H alloys with good properties is not straightforward. In poorly optimized processes, the addition of even modest amounts of Ge produces a large density of defect states in the mobility gap. The density of defects is generally found to increase exponentially with Ge content with an overall increase by a factor of ~3000 for a-Ge:H relative to a-Si:H for otherwise equivalent deposition conditions. This is true in all alloys studied to date, although some materials have lower defect densities overall across the range. One may understand why a-$Si_{1-x}Ge_x$:H shows a greater tendency to defect states when one considers the heats of formation for Si and Ge – 3.8 and 3.4 eV per molecule (78.44 and 88.04 kcal g^{-1} mol^{-1}), respectively. Thus, Si has ~10% stronger bonds that are, consequently, less inclined to strain or break. The resulting higher broken and distorted bond density in a-Ge:H explains the greater sensitivity to deposition conditions and the greater defect density. Indeed, in a-(Si,Ge) alloys, the majority of dangling bonds are from Ge atoms (90-99%). A delicate touch is required on the deposition process controls to obtain an equivalent defect density across the alloy range. Recipes for the production of a-$Si_{1-x}Ge_x$:H without significant increases in mobility edge width (a measure of material quality) have been obtained (although gap state densities still increase). Currently, alloys with Urbach energies of 45-50 meV, consistent with the values for unalloyed a-Si:H can be produced routinely with care and are sufficient for applications such as photovoltaic solar cells. Thus, a-$Si_{1-x}Ge_x$ has become an important alloy for band-gap engineered devices. Still, it is a difficult material to work with.

A-$Si_{1-x}C_x$:H alloys have also been explored as a possible means to increase the mobility gap. Indeed, very large increases in mobility gap have been achieved, at the price of massive defect densities. For example, an increase in mobility gap from 1.75 to 3 eV was obtained. [17] However, the increase in gap was highly dependent upon the hydrogen content of the alloy. It is not surprising that a-$Si_{1-x}C_x$:H alloys behave very differently from a-$Si_{1-x}Ge_x$. The crystalline Si-Ge alloy is completely miscible over the entire composition range so the bond energies and bonding configurations cannot be very different, nor is there a strong tendency to order the mixture to form a compound. Si and C, however, react to produce SiC, with little solubility of excess Si or C in the elemental Si or C or in the SiC compound. Consequently, it should not be unexpected that a-(Si,C) alloys show a large preponderance of Si-C bonds in preference to Si-Si or C-C bonds. A detectable minority of C-C bonds also occurs for C contents below 50%. This tendency to form a-SiC + a-Si two phase mixtures rather than forming a random alloy leads to a change in the energy band structure of the material, with large numbers of defect states due to localized disordering or variations in the material chemistry. To picture this, consider an a-SiC matrix with a large energy gap but including localized Si-Si bonds with low energies. Such defects would appear in the mobility gap of the final structure. Being first-nearest neighbor bonds, they would be present regardless of their state of local distortion and would be unaffected by local hydrogen content. The situation is not quite this bad, as one needs a cluster of at least five Si atoms to make a reasonably a-Si-like region. However, the tendency of this alloy to phase separate leads intrinsically to trouble and poorly performing alloy materials.

8.7 APPLICATIONS

8.7.1 Thin film transistors

Amorphous semiconductors are employed where crystalline substrates are unusable or impractical such as when a large number of discrete devices need to be integrated onto the surface of a material such as glass. Even when a single-crystal substrate could be used the maximum area is limited to the size of a single-crystal substrate, while amorphous semiconductors are easily adapted to large substrates. Devices on flexible surfaces and on materials that cannot be heated significantly (polymers for example) are also commonly constructed from amorphous semiconductors.

Amorphous semiconductors are used in switching devices (mostly field-effect transistors), which must be made reliably on polycrystalline or amorphous surfaces, and solar cells (essentially simple diodes), which must be made in extremely large areas at the lowest possible cost on the least expensive substrate (glass for example) to have an acceptable price. These devices function well because for light-absorption and diode operation the mobility gap of an amorphous material acts essentially as does a normal semiconductor energy gap. Luminescent devices are not made from a-Si:H or other inorganic amorphous semiconductors because these do not emit photons at a well-defined energy, nor at all efficiently. Amorphous organic materials are used as light emitters, as discussed in the next chapter.

The most common application of amorphous semiconductors is in thin film transistors that drive "active matrix" liquid-crystal displays or large-area optical sensors. In display devices, electrodes control the orientation and/or twist of polar organic molecules in the liquid crystal. These, in turn, determine the polarization of light passing through the display. To improve the speed of the display and to reduce crosstalk between pixels or areas, transistor switches need to be integrated into each pixel. Because each pixel requires a separate transistor for control, and because the pixels are distributed across the entire display, the transistors must likewise be distributed. To fabricate all of these devices from crystalline bits, and position and contact each separately would be prohibitively complex and expensive. It is far easier to make transistors from amorphous Si on the glass surface of the display directly. This is how all current active-matrix liquid crystal displays are fabricated. In a video display, speed requirements are modest (switching times of a few

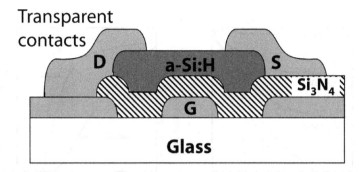

Figure 8.13: A schematic diagram of a typical thin film field-effect transistor using a-Si:H as the semiconductor. The configuration shown is known as the inverted staggered geometry, which is the most popular and highest performing structure. "Inverted" refers to the way the gate electrode and dielectric underlie the a-Si:H channel. "Staggered" indicates that the gate is on the opposite side of the semiconductor from the source and drain contacts. S, G, and D designate the source, gate, and drain contacts.

microseconds are sufficient). However, for most applications, power-efficiency is also important. Liquid crystal devices require very small amounts of power to operate. To prevent unintended turn-on and to minimize off-state power consumption, the maximum ratio of on-state to off-state current and the minimum off-state current are required. Amorphous Si TFT's meet these requirements well.

A second application for a-Si TFT's of growing importance is imaging systems acting as direct replacements for photographic film, especially in large-area medical systems such as x-ray instruments. These switch charge generated by local photosensors to amplifiers. This application also requires large ratios of on-state to off-state resistance and very high off-state resistances. As with liquid crystal displays, the speed of the device need only be fast enough to handle the sampling frequency of each pixel, typically a few milliseconds, or tens of kilohertz frequencies. Devices with low carrier mobilities and large trap state densities such as a-Si:H can still handle these switching rates adequately.

The typical a-Si:H thin film transistor (TFT) is a field-effect device, shown schematically in Figure 8.13. The electrodes are patterned on a glass substrate and the gate is covered by a dielectric, usually Si_3N_4. The active a-Si:H layer is added on top of the dielectric and the device is finished with source and drain electrodes. Application of a bias voltage to the gate controls channel conductivity as usual in field-effect transistors. These switch power onto and off of electrodes in displays quickly and efficiently.

More recent applications are beginning to require higher speeds, which a-Si:H devices with carrier mobilities of $< 1 cm^2 V^{-1} s^{-1}$ are not always capable of meeting. This is driving movement away from amorphous semiconductors to nanocrystalline

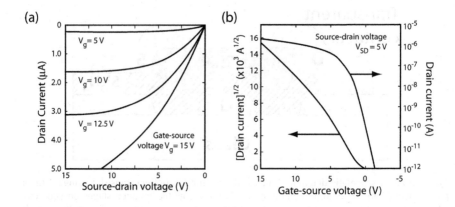

Figure 8.14: Characteristics for two a-Si:H thin film transistors. (a) The source-drain current as a function of source-drain voltage at several gate voltages for a transistor described in Powell [1989, Ref. 18] and (b) the source drain current as a function of source-gate voltage for a device described in Street [1998, Ref. 19]. Plotting data from [Powell, Ref. 18] on figure (b) shows a linear rather than quadratic behavior but a generally similar curve on the logarithmic scale. Note: compare these graphs with Figure 9.12 for an organic TFT.

materials with higher mobilities. The usual trade-off of converting to a nanocrystalline film is that higher mobilities are accompanied by higher leakage (off-state) currents. Therefore, there is still an important market for a-Si:H switching devices.

Typical results for a-Si:H TFT device characteristics are given in Figure 8.14. The critical parameter for current capacity is the width of the active region (in and out of the plane of Figure 8.13) relative to the length from source to drain. Higher widths at a given length increase current capacity at a given drive voltage and decrease on-state resistance. Typical width-to-length ratios are of the order of five. One of the features of the performance data evident in Figure 8.14 is the very high off-state resistance and high on-to-off state resistance ratio. a-Si:H TFT's can be produced with off-state resistances typically exceeding 10^{14} Ω and on-to-off current ratios exceeding 10^8. These properties are of great importance to the success of a-Si:H devices in display and imaging systems. The shape of the subthreshold current (the low current, low bias side of Figure 8.14b) can be modeled to provide an understanding of the density of states in the energy gap and the nature of the band tails in the material from which the TFT is produced.

a-Si:H TFT's are not without problems. Due to generation of carriers from trap states, current persists, decaying roughly inversely with time, after the device is nominally switched off. Therefore, the resistance appears to increase or the reverse leakage current appears to decrease as time goes on. This makes the device switch slowly. Slow switching is not a major problem for most applications where speed is not critical. More significantly, the threshold voltage for turn-on of the device can vary over both long and short time scales, from device to device, and, most unfortunately, can have a history/memory effect. In the latter, injection of carriers results in increases in dangling bond density (the Staebler Wronski effect). These midgap states trap charge and can affect device capacitance, switching rate, and threshold voltage. This is a serious problem for device designers. In addition to charge-injection-induced increases in trap density, normally present and unchanging traps have history and time-dependent charges. In other words, trapped charge can take a significant amount of time to escape a trap. This leads to additional variations in threshold voltage. Finally, defects in the Si_3N_4 dielectric can trap additional charge and modify the threshold voltage. All of these defect behaviors lead to considerable variability in resulting devices, which can have a substantial impact on the yield and performance of large-area devices.

The lesson of the thin film transistor is that amorphous semiconductors work well when the demands on the performance of the device are not great and where light emission is not important. When such requirements are coupled with restrictions on the surface on which the material is to be applied, amorphous semiconductors can become essential. The major question at this point is becoming – will organic materials replace inorganic materials in amorphous thin film devices?

8.7.2 Solar cells

Solar cells are diodes which, when exposed to light, produce an electric current. The concept is very similar to operation of a light-emitting diode in reverse, and was discussed in Chapter 6 for crystalline materials. Here we turn to non-crystalline solar cells. A-Si:H is the most widely manufactured material for thin film solar cells, although other materials are rapidly approaching it in production volume, because it can be deposited by relatively straightforward techniques onto glass or other materials coated with a transparent conductor. This is much less costly than single-crystal devices such as GaAs-based solar cells. As noted above, it is difficult to dope a-Si:H p or n type. Therefore, most a-Si:H solar cells are p-i-n structures. In other words, they consist of a thin, highly p-type region, a thick undoped region, and another thin n-type region. This is good because the p and n regions can be optimized for high doping, while the i-layer can be adjusted for best carrier collection. Generally, the deposition conditions under which these two behaviors are obtained are very different.

A typical a-Si:H single-junction solar cell along with its operating characteristics are shown schematically in Figure 8.15. In a crystalline solar cell the optimal energy gap for maximum efficiency is between 1.4 and 1.5 eV. In a-Si:H devices the optimal value may be somewhat higher, closer to 1.8 eV, if the device does not collect current well. Fortunately, it is easy to produce a-Si:H with a mobility gap energy near the optimal value.

A significant advantage of a-Si:H and alloys with Ge and C is that these are direct-gap materials, unlike their parent crystalline semiconductors. Consequently, they have at least 100 times higher optical absorption coefficients than the equivalent crystalline materials. The benefit is that only 1% of the thickness of the crystalline devices is required to absorb an equivalent amount of light. At the same time, one needs carriers to move only 1% as far as in crystalline Si or Ge before they reach a contact. Therefore, the electrical quality of the material may be much lower and carriers can still be collected. For a-Si:H the very high absorption coefficient allows all light to be collected in ~1μm of material. This means it needs only a relatively thin film of a-Si:H to produce a useful device. Hence the name thin film solar cells.

In the simulated device shown in the Figure 8.15 the current/voltage curve under light exposure has a lower resistance in forward bias than the curve without light (the light and dark curves cross). This crossover is the signature of a photoconductive response in the device – under light the amorphous semiconductor becomes more conductive. The slope of the curve at zero voltage can indicate a shunt in the junction allowing current to pass in spite of the resistive response of the diode. In the simulated device, however, this simply indicates a voltage-dependent collection of current. If there were a real shunt in the device then the dark current curve would also show the same or nearly the same resistance and hence current as a function of voltage.

If the absorber material is of high quality, primarily meaning that it has relatively few defect states in the gap, then the built-in electric field will transition continuously and uniformly from one contact to the other through the i-region. This maximizes the

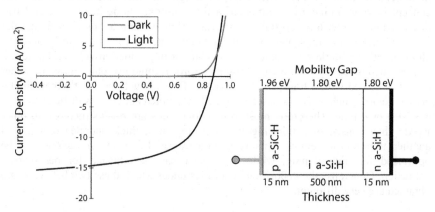

Figure 8.15: Simulated current/voltage curves calculated using the AMPS computer code [20] based on a standard simulation supplied with the code. Notice that the light curve crosses the dark one indicating that the device exhibits photoconductivity.

field assisting collection of carriers where they are generated and produces the best solar cell. If the defect density is high, as in poor material, the field is localized near the contacts. Field localization has several effects. The two most significant are that the electic field in the bulk of the i-layer is reduced, leading to less current generation by the device, and that in forward bias without photogeneration (in the dark) current injection at the contacts becomes space charge limited.

This latter point deserves emphasis. As we will see in organic devices, any time that carriers are injected from a contact into a very-low-mobility material they will tend to pile up near the contact (creating a "Coulomb blockade"). As their mobility is low, they may be unable to diffuse away into the bulk of the low-mobility layer as fast as they are injected. The resulting pile-up creates a field near the contact that tends to reject additional injection, effectively increasing the contact resistance at high current levels. This condition is known as space-charge-limited injection, and also occurs at heated cathode surfaces under some conditions. Two devices with and without space charge limited injection are compared schematically in Figure 8.16.

If the solar cell is produced from good material, mobilities are relatively high as are carrier diffusivities. Therefore, space-charge limitations are insignificant and the device works well. If the carrier mobility is low then carrier collection is poor

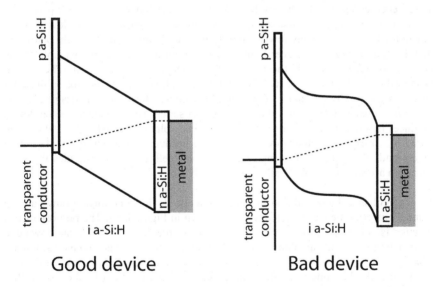

Figure 8.16: Band edge diagrams for good and bad a-Si:H solar cells similar to the one in Figure 8.15 showing the effect of increased defect density in the i-region. The p-type region is shown with a higher gap to prevent electrons from escaping to the transparent contact.

and the device yields both low currents and low voltages. One may reduce contact injection problems by using a nanocrystalline rather than an amorphous semiconductor at the contact. The nanocrystalline material has a higher mobility and is easier to dope heavily.

One of the greatest limitations to the application of a-Si:H in solar cells is the Staebler-Wronski effect, mentioned above. This is a light- or current-injection-induced enhancement in defect state density in the material. Because performance of the solar cell is reduced by increasing the defect density, the solar cell degrades during operation. Atomic-scale mechanisms for this degradation continue to be hotly debated but are most likely the result of a rearrangement of hydrogen atoms or a reorganization of dangling bond states such that the dangling bonds become more electrically active. Although the device performance recovers to nearly its initial value with proper treatment (usually annealing at moderate temperatures will suffice), the degradation reduces the photovoltaic conversion efficiency by up to two percent (for example, from ten to eight percent efficient). Even more annoying is that better devices are more susceptible to this phenomenon because the poor devices have so many defects to begin with. Products can only be specified to operate at the degraded level, even though they may initially produce higher powers.

The current technology for amorphous solar cells involves three diode junctions in series, each collecting the energy from a successively longer wavelength portion of the solar spectrum as one passes from the top to the bottom of the device. Each junction has a progressively lower mobility gap and is transparent to light below the mobility gap energy. Whatever light passes through one device is available to the next, if its energy exceeds the mobility gap energy of that material. The current produced by each junction is set by the number of photons in the portion of the solar spectrum that it absorbs and its collection efficiency. This current is designed to be the same for each device since they are wired in series. The voltage that each device produces is some fraction of its mobility gap. The voltages add when the devices are wired in series. The output power is then the product of current and voltage at a given load. The concept of a multijunction solar cell and how it divides the solar spectrum are shown schematically in Figure 8.17.

To accomplish voltage addition in a multijunction device, one must arrange an n-p "tunnel" junction between each device, as shown in Figure 8.17. The purpose of this junction is not to produce a working diode. Rather the diode should break down due to tunneling to provide an ohmic contact from device to device and convert smoothly from one type of majority carrier to the other, forcing all minority carriers to recombine. At the same time, it is transparent, which would not be the case with a metal-to-metal contact. It has been primarily the progression from single to double to triple junction devices that has been responsible for the gains in efficiency of a-Si:H solar cells in recent years.

Selection of the three energy gaps varies and the optimal values depend upon the performance of each junction and the light conditions for which performance is to be optimized. The overall objective of this design is to ensure that each device produces as nearly the same current as possible at the maximum power output voltage. Values shown in Figure 8.17 are schematic and not representative of a particular device. Multijunction devices are also commonly used in single crystalline solar cells based on III-V type semiconductors and are being explored for use in polycrystalline thin film devices. It should be noted that each device in the structure must be a good solar cell (at least in terms of current generation) if the multijunction solar cell is to be more efficient than a single-junction device of the same quality. If one device is

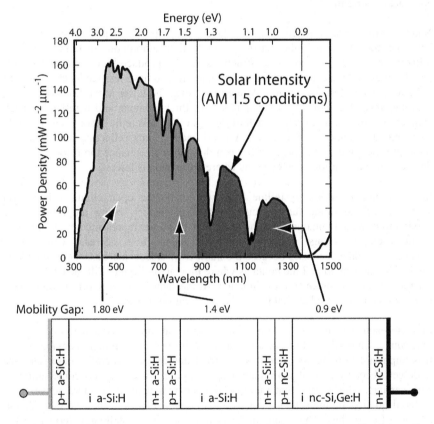

Figure 8.17: A schematic diagram showing how a triple junction solar cell might collect current from photons in different parts of the solar spectrum. Light enters the device from the left through the wide-gap solar cell and portions of the spectrum penetrate into the underlying devices.

poor, producing little current, then for the other solar cells to drive their output current through it they have to forward bias the bad device. This leads to net power dissipation in the bad junction rather than net power production.

The requirement for progressively lower energy gaps is met in common structures by two layers deposited with different H contents, and consequently having different gaps, adjusted as in Figure 8.6, and a third layer consisting of a Si, Ge alloy. Other devices are constructed using a-Si:H only for the wide-gap layer with Si, Ge alloys for both narrower gap regions. Still another approach is to produce nanocrystalline layers for a narrow-gap junction in the device. The latter is one of the most popular approaches currently.

Nanocrystalline materials for absorber layers (as opposed to just contact regions) have the advantage of higher carrier mobility than amorphous materials. Devices utilizing them for the narrow-gap structures tend to produce higher currents (fewer defect states are present to mediate recombination) and have less tendency to light or forward-bias-induced degradation. At the same time, the lower absorption coefficient, resulting from the indirect gap, requires both thicker layers and a reflective back contact to scatter and return as much of the light that fails to be absorbed as possible back into the nanocrystalline layer. A good back contact reflecting and scattering light into the nanocrystalline absorber can dramatically enhance the response of the device, especially at long wavelengths where absorption is lowest.

In single-junction microcrystalline Si:H solar cells the open circuit voltage and sharpness of the diode turn-on curve (the "fill factor", being the ratio of the maximum power obtainable from the device to the power represented by the product of open circuit voltage and short circuit current), both decrease with device thickness. Voltage decreases because the number of defects increases between where carriers are generated and the contacts where they are collected. Fill factor decreases because of series resistance in the junction. At the same time the short-circuit current increases as the material available for photogeneration increases.

Because each device in a multijunction structure absorbs only a part of the solar spectrum, each produces less current than an optimized single junction device would. It is important to remember that lower-current devices are generally more sensitive to defects, because recombination and trap states are less likely to become saturated with carriers. This means that multijunction devices tend to be more sensitive to material quality. In amorphous devices it also means that light-induced degradation is more of a problem. This is another reason to use nanocrystalline materials for the narrow-gap device.

8.8 SUMMARY POINTS

- Coordination of atoms in inorganic amorphous semiconductors is usually either two-fold (chain-like structure) or four-fold (tetrahedron based structure).
- Tetrahedron-based structures consist of randomly-organized networks of tetrahedrally-coordinated atoms. Bonding distortions are primarily in bond angle rather than bond length.
- Amorphous semiconductors must have direct mobility gaps.
- Highly distorted and dangling bonds result in a continuum of states in the energy gap. These come primarily from the band edge density of states of the equivalent crystalline structure.
- Defect-related states in the energy gap are increasingly widely separated in space as the distortion increases and they move farther into the gap. At some point the separation leads to a sudden decrease in carrier mobility among the defect states.
- Therefore, there is a "mobility gap" equivalent to an energy gap in a crystalline material which functions, for device modeling purposes, exactly as does a true energy gap in a crystalline device.
- Band tails resulting from an exponentially decreasing density of states are observed around the mobility edges. Wider tails with higher density of states result in lower-performance devices.
- Discrete states distributed across the gap occur. Higher densities of these states lead to lower carrier mobilities and higher minority carrier recombination and trapping rates.
- Dangling bonds tend to result in states near the middle of the gap. Hydrogen atoms may effectively passivate these bonds moving the states out of the mobility gap.
- Hydrogen concentration may be used to control the magnitude of the mobility gap in a-Si:H. Increasing hydrogen increases the mobility gap energy.
- Defect states in the gap make doping difficult.
- Doping may be improved by using nanocrystalline material for the heavily doped regions.
- Optical absorption has three distinct regions associated with: states outside the mobility gap, band-tails, and discrete states.
- Mobility gaps vary with temperature differently from crystalline material energy gaps.
- Generally one observes "dispersive transport" in which the mobility for each carrier is different with some carriers moving very well but with the average carrier moving very poorly.
- Deposition of a-Si:H and related materials is most commonly performed by chemical vapor deposition enhanced with a plasma or hot wire. Sputtering is also used.

- Doping is accomplished by adding B, As, or P to a-Si:H and related materials. Doping efficiency is generally low requiring very high concentrations of dopant to move the Fermi level in the mobility gap. Better materials are more easily doped.

- Nanocrystalline materials may be produced controllably by modification of deposition conditions. These have narrower and in the case of Si-Ge alloys indirect energy gaps.

- Carrier mobilities are higher and absorption coefficients lower in nanocrystalline indirect-gap materials.

- a-$Si_{1-x}Ge_x$:H alloys can be produced by codeposition of the elements. These have lower mobility gaps than a-Si:H and much higher defect state densities (increasing exponentially with x).

- a-(Si,Ge,C):H can be produced with C increasing the gap.

- Generally adjusting H content is more effective for gap increase than adding C because of additional defect states added with alloying with C.

- Little band-gap bowing occurs in amorphous semiconductor alloys and most band offset in Si-Ge alloys occurs in the antibonding (conduction) band.

- Low mobility materials often suffer space-charge-limited carrier injection at contacts in which contact resistance increases at high current densities.

- The Staebler-Wronski effect is the result of an increase in states in the mobility gap and leads to lower carrier mobilities, higher trapping and recombination rates, and worse device performances. It is the result of minority carrier injection into the amorphous material either through forward bias carrier injection or through optical generation by light.

- Thin film transistors (TFTs) are used for display applications in which large numbers of transistors are needed, distributed across a non-single-crystalline substrate.

- TFTs are generally field-effect devices in which only the channel is an amorphous semiconductor.

- a-Si:H TFTs have high off-state resistances and high on-to-off state current ratios.

- Switching speed in a-Si:H devices is generally related to carrier mobility and trap state density leading to a preference for nanocrystalline materials.

- a-Si:H solar cells are interesting as they can be produced in large areas on inexpensive substrates by simple processes.

- Amorphous solar cells are usually p-i-n structures with high doping levels in the p and n regions and the i-region optimized for higher mobility and low gap state density.

- Multijunction solar cells divide the solar spectrum into segments of increasing energy. Each junction absorbs a lower portion of this spectrum. These devices have higher efficiencies if each junction performs well but are more sensitive to defect state density.

8.9 HOMEWORK

1) Explain briefly why a-Si:H has a much higher absorption coefficient near the mobility gap energy than crystalline Si has.

2) Explain quantitatively why optical absorption

   a) increases as the square of the photon energy above the mobility gap.
   b) decreases exponentially with energy below the mobility gap energy.

3) If the defect density is constant across the majority of the mobility gap below the band tail region why does the absorption coefficient decrease as the photon energy decreases? What would you expect the mathematical dependence to be?

4) If the most common minority carrier in an amorphous material does not move, how can current be transported?

5) Describe the three mechanisms of carrier transport in amorphous semiconductors.

6) What is the Staebler-Wronski effect and why is it important in amorphous solar cells?

7) What is the best way to adjust the mobility gap in an amorphous semiconductor and why is this superior to alloying?

8) Why is band gap bowing not observed significantly in a-Si$_{1-x}$Ge$_x$:H?

9) Why is the mobility gap in a-Si:H greater than the energy gap in crystalline Si?

10) How does hydrogen help improve the electrical properties of a-Si?

8.10 SUGGESTED READINGS AND REFERENCES

Suggested Readings:

Brodsky Marc H., *Amorphous Semiconductors* in *Topics in Applied Physics*, volume 36. Berlin: Springer-Verlag, 1979.

Cohen J. David "Light-induced defects in hydrogenated amorphous silicon germanium alloys." Solar Energy Mater. & Solar Cells 2003; 78: 399-424.

Conde J.P., Chu V., Shen D.S., Wagner, S. "Properties of amorphous silicon/amorphous silicon-germanium multilayers." J. Appl. Phys. 1994; 75: 1638-1655.

Willardson R.K., and Beer, Albert C., *Semiconductors and Semimetals*, v. 21: Hydrogenated Amorphous Silicon. New York: Academic Press, 1984.

N.F. Mott, "Electrons in Non-crystalline Materials", in *Electronic and Structural Properties of Amorphous Semiconductors*, P.G. Le Comber and J. Mort, editors. (Academic, New York, 1973), p.1.

Zallen, Richard, *The Physics of Amorphous Solids*. New York: Wiley, 1983.

References:

[1] Kramer B and Weaire D., "Theory of electronic states in amorphous semiconductors," in Brodsky M.H., *Amorphous Semiconductors* in *Topics in Applied Physics*, v. 36. Berlin: Springer-Verlag, 1979.

[2] Lecomber, P.G. and Spear, W.E., "Doped Amorphous Semiconductors," in Brodsky M.H., *Amorphous Semiconductors* in *Topics in Applied Physics*, v. 36. Berlin: Springer-Verlag, 1979.

[3] Dawson, R.M.; Li, Y.; Gunes, M.; Nag, S.; Collins, R.W.; Bennett, M.; and Wronski, C.R.; "Optical properties of the component materials in multijunction hydrogenated amorphous silicon based solar cells." Guimaraes, Leopoldo, editor, *Proc. 11th European PV Solar Energy Conference and Exhibition*, Montreux, Switzerland, October, 1992. Chur, Switzerland: Harwood Academic Publisers, 1993: 680-3.

[4] Staebler, D.L. and Wronski, C.R., "Reversible conductivity charge in discharge-produced amorphous Si." *Appl. Phys. Lett.*, 1977; 31: 292-4.

[5] Nagles P. "Electronic transport in amorphous semiconductors," in Brodsky M.H., *Amorphous Semiconductors* in *Topics in Applied Physics*, v. 36. Berlin: Springer-Verlag, 1979.

[6] Mott N.F. "Conduction in non-crystalline systems. IV. Anderson localization in a disordered lattice." *Phil. Mag.*, 1970; 22: 7-29.

[7] Mott N.F., "Conduction in non-crystalline materials. III. Localized states in a pseudogap and near extremities of conduction and valence bands." *Phil. Mag.*, 1969; 19:835-52.

[8] Zallen, Richard, *The Physics of Amorphous Solids*. New York: John Wiley & Sons., 1983.

[9] Paul W., Chen J.H., Liu E.Z., Wetsel A.E., Wickboldt P., "Structural and electronic properties of amorphous SiGe:H alloys." *J. Non-cryst Solids*, 1993; 164-6: 1-10.

[10] Unold T., Cohen J.D., and Fortman C.M. "Electronic mobility gap structure in deep defects in amorphous silicon-germanium alloys." *Appl. Phys. Lett.*, 1994; 64: 1714-6.

[11] Feng G., Kaytyar M., Yang Y.H., Abelson J.R., Maley N., "Growth and structure of microcrystalline silicon by reactive DC magnetron sputtering." Thompson, M.J.; Hamakawa, Y.; LeComber, P.G.; Madan, A.; Schiff, E.A., editors, *Amorphous Silicon Technology – 1992*, Proc. Mater. Res. Soc., v. 258, Pittsburgh, Pennsylvania: Materials Research Society, 1992: 179-84.

[12] Based on data from Cody, G.L. "The optical absorption edge of a-Si:H" in Willardson R.K., and Beer, Albert C., *Semiconductors and Semimetals*, v. 21: *Hydrogenated Amorphous Silicon, Part B: Optical Properties*, Jacques I Pankove, ed. New York: Academic Press, 1984.

[13] Connell, G.A.N. "Optical properties of amorphous semiconductors" in Brodsky M.H., *Amorphous Semiconductors* in *Topics in Applied Physics*, v. 36. Berlin: Springer-Verlag, 1979.

[14] Smith F.W., "Optical constants of a hydrogenated amorphous carbon film." *J. Appl. Phys.*, 1984; 55: 764-71.

[15] Unold, D.; Cohen, J.D.; and Fortmann, C.M. "Electronic mobility gap structure and deep defects in amorphous silicon-germanium alloys", *Appl. Phys. Lett.*, 1994; 64: 1714-6.

[16] MacKenzie, D.W. and Paul, W., "Comparison of properties of a-$Si_{1-x}Ge_x$:H and a-$Si_{1-x}Ge_x$:H:F" in Madan, A.; Thompson, M.; Adler, D.; and Harnakawa, Y., editors, *Amorphous Silicon Semiconductors – Pure and Hydrogenated*. Materials Research Society Symposium Proceedings 1987; 95: 281-92.

[17] Conde, J.P.; Chu, V.; Shen, D.S.; Wagner, S., "Properties of amorphous silicon/armorphous silicon-germanium multilayers." *J. Appl. Phys.*, 1994; 75: 1638-55.

[18] Powell, M.J. "The physics of amorphous-silicon thin-film transistors." *IEEE Trans. on Electronic Devices* 1989; 36: 2753-63.

[19] Street, R.A. "Large area electronics, applications and requirements." *Phys. Statu Solidi a* 1998; 166: 695-705.

[20] McElheny, P.; Arch, J.; Lin, H.; and Fonash, S. "Range of validity of the surface-photovoltage diffusion length measurement: a computer simulation." *J. Appl. Phys.*, 1988; 64: 1254:65; and Zhu, H. and Fonash, S.J. "Computer simulation for solar cell applications: understanding and design" and references therein. *Amorphous and Microcrystalline Silicon Technology-1998* Schropp, R. et al. editors, Proc. Mater. Res. Soc., v. 507, Pittsburgh, Pennsylvania: Materials Research Society, 1998: 395-402.

Chapter 9

ORGANIC SEMICONDUCTORS

One of the most exciting opportunities in optoelectronics currently is devices based on organic materials. These have many advantages, primarily: lower-technology processing with less sensitivity to processing environment (but many are very air sensitive), flexibility, and the opportunity to apply the enormous power of organic synthesis to tailoring the properties of the materials to specific applications. Furthermore, organics can emit light directly as do conventional cathode-ray-tubes and plasma display panels, rather than relying on back-lighting systems such as are used in liquid-crystal displays. One can imagine these technologies leading to poster-sized televisions which can be rolled up and stored in mailing tubes, or unrolled and thumb-tacked to a wall. The materials are already being applied in compact lightweight, power-efficient light emitting devices in small areas such as cell-phone displays. The primary problem with all organic devices is stability. When carriers are injected into these materials, sometimes a molecule falls apart. This does not need to be very common for the device to degrade significantly over relatively short operating times. This chapter considers the options for organic semiconductors and how they are applied.

9.1 MATERIALS OVERVIEW

Organic materials used as active layers in optoelectronics can be divided into two major classes, molecular and polymeric. Molecular organic electronic materials are relatively small molecules, typically with a non-repeating structure – in other words, they are monomers. Details are given in Section 9.1.2. Polymer organic electronic materials, as with other polymers, consist of linked chains of single monomers (simple polymers) or strings of two or more monomers (copolymers). All electronic organic materials, whether monomer or polymer, contain a backbone of "conjugated" bonds, as described in detail in the next section. This structure is essential to achieving conductivity in the molecule. In general, polymers are more soluble than small molecules in common solvents (such as the polychlorinated methanes and tetrahydrofuran [THF]). Thus, polymers are often deposited by spin-coating from a solution, while small molecules are generally evaporated in vacuum. The possibility of functionalizing monomer units or copolymerizing from different monomers allows a single polymer to combine properties of various monomer materials. To achieve the same effect in small molecule structures, several must be co-deposited or otherwise mixed. Likewise, the possibility of producing branched, cross-linked, or network polymers allows greater tailoring of the mechanical properties and greater control of the tendency to crystallize. Both polymer and molecular devices often contain multiple layers to optimize carrier injection into the device and to control the efficiency of recombination of carriers resulting in optical emission. A further advantage of polymers is that thinner active layers may be produced without pinholes and other defects typical in evaporated small molecule films. This can allow operation of a device at lower voltages with the same resulting field. In the following sections, we will consider the structure of selected molecular and polymeric electronic materials. Following this is an overview of typical organic devices. The remainder of the chapter then considers associated issues such as contact materials, and ways of improving the performance of the devices.

9.1.1 Conjugated organic materials

The basic structure of a conducting or semiconducting organic material is shown in Figure 9.1. The essential feature is a continuous series of double bonds that runs the length of the molecule. Molecules with this structure are known as "conjugated". Note that each C atom in the backbone has one bond to each of its neighbors and one bond forming the double-bonded backbone. Because each C atom supplies one bond to this double bonding structure, one of the two adjacent bonds, on average, is a double bond, while the other is a single bond. It does not matter, in principle, which is the double bond and which the single bond. Another way to look at the situation is that each C atom has a half-filled p state available for π bonding, thus the double bonds can be represented as a π-bonded molecular orbital running down the length of the carbon backbone (see below for more discussion). Note that each C atom also has one bond left over for attachment of functional groups, usually a H atom. This opportunity to add functional groups makes conjugated polymers highly adjustable.

The molecular orbital structure of a portion of a trans polyacetylene molecule (Figure 9.1) is shown in Figure 9.2. The carbon atom bonding is hybridized as was discussed for semiconductors in Chapters 2 and 5, with $s+p_x+p_y$ atomic orbitals (Figure 9.2b) forming the sp^2 triangular planar molecular orbital structure and

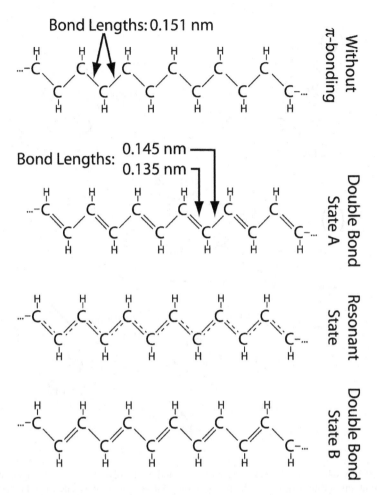

Figure 9.1: A schematic diagram of the simplest fully-conjugated polymer molecule, trans polyacetylene. Three possible configurations are shown for this molecule. In addition, the structure of the molecule in the absence of π-bonding is shown at the top, giving a comparison of bond lengths.

leaving a lone unhybridized p_z orbital projecting along the orthogonal z-axis. The sp^2 hybrid orbitals form conventional σ bonds (Figure 9.2c). (Recall that the bond axis lies along the bond direction in a σ bond.) These bonds have fully paired electrons in their bonding states and empty antibonding states, resulting in a very strong and stable covalently-bonded molecular backbone. The unhybridized, half-filled p_z orbitals form the π bonds (Figs 9.2c and d). (Recall that in π bonds the bond axis is perpendicular to the bond direction.) The energy of the resulting molecular orbitals

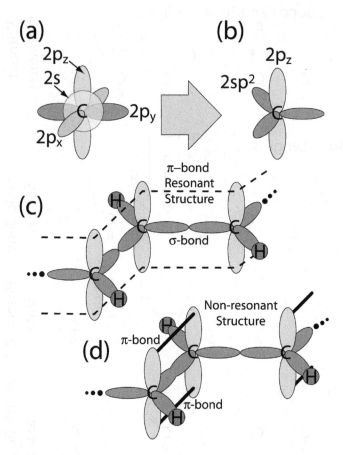

Figure 9.2: A schematic of the hybridization and bonding in poly-ene structures. (a) shows the valence atomic orbitals of C, (b) the hybrid molecular orbitals in poly-enes, (c) the resonant backbone structure of a poly-ene, and (d) the structure resulting from the choice of one of the two possible double bonding structures.

and the filling of these states with electrons is shown schematically in Figure 9.3 for octatetraene (polyacetylene with only 8 C atoms in the backbone).

Adding more C atoms results in more states near both the lowest unoccupied molecular orbital (LUMO) and the highest occupied molecular orbital (HOMO) states shown. Because each C atom brings in, on average, half filled states, the molecular structure fills all of the orbitals for which bonding dominates without filling any of the orbitals with primarily antibonding character as in other semiconductors. Thus, the bonds are very strong. Furthermore, because the π bond is specific to a particular direction, the double bond is rigid and the molecule is not free to rotate the bond. This leads to rod-like molecular structures. In contrast, flexible molecules have single-bonded backbones such as in polyethylene. (Single bonds are free to rotate around their bond axes.) As with inorganic semiconductors, there is an energy gap between the HOMO and LUMO states. It is this gap that makes most polymers (conjugated or not) insulating or semiconducting.

The situation in trans polyacetylene, and in other linear trans poly-ene's is, unfortunately, more complicated than the picture represented by Figure 9.1 would suggest. The simple symmetric structure with a π-bonded backbone (Figure 9.2c) (determined from a Hückel calculation for example), does not account for electron correlation effects. With correlated electron motions, each electron interacts with all other electrons rather than behaving as if it were the only electron present. Electron-electron correlations alter the state energies and exchange one state with another as, for example, the LUMO. This is unfortunate because the LUMO is transformed from a symmetry different from that of the HOMO to the same in most cases. This simple

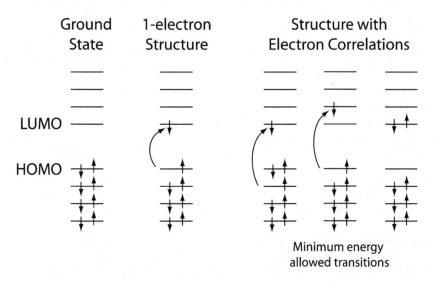

Figure 9.3: Shows the molecular orbitals for trans octatetraene ignoring, hydrogen-carbon bonds, and the filling of these orbitals with electrons. Both possible transitions that would result from a 1-electron model of the structure and the actual result including electron correlation effects are also shown, indicating that the lowest energy optical transitions are not from the HOMO to the LUMO states. Only a two-electron transition is allowed between these states for a photon-only process. Reprinted with permission from Brédas, Jean-Luc et al. "Excited-state electronic structure of conjugated oligomers and polymers: a quantum-chemical approach to optical phenomena". *Accounts in Chemical Research* 1999; 32:267-276. [1] Copyright 1999 American Chemical Society.

statement has profound consequences for an optical material. In a single photon process, quantum mechanical selection rules forbid transitions between states of like symmetry. The electron correlation effect changes the HOMO-LUMO transition from allowed to disallowed. The effect is similar to converting an inorganic semiconductor from direct to indirect. While optically-active transitions still exist within the molecule, they are no longer the lowest energy transitions. Thus, electrons and holes in their lowest-energy states in organic molecules cannot normally recombine radiatively in simple molecules such as trans-polyacetylene, see Figure 9.3.

Repeating this important point, it is possible for organic molecules to behave as do *indirect-gap* semiconductors, reducing the light emission probability.

In addition to electron correlation effects, there are structural asymmetries that must be considered. As mentioned above, in a simple molecule such as the trans polyacetylene the structure has exactly one double bond per two bonds along the chain. In a perfectly symmetric structure, either of these bonds could be the double bond or the structure could resonate (Figure 9.1), switching the double bond from one side to the other so fast as to be indistinguishable. However, the electron density in a double bond is higher than in a single bond and the consequent bond length is smaller. Thus, picking one of the double-bonding states of the molecule (state A or B in Figure 9.1) defines a short double bond. This, in turn, specifies that the next bond should be longer, the following bond shorter, and so on down the molecular backbone. It further turns out that the non-resonant states have lower energies than the resonant state (shorter overall bond lengths in the non-resonant state). In the resonant state the molecule is completely symmetric and would be a metal with the HOMO and LUMO at the same energy. Breaking the resonance to choose a specific double-bond pairing leads directly to the HOMO/LUMO gap. This stabilization has important consequences for the behavior of the molecule, as we shall see below.

The symmetry breaking of the bond lengths also applies to more complex molecules such as poly-paraphenylene vinylene (PPV). The three possible double bond configurations of this molecule are shown schematically in Figure 9.4.a-c.

One of the most important concepts in π-bonded molecular orbital (MO) structures is that of nodes. MOs are generally understood as being made up of linear combinations of atomic orbitals (AOs), as discussed in earlier chapters. Any such combination will include positive and negative terms. (See, for example, Equations 2.5 and 5.11.) The situation is generally represented schematically as shown in Figure 9.5. Here, a "+" lobe above the molecular backbone represents the opposite sign in the MO sum from terms with a "-" lobe above the molecular backbone. Mathematically and physically, AOs for given atoms being added together in this manner are identical. Thus, the sum of two AO's gives a high electron density between the atoms (bonding) while a difference of AOs gives a node, resulting in zero electron density at the bond center

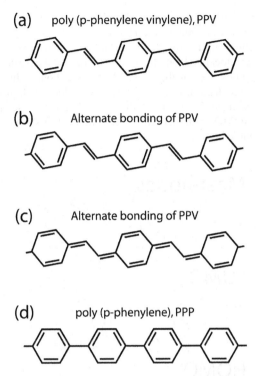

(a) poly (p-phenylene vinylene), PPV

(b) Alternate bonding of PPV

(c) Alternate bonding of PPV

(d) poly (p-phenylene), PPP

Figure 9.4: Structure and three possible double-bonding arrangements for PPV (a-c) and the structure for PPP (d). Note that PPV structure (c) has a higher energy than structures (a) and (b). Thus, (c) represents an excited state.

(antibonding). For a given combination of AOs there will be a given number of (regularly spaced) nodes.

Each combination of sums and differences of AO's down the π-bonding structure represents one electronic state of the system that can accommodate exactly two electrons. The resulting MOs include both bonding and antibonding character. The overall bonding/antibonding behavior of all but two molecular orbitals is somewhere between a pure bonding state (zero nodes) and a pure antibonding state (one node per bond). In Figure 9.5, the four structures shown are the highest and lowest energy states along with the HOMO and LUMO states for trans polyacetylenes. Notice that these structures are highly symmetric with well defined wave-like alternations of AO sign and regular spacing of nodes. This is because the resulting MOs are wave-like as was the case for Bloch wave states in solids, discussed in the free and nearly free electron models of inorganic semiconductors (Chapter 2). One can see that the energy of an MO increases as the wavelength of the node pattern decreases, exactly as it does with increasing wavevector for free electrons. A striking feature of the MO polarities and nodal wavelengths in the HOMO and LUMO states shown in Figure 9.5 is that the structures are nearly identical. The only difference is that the nodes in the HOMO state are localized on the single bonds of the structure while in the LUMO state the nodes are localized on the double bonds. In other words, the electron density for an electron in the HOMO state is localized on the double bonds while the LUMO electron density is on the single bonds. Consequently shifting an electron from the HOMO to the LUMO state (Figure 9.3) shrinks the single bonds and expands the double bonds.

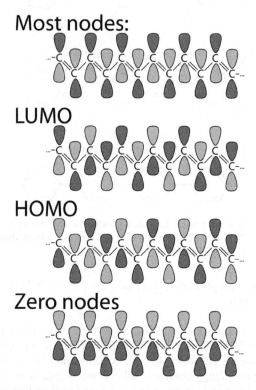

Figure 9.5: The lowest and highest energy nodal structures possible in trans polyacetylene along with the HOMO and LUMO states. The more nodes occur in the structure, the higher the energy of the linear combination of atomic orbitals. The HOMO and LUMO states divide at the point where half of the states have been filled. Thus it is perhaps not surprising that every other bond would have a node at this point. Reprinted with permission from Brédas, Jean-Luc et al. "Excited-state electronic structure of conjugated oligomers and polymers: a quantum-chemical approach to optical phenomena". *Accounts in Chemical Research* 1999; 32:267-276. [1] Copyright 1999 American Chemical Society.

At the same time it increases the electron density on the single bonds and reduces density on the double bonds. This coupling of atomic positions (shifted through atomic vibrational modes or phonons) with electron modes is the same type of coupling that occurs in indirect gap semiconductors. It is immediately evident that molecular design for optical activity will require a minimization of the molecular distortional changes associated with a transition from the HOMO to the LUMO states. In long molecules such as those shown in Figs 9.1 and 9.5, the coherence distance of distortions associated with a charge in an excited state do not propagate over the entire molecule. It is estimated that the excited state affects 2-2.5 nm of the chain. [2] The bond distortion pattern of the excited section is interpolated to the distortion pattern of the unexcited section through a resonant region.

Molecules such as poly (p-phenylene vinylene), Figure 9.4.a, or poly (p-phenylene), Figure 9.4.d, contain more phenyl rings and fewer linear chain segments than trans polyacetylene and similar molecules. The phenyl rings have a weaker distortion pattern and are often considered truly resonant structures. Consequently, there are smaller changes of the structure for HOMO to LUMO transitions, and the optical activity of the molecule is higher. Similar distortion wave phenomena occur in small molecules during excitation. We will see, for example, that the dye molecules discussed in the molecular optoelectronic materials Section 9.2, following, are made up primarily of linked functionalized ring structures rather than π-bonded chains and are highly efficient in radiative exciton recombination.

9.1.2 Ionized organic molecular structures

The above discussion shows that the bonding structure of fully conjugated organic molecules leads to a strongly bonded insulating state, as is the case in cold semiconductors. We are now prepared to consider what a free electron or hole might look like in an organic molecule. A free electron in a semiconductor exists in an antibonding state. Likewise, a free electron in an organic molecule usually occupies the lowest energy predominantly antibonding state (the LUMO). Occupying this state eliminates a double bond (both the bonding and antibonding states are filled on one atom). This leaves a negative charge on one atom and an unpaired electron on the other. The unpaired electron is known as a soliton or a free radical and is typically distributed over a moderate distance (1-2 nm) on the carbon backbone. Likewise, removing an electron (creating a "hole") removes one of the electrons forming a double bond and breaks the bond, leaving a positive charge behind on one atom and a soliton on the other atom.

Charged states on molecules can also be viewed as the result of oxidation or reduction processes. To see how charge states can be developed by these reactions, consider a misalignment of the double bonds on the backbone of polyacetylene, leading to an unpaired electron, as shown in Figure 9.6a. This soliton or free radical and has no net charge. However, the extra electron (half-filled dangling bond) is highly reactive. Oxidation then can remove this electron, leaving a positive residual charge (Figure 9.6b), while reduction can leave an extra electron behind, pairing the

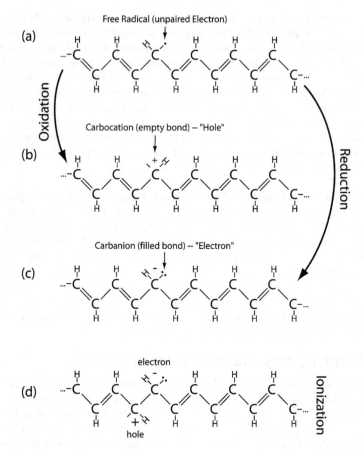

Figure 9.6: Bonding configurations for imperfectly bonded states of trans polyacetylene. (a) Shows the result of a reaction leaving a reduction in conjugation (a mismatch in the double bonding structure, known as a free radical. These are highly reactive sites and not stable in bulk solids. (b) Shows the organic molecule equivalent of a hole in an inorganic material, (c) an electron, and (d) an electron hole pair.

electrons on the dangling bond (Figure 9.6c). These charged defects are free to diffuse along the backbone of the molecule. This process of oxidation or reduction of the backbone is a standard method for achieving conductivity, effectively doping the polymer.

The negatively charged atom (Figure 9.6b) is referred to as a **carbanion**, while the positively-charged atom (Figure 9.6c) is a **carbocation**. Either of these structures (or both) can occur on a conjugated molecular backbone. As one can see by a close examination of the bonding of the molecule, both of these cases break the alternation of the double bonds, leading to a situation in which one atom has an unsatisfied desire to form a double bond. Consequently, the molecule reorganizes its bonding to maximize the number of double bonds on the backbone (Figure 9.6). One can quickly establish however, that in most cases the electron count does not work out. In fact, this will always be the case when one transfers a single carbocation or carbanion from one molecule to another and always creates not only a charged defect but also a soliton. When two defects are present on opposite sides of the molecule the two corresponding solitons can join to reconstitute a double bond on their own leaving a separated carbanion/carbocation pair on the molecule without additional partially filled dangling bonds.

The equivalent of an electron-hole-pair or exciton results from a carbocation lying on a site on the backbone of the molecule adjacent to a carbanion. This would also be the result of breaking a double bond where both electrons of the double bond went to one atom of the double bonded pair, leaving a negative charge on that atom and a positive charge on the other atom (Figure 9.6d). A carbocation-carbanion pair on adjacent C atoms is particularly favorable for two reasons. First, the opposite charges of the carbocation and carbanion experience an electrostatic attraction, as in an exciton in an inorganic semiconductor. Second, the bonding distortion pattern on the nearby π-bonds is less affected than the distortion pattern for separated defects. These contribute to a strong binding of the carbocation to the carbanion. In solar cells and photodetectors where charge separation following light absorption is required, this binding energy reduces the efficiency of current generation and increases the recombination rate. In light-emitting devices, the exciton binding energy coupling the carbocation to the carbanion lowers the energy of the emitted photons relative to the energy of a pair of dissociated defects. Exciton binding energies in organic molecules have been estimated to range from 0.1 eV to as high as 1.0 eV. [1] These very high binding energies have the advantage for light emitters that once an exciton forms, radiative recombination of the charges, reconstituting the double bond, is likely. This tends to reduce loss of free carriers in the thin emissive organic layers and increase their light emission efficiency per unit current injection (see the discussion of light-emitting devices below).

An interesting feature of carbocation/carbanion pairs is that when the charges separate, they remain on the side of the backbone of the carbon chain on which they began. This is the result of the need to maintain the double bonds on the remainder of the structure and the bonding distortions associated with the charged defect. Only when the two are on opposite sides of the molecule can the double bond be reconstituted by transfer of the electron. When the double bond does reform, a large amount of energy is released as an electron is transferred from a filled antibonding state to an empty bonding state. This extra energy can be released as heat or light as for inorganic light emitters. The possibility of light emission is the basis of organic optical devices. The situation described above is summarized schematically in Figure 9.7.

Note that all states in these molecules are resonant states, as with Bloch waves and so are more delocalized than this description suggests.

There is a temptation to discuss organic electronic materials and devices in the same terms as for inorganic materials and devices. Thus, the HOMO state becomes the valence band edge, the LUMO state is the conduction band edge, carbocations become holes and carbanions become electrons. With these substitutions, the discussion in broad terms is identical to that of Chapters 2 and 3. However, this picture is inaccurate and deceptive in a number of ways. Recent calculations, for example those by Köhler and collaborators [3] have shown that molecular orbitals on one carbon atom interact only with the nearest few repeat units on a polymer such as PPV and only weakly from molecule to molecule. Interchain electronic interactions are further complicated by the irregular structure of the organic matrix (although local crystallization of some materials can have a strong influence on this situation).

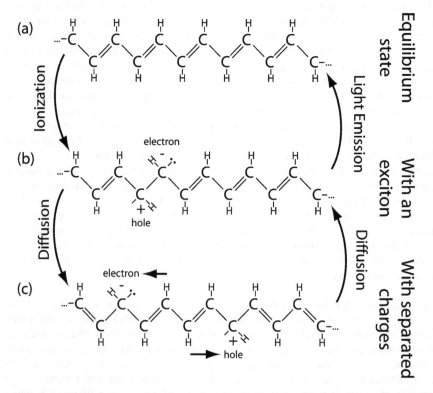

Figure 9.7: Schematic diagrams showing the transition from the equilibrium state of trans polyacetylene to states with an exciton and two separated carbocation and carbanion charges. (a) The equilibrium state, (b) the molecule with a carbocation+carbanion pair bound together as an exciton, and (c) with the two charges separated.

This topic is a matter of intense current research and remains to be fully resolved. Nonetheless, it is clear that the interactions of bonds are weak enough to produce little broadening of the molecular orbitals into bands and little delocalization of the bonding structures. This lack of interaction is, in part, responsible for low carrier mobilities in these materials. Consequently, it is inaccurate to think of organic electronic devices in terms of bands and free carriers and preferable to stick with a more traditional physical chemistry lexicon based on HOMO and LUMO states and localized oxidation and reduction of the bonding structure to produce carbocations and carbanions. One can go too far in this direction as well. Significant broadening in the optical emission properties of organic molecules occurs as one transitions from light emission of single molecules in solution to emission from solids composed of many molecules. This indicates some level of broadening and band formation. Nonetheless, the band-like behavior is much less significant than in inorganic semiconductors and the dominant behaviors are intramolecular rather than intermolecular except in the case of charge transport where interchain transfer is critical. In spite of the admonition above that it is inaccurate to refer to carbanions and carbocations as "electrons" and "holes" in organic materials, the remainder of this chapter will generally use these oversimplified terms for convenience. In most cases, the distinction is not significant.

It is interesting to note that in general the mobility of holes (carbocations) is higher than the moblity of electrons (carbanions) as there is a greater overlap from one molecule to another in the HOMO bonding states than in the LUMO antibonding states. [1] The overlap between molecules is strongest as the distance between the chains shrinks. Unfortunately, this has the simultaneous effect of rehybridizing the bonding structures to increase the energy of the lowest-energy optical transition relative to the HOMO-LUMO gap in some molecules (as was the case in poly-acetylene). Consequently, *optical emission is quenched as the distance decreases in some molecules.* This is primarily a problem when the long axes of the molecules lie parallel to one another as in organic crystals. Thus, organic crystals often will give a relatively high carrier mobility and a relatively low luminosity. Such materials are beneficial to organic transistors that rely on high mobility of carriers for improved performance, while amorphous structures provide better performance in optical devices. Indeed, we shall see below that the efficiency of luminescent devices degrades if the organic matrix crystallizes. In spite of these trade-offs, in the some optimal materials, both good mobility and high luminescence (quantum efficiencies approaching one) is obtained. The fact remains though that typically optical luminescence and carrier transport are inversely related.

9.2 OVERVIEW OF ORGANIC DEVICES

Before turning to details of specific organic materials used in devices, it is helpful to examine the typical structure of such devices. Organic devices are similar to conventional devices in some ways but dramatically different in others. Organic materials have exceptionally low carrier mobilities relative to conventional inorganic crystalline materials (and even compared to a-Si:H in many cases). Doping levels are generally relatively low as doping requires oxidation or reduction of the organic material. Thus, many devices, and in particular light-emitting devices, do not have a p-n homojunction nature as in inorganic devices. Rather, they resemble the p-i-n structures in amorphous Si solar cells discussed in Chapter 8. Thus, the built in or applied field generally is dropped across the entire device. In other words, there are no depletion regions – the entire device is a depletion region as in a capacitor. The low carrier mobilities require relatively high fields to obtain adequate current injection and conduction through the organic layer. This is reflected in the relative thinness of the final devices and the high fields at which they often are forced to operate (10^5 V cm^{-1} or more).

9.2.1 Light emitting devices

The basic structure of all organic light-emitting devices (OLEDs) is shown schematically in Figure 9.8. The structures are usually deposited on glass substrates coated with indium-tin oxide (ITO), a transparent conducting oxide, as anode. The active layers consist of one or more polymer or molecular films, generally well below 1 μm in thickness. If a single layer is used, that layer must transport both holes and electrons, accept both efficiently from the contacts, and emit light by carrier recombination. This combination of properties has been difficult to achieve in a single material. Therefore, both polymer and molecular materials adapted to act primarily as electron transport layers (ETLs) or hole transport layers (HTLs) are used. Light emission is usually incorporated into one of these materials either by addition of dye molecules or by design of functional groups and the basic structure of the ETL or HTL itself. Additional layers may also be included to improve carrier injection. In some cases (usually in molecular organic devices) a layer optimized to trap carriers of both types and promote radiative recombination is included. Finally, the device is completed by deposition of a cathode metal, which is responsible for electron injection into the ETL. The anode is usually incorporated into the substrate. Specific materials used for many of these layers and issues associated with them are discussed in later sections. Additional details may be found in the literature.

The emission intensity from current devices similar to the one illustrated in Figure 9.8 deteriorates with time. The absolute efficiency of the device (optical power out per Watt of input power) also varies significantly from device to device. The major factors limiting the performance of an organic light-emitting diode (OLED) have been discussed in detail by Patel et al.. [4] The four primary limitations to total optical emission intensity are as follows.

(1) An imbalance in the carbocation and carbanion injection rates on the two sides of the device. This causes net transport of one type of carrier completely through the structure to the other electrode. This is inefficient, as the transported current generates no light. It results from dramatic differences in the electrical properties of the two contacts and the absence of adequate barriers to transport of either or both types of carrier through the entire device.

Device Schematic

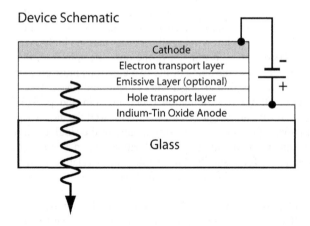

Band Edge Diagram

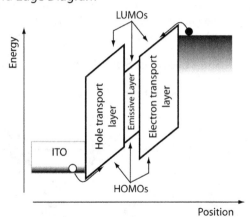

Figure 9.8: A schematic diagram of a typical organic light-emitting device along with a band-edge diagram for the device under forward bias (for light emission). The hole transport layer (HTL) has a relatively high HOMO energy and accepts and transports carbocations relatively easily. A potential barrier between the HTL (or emissive layer) and the electron transport layer (ETL) blocks further transport of the holes. Likewise, the ETL accepts and transports carbanions relatively easily. Emitted light escapes through the indium tin oxide (ITO) anode and the glass.

(2) The electronic state distribution of the excitons trapped in the material affect recombination efficiency. An excited configuration of a molecule consists of an excited electron in a LUMO state paired with a hole in a HOMO state. The hole,

however, is the absence of an electron in a state that normally holds two electrons. Thus, effectively, the high-energy electron must mesh well with the electron in the half-filled HOMO into which it would fall. In order that the high-energy electron can relax into a low-energy state, it must have the opposite spin from the other electron in the low-energy state and the spins of the two electrons must precess with opposite phase. There are four possible arrangements of spin and phase, (see Figure 9.9) but only one of these allows recombination (opposite spin and phase). Thus, recombination is allowed for one (singlet) state and disallowed for three (triplet) states. The relative populations of carbocation/carbanion pairs in the singlet and the three triplet states controls the recombination rate. Unfortunately, in some organic materials, electron correlation effects stabilize the triplet states relative to the singlet states, reducing the efficiency of radiative recombination.

The solution to this is to add molecules incorporating high mass metals such as rare earths. These elements assist in flipping the electron spin and phase in the triplet state, transferring the exciton to the singlet arrangement, allowing recombination and greatly increasing radiative efficiency. Dye and spin-mediating molecules and their design are discussed briefly below. In long-chain molecules the situation is softened somewhat, resulting in a higher population of singlet states (the energy of the singlet is reduced in the polymer), although the mechanism is not known.

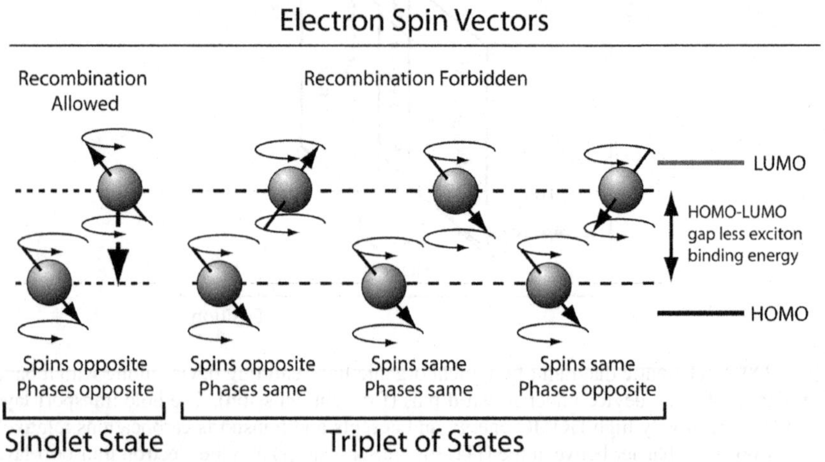

Electron Spin Vectors

Figure 9.9: A schematic representation of the various relative wave function relationships between electrons in a bound exciton. Only when the electrons have opposite spin and opposite precession phase can recombination occur based on quantum mechanical selection rules.

(3) The efficiency of radiative recombination decay of the singlet state as compared to possible non-radiative processes controls luminescence efficiency. The most common reason for lowering of the radiative recombination rate is diffusion of impurities into the organic layer, usually from the contacts. Recombination of excitons at a contact also lowers efficiency. Increasing the organic layer thickness and providing barriers to excitons diffusing to the contacts can improve this.

(4) Escape of light from the device is required for it to be useful. Light is given off isotropically in most cases but it escapes only through the transparent front contact. Inward-directed light must be reflected toward the transparent contact. At the same time, light moving outward must not be reflected back into the device at interfaces. Reflection can be reduced by standard optical film techniques such as by use of anti-reflection coatings. Reflection from the back of the device may be enhanced by choice of the back contact metal.

Resulting devices have achieved relatively good performances in a startlingly short time. An illustrative comparison of device performances is given in Figure 9.10 (after Roitman et al.). Recent results continue to increase efficiencies, luminosities, and lifetimes, although the rate of further increase is slow. In some cases, internal quantum efficiencies have approached 100% (all injected current results in photon generation). However, this efficiency generally decreases dramatically as power levels are increased where more non-radiative processes become active. Furthermore, the performance of devices generally degrades with time.

The degradation rate increases as the device is driven harder. Thus, the higher the current injected into the device, the shorter its lifetime. Lifetimes also vary depending upon the wavelength of operation and the contacts and other materials used. These effects are discussed in detail below.

9.2.2 Transistors

Field effect transistors have been demonstrated based on both molecular and polymer organic semiconducting materials. [See References 5 and 6, respectively, for examples of these devices] A typical device, such as that illustrated in Figure 9.11 consists of an organic semiconductor coated onto a dielectric such as SiO_2 and making use of an underlying gate electrode for conductivity control. Source and drain contacts underlie the edges of the organic channel, as shown in the figure. Carrier mobilities in molecular organic TFTs exceeding 2 cm^2 $V^{-1}s^{-1}$ with on/off current ratios in excess of 10^8 have been achieved. In the device reported by Jackson et al., [5] the organic semiconductor was pentacene (see Section 9.3, below).

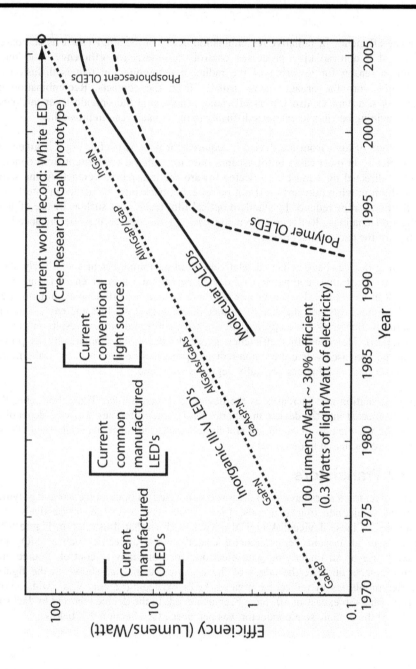

Figure 9.10: Efficiency of solid-state light emitters as a function of time. Ranges for current commercial devices are also shown. Data based on references 2-4 in Chapter 1.

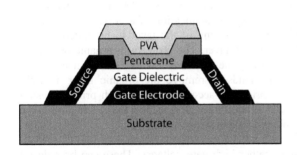 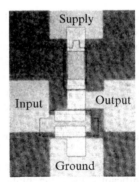

Figure 9.11: A schematic diagram of the elements and structure of an organic TFT (left) along with an image of an actual device. The PVA serves to encapsulate the device. With permission from: Jackson, T.N.; Lin, Y-Y.; Gundlach, David J.; and Klauk, H.; "Organic thin-film transistors for organic light-emitting flat-panel display backplanes." *IEEE J. Sel. Topics In Quantum Electronics* 1998; 4: 100-4. Copyright [1998] IEEE.

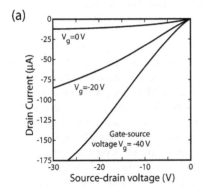 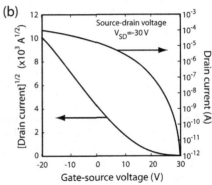

Figure 9.12: Device characteristics for a pentacene TFT. With permission from: Jackson, Thomas N.; "Organic thin film transistors-electronics anywhere" *Proceedings of the 2001 International Semiconductor Device Research Symposium.* Copyright [2001] IEEE.

To achieve high carrier mobilities, highly organized molecular or polymer layers are necessary as chain-to-chain or molecule-to-molecule charge transfer generally limits the mobility. High levels of crystallinity assist such transfers. Remember that this is the opposite of the case for organic luminescent materials where an amorphous structure is preferred. Treatment of the substrate and careful control of the deposition conditions are generally necessary to obtain the highest crystallinity. In addition, very regular molecules such as pentacene are needed for such layers. An example of the current/voltage characteristics of a typical resulting TFT is shown in Figure 9.12.

One of the complications in fabrication of organic electronic devices is patterning of layers. In standard microelectronic processes, application of a polymer photoresist, exposure and development of the resist, etching of the layer to be patterned, and removal of the resist are used to pattern a layer. The problem in organic devices is that the layer to be patterned is essentially the same material as the resist itself. Thus, any method of removing the resist tends to remove the organic layer as well. Consequently, patterning has been primarily accomplished through printing (using ink-jets or other methods) or by evaporation through a shadow mask. This limits the size of the devices and the possible processing steps used.

The devices described above have rather high gate voltages. These can be reduced significantly by using higher dielectric-constant gate insulator materials such as barium zirconate titanate. For an example, see Dimitrakopoulos et al. [8]

9.3 MOLECULAR OPTOELECTRONIC MATERIALS

As noted at above, organic semiconductors can be divided into two major classes of materials, small molecules and polymers. This section reviews some of the more common small molecules currently used in organic electronics. The following section considers the polymers.

The small molecular organic semiconductors typically consist of phenyl or napthyl groups coupled together. The electronic structures of these molecules differ from one molecule to another in the coupling of HOMO and LUMO states within the molecule and from molecule to molecule. The differences provide relatively good conduction for either carbanions or carbocations. Thus the molecules are often described as electron (ETL) or hole (HTL) transmitting layers, respectively. In general, hole transmitters have relatively high HOMO energies while electron transmitters have relatively low LUMO energies. More importantly, carbocations ("holes") are distributed broadly across the molecule in HTL's while carbanions ("electrons") are localized. The converse is true in most ETL's. The most common electron transmitter is tris-(8-hydroxyquinoline) aluminum (Alq$_3$). There are a variety of hole transmitters with generally similar structures. A typical example is *N,N'-di(naphthalene-1-yl)-N,N'-diphenyl-benzidine* (NPD). The structures of these molecules are discussed in detail in the following Sections, 9.3.1 for ETL's and 9.3.2 for HTL's. These two molecules are used commonly in light emitting devices. In general, a different set of molecules is used for switching devices (FET's for example), which are optimized for carrier mobility and conduction properties at the expense of light emission. Two of the more popular small molecules in current use for transistors are pentacene and α-sexithienyl. These are considered in detail in Section 9.3.4. Finally, in light-emitting devices small dye molecules optimized for radiative recombination of carbocations and carbanions are added to enhance efficiency as discussed in Section 9.3.3.

Molecular organic semiconductors are generally deposited by vacuum evaporation as they are typically insoluble and thus difficult to spin-cast or ink-jet print, but are stable and have reasonable vapor pressures and are therefore evaporable. Vacuum evaporation is more complex than spin-casting but provides quite pure materials with well-controlled structure and layer thickness. It is also straightforward to produce multilayer structures by evaporation.

9.3.1 Molecular electron transporters

The near ubiquitous choice for the electron transporter molecule in OLEDs and related devices is tris-(8-hydroxyquinoline) aluminum (Alq$_3$), shown in Figure 9.13. This molecule has proven highly reliable and durable and is nearly always used in contact with the cathode. Alq$_3$ has been found to be the most stable electron transporter molecule in conjunction with the low-work-function cathode metals

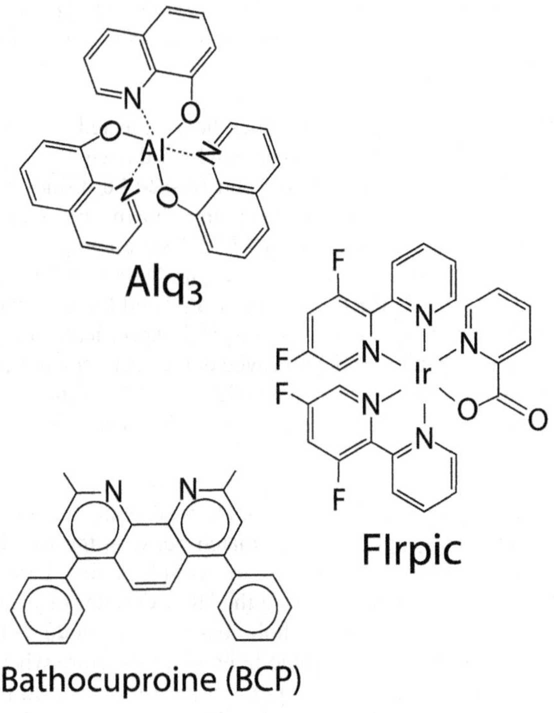

Figure 9.13: The molecular structure of three successful ETLs. Tris-(8-hydroxyquinoline) aluminum (Alq$_3$) is the most common electron transporter molecule in organic electronics and is used in contact with the cathode metal in most devices. BCP and FIrpic are ETLs that provide a barrier to hole and exciton transfer from the anode side of the device into the Alq$_3$.

(including such reactive species as Al and Li) and is relatively resistant to chemical attack. Furthermore, the substantial energy of the HOMO and LUMO states compared to the vacuum level makes electron injection into Alq_3 relatively easy. Other molecular electron transporters have been developed for specific purposes (mostly as hole transport barriers) but are generally used in conjunction with Alq_3 at the cathode.

Numerous theoretical studies have considered the electronic structure of the Alq_3 molecule. See, for example, Zhang and Lee. [9] Calculations suggest that electrons added to Alq_3 molecules are relatively delocalized, and are primarily found on the O, N and Al atoms. The distribution of the charge across the center of the molecule contributes to the high electron mobility. Furthermore, the electronic structure of the Alq_3 molecule is largely unchanged by adding an electron to the LUMO state. By contrast, adding a positive charge to the HOMO results in significant alterations in bonding and the positively-charged state is relatively localized on one of the quinoline units. [9] *This suggests that the mobility of holes will be lower than that for the more generally-distributed electrons and explains why Alq₃ is not a good hole transporting material.*

Several other ETLs are used, primarily to block holes injected from the anode from passing through devices. Two of these, 2,9-dimethyl 4,7-diphenyl 1,10 phenanthroline (also known as bathocuproine or BCP) [10] and (bis(2-(4,6-difluorophenyl)pyridyl-N,C2')iridium(III) picolinate) or FIrpic [11] are shown in Figure 9.13. These molecules are primarily designed to have a LUMO state energy similar to that of Alq_3 but to have a larger HOMO/LUMO energy gap. Note how FIrpic has a central structure similar to Alq_3. The distinction between BCP and FIrpic is the ability of the hole barrier to block triplet-state excitons. FIrpic is specifically designed for this purpose and has been shown to provide improved device performance as a result. [11] The range of materials which have been considered as ETL's includes even carbon nanotubes and carbon nitrides. However, the materials listed above are the most successful.

In summary, the most important properties of ETL's, generally achieved in the above materials are as follows. (1) A good energy match between the LUMO of the ETL and the cathode Fermi energy is designed to provide a low barrier to electron injection. (2) A high barrier to holes entering the HOMO states or excitons diffusing from the anode side of the device blocks hole loss to the cathode. (3) A distributed density of negative charge with little change in bonding structure when the molecule contains an extra electron enhances electron mobility. (4) Localized positive charge and significant changes in bonding in positively-charged molecules reduces hole mobility. (5) The ETL material, as with other insoluble molecular materials, must be easily evaporable and should not crystallize. (6) ETLs should be unreactive in operating devices.

9.3.2 Molecular hole transporters

The choice of hole transporting molecule is not as clear cut as for the electron transporting layer. Examples of HTLs include diphenyl-biphenyl and related compounds such as N,N-bis(1-naphthyl)-N,N-diphenyl-1,1-biphenyl-4,4-diamine (NPD) shown in Figure 9.14. As with ETL's, the primary objective in design of a hole transport material is to obtain a delocalized hole density across an entire positively-charged molecule while a localized electron would occur in a negatively-charged molecule. In addition, a lower binding energy of the HOMO and LUMO states relative to the vacuum level make hole injection easier in these materials.

A key issue in hole transporter molecule stability has been found to be the degree of crystallinity of the molecules and the grain size once crystallization occurs. Reduction

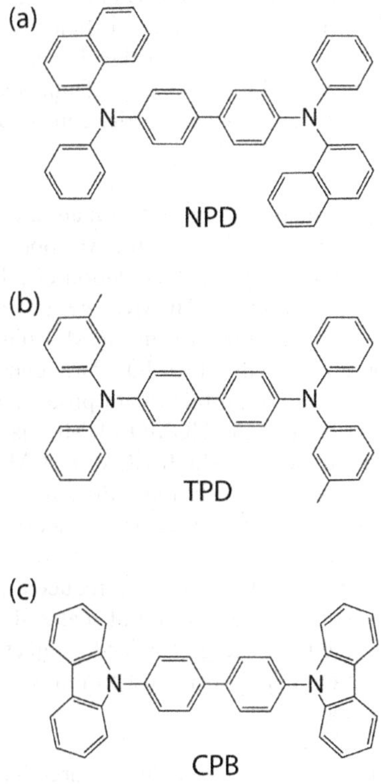

Figure 9.14: Shows the chemical structure of three hole transporter molecules, (a) *N,N'*-diphenyl-*N,N'*-bis(3-methylphenyl)-1,1-biphenyl-4,4'-diamine or NPD, (b) *N,N'*-di(naphthalene-1-yl)-*N,N'*-diphenyl-benzidine or TPD, and (c) 4,4'-N,N'-dicarbazole biphenyl or CPB.

in tendency to crystallize can be accomplished by reducing molecular symmetry (necessary for close molecular packing), increasing the molecular weight, and increasing the glass transition temperature. All of these differences occur in NPD compared to a similar molecule TPD (Figure 9.14b). TPD has a methyl-phenyl group in place of the napthyl groups of NPD. The glass transition temperature of TPD is 63°C as compared to 95° for NPD. NPD is also more rigid. Both effects reduce the self-diffusion coefficient and tendency to crystallize for NPD.

In addition to crystallization effects, degradation of the devices is increased with an increased barrier to hole injection from the anode. This is a persistent problem in OLED's. There are no metals with high enough work functions to make ohmic contacts to the HOMO states in typical organic compounds. Transparent conducting oxides (TCOs) are the best hope and most commonly used. N-type TCOs result in high barriers to hole injection but are common and easy to produce. P-type TCOs are rare, far less conductive, and little explored to date. The high barrier with n-type TCOs requires a large electric field to induce sufficient carrier injection. This, in turn, requires a large bias voltage to turn on the device. The large bias is inconvenient but more significantly the kinetic energy of injected carriers increases with increased bias and raises the probability of a carrier causing damage to the molecular structure of the HTL.

Charge-injection damage is a major source of degradation of OLEDs. Clear evidence of the effect of injection barrier on device lifetime was obtained by Adachi et al. [12] as shown in Figure 9.15. In this study the degradation of a large number of possible hole transport layers having a variety of HOMO energies were studied. A roughly linear decrease in lifetime with an increase in HOMO binding energy was found, demonstrating the importance of the HOMO state energy in charge injection degradation. This also illustrates the problem with producing blue-emitting OLEDs. To produce a blue-emitting device, the HOMO/LUMO gap must be at least in the blue spectral energy range. Achieving a high HOMO/LUMO gap generally requires a deep HOMO state. It will be necessarily more difficult to inject holes into such a state and the consequent lifetime of the device will be shorter.

Fortunately, charge injection degradation can be reduced at least in some devices. One common method is to introduce a cascade of materials with increasing HOMO binding energies, breaking the total energy barrier for injection of holes into a series of smaller steps. Many different materials are used in these cascades. Some examples appear elsewhere in this chapter.

Particularly attractive materials for assisting charge injection are the metal phthalocyanines. It has been shown that the introduction of one or two metal phthalocyanine layers between the ITO transparent anode and the hole transport layer can dramatically increase the device lifetime at a given luminosity. The most commonly-used of these molecules is copper phthalocyanine (CuPc), shown in Figure 9.16. In addition to mediating hole injection, it is thought that diffusion of

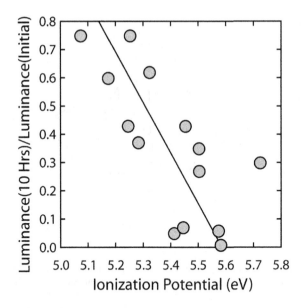

Figure 9.15: Shows the lifetime of OLED devices based on various compounds with a variety of ionization potentials. The lifetime here was determined as the output of the device after 10 hours relative to the initial luminance. Reused with permission from Chihaya Adachi, Kazukiyo Nagai, and Nozomu Tamoto, Applied Physics Letters, 66, 2679 (1995). Copyright 1995, American Institute of Physics.

atoms from the ITO into the HTL can degrade performance. The CuPc molecules appear to be relatively insensitive to this type of damage.

Additional evidence for the effect of metal-Pc molecules on the barrier to charge injection at the anode has been obtained by Zhu et al.. [13] They observed in ITO/MPc/TPD/Alq$_3$/Mg-Ag devices that the operating voltage decreased in the sequence H$_2$Pc > SnPc > VOPc > NiPc > PbPc > CoPc > FePc > CuPc > ZnPc. This trend is consistent with a corresponding change in the energies of the HOMO states of these compounds. Thus, substitution of the metal ion in the compound provides a direct control of the injection barrier. Of the various phthalocyanines listed above, CuPc has been found to have the best overall combination of low charge injection barrier and resistance to chemical reactions and other degradation mechanisms. Note that the device requires the TPD or other HTL to block electron injection from the Alq$_3$ into the MPc. Attempting to produce a device without the HTL results in a high current but a low luminescence due to a failure of moving charges to meet and recombine radiatively. [14]

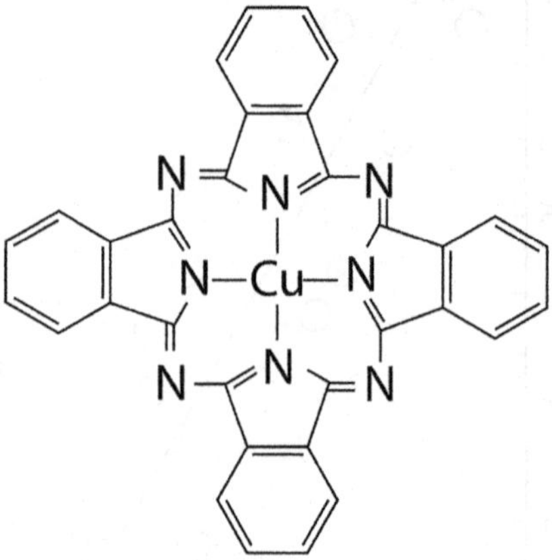

Figure 9.16: Shows the chemical structure of the copper phthalocyanine molecule. Other MPc molecules referred to in the text are similar with different metal ions in the core of the molecule.

9.3.3 Dye molecules

The overall performance of OLEDs can be dramatically improved by the use of dye molecules. These small molecules efficiently trap electrons and holes (ie: carbanions and carbocations) and have high radiative recombination rates relative to non-radiative decay processes. One can quickly see why this would be effective in a device as the dye molecule can be optimized for recombination, while the HTL and ETL layers can be optimized for carrier transport, carrier injection and chemical stability without having to worry about radiative recombination. From the discussion in Section 9.2, it is known that many molecules exhibit low radiative recombination rates due to the symmetries of their HOMO and LUMO orbitals, resulting in disallowed optical recombination pathways. The "solution" to the problem is to dissolve dye molecules optimized for emission in one or more of the transport layers.

In a full color display, it is necessary to have efficient light emission from individual devices with different colors (red, blue, and green). Dye molecules allow the color of the light emission to be tailored by changing the dye without affecting the composition of the remainder of the device. This means that the HTL and ETL layers and any buffer layers surrounding the contacts can be optimized for high device performance without worrying about the resulting wavelength of emitted light. The only requirement is that the HTL and ETL have a higher HOMO/LUMO energy gap than the highest energy photons to be emitted (blue). This requirement is not trivial, as the stability of molecules generally decreases as the HOMO/LUMO gap increases. Common matrix materials for the HTL and ETL are TPD and Alq$_3$, which have HOMO/LUMO gaps of 3.1 eV for TPD and 3.3 eV for Alq$_3$ in the green portion of the spectrum. The problem may be finessed somewhat by adding higher-gap barrier layers such as BCP between the Alq$_3$ and the emissive layer. The ability to change the light emission wavelength without changing the remaining materials in the devices certainly simplifies manufacturing. Dye molecules generally emit in a narrower range of wavelengths than do the HTL and ETL layers. Narrower emission lines mean purer colors. More saturated (very red, blue or green) light emission results in more vivid color displays. Finally, because dye molecules increase light emission efficiency, less drive current is required. This increases device lifetimes.

Based on the above, criteria important in the selection or design of a dye molecule include:

1. The dye should trap free carbocation and carbanion defects from the surrounding matrix and efficiently convert these to excitons.

2. These excitons should remain on a single dye molecule, rather than being distributed over multiple molecules. (The carbocation and carbanion should both be on the same molecule.)

3. The excitons should decay radiatively. To accomplish this the exciton lifetime should be relatively short.

4. Dye molecules should be designed to make use of energy in triplet states as well as singlet states for luminescence.

5. There should be a high solubility of dye molecules in the matrix.

6. The dye color should be strongly saturated and the emission line should be sharp.

7. The dye molecule should be chemically stable in the excited state (i.e. when an exciton is present).

In a typical high-efficiency OLED, see Figure 9.17, the majority of the formation of excitons and light emission occurs in the ETL or in a special recombination layer material between the ETL and the HTL (CPB in the example shown in Figure 9.17). Generically, the HTL is less favored because the exciton energy is lower (excitons are more stable) in ETLs compared to HTLs. If we ignore the possibility of a recombination layer and focus on a simple HTL/ETL heterojunction device, the general exciton formation process is as follows. Holes are injected from the HTL into the ETL. These holes are not highly mobile and soon encounter either a dye molecule or an electron. In the former, the hole is trapped on the dye molecule and awaits the arrival of an electron. In the latter case, an exciton is formed immediately. This exciton can decay radiatively. However, the host is not optimized for such a process. Therefore the exciton lifetime is relatively long. For sufficient dye concentrations the exciton will diffuse to and become trapped on a dye molecule before it decomposes or decays in the host ETL.

The details of the mechanisms of carrier and exciton trapping by the dye are beyond the scope of this text but may be found in the references. It is sufficient here to state that the efficiency of trapping by the dye molecule depends upon its shape and

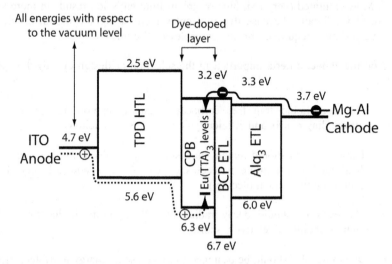

Figure 9.17: A band-edge picture of an efficient tested OLED device biased for flat bands. This device contains both TPD and Alq₃ hole and electron transport layers, a CPB hole transport layer (see Figure 9.14) hosting Eu(TTA)₃ dye molecules, and a 2,9-dimethyl-4,7-diphenyl-1,10-Phenanthroline (BCP) buffer layer. The buffer layer serves to prevent injected holes from entering the Alq₃. The TPD/CPB junction acts as an electron barrier. Redrawn with permission from Chihaya Adachi, Marc A. Baldo, and Stephen R. Forrest, Journal of Applied Physics, 87, 8049 (2000). Copyright 2000, American Institute of Physics.

charge distribution; the shape, charge distribution and organization of the host; and the host/dye bonding interactions. These control the distribution of dye molecules, their orientation relative to the matrix molecules, and the interaction of the electronic states among the molecules. The effectiveness of a dye at trapping carriers can be related to a quantity known as the Förster radius (~3 nm for most dyes). [16] Therefore, dye design criterion (1), above, can be represented in part as a requirement for the dye to have a large Förster radius.

Dye molecules have been shown to greatly increase the photon output of OLEDs, as shown in Figure 9.18. However, there is a strong peak in the luminescence as a function of dye concentration. At very low concentrations a minority carrier may not encounter a dye molecule before it recombines, so the contribution of the dye to luminescence is small. Initially, increasing the dye content increases the rate at which charges are trapped and the rate at which excitons are formed. The peak in luminescence will occur where virtually every minority carrier is trapped on a dye molecule before it can decay in the matrix. The average spacing between the dye molecules at the peak efficiency is near the Förster radius.

At higher concentrations the luminosity decreases gradually for several reasons. A pure dye molecule matrix is not an effective light emitter. At high dye concentrations, carbocation/carbanion excitons become delocalized over two dye molecules. This state is not strongly bound (indeed, it is not a ground state of the system) and the carbocation and carbanion can separate relatively easily. The delocalized exciton also tends to decay more frequently by non-radiative processes, reducing luminescence. Thus, it is evident that there is a positive influence of the presence of the non-dye matrix. Dye molecules are often not very soluble in the organic matrix and phase separation can occur as the concentration increases. Because dye molecules are not effective emitters in groups, the resulting aggregates of dye species show decreased luminosity. At high dye concentrations, the radiation spectrum also broadens, producing a less pure color and giving a more pastel hue to the device.

Based on the above discussion, we can see that the ideal situation is to have individual dye molecules separated from each other by the host matrix but at a sufficiently high concentration that no recombination occurs in the host. The matrix supplies carbocations and carbanions to the dye, which traps and efficiently recombines them radiatively. The most effective method for trapping charges and forming excitons is for the dye molecule to have a lower HOMO/LUMO gap energy and to have the HOMO lie above and the LUMO below the corresponding states of the host material. This situation is shown schematically in Figure 9.17. However, this also implies a larger HOMO/LUMO gap in the matrix for a given emission energy, which reduces device stability and increases contract problems.

Because the host matrices are typically either electron or hole transmitters and relatively poor at transmitting the other type of charged defect, most of the recombination occurs near the heterojunction, and usually on the ETL side. Therefore, the most critical location for dye molecules is generally in the ETL near the heterojunction. The situation can be optimized by providing a thin recombination layer between the ETL and the HTL in which the dye is located. This layer can be designed for high dye solubility and effective transfer of charges to the dye molecules, rather than for stability with respect to contacts or effective hole or electron transport properties. Such a layer is shown in the device in Figure 9.17.

Additional advantages of using dye molecules are that they decrease the tendency of the host matrix to both crystallize and agglomerate. In agglomeration, a layer forms in clumps or hillocks rather than remaining uniform and flat. Such clumping concentrates current in the thin regions where the resulting fields are higher. This causes rapid degradation of the device.

A large number of molecules have been developed for use as dyes in OLEDs. For a detailed discussion see Shoustikov et al. [16] or the less detailed Chihaya et al. [18] or Kanno et al.. [19] Examples of the more popular molecules or those that provide relatively saturated colors (as of this writing) are shown in Figure 9.19. The most common way to represent the color of a dye molecule is on a chromaticity diagram.

The location of the dye molecules with structures given in Figure 9.19 on such a diagram is shown in Figure 9.20. Molecules luminescing near the corners of the triangular curve in the diagram provide the most strongly saturated red, blue, and green colors. Points closer to the center of the diagram provide more pastel colors.

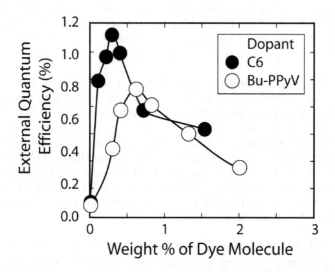

Figure 9.18: Efficiency of OLED's formed with coumarin 6 (C6) [see Figure 9.19] and poly(3-n-butyl-p-pyridyl vinylene) (Bu-PPyV) dyes in a polyvinyl carbazole (PVK) hole transport layer. The remainder of the device included an electron transport layer and ITO and Mg-Al electrodes. With permission after Shoustikov, Andrei A.; You, Yuijian; Thompson, Mark E.; "Electroluminescence color tuning by dye doping in organic light-emitting diodes." *IEEE J. Sel. Topics in Quantum Electronics* 1998; 4: 3-13. Copyright [1998] IEEE

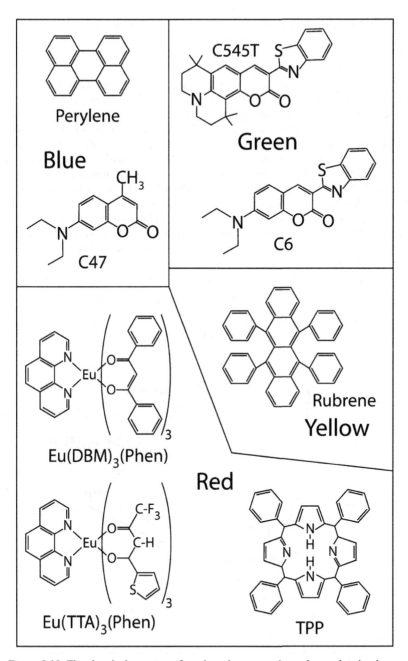

Figure 9.19: The chemical structures for selected compounds used as molecular dyes.

For a full color display, luminescence near the corners of the diagram is optimal. For an organic "light bulb", emission near the center of the diagram (white emission) would be acceptable or even preferred.

In recent work, a significant improvement (roughly a factor of three) in luminescence efficiency was obtained by adding rare earth transition metal complexes to the device. These exhibit efficient phosphorescence (a slower form of fluorescence which does not require spin conservation). The high electron density in the rare earth mediates electron spin transitions and mixes the singlet and triplet states, allowing conversion of triplet to singlet states. What makes the situation more efficient is co-doping of a host matrix with both a phosphorescent and a fluorescent material. (The fluorescent material is a standard dye molecule such as those shown in Figure 9.19.) The phosphorescent material captures excitons from the host, and passes the singlet state of the exciton to the fluorescent material. Thus the concentration of singlet states in the fluorescent material is enhanced significantly, along with the luminescence intensity of the device. Several examples of such systems have been described recently. For example, M.A. Baldo et al. discuss the use of $Ir(ppy)_3$ (see Figure 9.21) as a phosphorescent spin mediator leading to luminescence of a separate dye

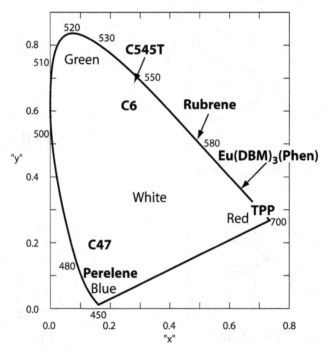

Figure 9.20: A chromaticity diagram for some of the more common dye molecules shown in Figure 9.19. Note the lack of a really saturated green dye.

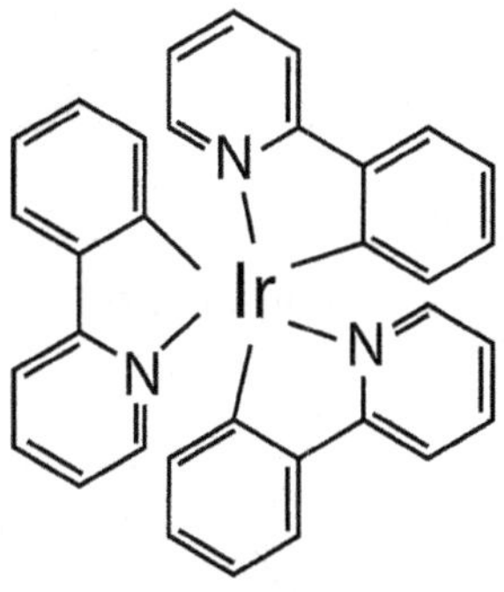

$$Ir(ppy)_3$$

Figure 9.21: The chemical structure of Ir(ppy)$_3$ used as a phosphorescent mediator converting triplet state excitons to singlet states.

molecule. [20] These rare-earth complexes are now used ubiquitously both as dyes themselves and as triplet-singlet mediators for other materials. As dyes they tend to produce wide emission spectra and are used primarily with other compounds to produce white-emitting devices.

9.3.4 Molecules for thin film transistors

Organic thin film transistors have achieved performances in excess of those of the amorphous Si devices described in Chapter 8. The highest performance devices have been based on molecular rather than polymer species. The best overall results have been obtained with pentacene (see Figure 9.22a), as described in Section 9.2.2. However, α-sexithienyl (see Figure 9.22b) has also been used successfully. Both of these molecules have a tendency to form relatively large crystalline regions with grain sizes up to one micron estimated from atomic force microscopy analysis. Crystallinity is essential to high carrier mobilities. Unfortunately, there is a cost in weak bonding of the film to the substrate when the film is deposited under conditions yielding good crystallinity. To improve adhesion in the presence of high crystallinity, Jackson and coworkers have shown that a self-assembled monolayer of octadecyltrichlorosilane between the pentacene and the SiO$_2$ underlayer improves the performance of the devices. [5] The field-effect mobilities obtained in pentacene devices of 2 cm^2 V^{-1}s^{-1} is in the range anticipated for the upper limit to mobilities for such materials. [21] This indicates that the well-ordered polycrystalline pentacene layers obtained are not primarily limited by their microstructure. Indeed, while most organic materials exhibit thermally activated hopping conduction, the highest mobility values in pentacene suggest a transition to temperature-independent band-like conduction. Perhaps this is not surprising if the molecules are crystallized and lie very close to one another such that their molecular orbitals can overlap to form bands.

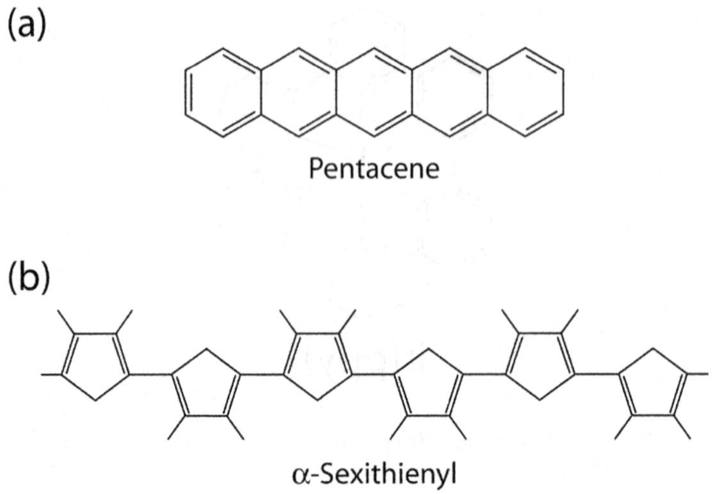

(a)

Pentacene

(b)

α-Sexithienyl

Figure 9.22: Shows the chemical structures of the two most popular molecules for organic thin film transistors, as discussed in the text.

One of the significant problems facing molecular organic TFTs is the formation of coulomb blockades at Au source and drain contacts. These have been described by Schoonveld et al. [22] and can have a significant effect on the apparent mobility of carriers in the device. The solution to this problem almost certainly involves more careful selection of the contact metallization materials.

9.4 POLYMER OPTOELECTRONIC ORGANICS

In spite of the generally higher performances achieved in light-emitting and thin film transistor devices with molecular organic materials, organic polymers have generally been the choice for demonstration of larger scale devices and product manufacturing. Polymers have superior processing options primarily because they are relatively soluble in organic liquids (at least as compared to the molecular species described above). This makes spin-coating, stamp printing, and ink-jet printing possible. The latter two, in particular, allow fabrication of devices without relying on lithography – a significant benefit. Most of the issues related to polymer electronic materials design are similar to those for molecular species. However, polymer design also includes significant work on solubility. Some of the more popular molecules for light emitting and thin film transistors are discussed below, along with general aspects of their design.

The most significant optoelectronic requirements for improving polymer devices currently include reducing the turn on voltage, lowering operating current by improving efficiency, and balancing electron and hole injection into the device. In Section 9.3, we saw that the introduction of barriers to specific carriers through control of the electron affinity of the molecules can effectively prevent carriers of either type from crossing the entire device. Furthermore, we saw that similar control of orbital energies can reduce carrier injection barriers. This lowers the voltage necessary for a given level of current injection and hence the input power. Likewise, increasing carrier mobility reduces this voltage by lowering the resistance of the device. Finally, more efficient luminescent recombination mechanisms are required to reduce injection currents at given luminosities. Increased luminosity is achieved by tailoring the polymers by the addition of specifically designed efficient fluorescent materials or functional groups. As with molecular species one may also dope the matrix with molecular species such as Ir(ppy)$_3$ or related polymeric materials to harvest energy in triplet states as discussed in Section 9.3.3. For switching devices, carrier mobility and material stability are important as for molecular species.

The above requirements can be summarized in the following general design goals for polymeric optoelectronic materials:

1. Control of HOMO and LUMO energies relative to the vacuum level.

2. Control of crystallinity and molecular organization for improved carrier transport characteristics.

3. Achieving a large range of ratios of carbocation and carbanion mobilities to provide effective electron and hole transporting materials capable of resisting transport of the other species.

4. Control the luminescent wavelength and maximized radiative relative to non-radiative recombination of excitons.

5. Maximization of solubility of the polymer and enhancement of other aspects of processability.

Molecules exhibiting particularly desirable properties in these respects are discussed below. Issues directly related to carrier injection and stability at contacts are deferred to Section 9.5. As we shall see, all of these properties may be controlled by addition of functional groups to the sides of the fully conjugated polymer chains.

9.4.1 Polymers for organic light emitting devices

In polymers, less distinction is drawn between primarily electron-transporting materials and primarily hole-transporting materials. Indeed, some molecules can be functionalized to accept carbanions efficiently in one region and carbocations in another. Because the molecules are continuous strands the two ions can move along the chain backbone without necessarily requiring interchain transfer. Self-assembled functional blocks in the molecule can also reduce the need for multiple molecules. In spite of this, at least two polymer layers are generally used in devices to optimize electron injection from the cathode and hole injection from the anode. Because of the smaller level of distinction between the materials, this section discusses both electron and hole transmitting polymers together.

The quintessential polymer optoelectronic material is poly-paraphenylene vinylene, PPV, with the structure shown in Figure 9.4. It is primarily used as an electron-transporting material but can be used in single-layer devices as well. A better description might be to say that PPV is used primarily on the cathode side of two-layer devices and acts as the primary recombination layer. Pure PPV is a rigid rod polymer with a strong tendency to crystallize and very low solubility. Therefore, it must be deposited in another form and polymerized on its intended substrate. This is extremely inconvenient from a processing standpoint. Therefore, there has been a strong driving force to engineer PPV to improve its properties. The currently most popular electron transporting (or cathode side) molecular organic semiconductors are derivatives of PPV based on functionalizing the vinyl or phenyl groups to meet the design criteria outlined above.

One of the most successful of the altered molecules is poly(2-methoxy-5-(2'-ethylhexyloxy-)*p*-phenylenevinylene), or MEH-PPV. The basic structural unit of this molecule is shown in Figure 9.23. MEH-PPV has acceptable solubility in common solvents such as tetrahydrofuran (THF). MEH-PPV emits in the red-orange spectral range (which can be further modified as described below). While such colors are potentially useful in producing a full-color display, MEH-PPV is not satisfactory as a host for other fluorescent molecules. The 2.3 eV HOMO/LUMO gap [23] of PPV likewise prevents it from fluorescing above the green portion of the spectrum. Typically, further functionalization of MEH-PPV lowers the HOMO/LUMO gap, red-shifting the resulting emission and further restricting the range of applicability. This has contributed additional impetus to explore other non-PPV-based materials.

In recent years, materials based on polyfluorenes (PFOs), see Figure 9.23, have become increasingly popular as blue light emitters for the cathode side of OLEDs as they provide strong emissions in the 380-420 nm wavelength range. [Note that polyfluorene contains no fluorine atoms. It is based on the *fluorene* monomer, a carbon ring structure (Figure 9.23a).] The fluorene monomer can also be copoly-merized with other monomers to alter its properties (in particular the fluorescent wavelength). As with the PPV materials, side chains can be added to modify the solubility and optoelectronic properties. In Figure 9.23a, two functional side groups on the PFO are designated by "R". These can be as small as hydrogen atoms or as large as massive side chains. The most common side chains are short alkanes. The

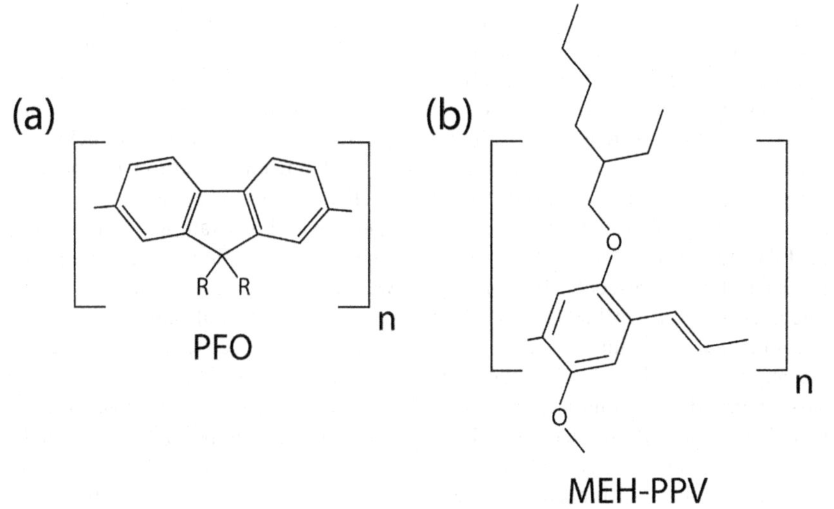

Figure 9.23: Chemical formulas for two molecules frequently used as electron-transporting and recombination layers in polymer light-emitting devices.

power of synthetic organic chemistry has been demonstrated effectively in the design and modification of light-emitting molecules such as PPV and PFO. As examples, let us briefly consider selected modifications of PPV in more detail.

The functionalization of PPV has been shown to allow control of electron affinity, HOMO/LUMO gap, and solubility. For example, the addition of alkoxy side groups, converting PPV to MEH-PPV increases solubility significantly. A difference in peak fluorescence wavelength is also obtained depending upon whether the side chain is an alkane or an alkoxy group (ie: whether the unit binding the side chain to the aromatic ring is a CH_2 or an O atom). The electron donating alkoxy groups result in a greater red shift in the resulting light emission. Likewise, electron donating or electron withdrawing groups can be added to the vinyl linkages to further modify the emission wavelength of the conjugated polymer backbone. For example, addition of electron withdrawing cyano (-CN) groups reduces the HOMO/LUMO gap and increases the electron affinity. This has been found to improve the luminescence of resulting devices significantly through changes in the efficiency of charge injection. (See, for example, N.C. Greenham, et al.. [24]) The cyano addition has a different effect depending upon whether the group is added closer to or farther from the alkoxy chains. The primary effect is through controlling twisting distortions in the π-bonding backbone of the molecule. The details of such polymer engineering are beyond the scope of this book. A discussion and detailed references concerning the engineering of side chain chromophores on PPV and their affect on the luminescent properties of the material may be found in van Hutten et al.. [25]

Fortunately, the design of such molecules can be aided significantly by quantum chemical calculations on model molecular segments (see, for example, Bredas et al. [1]). These methods permit effective prediction of the effect of specific side-group additions at specific locations on the optoelectronic properties of isolated molecules in both ground and excited states. The ability to predict the behavior of excited states is important as luminescence results from relaxation of these. The energy of excited states is closely coupled to changes in molecular conformation or organization upon excitation as noted in Section 9.1. For example, results suggest that PPV derivatives can become more flexible upon excitation, which can strongly influence the fluorescent efficiency and wavelength. Likewise, the stacking of molecules strongly influences the overlap of electron orbitals and contributes to intermolecular state luminescence. [25] This is in some senses unfortunate, as in amorphous molecular films a wide variety of chain packing arrangements occur and leads to broadened emission spectra and less saturated colors. Furthermore, it is difficult to assess the collective behavior of an amorphous polymer network from theory. The morphology and nanostructure of the resulting material becomes important in this case. Indeed, it is the interplay of nanostructure and properties that, in the end, makes a detailed prediction of the properties of the materials most difficult.

In addition to chemical methods used to design homogeneous polymers, all of the options of copolymerization and surface treatment have been applied to design of organic layers for optoelectronics. Subunits in the polymer chain can be optimized to accept electrons or holes or for recombination of excitons. Surface treatments can be used to orient the molecules on the substrate in the correct direction for optimal performance. These treatments can also be used to control the crystallinity or molecular nanostructure of the organic layer to some extent (or to a large extent in some cases), as was the case for pentacene self-assembled monolayers described above. Such modifications can be applied in both electron and hole transporting materials, which again blurs the boundary between the materials. Methods for engineering polymer subunits and copolymerization can be found in M.A. Fox. [26]

The results of copolymer engineering can be observed in the control of emission wavelength in PFO-related materials. PFO alone emits in the blue spectral region. To provide longer wavelength materials, PFO has been copolymerized with PPV-like (for red) or benzothiadiazole (BT) (green) units. An example of the resulting polymers, the performance of the devices, and full references may be found in I.D. Rees et al.. [27] Changes to the work function also occur with copolymerization. For example, PFO has HOMO and LUMO levels ~5.8 and 2.6 eV from the vacuum level, respectively, while the BT-PFO copolymer has corresponding values of 5.9 and 3.1 eV, lowering the injection barrier for electrons into the material. [27]

Although one can produce polymer OLEDs without an HTL, the performance of the device improves dramatically if a material specialized for hole injection from the anode is used. The HTL reduces the device operating voltage, increases lifetime, and

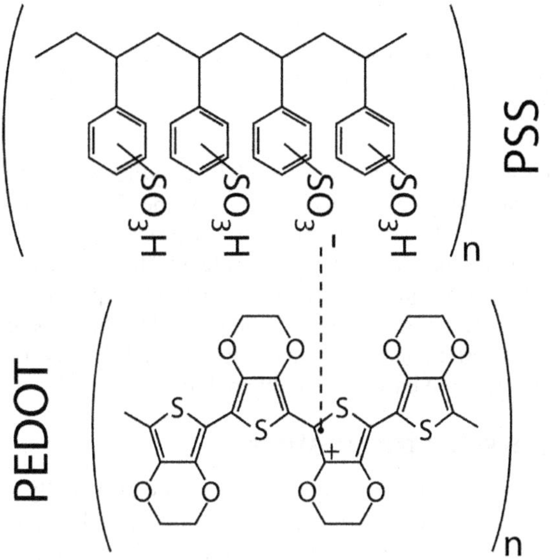

Figure 9.24: The chemical formula for the two components poly(3,4-ethylene dioxythiophene) (PEDOT) and poly(styrene sulphonic acid) (PSS) making up the PEDOT:PSS polymer blend used in contact with the anode and for hole transport in polymer OLEDS. Note that the two molecules bind together strongly as a double bond is broken in the PEDOT units and an electron is transferred to the PSS, resulting in release of a hydrogen atom (which may bond with the reactive radical site of the double bond. The molecules are held together, in part by the attraction of the ionic charges.

results in a better balance of hole and electron currents. The most popular anode-side material currently is poly 3,4-ethylene dioxythiophene (PEDOT), normally used in a polymer blend with poly (styrene sulphonic acid) (PEDOT:PSS). The structures of these molecules are shown in Figure 9.24. Note that PEDOT and PEDOT:PSS are relatively insoluble in their conventional forms and must be synthesized after precursors are deposited. Poly(aniline) or poly(thiophene) have also been popular for anode-contacting layers. These materials are highly conductive and optimized for efficient hole injection.

Comparisons of polymer OLEDs generally show an order of magnitude improvement in light emission efficiency for PPV-based devices when the surface of the ITO anode is treated with oxygen plasmas or acid cleaning methods (which modify the effective work function of the ITO surface). However, a thousand fold improvement results from the use of hole injection layers such as PEDOT:PSS.

Some PFO-like materials can also act as acceptable hole-injection layers as well as recombination layers fluorescing in the blue. An example of such a material is poly(9,9-dioctylfluorene-*co*-bis-*N,N*8-(4-methoxyphenyl)-bis-*N,N*8-phenyl-1,4-phenylenediamine) [PFMO], shown in Figure 9.25. [28] These modified PFO's have lower HOMO and LUMO energies relative to the vacuum level and therefore have lower hole-injection barriers.

Many other materials have been considered in polymer OLEDs. The results are reflected in the dramatic gains in performance of the devices as shown in Figure 9.10. All of these materials combine multiple functions (as compared to molecular materials where each separate constituent molecule has a specific purpose). The molecules are designed by quantum chemical methods supplemented with a considerable experience base among synthetic chemists. The final materials represent a compromise between performance and manufacturability. The most successful materials in manufactured devices are based on soluble polymers.

9.4.2 Polymers for transistors

The polymers used for organic thin film transistors are chemically similar in general to α-sexithienyl and include materials such as poly(3-hexylthiophene) (P3HT) and related compounds (see Figure 9.26). This material has functional groups incorporated onto the thiophene units as shown in the figure. However, there are two possible arrangements of these units, known as "head-to-head" and "head-to-tail", see Figure 9.26. Polymers consisting of the head-to-head arrangement can stack much more regularly and produce higher carrier mobilities in solution cast layers. Therefore, the head-to-head moieties are preferred for polymer TFTs. For a discussion of devices based on these materials, see Z. Bao et al.. [29]

As in the case of OLEDs, the primary driver for considering polymeric materials for OTFTs is the possibility of rendering them soluble by addition of appropriate side chain functional groups. The orientation of the molecules in the resulting films

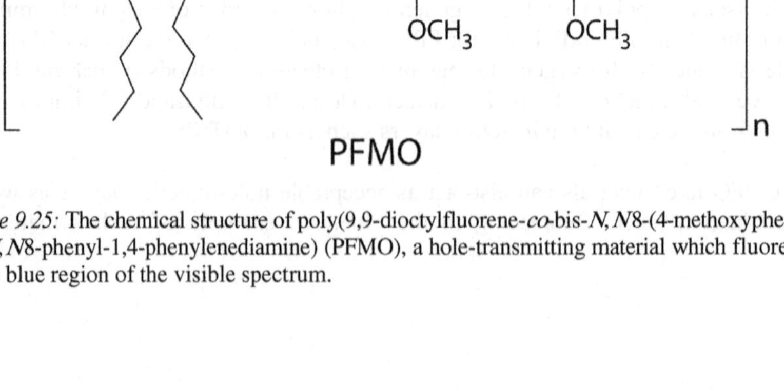

PFMO

Figure 9.25: The chemical structure of poly(9,9-dioctylfluorene-*co*-bis-*N, N*8-(4-methoxyphenyl)-bis-*N, N*8-phenyl-1,4-phenylenediamine) (PFMO), a hole-transmitting material which fluoresces in the blue region of the visible spectrum.

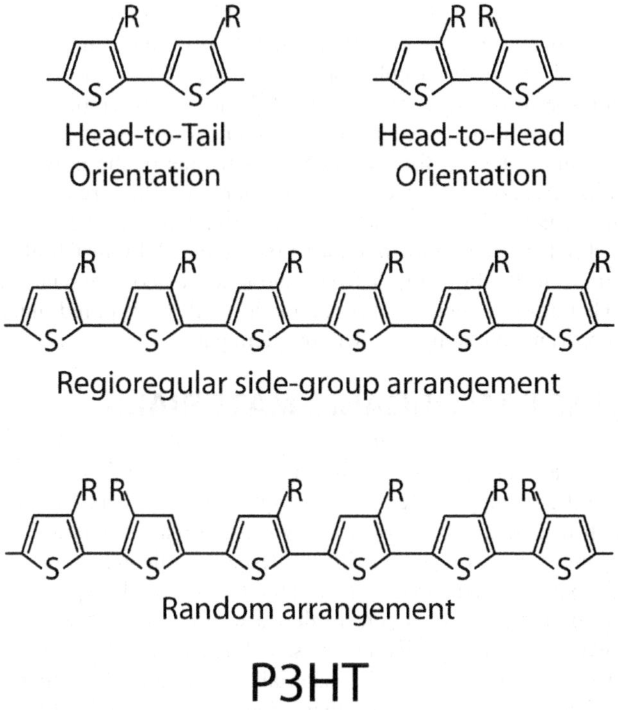

Head-to-Tail
Orientation

Head-to-Head
Orientation

Regioregular side-group arrangement

Random arrangement

P3HT

Figure 9.26: The chemical structures of poly(3-hexylthiophene) (P3HT) showing the head-to-head, and head-to-tail arrangements of functional groups. The head-to-head arrangement can be present in large blocks in a molecule (regioregular) or the two arrangements can be mixed randomly.

depends upon the solvent from which the material is cast. For P3HT, the best organization has generally been obtained using chloroform. The organization of P3HT layers on surfaces can also be influenced by surface treatments. When the layers are deposited on SiO_2 surfaces, stronger interactions and better organization can be obtained when the hydroxyl termination of typical SiO_2 is replaced with an organic surface termination. Such a surface generally improves carrier mobilities in P3HT on SiO_2.[30]

The architecture of devices based on polymer organic materials are similar to those described for molecular organic devices (see Figure 9.11), although devices using polymer dielectrics have also been produced. The thiophene compounds and most materials used for OTFTs are p-type (majority hole conductors) and the devices generally operate in enhancement mode. In other words, the conductivity is increased by the applied bias in the device. The performances of the polymer semiconductors in these devices are generally lower than for their molecular cousins by nearly two orders of magnitude. The highest field-effect mobilities are typically 0.1 cm^2 V^{-1}s^{-1} with device on/off current ratios of 10^6. [30] The difference in performance of the polymers relative to the molecular species is significant in that the polymers currently under-perform the competing a-Si devices now employed in flat panel displays, while molecular device results are comparable or even superior to a-Si devices. The primary reason for the underperformance of the polymers is the lack of well-organized crystals. Consequently, carrier mobilities remain very low and limited by interchain hopping rather than intrachain transport. As for amorphous silicon based thin film transistors, polymer-TFT's can exhibit instabilities. Unlike the a-Si TFT's, these instabilities result from interaction of contaminant gases with the molecules. Therefore, a high-quality hermetic seal is required under inert gas conditions to prevent contamination with problem gases.

9.5 CONTACT TO ORGANIC MATERIALS

As noted above, a major problem facing designers of organic electronic devices is the contacts to the organic semiconductor, especially when the HOMO-LUMO gap is large (which it generally is). Contact to wide-gap materials is a generic problem for all electronic devices – contacting wide-gap semiconductors is much more difficult than contacting narrow-gap materials. Consequently, special care needs to be devoted to the materials selection. In the case of organic materials, as compared to conventional inorganic semiconductors, a further complication is the relative instability and reactivity of organic materials as compared to inorganics. Some organic materials are more sensitive than others. For example, PPV degrades by oxidation following charge injection more rapidly than does Alq$_3$. The issues in cathode and anode design differ significantly and are therefore considered separately in the following sections.

9.5.1 The cathode contact

Charge injection from the cathode into the organic layer can be viewed as a process of thermionic emission combined with tunneling of electrons from the metal into the organic material, as shown schematically in Figure 9.27. However, the situation is complicated by the low mobility of carriers in the organic material. Electron injection at the cathode results in reduction of the organic molecule and formation of carbanions, as described in Section 9.1.1. One can rapidly reach a point of so heavily reducing the organic molecule locally that further charge injection is inhibited pending diffusion of the carbocations away from the cathode. At the same time, the accumulation of negative charge reduces the electric field driving charge injection and contributes to a back diffusion of charge into the cathode. Over-biasing the system to encourage charge injection and diffusion, however, provides increased opportunities for creating energetic charges, stimulating chemical reactions and reorganizations, and consequently to materials degradation.

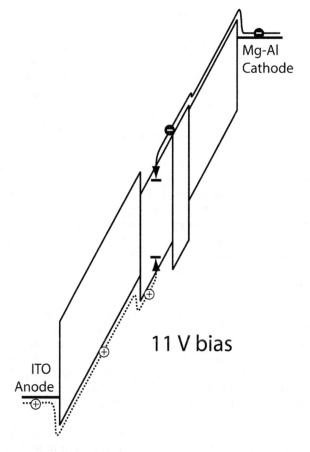

Figure 9.27: Shows the effect of bias on the band-edge diagram sketched in Figure 9.18, assuming that all organic compounds are equally conductive (and therefore the electric field in the organic layers is equally distributed).

Traditional cathodes use low work-function metals such as Ca. These have energies comparable to the LUMO energy into which electrons are injected. Any such low work function metal is naturally highly reactive, rendering the contact intrinsically unstable. Not surprisingly, reaction of the Ca electrode with the organic material degrades the device. It has been found that some preliminary oxidation of the electrode material is necessary prior to formation of the organic layer to avoid damaging reactions. Unfortunately, the cathode is not always the first layer formed and so pre-oxidation is not straightforward. When the cathode is the last layer produced, deposition of a reactive metal such as Ca often yields a non-functioning device. As a consequence, most devices are based on the more stable Mg or Mg-Al alloys. These provide an acceptable contact material with adequate stability.

An interesting alternative cathode, ZrC was explored by Sheats et al.. [31] ZrC has a low work function (between Ca and Al) and is unreactive. For example, it is highly stable in air for long times. Like many other transition-metal carbides, it has a good electrical conductivity. These factors make it appear a nearly ideal contact material. Indeed, good electrical results were obtained for ZrC cathode contacts to MEH-PPV. Unfortunately, the devices showed low luminescence efficiency, possibly because of defect formation near the contact in the organic layer. Likewise, the lifetimes of the ZrC based devices were relatively poor. Overall, this has led to the abandonment of ZrC as a contact material. Further research might find a solution to the observed problems.

The tendency to form an electrostatic dipole layer at the cathode junction that resists charge injection should be counteracted by use of a very low work function metal (with Fermi energy above the LUMO of the organic) that will supply electrons to the LUMO. This should spontaneously reduce the conjugated structure of the molecule and lead to a negative charge on the organic molecules near the contact. The large number of potentially reducible double bonds in the organic material restricts this dipole to the interface. Thus, no depletion region is found associated with the heterojunction.

In some cases, the reverse situation is found, in which electrons spontaneously transfer from the organic material to a higher work function metal. Again, there is no observable depletion layer and a positive charge occurs on the surface of the organic material. Both of the above cases result in interface dipoles. It is also possible to find no spontaneous charge transfer across the interface and no resulting dipole when the Fermi energy of the metal lies in the energy gap of the organic material. However, in practice, surface and interface states result in dipoles, even in these situations. [32]

It should be possible to create a favorable dipole at the contact capable of promoting charge injection by the application of an appropriate material in the interface. For example, an ionic (polar) compound could be inserted in an orientation in which the dipole of the ionic material would cancel out the interfacial dipole. This is one possible mechanism behind the success of LiF as an interfacial layer between Al and many organic compounds. Indeed, a thin layer of most alkali fluorides, oxides, or sulfides (in other words, strongly polar compounds) is found to lower the charge injection barrier, although LiF is the most popular. A discussion and detailed references may be found in N.K. Patel et al.. [4] A similar improvement has been observed by the use of elemental Li metal in contacts to Alq_3. This allowed a transparent cathode to be fabricated joining Alq_3 to ITO through layers of 2,9-dimethyl-4,7 diphenyl-1,10-phenanthroline and elemental Li. [33] However, the reactivity and diffusivity of the Li may make such contacts relatively less stable over the longer term and their ultimate performance remains to be determined.

9.5.2 The anode contact

In a typical light-emitting device, at least one of the contacts must be transparent to allow the emitted light to escape. In the case of OLEDs, this is usually the anode. In general, the anode has been fabricated from indium tin oxide, $(In_2O_3)_{1-x}(SnO_2)_x$, [ITO]. This material has a high electrical conductivity and a high transparency to visible light, although it absorbs substantially in the infrared due to the high concentration of carriers. ITO can also be viewed as a degenerately-doped n-type wide-gap semiconductor. It is commonly used as a transparent electrode in many devices including current flat-panel displays. The problem is then to couple this material to the organic layers into which holes are to be injected. ITO is used on the HTL side of the device because it has a relatively high work function. It is not usable as a cathode material, in spite of its heavy n-type doping, because of the barrier to electron injection into the LUMO state that would result for most organic compounds.

Anode contacts to hole transporting molecular layers were discussed briefly in Section 9.3.2. To reiterate, it is often found that metal phthalocyanine compounds effectively mediate the injection of holes from ITO into hole transport layers. (One could also say oxidation of carbocations in the HTL with net transfer of an electron from the HTL into the ITO.) Some experiments have suggested that multiple layers of phthalocyanines based on different metals, and hence having different HOMO state energies, can further reduce the injection barrier. The basic method in use in these cases is to provide a staircase of small barriers up which a hole can climb more easily than would be the case for a single large step. The low mobility of holes in the organic interlayers comes in handy here as it reduces the rate at which holes return to the contact.

For polymer-based devices, conducting polymers such as polyanaline and polythiophene have been used extensively as anodes. These are naturally p-type (hole conducting) materials produced by oxidation of conjugated molecules and thus are naturally suited to mediating the hole injection problem. However, these materials are not ideal. The polymer devices have generally lagged behind the molecular-based devices, probably in part, because more intensive study and improvement of contacts has been pursued for the molecular materials.

Another source of trouble at the anode is that ITO may be degraded by extended operation in organic devices. It is thought that the compound decomposes and that In ions migrate into the organic layer (possibly accompanied by oxygen, which would cause damage as well). The In ions can be reduced to In metal as well with potential for developing localized shorting through the thin organic layers. This process is presumably driven electrochemically by the device bias voltage, as it occurs even in cases where the organic compound is known to be stable in contact with ITO under zero-bias conditions. Further evidence of an electrochemical source of the problem is found in the fact that coating the ITO with a conducting polymer such as polyaniline or polythiophene eliminates the decomposition. Presumably, the conducting polymer reduces the electric field present near the surface of the ITO and consequently eliminates the electrochemical driving force for decomposition.

9.6 DEFECTS IN ORGANIC MATERIALS

It may not be surprising that defects in organic materials are just as much of a problem as in inorganic materials. Upon further reflection, it may be surprising that defects are not completely prohibitive to the operation of organic devices. Defects in organic materials can be divided into primarily chemical and primarily structural defects. Chemical defects might be materials with incorrect functional group attachments. For example, if some of the side chains in MEH-PPV failed to form or were replaced by other species a local chemical defect would result. Likewise, a hydrolyzed or otherwise destroyed double bond would destroy the conjugation of the molecule. Organic synthesis inevitably results in local chemical defects of this type. It should not be surprising that such defects could trap carriers, bind excitons, block carrier and exciton motion, and encourage non-radiative exciton recombination. Chemical defects can be expected to increase over time due to the presence of excitons and other reactive sites on molecules. For example, a water molecule can be decomposed in the presence of an exciton on a -CH=CH- conjugated structure to yield a –CH$_2$-CHOH- single bonded structure. This is bond hydrolysis referred to above and is a major reason why water is a problem in organic electronic devices. We will not say more about chemical defects as they are simply the result of unintended chemical reactions and destroy the structures we were relying on to achieve a given behavior in the organic compound. Some chemical defects can be reduced or their formation prevented by more careful synthetic approaches with higher yields, approaches designed specifically to avoid particularly harmful results, better purification of the resulting materials, better removal of problem species such as O$_2$ and H$_2$O from the final device layers, and better seals to protect the finished structure from contamination by reactive species. Discussion of the effect of chemical defects may be found in many sources, for example in Nguyen or Grozema in the recommended readings.

A more subtle type of defect is a structural defect in an organic molecule. The more commonly recognized structural defects include crystalline defects when the organic material crystallizes, chain ends in polymers, particularly problematical geometries in which two molecules might meet or cross, and physical deformations. These defects are intrinsic regardless of the skill of the synthetic chemist, and cannot be as easily detected or eliminated.

For example, a simple bend in a conjugated molecule will affect the HOMO and LUMO states dramatically. Consider a conjugated organic molecule such as trans polyacetylene (Figure 9.1). If one of the single bonds is allowed to rotate through a moderate angle then it becomes immediately impossible to transfer the double bond to the site of that single bond, because the double bond requires a very specific orientation of the sp^2 hybrid orbitals (Figure 9.2). The presumption in the description of diffusion of carbocation and carbanion defects was that the local double bonding structure could be rearranged at will. Therefore a twist of any of the single bonds, which would be one way of bending trans polyacetylene would restrict movement of the carriers. Theoretical models of torsional deformation of conjugated organic molecules confirm this picture. Although charges can move through such a structure, their mobility is reduced.

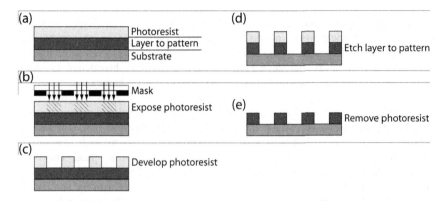

Figure 9.28: The basic steps in conventional photolithography. The approach begins with deposition of a photoresist material on the surface of a wafer coated with a layer to be patterned. This process is for a "positive" photoresist that increases its solubility when exposed to light (arrows) [step (b)]. The photoresist development step removes the exposed resist. The problem when patterning organic materials is step (e) in which the surface is typically exposed to an oxygen plasma that would remove both the photoresist and the organic layer to be patterned.

Returning to the LCAO discussion of Chapter 5, it is also hopefully apparent that any distortion of the geometry that would normally lead to an ideal bond will raise the energy of the bonding states and lower the energy of the antibonding states. Therefore, any bend in the backbone of a conjugated organic molecule must, by distorting the bonding patterns giving rise to the HOMO and LUMO states, reduce their energy difference locally. Therefore, a bend in a conjugated molecule, even if it does not involve a chemical defect, should bind excitons locally and reduce their energy. As we saw with inorganic materials, light emission is generally localized at such defects. Therefore, the only way to achieve the excitation properties of an ideal organic molecule is to assure that it never distorts. This is formally impossible to achieve for finite temperatures and can be made worse by structures that favor distortion of the molecules. Such distortions would also be natural in any non-crystalline material. Localized luminescence of organic molecules at bends in their structure has been observed by scanning probe techniques although whether the bends are caused by physical or chemical processes is not clear. The nature of physical and chemical defects is a matter of current debate and study. We will not say more here except to reiterate the essential point that **any physical or chemical distortion of an organic molecule will lead to a local trap for carriers or excitons and will degrade the optoelectronic properties.**

9.7 PATTERNING ORGANIC MATERIALS

Although this book does not consider lithography and other issues related to patterning of microelectronic materials in general, patterning organic materials is sufficiently distinct because of the problems involved that it deserves some mention. The patterning methods that have resulted have since been broadened to include application to other materials including conventional semiconductors.

The traditional method of patterning electronic materials, outlined schematically in Figure 9.28, begins by forming a uniform layer of the material across a substrate surface. This is then coated with a polymer known as a photoresist, generally by pouring it onto the wafer surface and spinning the wafer at a high speed. This results in a nice uniform photoresist coating across the substrate. The polymer is exposed to a pattern of light that either increases or decreases its solubility in some solvent. The higher solubility material is removed by the solvent, leaving the remainder behind. So far so good. To this point the process is compatible with organic materials. The next step is to remove the layer to be patterned which is not protected by the photoresist. Now comes the problem. The photoresist is typically an organic polymer. Therefore if the material to be patterned is also an organic polymer it is difficult to achieve a selective removal of the desired layer without also removing the photoresist. Even if one could do that the photoresist would remain as a coating on the protected parts of the layer to be patterned. In inorganic pattern generation the next step would be to expose the entire wafer to an oxygen plasma. Such a plasma very rapidly removes any organic material from the surface, leaving a nice clean patterned layer. Again, this is a problem when the layer of interest is itself a polymer. The problem of photoresist removal is an even greater problem than the selective etching of the layer to be patterned because the exposure process rendering the photoresist relatively insoluble makes it much more difficult to remove by design.

Various methods have been developed to deal with the problems with conventional lithography outlined above. Those described here are shown schematically in Figures 9.29 and 9.30. Some are adaptations to the traditional lithography approach that attempt to make it usable with a polymer underlayer. For example, one can insert a metal protective layer under the photoresist and use the photoresist to pattern the metal. That metal then acts as the mask for removal of the underlying organic material. Typically one can choose a metal that can be etched away at the end without affecting the underlying polymer that was to be patterned. This approach is cumbersome and if the metal is not sufficiently protective the underlying layer may be damaged during removal of the photoresist.

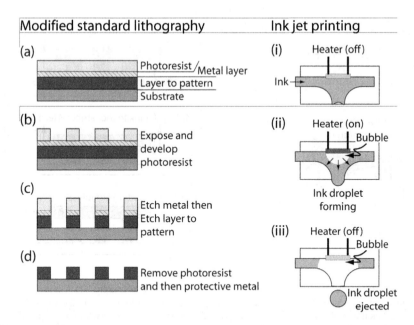

Figure 9.29: Two methods for conducting lithography on a material not compatible with conventional methods. Left: use of a metal layer to protect the material to be patterned from the process used to remove the photoresist. The protective metal is removed after the photoresist is removed. Otherwise the process is similar to that shown in Figure 9.28. Right: ink-jet printing has been used to print materials that can be dissolved or suspended to form an ink. The figure shows the operation of a print head. (i) Ink flows into the channel in the print head against the heater and opposite the print nozzle. (ii) The heater is turned on boiling the ink locally. The resulting bubble ejects a small droplet of ink (iii) and the heater is turned off, allowing the bubble to collapse and new ink to be drawn into the print head.

A second approach, which is attractive because it is a direct-write method, is to synthesize the material for the layer to be patterned in soluble form and then to dissolve it and write it directly onto the substrate in patterned form using an ink-jet printer [see, for examples, References 34-36]. These printers have remarkable resolution (patterns below 10 micron feature sizes can be achieved) and can directly write the material in question on the wafer without the need for subsequent patterning with a photoresist. This direct write approach means that no photomask is required and the pattern can be changed at any time by reprogramming the ink-jet printer. This method has been used for printing organic light emitting diodes devices with considerable success.

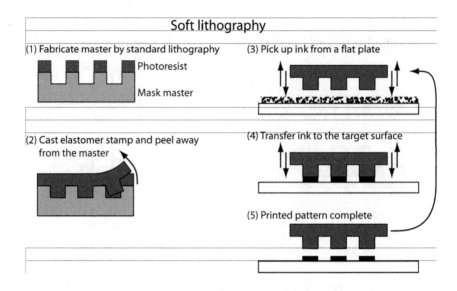

Figure 9.30: A schematic diagram of the basic steps in soft lithography. (1) A mask master is produced by standard lithography methods (see, for example Figure 9.28). (2) An elastomeric material such as polydimethylsiloxane is poured over the master and, with proper processing, covers the master conformally. The resulting stamp is peeled away from the master. The material to be printed, for example an ink, is picked up with the stamp (3) and transferred to the target surface (4). The ink should be designed to adhere well to the target and only weakly to the stamp so that the stamp comes away clean from the surface (5) leaving a complete and defect-free pattern. The stamp can be re-inked and used repeatedly.

The most dramatic improvement that is now being applied to patterning of a wide variety of materials from Si to polymers to metals is "soft lithography", known also by various names other, the most intuitive of which is "micro contact printing". A more detailed discussion and references may be found in Rogers and Nuzzo in the recommended readings. At its heart this is classical printing technology in which a raised pattern is produced on a stamp, the stamp is inked, and the ink is printed onto a surface. Amazingly, this process has been shown to be capable of printing even nanometer sized features, although the practical limit for the method is probably between 0.1 and 0.2 microns.

A typical process sequence for microcontact printing is shown in Figure 9.30. There are various implementations of this approach. We will consider a typical and relatively generic method here. The first step in the process is generally to create a master pattern in a hard material such as Si or quartz using conventional lithographic methods outlined above. For the highest resolution patterns it may be necessary to use electron-beam lithography. One then casts a layer of flexible rubbery polymer uniformly across the patterned master. A typical material used for this casting is polydimethylsiloxane (PDMS). The PDMS layer can then be peeled off of the master leaving a flexible polymer stamp containing the imprint of the master pattern. The remarkable aspect of this transfer is that the substrate pattern can be faithfully transferred to the PDMS even for nanometer feature sizes. It is useful for polymer lithography particularly because the only requirement is the ability to tailor the adhesion between the various layers that are put in contact with each other. It is not necessary to use an aggressive solvent or oxygen plasma to remove a polymer during this process.

The material to be printed is then transferred to the PDMS stamp by one of various methods. For example, a metal could be evaporated onto the PDMS stamp or a polymer or organic molecule could be applied to the stamp by precipitation from a solution. The simplest method is to deposit the ink on a flat plate and to pick up that ink on the stamp by simply pressing the stamp down on the surface. One then inverts the stamp and adheres the coating on the PDMS to a substrate of choice such as a Si wafer. The PDMS stamp is then peeled away, leaving the printed material on the substrate. To accomplish this it is necessary that the PDMS faithfully adopt the shape of the master, it must peel away from the master cleanly, it must pick up the ink effectively, and it must deposit that ink on the substrate. Of these the hardest to accomplish is the last. It is helpful if the ink will react strongly with the surface to form a strong bond and that the ink interact with the stamp only sufficiently to assure it is picked up initially. Specifically designed adhesion and release layers are commonly used, especially when transferring solid materials (see next) rather than inks.

Soft lithography is even more interesting because it can also transfer an active layer such as a thin single crystal of Si to any substrate surface desired. This is done by forming a brittle or soluble layer in the Si substrate at a fixed distance below the surface and below the active devices in the Si to be transferred. The stamp is attached to the active layer to be transferred and the detachment layer is used to remove the bulk of the substrate. The active devices are now attached to the stamp as in the printing process described above and can be transferred to another substrate. This has allowed creation of "flexible" Si electronic devices.

For more details on soft lithography the reader is referred to the recommended readings and references.

9.8 SUMMARY POINTS

- Organic devices are made from molecular (better performance) or polymer (more easily processed) materials. Molecular species have a single function normally and are blended when multiple behaviors are required. Polymers may be multifunctional based on copolymerized sections.
- Molecular species may be designed to have high crystallinity, leading to high carrier mobilities and good OTFT performances. Molecular materials have lower crystalline quality and lower mobility in general.
- Organic electronic materials are fully conjugated (they have a continuous double bond structure running through their structure). The π bonds associated with conjugated structures are rigid (they do not rotate). Conjugated molecules may conduct charges by breaking double bonds. Reconstitution of the broken bond may result in light emission.
- Any physical or chemical defect in an organic molecule will tend to lower carrier mobilities, trap charge carriers and excitons locally, will encourage non-radiative recombination, and will lower the energy of emitted photons in radiative processes. Many of these defects are intrinsic to the materials and many are more common for certain types of molecules or certain processing methods.
- Organic molecules may behave as indirect-gap semiconductors in that their minimum energy excitation may involve a change in physical structure. Transitions involving triplet states are forbidden. Only singlet states may recombine efficiently.
- Molecules having little conformational change associated with excitation are efficient light emitters.
- Coherence distances for defects on molecules are usually 1-2 nm. Conformational changes associated with excitations or defects are usually limited to this distance.
- Carbocations are positively-charged atoms and behave as do holes in inorganic materials and have higher mobilities in general than electrons (carbanions).
- Carbanions are negatively-charged atoms and behave as do electrons.
- Solitons are (for our purposes) uncharged dangling bond states (i.e. free radicals).
- Exciton binding energies in organic materials are generally high, 0.1-1 eV. They involve carbocation/carbanion pairs on adjacent C atoms.
- Limitations to OLED efficiency include: (1) imbalance in charge injection, (2) poor use of energy in triplet states, (3) low radiative recombination relative to non-radiative recombination, (4) problems with escape of the emitted light.
- Most molecules transmit electrons or holes better than the other charge and are know as electron transmitting (ETL) or hole transmitting (HTL) layers, respectively.

- Transmitted charges tend to be more dispersed across the structure of the molecule while poorly transmitted charges are localized and are associated with larger bonding changes.
- HTL and ETL molecules are designed (1) for optimized energies of their HOMO and LUMO orbitals to match with contacts (for charge injection) or to provide barriers to the propagation of a given charge (usually also the poorly transmitted charge), (2) for good stability, (3) for resistance to crystallization if used in OLED devices.
- Dye molecules are used to provide efficient luminescence at a prescribed photon energy (wavelength). Dye molecules should (1) trap free carbocations and carbanions, (2) localize resulting excitons on a single molecule, (3) allow the exciton to decay radiatively in a short time, (4) use energy in both singlet and triplet states, (5) dissolve well in the matrix, (6) have a narrow emission spectrum, and (7) be chemically stable.
- Optimal dye concentrations place the molecules at a separation of roughly the Förster radius (1-2 nm).
- Some molecules (especially those incorporating rare earth or transition metals) are efficient converters of excitons from triplet states to singlet states.
- Cathode contact issues include the reactivity of low work function metals and the formation of dipoles at the cathode/ETL interface that interfere with charge injection. Polar compounds may sometimes be inserted into the interface to cancel the dipole and improve injection efficiency.
- The major issue at the anode is the barrier to charge injection because high enough work function metals are not available. This is dealt with by using a cascade of materials with increasing HOMO binding energy (farther from the vacuum level). In addition, electrochemically-driven reaction and decomposition of the ITO transparent conductor can occur leading to loss of performance in the device. This can be reduced by adding a conducting polymer in the interface.

9.9 HOMEWORK

1) Sketch the structure of trans-polyacetylene containing a carbanion (the organic equivalent to an "electron") and a carbocation (hole) as separate solitons, not as an exciton pair in to configurations: (i) with both defects on one side of the chain and (ii) with the defects on opposite sides of the chain. Briefly explain why the structure is more stable with defects on opposite sides of the chain.

2) What is the purpose of $Ir(ppy)_3$ in an OLED and why is this important?

3) What is required for an organic molecule to be capable of transferring charge (i.e. transferring carbocations or carbanions) along its length?

4) It is possible to produce an OLED with a single layer of material (for example ppv) between two electrodes. Explain briefly:

    a)   Why do the carriers (on average) not simply pass through the material to the other electrode? [There are two major factors in this behavior.]

    b)   Why is this a diode? (Why is it so much harder to inject current in one direction than in the other?) You may wish to accompany your answer with a sketch of the band edges/Fermi energies in the device.

5) When designing a molecular material for use in

    a)   an OLED (for example as a hole transporter), what is the advantage of choosing an asymmetric structure or one containing large side groups?

    b)   a TFT (for example as the channel), what is the advantage of a small and highly symmetric molecule?

6) Why is Alq_3 a better electron transporter than a hole transporter?

7) What is the primary problem in application of hole transport layers and how does CuPc improve this situation?

8) What is the ideal concentration of dye molecules in an OLED emission layer and why are higher or lower concentrations less efficient?

9) What is the primary advantage of polymer over molecular organic materials in OLED's?

10) If the cathode contact to an electron transporter layer is a Schottky barrier type contact, why is there no observable depletion region in the organic material?

11) What are the major issues related to the anode and cathode contacts to OLED's and how have these been reduced in operational devices?

12) Why are dye molecules producing saturated colors important to designing a better full color display based on OLED's?

13) Consider the device structure shown in Figure 9.17. For each layer shown give the primary purpose of the layer and the most important aspect of its properties that contributes to achieving this purpose.

14) If the mobility of a carrier in an organic material is 0.001 cm^2 V^{-1}s^{-1} and the effective mass of an electron is the rest mass, m_0,

   a) calculate the approximate time between scattering events. [Hint: 1 cm^2 V^{-1}s^{-1} = 10^{-4} A s^2 kg^{-1} = 10^{-4} C s kg^{-1}.]
   b) if the device is biased to 8V and is 8 nm thick, estimate the average drift velocity of the carriers.
   c) Given the results of (a) and (b), estimate the average distance a carrier moves between scattering events.

9.10 SUGGESTED READINGS & REFERENCES:

Recommended readings:

Ian D. Rees, Kay L. Robinson, Andrew B. Holmes, Carl R. Towns, and Richard O'Dell, "Recent Developments in Light-Emitting Polymers" in *MRS Bulletin*, 2002; 27: 451.

Any of various articles in a special issue of *Accounts of Chemical Research*, 1999; 32.

H. Bleier, "Kinetics of charged excitations in conjugated polymers – an example for the application of picosecond- and femtosecond spectroscopy," in *Organic Materials for Photonics: Science and Technology*, Z. Gerbi, ed. Amsterdam: North Holland, 1993: 77-102, and other articles in the same source.

Nguyen, T.P.; "Defect analysis in organic semiconductors." *Materials Science in Semiconductor Processing* 2006; 9: 198-203.

Ferdinand C. Grozema, Piet Th. van Duijnen, Yuri A. Berlin, Mark A. Ratner, and Laurens D. A. Siebbeles, "Intramolecular Charge Transport along Isolated Chains of Conjugated Polymers: Effect of Torsional Disorder and Polymerization Defects." J. Phys. Chem. B 2002; 106(32): 7791-5.

J.A. Rogers and R. G. Nuzzo, "Recent progress in soft lithography", *Materials Today*, 2005; 8: 50-6.

References:

[1] Brédas, Jean-Luc; Cornil, Jérôme; Beljonne, David; Dos Santos, Donizetti A; and Shuai, Zhigang; "Excited-state electronic structure of conjugated oligomers and polymers: a quantum-chemical approach to optical phenomena". *Accounts in Chemical Research* 1999; 32: 267-276.

[2] Cornil, J.; Beljonne, D.; Brédas, J.L.; "Towards a coherent description of the nature of the photogenerated species in the lowest-lying one-photon allowed excited state of isolated conjugated chains." *Synthetic Metals* 1997; 85: 1029-1030.

[3] Köhler, A.; dos Santos, D.A.; Beljonne, D.; Shuai, Z; Bredás, J.-L.; Holmes, A.B.; Kraus, A.; Müllen, K.; and Friend, R.H.; "Charge separation in localized and delocalized electronic states in polymeric semiconductors." *Nature* 1998; 392: 903-906.

[4] Patel, N. K.; Cinà, S.; and Burroughes, J. H.; "High-efficiency organic light-emitting diodes." *IEEE J. on Selected Topics in Quantum Electronics*, 2002; 8: 346-61.

[5] Jackson, T.N.; Lin, Y-Y.; Gundlach, David J.; and Klauk, H.; "Organic thin-film transistors for organic light-emitting flat-panel display backplanes." *IEEE J. Sel. Topics In Quantum Electronics* 1998; 4: 100-4.

[6] Sirringhaus, H.; Friend, Richard H.; "Integrated optoelectronic devices based on conjugated polymers." *Science* 1998; 280: 1741-4.

[7] Jackson, Thomas N.; Organic thin film transistors-electronics anywhere. *Proceedings of the 2001 International Semiconductor Device Research Symposium.* Piscataway N.J.: IEEE, 2001.

[8] Dimitrakopoulos, C.; Purushothaman, S.; Kymissis, J.; Callegari, A.; and Shaw, J.M.; "Low-voltage organic transistors on plastic comprising high-dielectric constant gate insulators." *Science* 1999; 283: 822-4.

[9] Zhang, R.Q. and Lee, S.T.; "Effect of charging on electronic structure of the Alq_3 molecule: the identification of carrier transport properties." *Chem. Phys. Lett.* 2000; 326: 413-20.

[10] See for example, Kijima, Yasunori; Asai, Nobutoshi; and Tamura, Shin-ichiro; "A blue organic light emitting diode." *Jpn. J. Appl. Phys.* 1999; 38: 5274-77.

[11] Adamovich, Vadim I.; Cordero, Steven R.; Djurovich, Peter I.; Tamayo, Arnold; Thompson, Mark E.; D'Andrade, Brian W.; and Forrest, Stephen R.; "New charge-carrier blocking materials for high efficiency OLEDs." *Organic Electronics* 2003; 4: 77-87.

[12] Adachi, C.; Nagai, K.; and Tamoto, N.; "Molecular design of hole transport materials for obtaining high durability in organic electroluminescent diodes." *Appl. Phys. Lett.* 1995; 66: 2679-81.

[13] Zhu, L.; Tang, H.; Harima, Y.; Kunugi, Y.; Yamashita, K.; Ohshita, J.; Kunai, A.; "A relationship between driving voltage and the highest occupied molecular orbital level of hole-transporting metallophthalocyanine layer for organic electroluminescence devices." *Thin Solid Films,* 2001; 396: 213-8.

[14] D. Hohnholz, S. Steinbrecher, M. Hanack, "Applications of phthalocyanines in organic light emitting devices." *J. Mol. Struct.,* 2000; 521: 231-7.

[15] Adachi, Chihaya; Baldo, Marc A.; and Forrest, Stephen R.; "Electroluminescence mechanisms in organic light emitting devices employing a europium chelate doped in a wide energy gap bipolar conducting host." *J. Appl. Phys.,* 2000; 87: 8049-55.

[16] Shoustikov, Andrei A.; You, Yuijian; Thompson, Mark E.; "Electroluminescence color tuning by dye doping in organic light-emitting diodes." *IEEE J. Sel. Topics in Quantum Electronics* 1998; 4: 3-13.

[17] Wu, C.; Sturm, J.C.; Register, R.A.; Tian, J.; Dana, E.P.; and Thompson, M.E.; "Efficient organic electroluminescent devices using single-layer doped polymer thin films with bipolar carrier transport abilities." *IEEE Trans. on Electron Dev.* 1997; 44: 1269-81.

[18] Adachi, Chihaya; Thompson, Mark E.; and Forrest, Stephen R.; "Architectures for efficient electrophosphorescent organic light-emitting devices." *IEEE J. on Selected Topics in Quantum Electronics,* 2002; 8: 372-7.

[19] Kanno, Hiroshi; Hamada, Yuji; Takahashi, Hisakazu; "Development of OLED with high stability and luminescence efficiency by co-doping methods for full color displays." *IEEE J. on Selected Topics in Quantum Electronics* 2004; 10: 30-36.

[20] Baldo, M.A.; Thompson, M.E.; and Forrest, S.R.; "High-efficiency fluorescent organic light-emitting devices using a phosphorescent sensitizer." *Nature* 2000; 403: 750-753.

[21] Pope, M. and Swenberg, C. E., *Electronic Processes in Organic Crystals and Polymers,* 2nd ed. New York: Oxford University Press, 1999, pp. 337-340.

[22] Schoonveld, W.A.; Wildeman, J.; Fichou, D.; Bobbert, P.A.; Van Wees, B.J.; and Klapwijk, T.M.; "Coulomb-blockade transport in single-crystal organic thin-film transistors." *Nature* 2000; 404: 977-80.

[23] dos Santos, D.A.; Quattrocchi, C.; Friend, R.H.; Brédas, J.L.; "Electronic structure of poly-paraphenylene vinylene copolymers: the relationship to light-emitting characteristics." *J. Chem. Phys.* 1994; 100: 3301-6.

[24] Greenham, N.C.; Moratti, S.C.; Bradley, D.D.C.; Friend, R.H.; and Holmes, A.B.; "Efficient light-emitting diodes based on polymers with high electron affinities." *Nature,* 1993; 365: 628-30.

[25] Van Hutten, P.F.; Krasnikov, V.V.; and Hadziioannou, G.; "A model oligomer approach to light-emitting semiconducting polymers." *Acc. Chem. Res.* 1999; 32: 257-65.

[26] Fox, M.A.; "Fundamentals in the design of molecular electronic devices: long range charge carrier transport and electronic coupling." *Acc. Chem. Res.* 2999; 32: 201-7.

[27] Rees, I.D.; Robinson, K.L.; Holmes, A.B.; Towns, C.R.; and O'Dell, R.; "Recent developments in light-emitting polymers." *MRS Bulletin,* 2002; 27: 451.

[28] Campbell, A.J.; Bradley, D.D.C.; Antoniadis, H.; Inbasekaran, M.; Wu, W. W.; and Woo, E.P.; "Transient and steady-state space-charge-limited currents in polyfluorene copolymer diode structures with ohmic hole injecting contacts." *Appl. Phys. Lett.* 2000; 76: 1734-6.

[29] Bao, Z.; Dodabalapur, A.; and Lovinger, A.J.; "Soluble and processable regioregular poly(3-hexylthiophene) for thin film field-effect transistor applications with high mobility." *Appl. Phys. Lett.* 1996; 69: 4108-10.

[30] For details, see Dimitrakopoulos, C.D. and Mascaro, D.J.; "Organic thin-film transistors: a review of recent advances." *IBM Journal of Research and Development* 2001; 45: 11-27.

[31] Sheats, J.R.; Mackie, W.A.; Anz, S.; and Xie, T.; "Polymer electroluminescent devices with zirconium carbide cathodes." *Proc. SPIE,* 1997; 1348: 219-27.

[32] For details, see Shen, C. and Kahn, A., "The role of interface states in controlling the electronic structure of Alq$_3$reactive metal contacts." *Organic Electronics,* 2001; 2: 89-95.

[33] Parthasarathy, G.; Adachi, C.; Burrows, P.E.; and Forrest, S.R.; "High-efficiency transparent organic light-emitting devices." *Appl. Phys. Lett.* 2000; 76: 2128-30.

[34] Shun-Chi Chang, Jie Liu, J. Bharathan, Yang Yang, J. Onohara, and J. Kido, "Multicolor organic light-emitting diodes processed by hybrid inkjet printing" in *Advanced Materials,* 1999; 11: 734-7.

[35] Pede, D.; Serra, G.; and De Rossi, D; "Microfabrication of conducting polymer devices by ink-jet stereolithography" in *Materials Science and Engineering C*, 1998; 5: 289-91.

[36] Chabinyc, M.L.; Wong, W.S.; Arias, A.C.; Ready, S.; Lujan, R.A.; Daniel, J.H.; Krusor, B.; Apte, R.B.; Salleo, A.; Street, R.A.; "Printing methods and materials for large-area electronic devices" in *Proceedings of the IEEE* 2005; 93: 1491-9.

Chapter 10

THIN FILM GROWTH PROCESSES

Up to this point this text has focused primarily on materials themselves and not how to produce them. A major aspect of materials science is the control of the kinetic and thermodynamic conditions under which materials are produced to yield specific properties. This chapter and the ones that follow describe some of the ways semiconductor electronic materials are created as thin films. For comparison, the most popular method of production of bulk materials was covered in Chapter 4. Bulk wafers are useful as substrates but are impractical for many applications, especially where alloys are needed. In current technology, thin films constitute most of the active and passive layers that are used in electronic devices.

This chapter covers the common features of all vapor phase thin film growth techniques – the processes by which atoms land on surfaces, move about, leave the surface, and how surface atoms go on to produce complete films. As with other chapters in this book, whole texts have been written on the subject so this treatment reviews only the highlights. Following chapters will cover specific classes of processes. Subjects of this chapter and include adsorption, desorption, surface structure and energy and how they are related to surface diffusion and the evolution of morphology, and adhesion.

10.1 GROWTH PROCESSES

In many of the earlier chapters we have assumed the ability to create a desired series of thin layers on a thick substrate. Usually, when semiconducting, these layers are grown as high-quality single crystals because of the problems with defects described in Chapter 7. As one can tell from the discussion of the devices in Chapter 3, the layers must be very thin, sometimes only a few monolayers, with carefully controlled compositions and doping levels. Therefore, the missing element in the previous chapters is how one can practically produce thin high-quality crystal layers. All of the common growth methods involve one or more phase transformations along the path from an initial state, containers of individual reactants, to a final state – a film of desired composition and structure. It is the details of how these transformations occur that determine the quality of the resulting film. While atoms or molecules may be supplied from solid or liquid phases, the most common approach is to provide them through the vapor phase. Before describing vapor phase methods, it is worth a brief mention of the solid and liquid phase methods.

Solid phase growth is used where only structural phase transformations occur and long-range transport of atoms or removal of product species is not required. This is necessary because in solids atoms cannot move very far and the solid material must serve as its own atom source and sink. Solid phase growth is only applied significantly in current semiconductor processing for crystallization of amorphous materials or solid phase reactions, as in formation of silicides by reaction of a metal with silicon.

An example of crystallization of amorphous material is when a large number of atoms are added to solid semiconductors by ion implantation. In implantation, high-energy impurity atoms strike and penetrate the crystal causing damage to such an extent that it may convert the solid to an amorphous structure. Upon annealing the atoms reorganize to recreate the crystal by adding atoms to the surrounding lightly damaged single crystal one layer at a time. The basic process is shown schematically in Figure 10.1. Because an amorphous solid is not unlike a very viscous liquid, the process of solid-phase growth is not unlike a liquid phase growth in which the driving force for the phase transformation is crystallization. The major difference from an atom's perspective is the reduced atomic mobility in the solid compared to in the liquid. For implants in Si single crystals, solid phase epitaxy (regrowth) produces a better final doped material, when done properly, than a lower-dose implant that does damage but does not amorphize the solid.

In the case of solid-phase reaction to produce a silicide, a polycrystalline metal layer is deposited on a Si wafer. The structure is then heated, typically to between 250 and 650°C, depending upon the silicide to be grown. A reaction occurs at the interface producing the silicide. Once the metal-Si interface is completely covered with silicide (at least locally), continuing the reaction requires diffusion of either the metal or Si atoms through the intervening silicide layer. In most cases a wide variety of silicide compounds may form by reaction with Si. Thus, in the case of Ni, the commonly identified silicides are Ni_3Si, Ni_5Si_2, Ni_2Si, Ni_3Si_2, $NiSi$, and $NiSi_2$, as may be seen from a Ni-Si binary phase diagram.

With so many silicides to choose from one might then ask which forms first. An obvious possibility is that the most stable silicide (with the highest free energy of

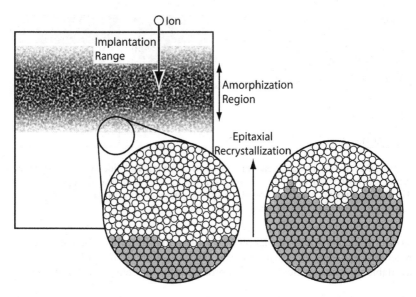

Figure 10.1: Shows the process of regrowth of a crystal from an amorphized region. For example, suppose that the right hand close-up represents the as-implanted material while the right hand close-up is after partial recrystallization. Regrowth of this type can result is a perfect crystal if done correctly.

formation at a given temperature) would nucleate first. However, one also requires reactants to produce a given product phase and low energy interfaces. Although exceptions occur, it turns out that in general the first phase to nucleate in silicide reactions is usually rich in the diffusing species, although it is not always the phase with the largest amount of that material. For example, Ni_2Si is the first phase to nucleate at a Ni-Si junction even though two more Ni-rich phases exist. In this case the Ni_2Si is the most stable (most exothermic heat of formation) of the Ni-rich phases. One may see how the nucleating phase might be rich in the diffusing species if nucleation is dominated by the law of mass action rather than by thermodynamics and if an excess of diffusing reactant accumulates before nucleation can occur.

After a phase is formed the reaction continues, either limited by the reaction rate at the interface where the diffusing species meets the stationary reactant, or limited by transport of the reactant through the silicide. The former does not depend upon the silicide thickness, so the thickness increases linearly with time, and exponentially with temperature (from the reaction rate). Diffusion-limited transport requires increasingly long-range movement the thicker the silicide becomes. Therefore, the rate decreases with increasing silicide thickness and the thickness of the silicide increases as the square root of time and exponentially with temperature (due to the increasing diffusivity). When the film becomes thick enough that reactant transport

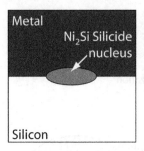  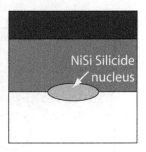

Figure 10.2: Shows the basic reaction process for forming a silicide by solid-phase reaction. A metal is deposited directly on a silicon layer and the couple is heated to permit the reaction. The first event is nucleation of one of the silicide phases. This grows thicker and slower until a new more stable phase has the chance to nucleate.

slows, a second more stable phase, if any, may nucleate. In the case of Ni-Si reactions, this leads eventually to NiSi and NiSi$_2$ formation if sufficient reactants are available. The basic silicide reaction process is shown schematically in Figure 10.2.

An interesting feature of the solid phase reaction to form silicides is that the diffusing reactants move by a vacancy mechanism. For this to occur it is necessary to move vacancies in the opposite direction as do the atoms. Thus vacancies are formed at the interface where reaction is occurring, diffuse through the silicide, and are eliminated at the other surface. However, this elimination does not occur instantly and the vacancies may penetrate a few nanometers into the diffusing reactant layer. When the moving species is Si, this means vacancies are formed at the metal/silicide interface, move to the Si and are eliminated at the Si/silicide interface. Their shallow penetration into the Si substrate can cause anomalously high diffusion of impurities in the Si near this interface, which may affect the doping profile there. When Si is the diffusing species, impurities in the Si, notably As, may be rejected by the silicide. Therefore, as the silicide grows the dopant concentration in the Si in front of the advancing silicide may be enhanced. Preferential incorporation of an impurity in the silicide is observed with other dopants leading to a reduction of dopant concentration in front of the silicide interface. For a detailed discussion of silicide reactions see Murarka in the recommended readings. For an example of dopant segregation during silicide formation see Reference 1.

Liquid-phase growth has also been used in thin film formation. We encountered an example in Chapter 4 – bulk crystal growth by the Czochralski process. Similar methods are available for thin film growth under the general name of liquid-phase epitaxy or LPE. These typically involve passing a seed crystal across the surface of one or more liquids that freeze new material onto the growing seed as in the Czochralski process. One can produce a multilayer structure by shifting from one liquid solution to another as growth proceeds. Although LPE produces good single crystal thin films at much higher rates than the more common vapor phase methods

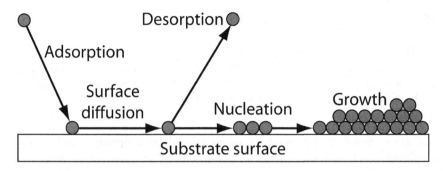

Figure 10.3: Events in a typical vapor-phase thin film growth process.

described below and in the following chapters, the options for control are relatively poor and the temperature is restricted to the melting temperature of the liquid being grown. Thus, it is not possible to grow materials as far from thermodynamic equilibrium, as is possible with vapor phase methods. In LPE, it is also common to remelt some of the substrate when switching from one material to another. This increases the minimum thickness of the layer being grown. Furthermore, remelting is not generally uniform and contaminates the growth liquid. Finally, the temperature is so high in general that diffusion is relatively rapid and abrupt interfaces become difficult to create.

For the above reasons, all of the commonly used thin film growth techniques supply atoms to the growing film from the vapor phase. The major differences among the techniques are the methods used to produce the vapor and the pressure of the chamber in which the growth takes place. In some cases, high-energy particles may also bombard the growing film and cause changes in the growth. The events necessary for film nucleation and growth from the vapor phase are shown schematically in Figure 10.3. Vaporization of species can be by physical or chemical means. Physical vapor deposition (PVD) processes involve evaporation by heat or sputtering by impact of high-energy particles with a target. PVD processes generally produce vapors of atomic or small molecular species. Chemical vapor deposition (CVD) methods use materials that are intrinsically gases under conditions in the reactor and are often supplied through pipes and valves from storage containers. In the following chapters we will consider heat-based vaporization (evaporation) methods, energetic particle based vaporization (sputtering), and CVD separately and in some detail.

Once the vapor has been produced, the steps needed to grow the film are condensation of the vapor on the growing surface (adsorption), surface diffusion of the reactants, reaction to form the film, coalescence of the reacted material into islands of atoms, and desorption (re-evaporation) of any product or unused reactant from the surface. The following sections discuss these steps in more detail.

10.2 GAS PHASE TRANSPORT

Before beginning the description of events on a growing surface, it is useful to consider a few important aspects of the behavior of gases briefly as this is essential to understanding supply of atoms to a surface from the vapor phase, to removal of products in CVD, and to a general understanding of the importance of vacuum conditions.

Gas atoms move at the speed of sound (for given temperature and pressure conditions) and undergo collisions periodically. For gas atoms near room temperature the mean free path λ between scattering events is given roughly by

$$\lambda = \frac{k_B T}{\sqrt{2}\,\sigma P} \qquad\qquad 10.1$$

where σ is the collision cross section, typically of the order of 10^{-16} cm^2 and P is the pressure. For T = 300 K and $\sigma \sim 4\times10^{-15}$ cm^2,

$$\lambda = 0.7 \text{ cm P}^{-1} \qquad\qquad 10.2$$

where P is the pressure in Pa (1 atmosphere = 1013 mBar = 1.013×10^5 Pa). At atmospheric pressure this would yield a mean distance between atomic collisions of ~70 nm. For a typical vacuum system with a source-to-substrate distance of 10-50 cm a pressure of 10^{-2} Pa is sufficient to allow a typical atom or molecule to transit from source to substrate without striking another atom. Higher energy atoms travel, on average, a greater distance. Based in equation 10.1, the distance would increase roughly linearly with kinetic energy. Thus, a 25 eV particle would move ~1000 times farther than a room temperature particle in the same gas.

In addition to mean free path, the flux of atoms striking a surface, or the approximate time between collisions on a particular atomic site on a surface, is important. Ideal gas theory yields a relation between flux and pressure of

$$F = \frac{P}{\sqrt{2\pi m k_B T}}, \qquad\qquad 10.3$$

where m is the mass of the gas particle, or

$$F = 2.63\times10^{17}\,\frac{P}{\sqrt{mT}} \text{ cm}^{-2}\text{ s}^{-1}, \qquad\qquad 10.4$$

where the mass is the molecular weight of the gas in AMU, the temperature is in K, and the pressure is in Pa. At 10^{-2} Pa the flux is roughly 3×10^{16} cm^{-2} s^{-1} for room-temperature nitrogen.

When a surface contains a typical atom density of 5×10^{14} atoms cm^{-2} then ~56 atoms strike each site per second at 10^{-2} Pa. Even if only ~2% of them stick to the surface this means the surface will be covered in only one second. For a pressure of 10^{-8} Pa, typical of "ultrahigh" vacuum, the time would be roughly 10^6 s or about 2 weeks. The significance of this time is that over a space of minutes the surface remains clean to within the detection limit of most analytical techniques. It also means that a film grown at a typical rate of one layer of atoms (a "monolayer") per second (1 ML s^{-1}) will have a maximum contamination due to the residual gas in the vacuum system of one part per million. Doing scanning tunneling microscopy (STM) sets a different standard in surface cleanliness as a single STM image may show one million atom sites on a surface. Hence, a visible contaminant may appear on the surface for every second the surface sits in the vacuum system at 10^{-8} Pa. Therefore, it is common to observe surface contamination by STM and very high quality vacuum is required for this analysis method. What saves most sensitive surface chemistry measurements is the fact that the average residual gas molecule that strikes a surface in ultrahigh vacuum does not stick. After all, if they stuck well they would be relatively easy to pump by creating a clean metal surface. This does work and is used in titanium sublimation pumps, but even when a surface is designed for pumping the sticking coefficient is very small. The process of atoms sticking to a surface is described in the following sections.

10.3 ADSORPTION

The first event in film growth, as suggested by Figure 10.3, is getting atoms or molecules needed for the film to adhere to the surface of the growing material and is known as "adsorption". The essential step in this event is to reduce the energy of the adsorbing species on the surface to below its energy in the vapor phase, preventing its escape. Typically, this happens in a series of stages.

Atoms in the gas phase have a kinetic energy of approximately k_BT where the temperature is that of the vapor source, if little scattering occurs in the gas phase, or is the temperature of the gas if scattering is common. If gas species experience an attractive interaction with the surface (bonding) they accelerate under the attractive force and gain kinetic energy. This is usually much greater than the gas phase kinetic energy. Without some other atom to transfer energy to and without an efficient process for making the transfer, the arriving atom would simply bounce off the surface most of the time, slowing as it returns to the gas but not stopping. This is why reactions cannot normally occur between pairs of atoms in the gas phase. Only when a third particle is present to take away the excess energy is a reaction possible without a reverse reaction being obligatory. In adsorption on a surface the process is much simpler as the surface itself is a reservoir of energy (in the form of heat) and momentum. The arriving particle may exchange these conserved quantities with the surface. Unfortunately, for crystal growth, this exchange does not generally come easily, even in the best of cases.

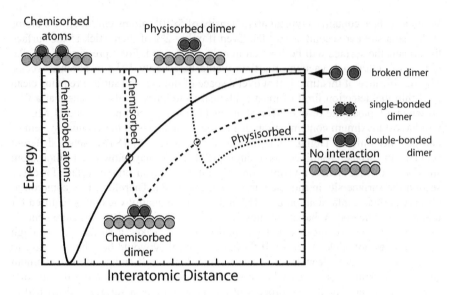

Figure 10.4: Shows an energy-distance diagram for several configurations of a molecule with a surface. Not all of these paths will occur for any given process of adsorption with a surface. In this example the molecule may make transitions from one of the various curves to another as reactions with the surface occur. The lowest-energy transitions are marked with open circles. For an example of a situation where chemisorption is very rare, consider how this figure would appear without the center of the three curves. In that case, transfer from the physisorbed to the chemisorbed case would add much more energy to the adsorbate than would be necessary to induce desorption.

For illustration, let us think of the problem of a diatomic molecule interacting with and adsorbing on a metal surface. The energy of this hypothetical dimer molecule as a function of distance from a surface is shown schematically in Figure 10.4. In the figure the curve with short dashes suggests to the energy-distance relationship for the dimer with unchanged bonding. Thus, if the dimer were O_2 it might be for the case where the dimer has a double bond. The dimer atoms may interact weakly with the surface, but no unpaired electrons are available to form a strong bond. Although both would like to react, the two oxygen atoms are strongly bound to each other and may not spontaneously reorganize their electrons to form a bond to the surface. The result is a van der Waals bond.

Once the molecule is weakly adsorbed it may acquire enough heat to escape back to the vapor (desorb) or it may be able to reorganize its electrons sufficiently to form stronger bonds to the surface. In Figure 10.4, two different reorganizations are hypothesized. These might, for example, correspond to a weaker and a stronger chemical bond to the surface. Each possible arrangement of atoms has its own energy vs. position curve. At a given position along a curve the molecule may be thermally excited such that it changes its orientation and bonding structure to conform to another curve, often with higher energy at that position. It is then free to relax along this curve to the minimum energy position. Such transitions can occur any time a molecule/atom is following a curve with a higher energy than another curve. Note that if one has a dimer initially, and the stronger bonds to the surface require breaking the bonds between the atoms, the energy of the molecule if it desorbed in this state would be higher than in the original gas phase (see Figure 10.4). This may reduce the chances of desorption from the more strongly bound configuration.

A critical issue is the energy necessary to make a transition from one curve to another. If that energy is higher than the energy binding the molecule to the surface in its initial state, then desorption (re-evaporation) is more likely than that it will become more strongly bound to the surface. For example, if the curve crossing points marked with open circles on Figure 10.4 lie above the energy in the vapor phase of the adsorbate in its current bonding configuration, then desorption is likely.

Additional transformations may be required before a molecule reaches its most stable bonding arrangement on the surface. For example, it may be necessary to break surface bonds as well as bonds in the initial gas molecule. Any of these may contribute to any of the energy barriers to adsorption. Therefore, there is generally a good chance that molecules will leave the surface rather than forming a more stable bond. A notable exception to this argument is single metal atoms landing on a metal surface. In this case, both atom and surface have unpaired electrons that can easily interact both to produce a bond and to dissipate the energy of adsorption. Therefore, metals adsorb strongly on metallic surfaces.

For a growth process where a given flux of atoms is arriving at a surface, as described by Equation 10.3, the flux of gas actually adsorbed, Γ, is simply this flux, F, multiplied by the probability, ϕ, for accommodation of the heat and momentum of adsorption.

$$\Gamma = F\phi \qquad\qquad 10.5$$

Note that ϕ is typically dependent upon the local chemistry and structure of the surface where the atom is attempting to adsorb. This is relatively constant for a given gas and a given surface. However, as adsorption continues the surface becomes covered with the adsorbate and the adsorption probability may change dramatically. The flux has not changed, only the chance of actually adsorbing. This is usually a consequence of a change in the energy-position relationships.

10.4 DESORPTION

To allow a thin film to grow it is necessary that the atoms not re-evaporate into the gas. Furthermore, in chemical vapor deposition processes some species must evaporate to continue the reaction. Therefore, it is important to also consider the process of desorption. At its simplest, desorption is simply the reverse of adsorption and may follow the same energy distance curve. For desorption of a single atom this simply requires the surface heat and momentum reservoir to supply enough kinetic energy in the correct direction that the atom may escape into the gas. In general the rate per unit area of this process will depend upon the areal concentration of atoms on the surface that are attempting to desorb, an attempt rate constant, and a Boltzman factor. The result looks very similar to a general reaction rate equation:

$$r_{des} = [A]^n \, \nu \, e^{-Q_{des}/k_B T},$$ 10.6

where [A] is the concentration of desorbing species on a surface, n is the order of the desorption process, ν is the attempt frequency, and Q_{des} is the heat of desorption (usually the same energy that was dissipated in the reverse process of adsorption). For a single atom desorbing into the gas n=1 while desorption as clusters of atoms generally produce larger values of n.

In many cases, the simple desorption of an atom into the gas is a relatively high energy, and hence relatively rare event. Figure 10.3 shows that the energy barrier may be reduced if two or more atoms associate into a dimer or small cluster before desorbing. This is because their energy in the gas phase is lower. The process is known as associative desorption. What is important about associative as compared to simple desorption is that its rate depends, via the law of mass action, on the square of the concentration of desorbing atoms on the surface (n=2) for a dimer desorption or n>2 for larger desorbing clusters. When this concentration is high, desorption is rapid. However, when the concentration is low associative desorption slows considerably. The desorption process is shown schematically in Figure 10.5.

The concentration dependence of associative desorption and the microscopic dynamics of that association forms the basis for operation of devices known as "cracking" sources. These are used to reduce the size of evaporant species leaving a gas source. For example, arsenic evaporates from its own solid as As_4 tetramer molecules. However, from most other surfaces it leaves as As_2. This is simply a consequence of the As concentration, which is high on the bulk arsenic solid and lower on other surfaces even when they contain most of an atomic layer of arsenic. Hence, the dependence of desorption rate on the fourth power of As concentration is not an issue on the bulk solid but prevents this form of desorption on other surfaces. Therefore, adsorbing As_4 clusters from a bulk arsenic source on a surface and allowing them to desorb back into the gas converts As_4 to As_2 (it "cracks" the As_4 clusters). Application of the cracking process to growth by evaporation is discussed in more detail in Chapter 11.

Most group V and VI elements desorb from their own solids as clusters of two or more atoms. In many cases clusters of up to eight atoms may be the dominant desorbing species. However, nearly all of these if adsorbed on another material such as graphite will desorb as dimers.

As discussed above, the desorption process often has a very different rate on different surfaces. For example, if we cool a Si surface and evaporate As onto it we may deposit a thick layer. Upon heating we can desorb this As which shows two distinct desorption behaviors, one associated with As desorption from on top of other As atoms and one for desorption from on top of Si atoms. A detailed example of how H desorbs from Si and how the surface coverages and energies binding the atoms to the surface is presented in the applications section at the end of this chapter.

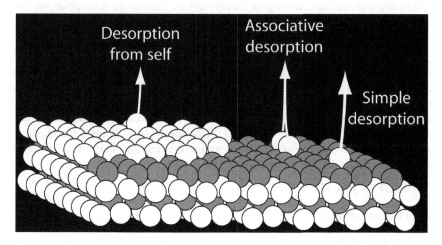

Figure 10.5: A schematic of several desorption processes discussed in the text.

In some cases a reactive site is needed to initiate desorption. For example, the native oxide on GaAs wafers, typically a gallium oxide, desorbs very abruptly. After considerable study it has been shown that the oxide desorption requires a nucleation step creating holes in the layer. These then expand rapidly as the desorption occurs. The desorption reaction can be nucleated by adsorption of a variety of species on the surface. The desorption temperature also depends somewhat on how the oxide is formed and its thickness. When desorption is nucleation-limited there may be a considerable delay in its onset if it is not significantly superheated.

10.5 STICKING COEFFICIENT & SURFACE COVERAGE

Typically during a growth process one has both adsorption and desorption of at least one species going on continuously. When this happens, conservation of atoms indicates that the system must come to a steady-state involving either a balance of adsorption and desorption without net growth of the film; or, if adsorption is faster than desorption, to net growth of the film. The remaining option, that desorption is faster than adsorption means net loss of adsorbate from the surface (the film will evaporate). Because desorption rate depends upon surface coverage, the desorption rate will fall as the surface loses material until it balances the adsorption rate or until the surface is clean.

To determine the surface coverage at steady state we balance the adsorption rate with the desorption rate. Setting equations 10.5 and 10.6 equal:

$$R\phi = \theta^n \, v \, e^{-Q_{des}/k_B T}, \qquad\qquad 10.7$$

where we have replaced the areal concentration of adsorbate [A] with the unitless fractional surface coverage $\theta = [A]/N_{surf}$ and N_{surf} is the number of surface sites per unit area. The flux F was also replaced with the average arrival rate (in units of inverse seconds) per site, $R=F/N_{surf}$. This puts the equation in area-independent units and eliminates unit difficulties when $n \neq 1$. Note that ϕ depends upon the chemistry and structure of the surface on which adsorption is occurring. This equation may be solved for θ^n to get the steady-state surface coverage:

$$\theta^n = R \, \phi \, \tau, \qquad\qquad 10.8$$

where we have defined

$$\tau^{-1} = v \, e^{-Q_{des}/k_B T} \qquad\qquad 10.9$$

with τ being the average residence time of an adatom on the surface. Steady-state adsorption and surface coverage is illustrated in Figure 10.6. As with the adsorption probability, τ will vary depending upon the type of site from which the atom is desorbing, resulting from changes in both v and Q_{des}.

In typical compound semiconductor deposition processes there will be a low-vapor pressure species that sticks well to the growing surface ($\phi \sim 1$ everywhere on the surface). This is the case for Ga in growth of GaAs. The other element has a high vapor pressure and adsorbs and desorbs moderately where the surface is covered with low vapor pressure element (e.g. Ga) and adsorbs poorly and desorbs rapidly where the surface is covered by high vapor pressure material (e.g. As). The low-vapor-pressure species determines the growth rate of the film while the high vapor pressure species is supplied in excess. Whatever does not find a low-vapor-pressure atom to react with and stick to desorbs leaving a stoichiometric compound on the surface automatically.

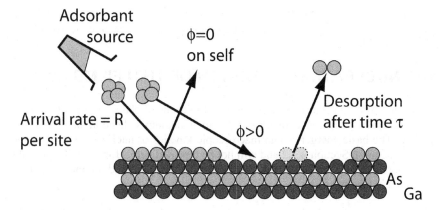

Figure 10.6: A schematic of a simplified version of the balance of adsorption and desorption on a GaAs surface.

Because only a portion of the flux of atoms that arrives at the film surface ends up sticking to and ultimately incorporated into the film, it is common to define the "sticking coefficient" and an "incorporation coefficient" referring to the probability that an atom will stick to and ultimately become part of the film, respectively. The distinction between these two is somewhat subtle. Normally they are the same. However, in a case where an impurity segregates strongly on the surface of the film as growth proceeds, the sticking coefficient may exceed the incorporation coefficient temporarily as the surface segregated layer accumulates, although ultimately at steady state they must match. In segregating species the surface concentration of that species is much greater than the bulk composition within the material. Atoms typically segregate to the surface when they are much larger than the host matrix atoms or when they reduce the surface energy (i.e., they are surfactants). Surface segregation is discussed in more detail in Section 10.7. The incorporation coefficient is generally a sensitive function of surface temperature, fluxes of other atoms arriving at the surface, and steady-state surface coverage for reasons described above. It is therefore a complex quantity and not easily predicted *a-priori*. However, it is the easiest to observe experimentally. One simply measures the bulk concentration of that element and calculates the fraction of the flux arriving at the surface that becomes incorporated into the film.

Because As is supplied as a tetramer when a standard evaporation source is used it has been found to be necessary to look closely at the actual site where the tetramer arrives to determine whether it will stick. One might imagine that an As tetramer would require as many as four open sites on the sample surface sufficiently close to each other to obtain a high sticking coefficient. In practice the number of sites required is closer to two but is significantly greater than one. A theory of sticking of multimeric adsorbates such as As_4 has been developed. For example, Madhukar and Ghaisas referred to the process as "configuration-dependent reactive incorpration" in their model. [2]

10.6 NUCLEATION & GROWTH OF THIN FILMS

Nucleation of a thin film is usually described in terms of classical nucleation theory resulting from coalescence of clusters from a random collection of atoms on a surface. The basic energetics and macroscopic kinetics of nucleation was described in some detail in Section 4.4.2 and will not be repeated here. However, it is useful to look more closely at the atomic processes that are involved in the phenomena described in Chapter 4.

When atoms are deposited on a film surface they produce what is generally termed a "lattice gas", meaning a low-density collection of free atoms diffusing at random among surface lattice sites, occasionally colliding with one another, and presumably occasionally sticking together. Thus, their behavior resembles a two dimensional gas. Single atoms generally move much faster than larger clusters. Therefore, atoms that are bound together are not part of the surface lattice gas. The density of the gas depends upon the rate at which atoms are supplied from the vapor phase by adsorption, lost by desorption, lost by incorporation into atom clusters or islands, or supplied by release from these clusters. In Figure 10.3 the surface atoms on the left side of the figure are part of the lattice gas.

Atoms in the lattice gas interact with others during their diffusion process by moving into adjacent surface sites. From such sites they may continue to diffuse but not into the occupied site of the other atom. Typically, when the atoms are in adjacent sites they are bound together somewhat to form a surface dimer. If while this state lasts another atom joins the dimer the cluster becomes a trimer, and so on. The process is shown schematically in Figure 10.7.

The probability of two atoms joining to form a dimer depends upon the square of the density of single atoms on the surface (n_1) and a Boltzman-type rate constant describing the probability of the atoms sticking together k_{12}. The rate constant incorporates both an effective surface diffusion rate for the moving species and a probability that the particles stick together. Both are temperature dependent. The chance of forming a trimer depends upon the product of a rate constant k_{23}, the density of dimers (n_2) and n_1. Similar terms can be written for higher order events. The backward reaction describing decomposition of dimers into monomers depends only upon the product of the decomposition rate constant and the dimer concentration, k_{21} and n_2, respectively. Decomposition of larger clusters depends upon the concentration of that cluster and a rate constant. A set of differential equations may then be written for the change in concentration with time of clusters of a given size, i, with the general form for a typical island of:

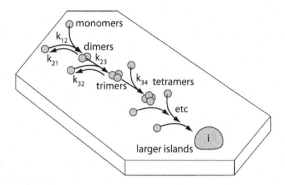

Figure 10.7: A schematic of the process of adatom agglomeration to form small and eventually large clusters.

$$\frac{dn_i}{dt} = \sum_{0<j<i-j} k_{j,i} n_j n_{i-j} - \sum_{0<m<i} k_{i,i-m} n_i + \sum_{\ell>i} k_{\ell,i} n_\ell \, . \qquad 10.10$$

The first term on the right describes coalescence of clusters of size j with those of size i-j. The second term describes decomposition of clusters of size i to sizes i-m and m. The third term represents formation of clusters of size i by break-up of larger clusters of size l. It is also necessary to add terms to account for deposition of atoms or molecules (for example, $F\phi_1$ as in Equation 10.5 added to the equation for n_1, for desorption (by subtracting a term such as $k_{des}(i)n_i$ from each equation), and by the addition of other terms such as may be necessary for a given situation. Note that the chance of sticking to the surface, ϕ, may be dependent upon the surface to which sticking occurs. Reminder: a species may stick very poorly to a surface covered by itself, as is the case for As sticking to GaAs at high temperatures, or may stick very poorly to a substrate other than itself, as is the case of CdS islands on a clean glass substrate where CdS sticks well only to existing CdS islands and not to the clean glass areas.

In practice, decompositions for m>2 (loss of more than one or two atoms at a time) are unlikely unless an energetic particle hits the cluster and breaks it up. Cluster aggregation where j>1 becomes increasingly important when the surface becomes heavily covered by clusters leaving little free space for the lattice gas, as discussed below. Actions involving monomers are generally by far the fastest and hence have the greatest values of k.

To complete the analysis of the equations one requires an estimate of the magnitude of the rate constants k. This estimate, carried out in detail, is complex and beyond the scope of this text. However, several observations on the nature of some of the contributing terms seems reasonable. First, the rate constants include a Boltzmann-like term based on the free energy change, $\Delta G_{j,i}$, for the system upon attaching/detaching an atom to a cluster or for merging two larger clusters. The energy change may be either endothermic or exothermic. This is multiplied by an "attempt frequency" which is related to the diffusivity of the moving species (usually the smaller of the two atoms, molecules or clusters involved in a given event). This attempt frequency also includes a Boltzmann-factor temperature dependence. Thus, the rate constants might be expected to have a general form such as

$$k_{j,i} = D_0 e^{-E_D / k_B T} \, e^{-\Delta G_{j,i} / k_B T} \qquad\qquad 10.11$$

where E_D is a diffusion activation energy and D_0 is the diffusion prefactor and could include other constants. The ΔG terms include, among other contributions, an energy such as that described by Equation 4.17 (Section 4.4.2), which is to say by the shape of the total energy curve as a function of cluster size in Figure 4.13. When clusters are well beyond the critical size there is almost no barrier to adding an atom and the diffusion term dominates the behavior, especially when coalescence of larger clusters is considered.

Equation 10.10 represents a potentially infinite set of equations (one for each value of i) in i unknowns. Therefore, the set can be solved for each value of n_i given a starting condition and all relevant rate constants. Such solutions have been carried out many times in the literature.

Because the rate constants are greatest for terms involving movement of monomers (single adsorbed atoms or "adatoms" on the surface), processes involving adatoms are generally the most important. This is particularly true initially, as no clusters exist. Since the first step depends quadratically on n_1, dimer formation is very sensitive to adatom density. Consequently, formation of dimers grows very rapidly. However, when the dimer formation rate constant is low, for example where adatoms do not stick to one another well, and where the decomposition rate is high, formation of dimers and consequently all larger clusters is slow. In this case nucleation is slow and requires accumulation of very high densities of adatoms (which also encourages desorption). When atatoms stick together well and dimers do not decompose rapidly, nucleation is fast and requires low adatom densities. If adatoms do not stick well to the substrate or if single adatoms desorb much more rapidly than atoms desorb directly from clusters, it may be difficult to accumulate enough adatoms to form an initial nucleus. This leads to a delay in nucleation. If one plots the thickness of a film as a function of time in such a situation and extrapolates to zero thickness, one finds that the extrapolation does not lead to zero time. Rather there is a nucleation delay before growth can begin as shown in Figure 10.8.

A rough estimate of the relative rates at which adatoms attach to and detach from clusters may be made by observing that both rate constants will include a Boltzmann term related to the binding energy of an adatom to the cluster. When the binding energy is high (relative to $k_B T$) the attachment rate constant will be much greater than the detachment rate. When low, the two rates will be more nearly similar. When

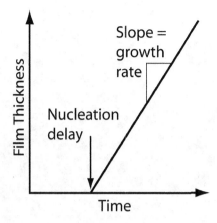

Figure 10.8: A schematic plot of film thickness as a function of time after exposing a substrate to a flux of film material. The plot extrapolates to the nucleation time. Note that this behavior requires desorption of adatoms during the nucleation delay. Otherwise, once nucleation begins all of the atoms deposited earlier would be available to contribute to the thickness.

binding is unfavorable, the dimer formation rate is low. Measurements of film morphologies in the very early stages of growth before islands coalesce as a function of temperature can provide a quantitative measure of the various energies through modeling, for example using Equation 10.10.

An example of the final microstructure of a real surface after the clusters have grown large is shown in Figure 10.9. It is hopefully clear from the figure that the clusters eventually come to occupy a large fraction of the surface. At some point any atom that adsorbs on the surface will meet and stick to a cluster before it sticks to another adatom. This turns off the nucleation process as no new dimers form and the pipeline of small clusters growing into larger ones that constitute the nucleation process is ended.

The end of nucleation actually begins as soon as stable nuclei develop because within some distance of any stable nucleus adatoms tend to be captured by that nucleus in preference to forming new dimers. This leads to a "denuded zone" around the cluster where new clusters are unlikely to form (see Figure 10.10 for a schematic and Figure 10.9 for an example). The greater the diffusion distance for adatoms, the larger the denuded zone around a cluster and the faster that cluster will grow. The most likely area for a new cluster to form is outside of the existing denuded zones. However, as these areas shrink the adatom density in them also decreases and nucleation there also becomes less likely. The formation of denuded zones and the preference for nucleation outside of the existing denuded zones is responsible for the relatively well-organized structure of clusters on the surface.

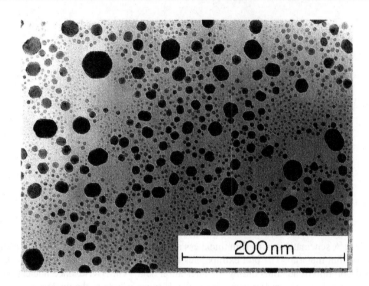

Figure 10.9: An example of a thin In film on an amorphous Si_3N_4 substrate grown by evaporation. See also Reference 3 for experimental details. Used with permission from M.-A. Hasan, S. A. Barnett, J.-E. Sundgren, and J. E. Greene, Journal of Vacuum Science & Technology A, 5, 1883 (1987). Copyright 1987, AVS The Science & Technology Society.

Even clusters that are very large and very stable (as indicated by Figure 4.13 and Equation 4.17) will lose atoms, $k_{i,i-1} \neq 0$. Although the most likely fate for these atoms is to reattach to the cluster that they left, inevitably some manage to escape and either form new nuclei or become attached to another cluster. This leads to a gradual process of exchange of atoms among clusters. The process can be rapid or slow in absolute terms or relative to the growth rate of the film. If it is rapid, with or without net growth, the clusters with the greatest stability gain atoms at the expense of the clusters with the lowest stability. This is because their adatom attachment rate relative to the detachment rate is larger than for less stable clusters. Thus, the small clusters, which are usually less stable, gradually disappear and the larger, more stable clusters grow. The exchange of atoms leading to growth of large clusters and shrinkage of small clusters is known as coarsening.

The process of coarsening is helped along if even relatively large clusters can move on the surface and hence can coalesce with one another. This movement is observed experimentally at sufficiently high temperatures. Cluster diffusion results from the fact that atoms may move around the perimeter of the cluster as well as leaving it to join the lattice gas. This results in random transport of atoms from one side of the cluster to the other and hence to movement of its center of mass. The larger the

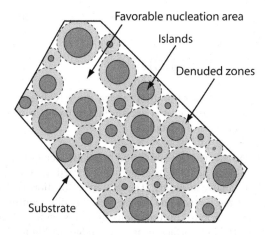

Figure 10.10: A schematic of the effect of capture of adatoms by large islands on nucleation. The denuded zones are areas relatively free of adatoms where further nucleation is unlikely. The white areas between the denuded zones may show nucleation, although the smaller these areas, the fewer adatoms are present on average, the less likely is nucleation, and the slower is growth of any island that does form.

cluster the more atoms must move in order for it to shift its position. Hence, larger clusters diffuse much more slowly than smaller ones. When two clusters touch they usually rapidly shift to an equilibrium shape (hemispherical or faceted in most cases) through movement of atoms around their perimeters.

As the clusters grow they may also coalesce without diffusion by simply coming into contact. Again, this is usually quickly followed by clusters reshaping themselves by diffusion across their surfaces. Coalescence eventually leads to complete coverage of the substrate with the growing layer. At this point there is a motivation to flatten the surface as much as entropy, atomic transport and surface energy differences from one plane of atoms to another permit. The smoother the surface, the less its area. Since the surface energy is always unfavorable, minimizing surface area is preferred. Ideally, the surface would be atomically smooth. However, entropy at finite temperatures prevents this from happening.

The result, near to an ideal overall orientation, is a surface consisting of relatively smooth terraces, occasional high points (clusters of adatoms), and various smaller structures including free adatoms. Any real surface will include surface steps that have a much larger radius of curvature than any other feature. As with surface islands on a foreign substrate, atoms may be transferred from small clusters to larger ones on a smooth surface. Ultimately they prefer to transfer to a surface step. This exchange process is shown schematically in Figure 10.11 and is the basic process of surface growth after coalescence of the film. If one stops growth and allows the surface to anneal, islands will typically disappear as atoms transfer to the (most stable) surface steps.

10.7 SURFACE DIFFUSION

Because growth mechanisms for thin films are so strongly determined by how atoms are transported across the surface, it is appropriate to pause to consider diffusion of atoms at the atomic scale in more detail.

Atoms on surfaces will always have preferred locations locally relative to underlying atoms. Sometimes this will be on a dangling bond, sometimes on top of an atom, and usually in a "hollow" between atoms. The preferred location is determined by the electronic structure of the surface and the possible bonding configurations. One can calculate the energy of various configurations of atoms on a surface, as shown, for example on the Si (001) 2x1 surface near a step edge, in Figure 10.12. This figure shows a map of the energy of a solid, calculated with a density-functional theory computer code [4] with an adatom placed at various locations across the surface. Different calculation methods may obtain different energy maps. However, the results illustrate the important phenomena involved in atom transport on a surface. The Si (001) 2x1 surface consists of rows of pairs of atoms (dimers) on top of a square grid of atoms below, as shown in Figure 10.12 (c). The dimers form because each atom on this surface has two dangling bonds sticking into the vacuum on opposite sides of the top atoms.

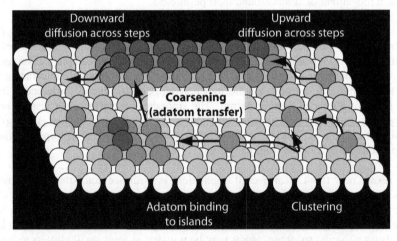

Figure 10.11: A schematic diagram showing typical routes of transfer of adatoms across a growing crystal surface among islands, between islands and surface steps, and across the steps.

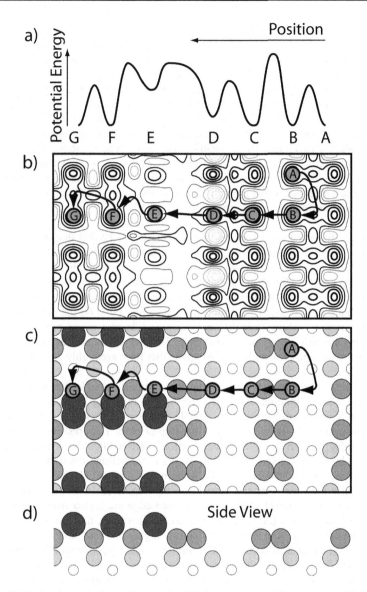

Figure 10.12: A schematic figure for atomic diffusion across a surface step on a Si (001) 2x1 surface. The approximate energy contour for the atom (a) following the path on the surface shown in the middle two portions of the figure is at the top. Regions of strongest binding of an adatom to the surface are shown with darker contours. A side view of the surface step (d) and a top view (c) as well as the full energy surface contour map (b) around the step edge are also given. (After results in Wang, Drabold and Rockett [4].)

By rocking the atoms on the surface toward one another bending the bonds connecting the top atom layer to the second layer, pairs of these dangling bonds may join to form new bonds. Thus the atoms on the surface, rather than having two dangling bonds each, have only one at the cost of distorting their bonds to the layer below. The dangling bond orientation in the surface plane rotates by 90° with the addition of a new layer of atoms. Thus, the rows of dimers also rotate with each step. This results in two types of surface step, one with the dimer rows on the upper terrace running parallel to the step edge, and one running perpendicular. The example shown in Figure 10.12 includes the latter type of step. When an atom diffuses on such a surface, its energy changes as it moves from one ideal site to another. Generally, on the energy landscape there will be maxima, minima, and saddle points. The atoms will typically reside at the minima and will move from site to site via saddle points as suggested by Figure 10.12.

One can see from the surface potential map [Figure 10.12 (b)] that [this simulation concludes that] the Si adatom prefers to bind to the dangling bond of one of the atoms in a dimer and moves along the dimer row through the channel between dimers from one dangling bond to the next. It is easy to switch from one side of the inter-dimer-row channel to the other. The hard step is to cross a dimer row from the dangling bond on one side of a dimer to the dangling bond on the other side. Thus, diffusion on this surface is highly anisotropic. Adatoms move rapidly along the dimer rows but only slowly across the rows. This has consequences for the resulting surface structure and growth. Islands of atoms that form are highly anisotropic.

When an atom encounters a surface step it is entirely possible that there will be a significant energy binding it to the step edge, repelling it from that step, or preventing it from crossing the step edge. The barrier to crossing the step may resist either upward or downward crossing of the step or both. The presence of a barrier to crossing steps, known as a "Schoebel-Ehrlich" barrier after the seminal works by groups led by these two scientists, [5,6] results in formation of new islands of atoms at step edges, surface roughening, and asymmetric growth. The example shown in Figure 10.12 indicates that the step is mildly repulsive to single adatoms. The binding energies at sites D and E along the step edge are higher than at sites on the upper or lower terraces (A, B, C, F, and G). Furthermore, the energy necessary to move from site F to site E is considerably higher than to move from F to G. Consequently, an adatom is most likely to return to the upper terrace from position F than to reach the step edge. Likewise, an atom approaching the step from below finds site D less attractive than site C and the energy barrier higher to move from site D to E, climbing the step edge.

The consequence of these predictions is that single adatoms will tend to stay away from this type of surface step and growth will not occur there until the dimer bound to the lower step edge is broken. The repulsion of adatoms by the step would also enhance nucleation of new adatom islands in the surrounding region because they have two directions to move on the open terrace but only one near the step. This results in a lower effective diffusivity and a higher adatom density around the step edge. The behaviors outlined here would normally result in a very rough growth surface. However, in the case of Si, the adatoms meet each other and form dimers on the open surface. It turns out that the dimers are themselves mobile and are the primary diffusing species participating in film growth. They move relatively isotropically and are less influenced by step edges. Still, the slow growth rate of the step shown in Figure 10.12 and the fast growth rate of the other step orientation leads to characteristic surface morphologies on the Si (100) surface.

The details of the dimer behavior are much less known than for single adatoms because it is difficult to do detailed *ab-initio* calculations similar to those above for a moving dimer. There are too many cooperative modes in which movement can take place. [*Ab-initio* means that the calculation does not rely on experimental data and the result is not a "fit". Rather, it is calculated from fundamental physics such as was outlined in Chapter 5. Indeed, the density functional *ab-initio* method used for Figure 10.12 is based on the LCAO approach.] The calculations are very time consuming for more than a few hundred atoms and so only a few configurations can be examined.

10.8 SURFACE ENERGY

Much of the surface structure observed in thin films grown from the vapor phase may be attributed to the surface energy of the material or the substrate or the energy of the interface between the two. Therefore, a discussion of surface energies is valuable to understanding thin film growth. Surface energy is highly dependent upon the number of dangling bonds on a surface and whether electrons in these dangling bonds are paired with others, whether the dangling bond is empty, or if the surface includes unpaired electrons. Unpaired electrons raise the surface energy significantly in covalently bonded materials.

As an example, consider the case of Si discussed above with reference to surface diffusion. The three common surface orientations for which it is most easy to purchase polished substrates are the "low-index" (100), (110), and (111) surface planes. The Si (100) surfaces in two structures (with and without surface reorganization ["reconstruction"] to reduce the number of dangling bonds) are compared to the unreconstructed Si (111) surface in Figure 10.13. Note that the simple, unreconstructed Si (100) 1x1 surface has two dangling bonds per atom, each half filled with electrons. The reconstructed Si (100) 2x1 surface has one dangling bond per atom but some of the surface bonds are distorted. Finally the stable unreconstructed Si (111) 1x1 surface has one dangling bond per atom, each half filled. Even the cases shown in Figure 10.13 are simplified because the Si (111) surface can undergo a further complex reconstruction to pair the electrons in the dangling bonds. This results in a surface with adatoms lying off proper lattice sites in the top layer in a "dimer-adatom-stacking fault" structure. [7] The Si (100) 2x1 surface includes one undesirable half-filled dangling bond per surface atom. This can

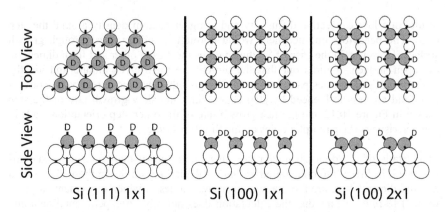

Figure 10.13: Shows the surface structure of three surfaces, the unreconstructed Si (111) and two forms of the Si (100), with and without the typical reconstruction. The (100) 1x1 surface is not observed experimentally because of the large energy gain due to reconstruction. At temperatures below ~700°C the Si (111) surface reconstructs. The "D" labels indicate the position of dangling bonds. On this surface, each is half filled with electrons.

be improved by tilting each dimer and changing the bonding such that one atom in the pair has more electrons in the dangling bond and one has fewer. This partially pairs electrons in one dangling bond while partially emptying the other. The tilt alternates along the dimer row, leading to a 2x4 surface structure (repeating every two lattice spacings in one direction and every four in the other).

The surface energies for the selected Si planes (with relatively low energies), as measured by Eaglesham et al.., [8] are 1.23, 1.36, 1.38, and 1.43 J m^{-2} for the (111), (100), (311), and (110) surface planes, respectively. Given enough atomic mobility during growth, the surface will adopt the lowest energy atomic arrangement. Ideally the surface will consist entirely of the lowest-energy planes. However, where these planes meet it may be better to produce a moderate energy surface plane to connect them. The equilibrium shape of an interior void in a solid will therefore typically consist of a mixture of relatively low-energy planes. Determination of the resulting void shape is by a Wulff construction. This is accomplished as follows (see Figure 10.14 for example). One draws, in three-dimensional space, planes at appropriate angles for their indices at a distance from the plot origin equal to their surface energy. The equilibrium shape of a void or gas bubble in a solid will be the shape of the space contained within all such planes. The construction also has relevance to the equilibrium shape of surface facets on a thin film. Concerning the example shown in Figure 10.14, no {111} or {311} plane lies perpendicular to the (100) and (010) planes simultaneously. Therefore, the planes lie at an angle to the drawing. To be correct, one must project the various planes onto a given surface in which the plot is being made. In the example, this tends to enhance the length of the {111} and {311}

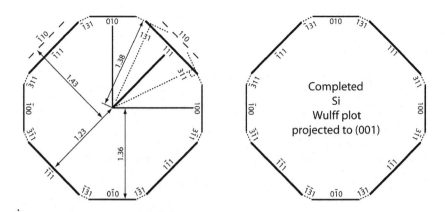

Figure 10.14: Shows the construction of a Wulff plot for Si based on data in Eaglesham, Reference 8. Note that because the four planes indicated do not lie perpendicular to a single plane, the perpendiculars shown in the figure are projected such that they are in the same plane. Therefore, the distances are the projected lengths on the diagram. The shape of the Wulff plot should be approximately correct based on the observed void pubished in Eaglesham et al. [8]

lines relative to the {100} sides. Note also that because the {110} planes have higher energy than the others, they do not form spontaneously in most of the void.

An example of the ability of a semiconductor surface to decompose during growth in an attempt to minimize the surface energy is shown in Figure 10.15 for the chalcogenide semiconductor $CuInSe_2$ grown by an evaporation-like process. [9] Although the average surface orientation for this film is {110}, the film has decomposed into close-packed planes of opposite polarities, equivalent to the {111} planes of Si in general atomic arrangement. A series of studies of the growth of this semiconductor shows such a strong tendency to form these planes, that most surfaces facet strongly to the low-energy orientation. Similar results can be found by examining voids within grains of $CuInSe_2$. Most other semiconductors do not show this strong a preference for one type of surface and a mixture of orientations would be expected both in voids and on free surfaces, as was the case for Si.

The surface energy of a solid can be considerably reduced by accumulation of a foreign atom on the surface. For example, if one considers the Si (001) 2x1 surface in Figure 10.13, if the top layer of Si atoms were replaced with a group V element such as As, an extra electron would be available to each surface atom. This would permit pairing of the electrons in the dangling bond without further reconstruction. Indeed, this is what is observed when Si is exposed to As. A strongly-bound monolayer of As forms on the surface and significantly lowers the surface energy. Any material which lowers the energy of a surface is known as a "surfactant".

CuInSe$_2$ grown on GaAs (110)

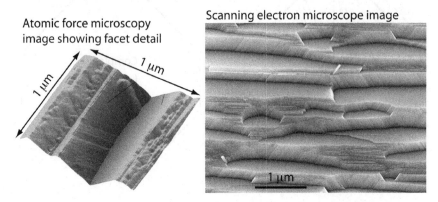

Atomic force microscopy image showing facet detail

Scanning electron microscope image

Figure 10.15: The surface of an epitaxial layer of CuInSe$_2$ grown on a (110)-oriented GaAs wafer. The surface has decomposed into two sets of facet planes with close-packed surfaces similar to the (111) in Si and other diamond-structure semiconductors. One of the two facet planes is rough because growth is occurring on that plane and the other is smooth.

Surfactants are problematic as dopants in semiconductors because there is a strong driving force for them to remain on the surface as a film grows, rather than being incorporated into the material. Therefore, the incorporation probability is much lower than the sticking coefficient, at least on an initially clean surface. For example, in Sb doping of Si it is common to observe a surface concentration more than six orders of magnitude greater than the bulk concentration.

In general, a fixed ratio of bulk impurity atom fraction to surface atom fraction is observed at a given temperature. Higher temperatures usually result in more incorporation as entropy favors mixing the atoms on the surface with those in the bulk. As one would expect, the extent of surface segregation decreases exponentially with temperatures when atom mobility is rapid. By contrast, when the impurity has a relatively low mobility, it will tend to be buried by new layers of atoms as they are deposited. This results in more segregation as the temperature increases due to an exponentially increasing diffusion rate with growth temperature. Decreasing segregation with increasing temperature thus indicates an equilibrium segregation ratio, while increasing segregation indicates a kinetically limited segregation process. Both are observed experimentally.

In addition to segregation driven by reduction of surface energies, strain energy can drive atoms to the surface. Thus, large atoms such as Sb in Si are favored on the surface because the surface can relax to accommodate them better than is possible in the bulk. One can calculate the energy of a misfitting spherical inclusion in the bulk of an elastic solid following the procedure of Eshelby. [10] The resulting energy is:

$$E_{elastic} = \frac{8\pi(1+\nu)}{3(1-\nu)} G\, r_{imp}^{3}\, \varepsilon^{2},$$
     10.12

where ν is Poisson's ratio and G is the bulk modulus for the matrix, r_{imp} is the radius of the spherical inclusion, and ε is the strain caused by that inclusion, see Figure 10.16. The strain is assumed to result from cutting a spherical hole out of the solid, inserting the misfitting inclusion into it, and allowing the system to relax. The change in radius of the space containing the inclusion is the strain.

This elastic energy drives segregation as do surface electronic considerations. Therefore, a large atom segregates to the surface of a solid very strongly in general. Small atoms can either segregate to a surface, driven by an ability to reduce surface energy, or may prefer to be in the solid. One may see how a driving force may exist to pull a small atom into a solid from the discussion in Chapter 5. As the lattice shrinks, if the bonding is otherwise relatively unperturbed, the bonding-antibonding interactions strengthen and the energy of the electrons in bonding states is lowered. Perhaps it is not surprising then to learn that B segregates preferentially into rather than out of Si.

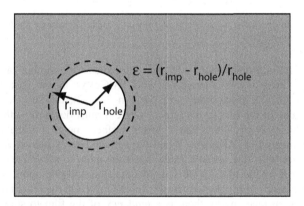

Figure 10.16: A schematic of the geometry of a misfit inclusion in a solid based upon which the strain energy is calculated.

10.9 MORPHOLOGY DETERMINED BY NUCLEATION

In addition to forming surfaces based on equilibrium energy relationships, non-equilibrium shapes may be determined by the process of nucleation and growth of the material. The nucleation process described in Section 10.6 leads to small islands on the surface. With sufficient atomic mobility these may have surface facets approximating the shape of the Wulff plot for that material. However, both the interfacial and surface energies and the surface lattice misfit also determine how the film grows.

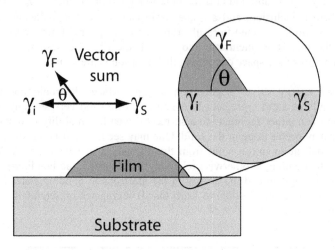

Figure 10.17: The geometry of an island of film on a substrate surface showing the contact angle and relevant energies.

Whenever a film is grown where one layer alternates with another (a superlattice), for example, GaAs layers alternating with (AlGa)As in the distributed Bragg reflector of a VCSEL laser diode (see Section 3.6.2.2), the surface energy must increase with one of the two layers grown, while it must decrease with the other. In the case of the example, GaAs has a lower surface energy for the typical (100) orientation compared to (AlGa)As. When the surface energy increases by replacing substrate surface with film surface, there is a motivation to minimize the amount of substrate covered by the film. In other words, the film will pile up in islands. This is why (Al,Ga)As tends to roughen when grown on GaAs. On the other hand, when the film lowers the surface energy there is a tendency of the film to spread across the surface, completely covering it, as for GaAs on (Al,Ga)As.

The equilibrium shape of a hemispherical island on a surface whose behavior is driven entirely by surface energy considerations will form an angle with the surface known as the contact angle, as illustrated schematically in Figure 10.17. This also applies to liquid droplets on surfaces, where the contact angle is particularly easy to observe. One may calculate this angle from the surface free energy change, ΔG_{surf} resulting from a change in the shape of the island, which changes the contact area, ΔA between the island and the surface:

$$\Delta G_{surf} = \Delta A(\gamma_i - \gamma_S) + \Delta A \gamma_F \cos(\theta - \Delta\theta) \qquad 10.13$$

where γ_i, γ_S and γ_F are the interface energy between the film and substrate and the surface energies of the substrate and film, respectively. The equilibrium condition is given in the limit of small ΔA by:

$$\frac{dG}{dA} = \gamma_i - \gamma_S + \gamma_F \cos(\theta) = 0 \qquad \qquad 10.14$$

assuming that in the limit of small ΔA, $\Delta\theta$ is also very small relative to θ. A brief examination of this formula with reference to Figure 10.17 will show that this looks like a force balance (vector sum) of the three surface energies where the film joins the substrate. It is a balance between the energy where the island is (both film surface and interface) and where the island is not (substrate surface). It is from contact angle measurements that much information about surface and interface energies are obtained.

The lattice mismatch between the film and substrate (discussed in detail in Chapter 7) also has an effect. Recall that strain energy increases with increasing thickness and that strain relief occurs when dislocations form at the film/substrate interface. This means that even when a film lowers the surface energy, it may still be energetically unfavorable to grow many layers of a relatively thin film. Rather, once the surface energy is reduced it may be favorable to pile the atoms up to take advantage of either local strain relief where a dislocation did form, or to reduce the strain energy by distortion of the island itself, as in Figure 7.23.

The combination of growth of a monolayer or two to reduce surface energy followed by island formation is referred to as the Stransky-Krastanov growth mode. Growth of islands without the thin layer to reduce surface energy (in the event that the new layer raises the energy) is Volmer-Weber growth, and a two-dimensional layer-by-layer growth mode is called Frank-Van der Merwe growth. The three modes are observed within ranges of strain and surface energy change as illustrated in Figure 10.18. These growth modes have been modeled in detail using molecular dynamics by Grabow and Gilmer with detailed discussion in Reference 11.

Finally, the orientation of the film atomic planes should be noted. When a grain in a thin film can match its lattice to that of the substrate, there is considerable motivation to do so as this greatly reduces the density of dangling bonds. Therefore, it is common to find an influence of a crystalline substrate on the texture (the tendency to develop a non-random average grain orientation) of a growing thin film. The ultimate example of this is epitaxy, where the film is fitted to the lattice of a single crystal substrate across an large area.

The most common observation of all is when the film cannot make any satisfactory alignment with the substrate and therefore effectively ignores it. This is also the case for deposition of a thin film on an amorphous substrate. In such a situation the film typically chooses to put its atoms as close together as possible to maximize bonding among them. This leads to a strong preference for nucleation of grains with a surface

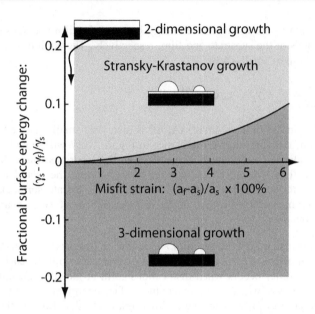

Figure 10.18: Shows the regions of misfit strain and surface energy phase space in which different growth modes are observed. At small misfit strains a film that lowers the energy of the surface produces smooth layer-by-layer growth. When the surface energy is raised by the film or when the strain energy is too large the growth is three-dimensional. As an example, the fractional surface energy change for growth of Si on Ge is +0.1, the misfit strain is 4%, and the growth mode is Stransky-Krastanov. The smooth layer is about three monolayers of Si before the growth switches to formation of islands. Ge growth on Si is by island growth without a smooth covering layer, as one might expect. Figure based on the analysis of Grabow and Gilmer.

that is a close-packed plane. This is why most thin films with a bcc crystal structure tend to show a (110) preferred orientation, while diamond and other fcc crystal structures tend to show a (111) preferred orientation. Note that this is not the reason for faceting of a surface such as was shown in Figure 10.15 as the facet planes, while close-packed, are not parallel to the substrate surface.

10.10 MICROSTRUCTURE EVOLUTION

The structure, morphology, and preferred orientation of a thin film is often very different from that of the nuclei that formed on the substrate surface, especially in the case of polycrystalline films. Before turning to that case, let us consider briefly the relatively simple case of the surface morphology of epitaxial layers. We saw in the previous section that when a film grows epitaxially it may roughen due to either surface energy or interfacial strain effects. However, the surface morphology may evolve significantly after the initial nucleation phase is complete. This may lead to either roughening of a relatively smooth film or to smoothening of a film that nucleated as islands.

A film that forms on a very smooth substrate and lowers the surface energy may grow as a very smooth layer itself. However, at either very high or very low temperatures, roughening may occur. At low temperatures atoms cannot move far enough from their initial adsorption point to reach the lowest point on the surface. Furthermore, if movement across surface steps is difficult then the more steps that form on the surface the shorter distance the adatoms can move. These factors lead to roughening of the surface as time goes on. If one stops growth and allows the surface to anneal, this roughness will gradually heal as atoms have time to assume their lowest energy positions. However, if roughness is allowed to accumulate, defects are often incorporated into the film. These defects increase the chances of further growth errors occurring. The defects may also collect together to produce grain boundaries. Thus, at sufficiently low temperature there will come a point where epitaxy breaks down and the film will become polycrystalline or even amorphous. In general, beginning from a perfect single-crystal surface, growth at a given temperature and rate can go on for a relatively fixed time before the transition to polycrystalline or amorphous material ensues. The result is a relatively well-defined layer of epitaxial material with a thickness known as the "epitaxial thickness" under the amorphous or polycrystalline layer.

Deposition at very high temperatures may also lead to surface roughening because the atomic mobility in the new layer is greater than in the underlying substrate. Now roughening is the result of entropy, which favors a rough surface to increase entropy in contradiction to surface tension that prefers a smooth surface. In some cases roughening can also be the result of a change in surface composition due to high temperature desorption of a species that is stable on the surface at lower temperatures.

10.11 RESIDUAL STRESS AND ADHESION

As a final brief note on thin film growth in general, a technologically important aspect is adhesion of the film to the substrate. Good adhesion normally requires the film to have the lowest possible stress, as any force applied to the interface will encourage decohesion. The most successful adhesives such as epoxy form a stress-free junction. Minimizing stress in a thin film is usually a matter of adjustment of deposition conditions. Two factors contribute to stress in a film, differential thermal expansion and deposition-related processes.

Deposition-related processes are the result of the mechanism of accumulation of the film on the substrate. A simple deposition process producing islands on the surface as in Figure 10.9 and 10.10 inevitably leads to shadowing of portions of the surface from the incoming flux as islands come together. The source of this shadowing is shown schematically in Figure 10.19. When atoms do not move sufficiently after

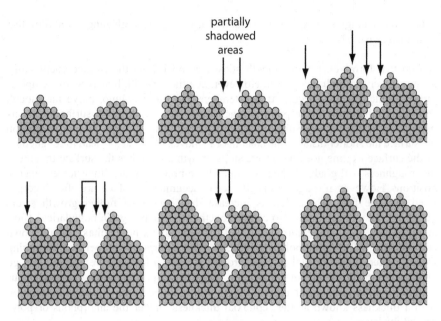

Figure 10.19: A schematic diagram of a thin film growing from a flux of atoms from directly above. Shadowing results from overhangs and protrusions in the growing material and leads to trapped voids in the films and in low-density grain boundaries.

deposition to fill gaps and shadows between coalescing islands, the result is narrow voids between columnar grains in the film, possibly nothing more than a somewhat low-density grain boundary. Because some atoms either make direct contact or form loose long range bonds between these columns, there is a bonding force that attempts to draw the columns together, while bonds to the lower portions of the column and its intrinsic stiffness tends to hold the columns apart. When the flux of atoms is at an angle to the surface these voids are both enlarged and directed toward the source of atoms, if the deposition process provides a flux from a single direction. A film that is moved continuously under a point or line source of atoms this may even cause the columns to wobble from side to side. The result of polycrystalline film growth at low temperatures is a tensile stress in the film that would cause its contraction were it not for the presence of the substrate. This often leads to tensile stresses in the films and possible adhesion failures.

When the film is bombarded during growth with energetic particles, atoms may be knocked into the intergranular voids, filling them even at relatively low temperatures. This knock-on process may drive the film into compressive stress if more atoms are pushed into lower layers of the film than it intends to accommodate in its preferred crystal structure. Thus, the residual stress may be adjusted if a

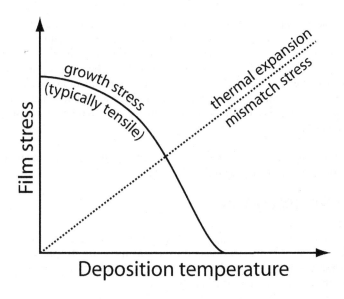

Figure 10.20: The effect of substrate temperature on film stress. Note that the thermal expansion mismatch stress can be either compressive or tensile while the growth stress is generally tensile at low temperatures. If the thermal expansion stress is tensile then the total stress is the sum of the two curves shown. If the thermal expansion stress is compressive then the total stress is the difference of the two curves.

mechanism is available to adjust the amount of energy and flux of particles bombarding the growing film surface (see Chapter 11).

The differential thermal expansion contribution to film stress is straightforward to understand. When a film is deposited it typically has only the growth stresses described above. However, if the deposition temperature is significantly different from room temperature (usually higher) then when heating is terminated after growth and the sample cools, the film usually contracts at a different rate than does the substrate. Because the substrate is usually very much stiffer than the film this differential thermal expansion stress is accumulated in the film.

The net stress in a film is the sum of the growth stress (typically decreased by increasing growth temperature) and the thermal expansion stress (typically increased by increasing growth temperature). The normal behavior is shown schematically in Figure 10.20. High stresses cause adhesion failures. On flexible substrates high stresses also cause curling of the substrate.

In addition to residual stresses, adhesion failures result from a lack of chemical reaction between the film and substrate, which leads to a weak interface; very smooth surfaces, which allows large forces to be projected onto the interface with no mechanical interlocking of the substrate and film; and contamination of the substrate surface. The latter is both most common and most easily dealt with. A combination of cleaning with solvents and ion bombardment of the sample surface prior to deposition can be highly effective in removing contamination. Surfaces are also roughened by ion bombardment, which can enhance adhesion. Finally, mixing the atoms across the interface on the scale of a few tens of atom layers can substantially enhance adhesion.

In production, deposition processes, surface cleaning and careful process control to minimize film stresses and maximize adhesion.

10.12 APPLICATIONS

The discussion in these examples makes significant reference to the growth techniques described in the following chapters as these are the practical implementations of the methods described in this chapter. Therefore, the reader is frequently referred to these chapters for more details.

10.12.1 Adsorption, desorption and binding of H to Si

A simple place to begin with some examples of concepts described in this chapter is hydrogen adsorption and desorption on the normal, reconstructed surface of Si (100) 2x1 (see Figure 10.13). This has been studied very extensively and no attempt is made here to review all of the literature. Rather, selected results will be considered as representative of the behavior of the material.

Hydrogen in its monoatomic form adsorbs very well on Si (100) 2x1. It has been found to produce two different surface phases, one consisting simply of termination of the dangling bonds on the (2x1) surface by H atoms – this phase forms easily and quickly, and a second phase where the Si-Si surface reconstruction bond is broken and two H atoms incorporate into the surface in its place. [12] This disrupts the reconstruction and results in an unreconstructed (1x1) surface. Both of these situations are shown in Figure 10.21. Typically formation of the (1x1) surface requires exposure of the initial clean Si (100) 2x1 surface to more than 1000 L of monoatomic H (more than 1000 atoms arriving per surface site).

Monoatomic H may be produced by passing H_2 gas by a heated W filament or is present, for example SiH_4, which is frequently used as a source gas for Si growth by chemical vapor deposition (CVD, see Chapter 12). When SiH_4 decomposes it leaves a surface covered with H behind. Indeed, it is this process that is in large part the reason for interest in H desorption from Si – it is a necessary step in CVD deposition of Si from SiH_4. The high sticking coefficient is not surprising perhaps since the Si-H bond is actually stronger than a Si-Si bond. Of course hydrogen is normally supplied as H_2 so one must also consider the energy of that bond when determining

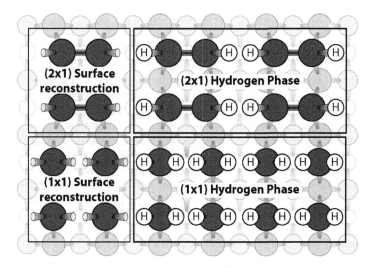

Figure 10.21: A schematic diagram of the Si (100) surface viewed from the top with the 2x1 reconstruction (top half) or with no reconstruction (bottom half). The latter is referred to as a 1x1 surface. The saturation coverage of each of these surfaces with H is shown to the right. The H positions are marked but are schematic and not determined quantitatively.

whether a reaction is endothermic or exothermic. However, in this case we assume that the H_2 is broken apart by another method. Monitoring the valence band structure during the process of H adsorption shows a loss of surface dangling bond states (see Figure 7.29) and an increase in state density at ~10 eV binding energy. [12] Therefore, one can see the elimination of the surface dangling bonds directly.

By contrast, H_2 adsorbs much more weakly with a sticking coefficient on the clean surface following an Arrhenius type behavior. Bratu et al. found an activation energy of 700 meV and a prefactor of 0.1 to 0.01. [13] This behavior is consistent from 275 to 725°C in spite of a very different desorption rate in this range. The same group found that the H_2 sticking coefficient decreases as coverage increases but that previously-adsorbed H is partially catalytic to adsorption such that in addition to the normal effect of loss of available sites for adsorption, the activation energy for adsorption is decreased somewhat. After accounting for this decrease in activation energy the coverage dependence appears to be second order in available surface sites, indicating that H_2 adsorption requires two adjacent open sites for adsorption. For lower temperatures Bratu et al. found based on a review of models and experimental data of their own and of others that adsorption was phonon-assisted, while normal dissociative adsorption occurs at higher temperatures. The behavior of D_2 is similar to that of H_2 with a small increase in adsorption activation energy.

Having adsorbed H_2, one may then consider H desorption. A number of groups have studied this process by a variety of techniques. They generally find that the "monohydride" surface with the 2x1 reconstruction and one hydrogen per dangling bond leads to first order desorption with an Arrhenius behavior having an activation energy of ~2.52 eV and a prefactor of ~10^{15} s^{-1}. [14] H_2 desorbs more easily from the "dihydride" where the surface exhibits the (1x1) reconstruction (Figure 10.21). The activation energy is ~1.88 eV and the prefactor is ~10^{13} s^{-1} for this case and the desorption is second order. [14] Considering that two hydrogens are leaving together, it is surprising that the monohydride surface would exhibit first order desorption kinetics (i.e. the rate is linearly proportional to the H concentration on the surface). A mechanism for this first-order process was proposed by Sinniah et al. [15] in which the first step was formation of an excited Si-H* bond followed by reaction of the H* with a normal H atom to form a molecule, which then desorbed. Thus the two-step process was rate limited by a single atom event; hence the first-order behavior. This behavior is consistent with the contention by Bratu [13] that the desorption process is most strongly influenced by the surface vibrational modes.

The adsorption and desorption behaviors of Si (100) 2x1 described above are summarized in Figure 10.22.

For a substrate surface heated at ~2°C per second, most of the dihydride surface will desorb H_2 and convert to a monohydride surface at ~400°C, while the monohydride will desorb H_2 and yield a clean surface at ~520°C. Slower heating or a static anneal will allow to desorption at lower temperatures if sufficient time is available. Hydrogen-terminated Si (100) surfaces can also be created from normal air-exposed oxidized surfaces by dipping the wafer in HF. The strength of the Si-H bond is illustrated by the observation that this surface is stable in laboratory air under standard conditions for several minutes to several hours.

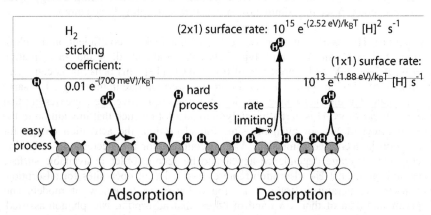

Figure 10.22: A schematic diagram summarizing the hydrogen adsorption and desorption mechanisms on Si (100) 2x1. The excited state for H_2 desorption from the 2x1 surface is indicated with a "*". H incorporation into dimer bonds is slow.

10.12.2 Surface processes in GaAs epitaxial growth

A more complicated situation than H on Si is GaAs. Indeed, this case is so complicated, having many differences as a function of growth properties, that here we will only summarize some of the behaviors and refer the reader to the literature for the details. We will consider only epitaxial vapor phase growth of the (100) GaAs surface by molecular beam epitaxy (MBE) (see Chapter 11). The basic surface is shown schematically in Figure 10.23. This surface is chosen here for illustration because it is the technologically important surface and the one that produces the best growth and film properties.

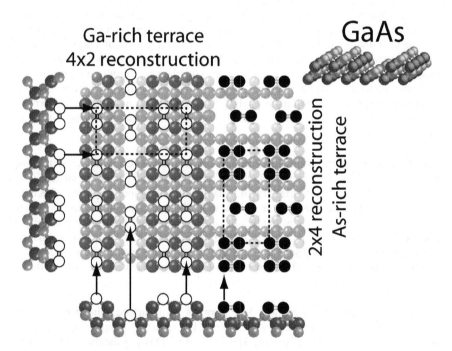

Figure 10.23: Shows the structure of the GaAs (100) surface. The larger darker circles represent As while the smaller lighter circles represent Ga. The top surface atoms in the reconstructed positions are highlighted in white (Ga) and black (As).

There are several points to note about the GaAs (100) surface. First, the surface reconstructs into dimer rows as does silicon (100) 2x1. However the arrangement of these dimers is more complex to account for the difference in chemistry of the anion and cation sites and to get the electron pairing that drives reconstruction correct. The structure removes half of the top layer of atoms and some of the second layer atoms to expose a third layer of the same chemistry as the top layer. These exposed atoms then reconstruct into dimers. The reconstruction is the same for a Ga-terminated surface or an As-terminated surface except that the structure is rotated by 90°. The reconstruction is periodic every second unit cell in one direction and every fourth in the other direction, hence the reconstructions are designated 4x2 and 2x4 for the Ga and As terminated surfaces, respectively.

The basic idea of the growth process is that Ga is evaporated from Ga liquid and arrives at the growing surface as atoms. As is evaporated from sources that produce either As dimers or tetramers. The condensing atoms provide the material from which the film grows. If the temperature is high (above ~650°C) As evaporates from the surface leaving it Ga rich unless a sufficient flux of As is available to counter-balance evaporation. Even at lower temperatures some evaporation occurs and some As is required. If evaporation is sufficiently rapid Ga will accumulate on the surface leading to formation of Ga droplets. This is very bad for growth of a good crystal. Ga can also evaporate from the surface but this process is essentially the same as for evaporation of Ga from its own metal surface (see the vapor pressure curves for Ga) and only happens when the surface is much hotter than during normal epitaxial crystal growth. Ga evaporation varies slightly depending upon the flux of As arriving at the surface at the same time. Normally one does not have to worry about Ga evaporation.

The balance between evaporation and condensation of Ga and As determines whether the surface is Ga rich or As rich. The question of whether the surface is Ga rich or As rich is critical to the growth behavior in many other ways and affects the ultimate quality of the grown layer significantly. To understand how the Ga/As ratio is set it is necessary to look at the details of adsorption and desorption primarily of As. Some of the behaviors are illustrated in Figure 10.24. In practice one can determine the cross-over from Ga rich to As rich by the transition from a (2x4) to a (4x2) reflection electron diffraction pattern (see Chapter 11) as one adjusts the As flux.

Arsenic evaporates as a tetramer, As_4, from its own surface. Therefore in MBE where simple effusion sources (see Chapter 11) are used to supply arsenic to the growth surface, the primary arriving species is As_4. This physisorbs weakly on the Ga-rich GaAs surface with a binding energy of ~260 meV [16] and does not adsorb detectably on a completely As-covered GaAs surface at normal GaAs growth temperatures. The weakly adsorbed tetramers can migrate across the surface with an activation energy of ~250 meV. [17] Note that the migration activation energy is very close to the adsorption/desorption energy so desorption is roughly as likely as surface migration. However, if the molecule can find a pair of open Ga surface sites it can chemisorb (without decomposing) with a higher binding energy of ~380 meV. [17]

It has been proposed and argued based on experimental results that decomposition of As$_4$ to deposit single As atoms on the surface requires multiple open Ga surface sites.

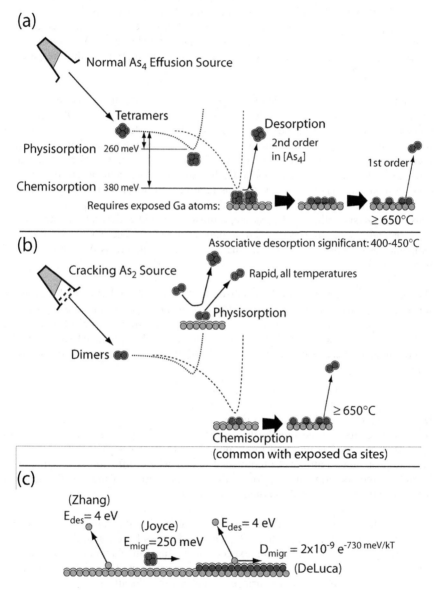

Figure 10.24: Some of the processes involved in (100) GaAs epitaxial growth from the vapor phase. Data for parts (a) and (b) are from Joyce [17]. Other references are to Zhang [18] and Deluca [19].

Joyce [17] proposed that four open Ga sites are required and that two As$_4$ molecules must adsorb and migrate such that these four sites that are adjacent and in a specific orientation to one another (later has been referred to as "configuration-dependent reactive incorporation" by Madhukar and others). These two As$_4$ molecules then react to produce four adsorbed As atoms and release an As$_4$ molecule. This configuration requirement makes the incorporation probability for As very low on a surface containing few open sites (exposed Ga) so adsorption in the presence of excess As is self limiting.

Understanding why it is difficulty to get As$_4$ to adsorb on GaAs immediately suggests a possible solution. When As evaporates from a surface that contains relatively few As atoms it comes off as As$_2$. One can design a cracking effusion source (Chapter 11) that provides As$_2$ to the growth surface rather than As$_4$. It was found that As$_2$ has a very high adsorption probability on Ga-terminated GaAs surfaces with both a weakly-bound physisorbed state and a strongly-bound chemisorbed state. [17] However, there is relatively little barrier to chemisorption if two adjacent open As sites may be found. Therefore, the As$_2$ incorporation process consists of physisorption on the surface, diffusion, and either desorption within a few microseconds or chemisorption on a pair of open surface sites. At moderate temperatures the physisorbed species may desorb associatively as As$_4$ but desorption is normally as As$_2$.

Having considered the adsorption and desorption of Ga and As on the growing surface, we can now briefly consider the growth process. Ga migrates across the GaAs surface anisotropically as might not be surprising from the anisotropic nature of the As- and Ga-terminated surface reconstructions. The preferred surface is the As-terminated where it has a diffusivity along the channels (parallel to the dimer bonds) of: [19]

$$D_{Ga}^{As-ter\min ated} = 2x10^{-9} \quad e^{-730meV/k_BT} \; cm^2 \; s^{-1} \qquad 10.15$$

Diffusion across the channels is roughly four times slower. On the Ga-terminated surface the diffusivity is roughly three six times higher. [20]. Therefore Ga on an open Ga-terminated region will move quickly until it comes to a surface step and moves up or down onto an As-terminated region where diffusion is much slower. Therefore, the As-terminated regions will accumulate most of the free Ga adatoms while As will only stick to the Ga-terminated areas.

Atomic diffusion on GaAs surfaces has been examined in the context of coarsening of surface islands on a flat surface. [21] They and others have found that coarsening does not proceed as one would expect for normal attachment/detachment and diffusion processes on surfaces. The data is consistent with a higher than usual concentration of adatoms on the surface, suggesting that the attachment rate of atoms to surface islands is relatively slow compared to the rate at which atoms leave the clusters. Likewise, it suggests that nucleation of new islands is even slower.

GaAs left to grow in its equilibrium crystal shape tends to produce {110} type facet planes. In some conditions growth even on a {100} type surface will produce such facets but normally the surface remains flat, as faceting to the {110} would increase the surface area more than enough to compensate for the improved facet energy. Therefore the surface tends to be flat. Growth proceeds by diffusion of Ga atoms onto the As-terminated regions. As they accumulate into islands they rapidly become covered with As atoms. Now we can see why there is an advantage in growth conditions that minimize As on the growth surface while maintaining the As-terminated structure. Too much As will tend to trap moving Ga atoms too easily and reduce their effective diffusivity. Too little As will allow the surface to convert to Ga-terminated. Once this occurs it is difficult to avoid droplet formation. Therefore, the optimal growth condition is just enough As_2 to maintain the As-termination but no more.

10.13 SUMMARY POINTS

- Solid phase epitaxy and film growth can only occur when reaction source materials need not diffuse far to react to form the final film, there should be no product species that need to escape from the reaction region.
- Reaction products may be dictated by nucleation or growth stages. Nucleation can depend upon the free energy of the phase forming, interfacial or surface energies, lattice misfit and strain energy, and what reactant is diffusing to form the new phase in the case of a solid phase reaction.
- Vapor phase film growth typically consists of adsorption, desorption, surface diffusion, nucleation, and growth steps.
- Atoms and molecules move substantial distances that increase with decreasing pressure and increasing energy of the moving species.
- The flux of gas particles crossing a plane is linearly related to gas pressure and inversely related to the square root of mass and average kinetic energy.
- The purpose of conducting growth processes and other experiments in ultrahigh vacuum is that the surface remains free of contamination for relatively long periods of time.
- Adsorption processes bond atoms to surfaces through weak interactions such as van der Waals bonds (physisorption) or strong chemical bonds (chemisorption). Each has a separate energy-distance relationship. Getting a molecule to bond to a surface may be limited by the energy and other barriers to transferring to a strong chemisorption configuration.
- Adsorption probability may depend strongly on the chemistry of the surface on which adsorption is to occur.
- Adsorption can depend upon the configuration of available sites on the surface.
- Desorption is a process of evaporating atoms from a surface into the vapor phase. Desorption depends upon the surface temperature, the number of atoms on the surface and how strongly they bind to that surface.
- Higher order desorption processes involve several atoms forming a cluster or molecule and desorbing as a unit. The rate of these processes generally scales with a power of the concentration of desorbing species where the exponent of the power is the number of atoms/particles of one type that must join together for desorption to occur.
- The number of atoms of a given type desorbing as a cluster from a surface may depend upon the chemical nature of that surface.
- The steady-state coverage of species on a surface depends upon their arrival rate, their adsorption probability, and their surface residence time.
- Thin film nucleation can often be described as a series of steps of attachment of atoms to each other to form a cluster and then attachment and detachment of atoms from the clusters. The attachment and detachment rates are affected by the energy change associated with adding an atom to the cluster.

- If single atoms do not stick together well or nucleation requires a large density of atoms and if isolated atoms on the surface desorb relatively easily a nucleation delay may result.
- Denuded zones in which no nucleation generally occurs surrounds existing nuclei on surfaces because the nuclei sweep the nearby area clean of atoms, which are required for further nucleation.
- Coarsening is a process in which atoms transfer between clusters. Because larger clusters are generally favored the transfer process is typically faster from small clusters to large clusters causing the small clusters to shrink and the large clusters to grow.
- Surface diffusion is faster than bulk diffusion and may be highly anisotropic due to surface structure.
- Energy barriers favoring or opposing transfer of atoms across surface steps and the energy binding or repelling an atom from step edges has a strong influence on surface morphology.
- Different surface crystal planes have different energies, often influenced by surface reconstructions and surfactants.
- Crystal growth may favor the lowest energy surface for thermodynamic reasons or the lowest energy or other surface planes may be favored because that plane grows more rapidly than others.
- Relative surface and interface energies determine the contact angle at which the film prefers to intercept the substrate surface.
- Thin films may form as two-dimensional films if they have a close lattice match to a substrate and if they lower the surface energy. (Lattice match may not matter if one is not growing the films as epitaxial layers.)
- Thin films generally nucleate as three dimensional islands when they raise the surface energy as they form or if they have a large lattice mismatch with the substrate or both.
- Films form by the Stransky-Krastanov mechanism (a flat layer across the substrate only a few atom layers thick followed by three dimensional islands) if the film lowers the surface energy but where there is a large lattice mismatch with the substrate.
- When a film has no specific reason to prefer one surface orientation over another on a substrate it tends to form surfaces consisting of close packed planes during film nucleation.
- The surface orientation after growth of a film for some time may change from that originally nucleated, for example by preferential growth of grains with other orientations.
- Residual stress in a growing film is typically tensile when the film has a columnar microstructure with small voids between crystalline columns.

- Residual stress may be made more compressive by bombarding the film with energetic particles as it grows.
- Differential thermal expansion affects residual stress in films.
- Segregation of impurities to surfaces may be driven by a desire to lower surface energy or because the impurity is much larger than the site into which it would need to incorporate in the bulk solid or both.

10.14 HOMEWORK PROBLEMS

1. Suppose that an evaporation system has a distance from an effusion source to the substrate of 30 cm. To prevent any significant scattering of gas during the movement of atoms from source to substrate you want at least three mean-free-path lengths over this path. Assume that the collision cross section is $\sigma \sim 4 \times 10^{-15}$ cm^2.

    a) Based on Equation 10.1, what is the maximum acceptable pressure in the evaporation system if the gas temperature is 300K?
    b) By the same method, what is the maximum acceptable pressure if one is evaporating Ga atoms at a temperature of 1000 K?

2. To grow a GaAs film at a rate of 1 ML s^{-1} you require a flux of 5×10^{14} Ga atoms and 2×10^{16} As$_4$ molecules arriving at the substrate surface per square centimeter per second. Assume that the Ga mass is 69.7 AMU, the As$_4$ mass is 300 AMU, and the source temperatures are 1000 and 500 K for Ga and As, respectively.

    a) Using Equation 10.4, calculate the beam equivalent-pressures to achieve the two fluxes needed for As and Ga.
    b) Using the cross section $\sigma \sim 4 \times 10^{15}$ cm^2, the temperature being the Ga temperature, the pressure is that of As$_4$ (assumed constant throughout the system), calculate the mean free path for Ga atoms passing through the As$_4$ vapor.
    c) Would you expect the As$_4$ vapor to scatter the Ga atoms significantly?

3. You are growing the hypothetical II-VI semiconductor CA by evaporation of A$_8$ octamers from an effusion source. Chemisorption requires open sites for all of the adsorbing A atoms. Suppose that the A$_8$ molecule chemisorbs as in the figure:

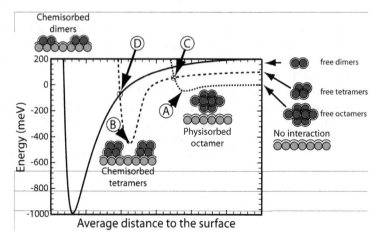

Where the energies at the marked points are:

     A: –25 meV
     B: –500 meV
     C: +50 meV
     D: –50 meV

Other important energies are:

Energy of a free octamer:	0	meV
Energy of a free tetramer:	+100	meV
Energy of a free dimer:	+200	meV
Energy of a chemisorbed dimer:	–1000	meV

Assume that the rates of processes go as concentrations multiplied by Boltzmann factors. For example, the rate of crossing the barrier for a tetramer to convert to adsorbed dimers is proportional to:

$$r_{TD} = C_T e^{-450 meV / k_B T} ,$$

where C_T is the concentration of surface tetramers. The rate of desorption of the tetramer would be:

$$r_{desT} = C_T e^{-600 meV / k_B T} .$$

a) Based on this assumption and the energies given and that there is no rate limitation to crossing from one curve to another at the crossing points, estimate the relative rate of octamers converting to tetramers relative to the number desorbing from the surface at the growth temperature of 400°C.

b) Likewise, calculate the relative rates (at equal concentrations) of:

     i.   Tetramers to dimers relative to tetramer desorption.
     ii.   Tetramer desorption to octamer desorption.
     iii.   Dimer desorption to octamer desorption.

c) Taking the result in (a) divided by the quantity one plus the results of (b.ii) and (b.iii), estimate the sticking coefficient for A_8 octamers on the growth surface. [Do the results in part (b) make a significant difference to the result?]

d) Recalculate the sticking coefficient at 200°C.

e) If the dimers desorb as tetramers from the surface, and if the dimers move together as dimers rather than moving as single adatoms, what would you expect the concentration dependence of desorption to be? [For example, the equation above for r_{desT} is proportional to the first power of tetramer concentration.]

4. You are depositing In metal on a glass slide.

   a) If you increase the temperature of the deposition would you expect the denuded zone around an island to increase or decrease in size.

   b) Suppose that the surface energy of the glass slide is 2.5 times greater than that of the In and that the interface energy is twice as great as the In. Calculate the contact angle you would expect for the In islands.

   c) If you were able to decrease the surface energy of the glass by a factor of two without affecting the interface energy through the application of a surfactant how would you expect the contact angle to change?

   d) Would you expect reconstruction of a semiconductor surface to raise or lower the surface energy? (Briefly justify your answer.)

5. You are depositing a thin film of InP on two different substrates, Si and Ge. The thermal expansion coefficients and lattice constants are:

	Thermal expansion Coefficient $°C^{-1}$	Lattice constant nm
Si:	2.6×10^{-6}	0.5431
Ge:	5.9×10^{-6}	0.5658
InP:	4.6×10^{-6}	0.5687

   a) Under growth conditions if the film is stress free, then upon cooling what is the sign of the stress of the InP grown on a Ge and on a Si substrate? (i.e. compressive or tensile)

   b) If the growth temperature is 600°C and if the film is strain free at this temperature, what is the magnitude of the strain at 20°C? (Strain is unitless.)

   c) If the strain in the film is determined by the lattice misfit at the growth temperature, which substrate would be the better choice as a substrate to minimize strain upon cooling? (Assume that thermal expansion strain directly offsets misfit strain.)

10.15 SUGGESTED READINGS & REFERENCES

Suggested Readings:

Murarka, S.P. *Silicides for VLSI Applications*. Orlando: Academic Press, 1983.

Adamson, Arthur W. *Physical Chemistry of Surfaces*, New York: Wiley, 1990.

Matthews, John Wauchope, editor, *Eptiaxial Growth*. New York: Academic Press, 1975.

Vossen, John L. and Kern, Werner, *Thin Film Processes* volumes I and II. New York, Academic: 1978.

Tu, King-Ning; Mayer, James W.; and Feldman, Leonard C.; *Electronic Thin Film Science for Electrical Engineers and Materials Scientists*. New York: Macmillan, 1992.

Venables, John A. "Atomic processes in crystal growth." *Surface Science* 1994; 299/300: 798-817.

Zangwill, Andrew *Physics at Surfaces*. New York: Cambridge University Press, 1988.

References:

[1]   Rockett, A.; Greene, J.E.; Jiang, H.; Ostling, M.; Petersson, C.S. "Dopant redistribution during the solid-phase growth of $CrSi_2$ on Si (100)." *J. Appl. Phys.* 1988; 64: 4187-93.

[2]   Madhukar, A. and Ghaisas, S.V. "Implications of the configuration-dependent reactive incorporation growth process for the group V pressure and substrate temperature dependence of III-V molecular beam epitaxial growth and the dynamics of the reflection high-energy electron diffraction intensity." *Appl. Phys. Lett.* 1985; 47: 247-9.

[3]   Hasan, M.-A.; Barnett, S.A.; Sundgren, J.-E.; Greene, J.E. "Nucleation and initial growth of In deposited on Si_3N_4 using low-energy ($\leq$300 eV) accelerated beams in ultrahigh vacuum." *J. Vac. Sci. Technol. A* 1987; 5: 1883-7.

[4]   Jun Wang; Drabold, D.A.; and Rockett, A. "Binding and diffusion of a Si adatom around type B steps on Si (001) c(4*2)." *Surf. Sci.* 1995; 344: 251-7.

[5]   Ehrlich, G. and Hudda, F.G., "Atomic view of surface self-diffusion – tungsten on tungsten" *J. Chem. Phys.* 1966; 44: 1039-40.

[6]   Schwoebel, R.L. and Shipsey, E.J. "Step motion on crystal surfaces." *J. Appl. Phys.* 1966; 37: 3682-6.

[7]   Takayanagi, K.; Tanishiro, Y. "Dimer-chain model for the 7*7 and 2*8 reconstructed surfaces of Si (111) and Ge (111)." *Phys. Rev. B* 1986; 34: 1034-40.

[8]   Eaglesham, D.J.; White, A.E.; Feldman, L.C.; Moriya, N.; and Jacobson, D.C., "Equilibrium shape of Si." *Phys. Rev. Lett.* 1993; 70: 1643-6.

[9]   Dong Liao and A. Rockett, (110) paper.

[10] Eshelby, J.D., *Solid State Physics*, ed. by. Seitz, F. and Turnbull, D. (Academic Press, New York: 1956) v. 3.

[11] Gilmer, G.H. and Grabow, M.H.; "Models of thin film growth modes." *J. of Metals,* 1987: 39: 19-23.

[12] Sakurai, Toshio and Hagstrum, Homer D.; "Interplay of the monohydride phase and a newly discovered dihydride phase in chemisorption of H on Si (100)2x1." *Phys. Rev. B* 1976; 14: 1593-6.

[13] Bratu, P.; Brenig, W.; Groß, A.; Hartmann, M.; Höfer, U.; Kratzer, P.; and Russ, R.; "Reaction dynamics of molecular hydrogen on silicon surfaces." *Phys. Rev. B* 1996; 54: 5978-91.

[14] Kim, Hyungjun; "H-mediated film growth and dopant incorporation kinetics during $Si_{1-x}Ge_x(100)$:B gas-source molecular beam epitaxy." Thesis, PH.D. University of Illinois at Urbana-Champaign, 1998.

[15] Sinniah, Kummar; Sherman, Michael G.; Lewis, Lisa B.; Weinberg, W. Henry; Yates, John T; and Janda, Kenneth C.; "Hydrogen desorption from the monohydride phase on Si(100)." *J. Chem. Phys.* 1990; 92: 5700-11.

[16] Atkins, P.W. *Physical Chemistry.* Oxford: Oxford University Press, 1978.

[17] Joyce, B.A.; "Molecular beam epitaxy." *Rep. Prog. Phys.* 1985; 48: 1637-97.

[18] Zhang, J.; Gibson, E.M.; Foxon, C.T.; and Joyce, B.A.; "Modulated molecular beam study of group III desorption during growth by MBE." *J. Cryst. Growth* 1991; 111; 93-97.

[19] DeLuca, P.M.; Labanda, J.G.C.; and Barnett, S.A.; "An ion-beam technique for measuring surface diffusion coefficients." *Appl. Phys. Letters* 1999: 74: 1719-21.

[20] Kangawa, Y.; Ito, T.; Taguchi, A.; Shiraishi, K.; Irisawa, T.; and Ohachi, T.; "Monte Carlo simulation for temperature dependence of Ga diffusion length on GaAs (001)". *Appl. Surf. Sci.* 2002; 190: 517-520.

[21] Braun, Wolfgang; Kaganer, Vladimir M.; Jenichen, Bernd; and Ploog, Klaus H.; "Non-Ostwald coarsening of the GaAs (001) surface." *Phys. Rev. B* 2004; 69: 165405-7.

Chapter 11

PHYSICAL VAPOR DEPOSITION

Physical vapor deposition refers to vacuum deposition methods that produce the source gas by evaporation, sputtering, or a related non-chemical method. Broadly, these methods transfer kinetic energy to atoms in a solid or liquid sufficient to overcome their binding energy. Evaporation refers to heating a material until the source atoms vaporize. Sputtering is a process of physical impacts transferring kinetic energy to atoms in a target. There are many related methods such as laser-ablation, which is similar to conventional evaporation but supplies energy to the surface locally by a laser beam rather than heating the entire material in an oven. Likewise, cathodic arc deposition is a sputtering-based process that uses a more localized, higher intensity glow discharge to bombard the target in a small region, rather than "sputtering" which refers to a more general bombardment of the target. This chapter describes the most common conventional evaporation and sputtering methods. Descriptions of the related techniques may be found in the recommended readings.

11.1 EVAPORATION

Evaporation includes a wide variety of techniques from simple resistive heating of a wire in a moderate vacuum to "molecular beam epitaxy" or MBE, where precisely controlled molecular beams are generated in an ultrahigh vacuum environment (less than $\sim 10^{-9}$ Pa or 10^{-14} atmospheres of residual gas pressure). In spite of the wide difference in appearance of these two techniques, they are, at heart, the same. A joke in the community is the interpretation of the acronym MBE as "megabuck evaporation" because MBE machines typically cost in excess of $1,000,000, while conventional evaporators are available for much less, typically only tens of thousands.

11.1.1 Basic system geometries

All evaporation systems include a vacuum chamber with associated pumps, and evaporation sources and their controlling electronics. Most also include methods for monitoring the flux of atoms generated by the sources and doors that cover the substrate and/or the sources during stabilization of the evaporation conditions or to modulate the beams of atoms striking the substrate without having to shut down the sources. Typical evaporation systems are shown schematically in Figure 11.1.

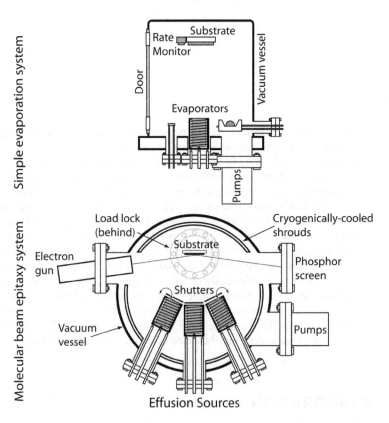

Figure 11.1: Typical evaporation system geometries. Both schematics are simplified and leave out typical components. However, the major parts are indicated. The major difference for the molecular beam epitaxy system is the increased emphasis on ultimate vacuum performance, often four to five orders of magnitude better.

One of the major differences between simple evaporation and MBE systems is the ultimate vacuum achieved by the system. As noted in Chapter 10, vacuum level affects both the chances of scattering of the beams of atoms leaving the evaporation sources and the contamination rate in the depositing film. Typically, a conventional evaporator would be based on a "high vacuum" system in which the ultimate pressure is 10^{-3} Pa (10^{-8} atm) to 10^{-6} Pa. This pressure range is sufficient to prevent gas scattering between the source and substrate for typical evaporation system geometries and is relatively easy and inexpensive to achieve. It also does not require, for example, extensive clean up efforts after the basic vacuum is established to remove adsorbed water and other contaminants from the components of the system.

In MBE the ultimate pressure is in the ultrahigh vacuum range below 10^{-6} Pa. For these pressures, very clean handling procedures are required for all parts in the vacuum system, special consideration must be given to the design of parts to allow efficient pumping of gas in small crevices such as screw threads, very high performance pumps must be used, often special highly-cooled surfaces known as cryoshrouds must be included inside the system, the samples must be inserted and removed through vacuum load-locks to prevent exposure of the interior of the system to atmospheric pressure as much as possible, and a long clean-up procedure must be followed after vacuum is established in the system. Hopefully it is clear from the length of this list of considerations for ultrahigh vacuum why it is used only when necessary. Such a clean environment is mostly to reduce contamination in growing films, as gas scattering is already minimal under high vacuum conditions.

An additional system geometry should also be mentioned, although it is not generally used with semiconductors – the linear or roll-to-roll coater. In these systems substrates are fed continuously into the deposition chamber from a roll of film or through a load lock or differentially pumped sample introduction system, and pass continuously across the evaporation sources. In a roll-to-roll coater the substrate is a long sheet of material, usually plastic, mounted on a roll in a sample introduction chamber and unloaded from a take-up roll in a second unloading chamber. There may be several evaporation chambers between the loading and unloading chambers. Often in roll-to-roll coaters it is inconvenient to shut down the source operation, so the sources must either have very large capacity or must be continuously refilled from feed stock added to the evaporator through a separate load lock system.

In evaporators a series of sources are typically located toward the base of the system, although they may be mounted to the side as well. It is very rare to evaporate downward because the evaporant is often a liquid at the operating temperature and would flow out of an inverted source. The substrate, because it is on the opposite side of the chamber facing the evaporation sources, is usually located near the top of the system and generally faces downward or nearly so.

The methods of loading samples into an evaporation system vary depending upon the complexity of the system and the ultimate base pressure desired. When an excellent ultimate vacuum is not essential, it is common that the entire side of the deposition system is a large door that opens to allow convenient access to the sample mounting position and to the sources, which may then be refilled as needed with each new sample loaded. This greatly reduces the capacity required in each source.

11.1.2 Sources

Evaporation sources range from simple resistively heated wires or metal foils to complex ovens or electron-beam heaters. The simple sources, see Figure 11.2, have limited capacity and small thermal mass, allowing rapid heatup and cooldown and reducing the time for a deposition. Additional examples and descriptions of how to choose among them may be found in the recommended readings such as the *Handbook of Thin Film Process Technology*. Because of their small capacity they must be refilled frequently, usually with each evaporation event. An accurate reading of the evaporant flux may be used to control the source temperature. However, a good control system is required for any sort of reasonable process management. Because of their small capacity and fast cycle time it is uncommon to preheat and degas the evaporants. Thus, the level of contamination in films deposited is generally relatively large. At the same time, the rapid deposition cycle usually results in low-quality films with small grain sizes and intergranular voids. Typical applications for resistively-heated open sources are for high-rate, low volume coatings with little sensitivity to either microstructure or contamination level. This might be for cases where a thin metal coating is needed and where thickness is not critical. The relatively low quality of the resulting films has generally caused these sources to become less popular in microelectronics applications.

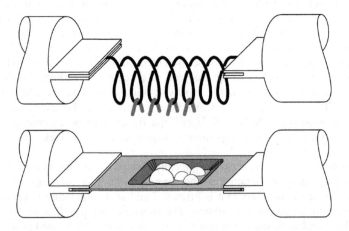

Figure 11.2: Open source evaporation boats (lower) and wire coils (upper figure). Wire coils are generally made of W although other low vapor pressure systems also work. Open boats are made in a wide variety of shapes and of a wide variety of materials including Ta, W, graphite, and other materials.

Much more common are the oven sources known as effusion cells. A typical example of such a source is shown in Figure 11.3. These consist of a crucible made of an inert material (usually boron nitride) containing the evaporant and surrounded by a refractory metal heater coil. This assembly is enclosed in a heat shield to reduce the power needed to operate the source and reduce the heat load on surrounding fixtures. The heat shield also increases the thermal mass of the oven so that it operates in a more stable manner over long periods. The temperature of the crucible is measured by a thermocouple contained within the heat shield and in conact with the base of the crucible. Because the environment within the heat shield is, to a good approximation, a black body, the thermocouple reads the outside temperature of the crucible quite reliably, usually to within 0.5°C.

A special case of the effusion cell is the Knudsen cell which has a relatively small opening from which evaporated material escapes. Inside the cell a constant gas pressure is established in equilibrium with the evaporant material (see Section 11.1.3.) For a sufficiently small opening the hole does not affect this pressure and the flux through the opening may be calculated based on Equation 10.3 or 10.4 and given the area, A, of the cell opening. Equations 10.3 or 10.4 multiplied by the opening of the effusion cell are known as the Knudsen equation. Although an exact mathematical description of a normal effusion cell with a wide opening is difficult, the qualitative behavior of the Knudsen equation is generally adequate to give an idea of the source flux.

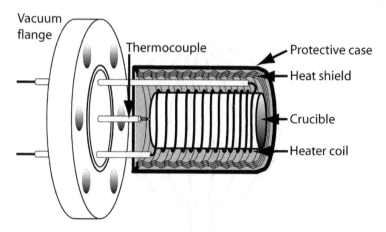

Figure 11.3: A simplified schematic sketch of a typical effusion cell. Omitted are the top cap that protects the heater windings from the evaporant as well as improving the black body environment of the crucible, the support clamps and details of the power feedthroughs, etc. The thermocouple junction is pressed gently against the base of the crucible but the black body environment assures an accurate temperature measurement. Note that the heater coils are more closely spaced near the front of the crucible to compensate for greater heat loss in that portion of the cell.

A more detailed analysis has been carried out by Luscher and Collins [1] taking into account the amount to which the cell is filled. If the cell is completely filled the evaporated material leaves the surface in a roughly cosine distribution. In other words, the flux relative to the surface normal scales as the cosine of the angle from that normal. This flux is plotted in Figure 11.4 on polar coordinates and appears as a circle. At 60° relative to the normal the flux has dropped to 50% of its maximum value. Because the area subtended by a given angular dispersion increases as the distance from the source, r, squared, the flux also decreases as r^{-2} at all angles. To estimate the uniformity of deposition, assuming a constant sticking coefficient, it is therefore only necessary to know the angle range and radii subtended by that substrate.

As the material in the crucible becomes consumed, because no scattering occurs in the gas, a geometric constraint on the flux results in forward focusing of the material. Therefore, the flux drops much more rapidly with angle than for the full cell and the uniformity of the coating on the substrate is decreased. This behavior is also shown in Figure 11.4 based on the calculations of Dayton [2] A detailed description may be found in Herman and Sitter. [3] The resulting nonuniformity would be a problem if it were not possible to correct for it. Fortunately, using a tapered crucible tilted relative to the vertical improves the results.

For such effusion cells, additional calculations are required. These were carried out by Yamashita et al. [4], see Figure 11.5. Tapering of the cell sides results in a more

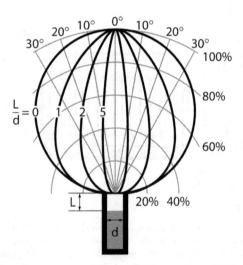

Figure 11.4: The flux distribution of atoms emitted by an effusion cell as a function of the depth in the cell of the evaporant surface relative to the diameter of the surface. [1]. Reprinted from Prog. Crystal Growth and Characterization, Vol 2, Luscher, P.E. and Collins, D.M.; "Design considerations for molecular beam epitaxy systems", pp 15-32, Copyright 1979, with permission from Elsevier.

uniform flux in the forward direction until occlusion by the sidewalls begins. This happens at a greater angle and more abruptly than for the untapered cell leading to the wide flux contour spacing near the center of the distribution and a narrower spacing near the occlusion edge. When the cell is then tipped relative to the vertical the basic flux pattern becomes distorted to one side as the deeper filling opposite this side is less strongly occluded than opposite the side filled more shallowly.

The flux patterns in Figure 11.5 are nearly ideal because the substrate may be rotated as it faces the source flange. Therefore, the high flux areas of the pattern are averaged with the low flux areas giving a net flux that is very nearly uniform within the angles at which the flux drops rapidly. This has resulted in achievement of high quality layers of well-defined thickness over 150 mm wafers in typical systems and over larger areas in more specialized instruments.

While there is an advantage to tapering and tilting the effusion cell for uniformity the usable volume of evaporant in the tapered cell is less. One cannot deplete the shallowest part of the evaporant charge beyond the point where the base of the

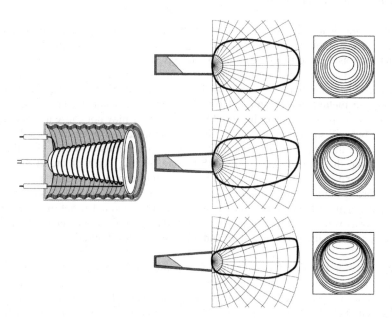

Figure 11.5: The flux patterns resulting from tapered effusion cells in both polar plots (center) and in flux contour plots (right). An isometric schematic shows the design of the tapered effusion cell with a flanged top to reduce creep of effusant out of the cell. Figure redrawn with permission from portions of Figure 3 in Tatuya Yamashita, Takashi Tomita, and Takeshi Sakurai, Japanese Journal of Applied Physics vol. 26 (1987), pages 1192-3. Copywrite 1987, Institute of Pure and Applied Physics.

crucible is exposed as this would lead to both a reduction in flux and a loss of uniformity. Therefore, typical crucibles are tapered only be a few degrees.

A final modification made to some crucibles is to include a flange around the top of the crucible rather than just ending with a simple opening. The purpose of the flange is to radiate heat away more rapidly than for the rest of the crucible, cooling the flange. When the evaporant wets the crucible effectively, as is the case for many metals, atoms may diffuse across the surface. Without the flange the diffusion carries atoms to the upper lip of the crucible, over that edge, and down the outside. Because the outside is hotter than the inside their diffusion rate increases leading to a net flux at constant concentration out of the crucible and down the heated exterior. The result is metal apparently creeping out of the crucible and getting into the heater area. The cooler flange rejects this tendency to creep for the same reason, a lower diffusion rate produces a flux back into the cell at constant concentration or, more normally, a concentration gradient decreasing rapidly across the surface of the flange. This effectively prevents escape of the material in the crucible. It is not generally required with high vapor pressure species such as metalloids because transport is more rapid through the gas phase than on the crucible surface and condensation in the hottest sections of the cell is unlikely.

Finally a practical note of caution concerning the operation of effusion cells is in order. When evaporating a material that wets the crucible, for example for common metals such as Al, the metal will adhere strongly to the crucible sides. This is also true when a reaction can occur or if the evaporant may dissolve some of the crucible material. When this happens one must be very careful when cooling the crucible through the melting temperature of the evaporant. The volume change associated with freezing often can crack the crucible. This may or may not cause an immediate problem but will surely lead to leaking of the evaporant charge when the cell is next raised above the evaporant melting point. The approaches used to deal with this are first to keep the effusion cells idling above the melting point of the evaporant if possible when not actively growing material, and second to cool the cells very slowly through the melting point of the material. A power failure can thus necessitate the replacement of the crucibles and their charges of evaporant if not of the effusion cell ovens as well. Therefore, it is important to have a reliable power source and/or backup power for the vacuum system and effusion sources if possible.

There are several modified forms of effusion sources that are used in certain process tools, especially in MBE machines, such as shown in Figure 11.6. For example, one may include a "gas cracker" which is simply a series of heated baffles or plates configured such that before a gas molecule can escape the source it must strike at least one of these surfaces. Group V and VI elements usually evaporate from their own solid/liquid surfaces as small clusters of four to eight atoms. The desorption rate equations tell us that the rate for such a process will depend upon a power of the species concentration. On the surface this concentration is constant and high. Therefore, evaporation of clusters is relatively easy. However, reaction of these large

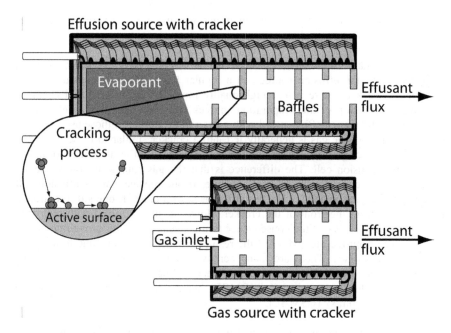

Figure 11.6: Cracking effusion sources for either a solid/liquid source (top) and for a gas source (bottom). The only difference is how the vapor to be cracked is supplied to the baffle section. Normal crackers typically have three baffles.

clusters on a growing surface is limited because one typically requires numerous open sites nearly adjacent to one another such that all of the atoms in the cluster have a place to deposit (see Section 10.12.2). A smaller cluster or an individual atom might be expected to show enhanced reactivity and reduced need for open adjacent sites. This is indeed what is observed. The purpose of the gas cracker is to break up clusters and render them more reactive.

The cracker is designed such that a multimer cluster leaving the effusion cell will typically adsorb on at least one surface before exiting the source. Cracking results from the relatively low concentration of the evaporant molecules on the baffle surfaces, as shown schematically in Figure 11.6. Comparing the rate of expected desorption for a small cluster relative to a large cluster, one might expect that the power-dependence of the desorption rate would become much more important in determining the desorbing species. Indeed, one finds that the desorbing species of group V and VI elements from a heated surface on which their adsorbed atom concentration is modest is generally the dimer. Dimers are far more reactive than the tetramers or octamers emerging from the effusion cell. If one further excites the gas by passing it through a glow discharge (similar to a gas plasma, see Section 11.3.4) the dimers can be cracked to monomers. This is not generally necessary and would typically require the addition of an inert gas to raise the pressure to the range necessary to maintain a discharge.

One of the primary disadvantages of MBE is that it is necessary to open the vacuum system to refill the effusion sources on a regular basis. This is a tedious process requiring a week or more before high quality material can be produced again. A popular modification of the evaporation sources that eliminates this problem is to replace the solid source effusion cell with a gas source similar to those used in CVD (see Chapter 12). The gas is generally fed into a cracker similar to those described above. The source gas decomposes within the cracker, resulting in a beam of atoms as from the effusion cell. The difference is that the gas source has other species mixed with the desired material. For example, if one replaces an As effusion cell with tube supplying arsine (AsH_3) gas at the back of a cracker, one still produces a beam of As_2 from the front of the cracker. However, this is mixed with hydrogen gas. Fortunately H_2 is of little consequence to most semiconductor growth processes and can even improve the quality of the resulting material by attaching to dangling bonds on the surface. There are many acronyms for this type of source and the resulting modified MBE process including "chemical beam epitaxy" (CBE), "metal-organic molecular beam epitaxy" (MOMBE) but the most generic is "gas source molecular beam epitaxy (GSMBE).

The final type of evaporation source that will be discussed here is the electron beam evaporator. The typical configuration for these sources is shown in Figure 11.7. Here a water-cooled crucible, typically made of Cu for high thermal conductivity, holds the evaporant material. To one side of the source is a heated filament biased to a high potential, typically thousands of eV, which emits an intense beam of electrons. With proper design the electric field lines direct the electrons outward relative to the filament. A strong magnetic field then deflects the outward electron beam in an arc that, for an appropriate electron beam energy, strikes the evaporant in the Cu crucible. By sweeping the acceleration voltage and with the addition of electrostatic plates to the sides of the beam, the electrons may be rastered across the surface of the evaporant charge to produce relatively uniform heating.

Electron beam evaporators can produce electron currents in excess of 1A at a high energy, resulting in many kilowatts of heating power. This is sufficient to balance the radiated energy and maintain the evaporant at a very high temperature, sufficient to evaporate even very low vapor pressure materials such as Ir, and Re. A normal effusion cell would fail at these temperatures because a simple radiant heater could not provide the heating power needed or because the crucible itself would evaporate. In an electron beam evaporator the evaporant effectively forms its own crucible because only the central portion of the charge is heated. The outer layer is in contact with the cooled Cu base and does not evaporate. Likewise, as long as cooling water is flowing through the Cu, its temperature does not rise to the point of allowing Cu evaporation. The result is a very clean evaporation source (because no contamination from the crucible occurs) that can operate effectively at a temperature of thousands of degrees. In cases where a reaction might occur between the evaporant and the Cu base or where one may use the electron beam evaporator for different materials over time, it may be desirable to add a conductive liner, usually made of graphite, between the Cu base and the evaporant.

One of the difficulties with electron beam evaporators is that it is difficult to reduce the electron accelerating voltage because the magnetic field requires a specific electron velocity to impact the charge in the crucible properly. Therefore, to reduce the heating power it is necessary to reduce the beam current. It is not always easy to obtain adequately stable low-temperature operation of an electron beam evaporator at low beam currents. Therefore, one generally uses electron beam evaporators only for low vapor pressure materials where the evaporation temperature exceeds perhaps 600°C. For lower temperatures an effusion cell is generally preferred.

There are various other versions of molecular beam sources such as high temperature effusion cells, modified to provide electron beam heating of the crucible, which we will leave for other authors to describe as they are relatively less used and are similar to the devices already described.

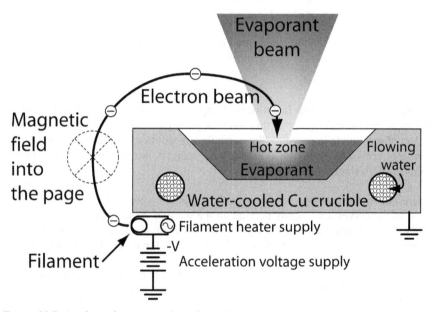

Figure 11.7: A schematic cross section of a typical electron beam evaporator. The magnetic field poles and consequently the magnetic field is out of the plane of the drawing. The high negative potential of the filament assists in electron emission and establishes the electron energy and point of impact on the crucible. Generally electrostatic deflection plates are added to sweep the electron beam into and out of the plane of the drawing.

Table 11.1: **Fits to Vapor Pressures of Selected Elements Based on Equation 11.2**

Element	p_0 (Pa)	Λ (eV)
K	5×10^9	0.94
As$_4$	5×10^{11}	1.10
Pb	1×10^8	2.00
Al	5×10^8	2.22
Ga	1×10^9	2.9
Ge	1×10^9	3.7
Si	3×10^9	4.5
Mo	1×10^{10}	6.7

11.1.3 Vapor pressure

The essential point in the operation of an effusion source is to know the vapor pressure of the evaporant material. Fortunately, in a major effort in the 1960's a group of researchers at the RCA laboratories headed by R.E. Hoing [5-8] measured and published vapor pressure data for virtually all of the elements in the periodic table. These data are reproduced in the data tables in the Appendix.

The general behavior of the vapor pressure curves can be approximated well by an exponential relationship between temperature and pressure:

$$p = p_0\, e^{-\Lambda / k_B T}, \qquad\qquad 11.1$$

where Λ is taken to be the latent heat of vaporization for the solid. Fitting some of the vapor pressure curves with this formula gives reasonable agreement over many orders of magnitude in pressure. For example, the values of p_0 and Λ for a few of the elements are listed in Table 11.1. Such an analysis is the basis of most heat of vaporization values for elements.

There are several points to note about the vapor pressure curves. First notice that most of the curves are roughly parallel on the scale of these plots. Based on the fits to the data it is apparent that the evaporation prefactor changes relatively little from element to element, while the primary changes are in the evaporation energy. This is a direct result of changes in the cohesive energy of the solid as reflected in the heat of vaporization. Note that the heat of vaporization is temperature dependent, which, in part, accounts for the curvature of the vapor pressure plots.

An additional point of note is that the vapor pressure does not correlate with the melting point of the solid. Thus, As boils (its vapor pressure exceeds one atmosphere) before it melts. Ga, by contrast, melts at ~27°C but has a negligible vapor pressure at this temperature. To evaporate Ga at a reasonable rate requires a temperature of many hundreds of degrees. Similar variability may be found for other materials. Therefore, the melting point is no guide. However, the boiling points, especially because of the relatively parallel nature of the vapor pressure curves, do provide a relatively good comparison from which one might infer a vapor pressure.

11.2 MONITORING DEPOSITION RATES

One of the major challenges in monitoring thin film deposition processes is determining the flux of atoms hitting the substrate surface. There are a number of methods, some of which probe the atoms in the gas phase, some that probe the atoms sticking to a probe surface, and some that measure what actually grows on the substrate surface. It is the latter that is most valuable as it measures what one is intending to form – a specific film on a specific substrate. The techniques considered briefly here are shown in Figure 11.8.

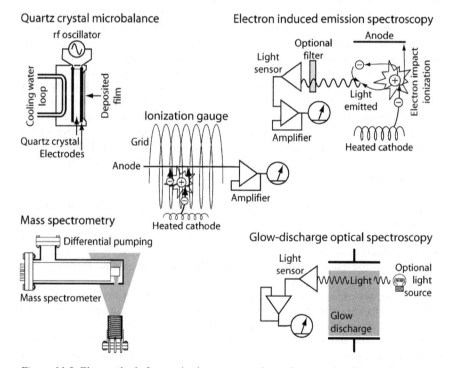

Figure 11.8: Five methods for monitoring concentrations of gas species. Many others exist.

11.2.1 Simple rate monitoring methods

Quartz crystal rate monitors, also known as quartz crystal microbalances, are one of the oldest and simplest methods. Quartz crystals have natural resonant vibrational frequencies, which are related to their mass. Quartz exhibits a piezoelectric response (see Section 2.3.1.2), which means that a physical vibration of the crystal is directly coupled to an electric field variation. Therefore, placing electrodes on opposite surfaces of a disk-shaped crystal and applying an appropriate radio frequency signal to them can excite one of the vibrational modes. This is also the basis for the use of precision quartz-crystal tuning forks as frequency standards for "quartz" electronic watches. The driving rf electronic circuit has no preference for a particular frequency. Therefore, it operates at the resonant frequency of the crystal. By reading this frequency one may estimate the mass of the quartz crystal. While a quantitative measure of the absolute mass of the crystal is moderately difficult to determine in this way, the change of mass associated with a change of frequency can be measured relatively precisely. Therefore, placing a quartz crystal resonator near the substrate in the thin film deposition system allows a measure of the flux of atoms striking and sticking to it. The major problem with approaches such as this is that the probe is not at the substrate temperature, nor does it generally have the same surface chemistry as the substrate. Therefore the probability of gas atoms sticking to the probe is different from the probability of sticking to the substrate. The net sticking probability includes both adsorption and desorption. As a result, the mass change of the quartz crystal is not an absolute measure of flux because not all atoms in the flux stick to the crystal and some that stick then desorb. Keeping the crystal cool (usually with circulating tap water) enhances the sticking coefficient but the result is still not a true flux measurement. This is not generally a major problem for coating processes that are not too sensitive. In recent years the quartz crystal microbalance has fallen somewhat out of favor because the precision demanded of thin film growth methods has increased dramatically.

A more precise group of methods measure the "beam equivalent pressure" in molecular beams near the substrate or the atom fraction of interest in the gas phase. There are several ways of doing this including electron impact emission spectroscopy (EIES), conventional ionization gauges, mass spectrometers, glow-discharge optical spectroscopy, and other methods. We will briefly consider these four in turn.

Electron impact emission spectroscopy and ionization gauges both begin the same way. Electrons are emitted from a hot filament (cathode) and are accelerated toward an anode by an electric field. Before reaching the anode (grid) they may strike a gas atom or molecule and ionize it. The probability of this event is directly proportional to the gas pressure. In an ionization gauge the ions are typically collected at a thin wire located in the field-free region in the center of the cylindrical anode grid (see Figure 11.8). The ratio of emission current from the cathode to detection current at the collector is directly proportional to pressure. Note that this relies on the ionization of the gas being constant. Gas atoms have significantly different ionization probabilities that depend upon the electron energy upon impact and the gas chemistry. Corrections to the ionization gauge pressure must be applied to account for the chemistry of the gas. An ionization gauge is normally calibrated to nitrogen gas. Correction factors of four to five are required for different gases. Note also that an ionization gauge has no way of indicating what gas is being measured (other than through changes in the correction factor). Therefore, there is no way to distinguish a partial pressure. An ionization gauge is a total pressure gauge. Knowing the pressure one may estimate the gas flux from classical thermodynamic formulas for gases such as in Equations 10.3 and 10.4.

The EIES system uses the same approach but rather than collecting the electrons or ions, one hopes that some of them will recombine and emit light. The light emission is directly proportional to gas pressure and has a characteristic set of wavelengths associated with each atom. Therefore, detecting the total emitted light gives a total gas pressure, while insertion of a filter on the detector can make the measurement sensitive to one species in the presence of another, making it more valuable than the simpler ionization gauge.

A third approach is to use a mass spectrometer for flux measurement. As with the ionization gauge, electrons flying from a cathode to an anode ionize gas atoms. These are extracted through an orifice by an electric field and enter a mass filter. Typical mass spectrometers use an electromagnetic field generated by application of radio frequency waves to four cylindrical metallic rods in a square pattern. The resulting quadrupole field allows some ion masses to pass through while rejecting others. The current at a detector is proportional to the partial pressure of a given species. This is the most sophisticated flux monitor for species in the gas phase. The mass spectrometer can independently measure atoms in the gas of different masses and can output a partial pressure for each. The pressures are sufficiently accurate that they may be used to control the sources directly.

It is uncommon to use a mass spectrometer for flux control in this way because the thin film deposits on the quadrupole rods and detector over time. Ultimately this will stop the mass spectrometer from functioning. Nonetheless, a mass spectrometer can be used for source flux control effectively, especially if the quadruopole detector does not face directly toward the flux and if the gas entering the ionizer passes through a small opening that limits the gas entering the ionizer. An example of this implementation is shown schematically in Figure 11.8.

The last gas pressure measurement system is based on optical spectroscopy. A gas emits, when ionized, and absorbs, when neutral, light on specific wavelengths as in the EIES technique described above. We can take advantage of this by shining white light through the depositing flux and measuring the absorption due to the gas. The absorption is directly related to the density of gas atoms and hence to the flux striking the substrate. Optical spectroscopy is a particularly well suited to the sputtering processes where the gas is ionized in the glow discharge as a natural part of the process. It emits light naturally so all that is necessary is to observe the intensity of the discharge at wavelengths characteristic of the species of interest. Both optical absorption and glow discharge optical spectroscopy work at high pressures so they are well suited to a variety of deposition conditions.

There are a number of other methods including ellipsometry that provide excellent measurements of either flux or thickness of a deposited film. Rather than attempt to cover all of these, we will content ourselves with one more example in the next section. Reflection high-energy electron diffraction is chosen because of the extraordinary capabilities of the technique and its relevance to other topics covered in the text.

11.2.2 Reflection high-energy electron diffraction

There is a complex but powerful technique based on reflection electron diffraction (RED) shown schematically in Figure 11.9. In this method a high-energy electron beam is incident at a low angle on the sample surface and reflects and diffracts onto a phosphor screen. The high energy beam explains why the method is also known as reflection high energy electron diffraction or RHEED. Several important features of the diffraction geometry are shown in Figure 11.10. Because the electron beam strikes the sample at a very low angle (generally less than 3°), it has very little kinetic energy in the inward direction. Consequently it only interacts strongly with the first layer of atoms.

From diffraction theory (see Section 4.1.2) we know that the reciprocal lattice point size depends upon the number of scatterers contributing to that point. A single layer of atoms yields a constant reciprocal space density along the direction perpendicular to that layer. Therefore the reciprocal lattice of a single plane of atoms produces diffraction rods perpendicular to the surface. The diameter of the rods is related to the regularity of the atoms in the plane, which also determines their horizontal spacing. From the Ewald sphere construction (Section 4.1.2) we can see that the Ewald sphere will cut a forest of regularly-spaced rods at points on an arc (see Figure 11.10). This will yield a pattern of spots – points on arcs with regular horizontal spacing, for an ideal flat surface. Notice that because the reciprocal lattice rods are

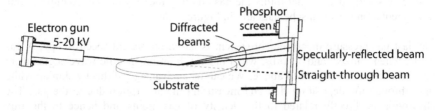

Figure 11.9: A schematic of the basic apparatus for conducting a RHEED experiment. Not shown is the remainder of the deposition system. This apparatus relies on at least high vacuum to protect the electron gun.

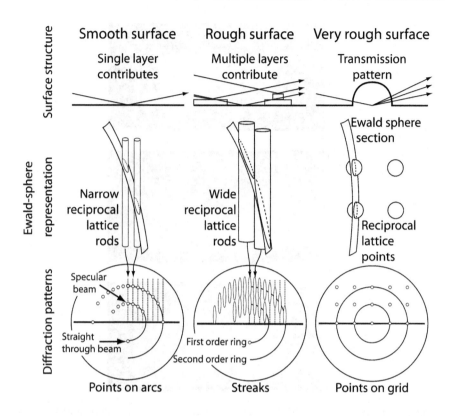

Figure 11.10: A schematic of the diffraction patterns in a RHEED experiment resulting from different surface structures. To simplify the drawing the broad rods of the rough surface are drawn as constant in width. Properly, for a two-level surface they have two non-zero Fourier coefficients and should oscillate as sine waves perpendicular to the surface. Therefore the rods should be wider and narrower along their length. See Figure 11.11 for a more accurate representation.

constant in magnitude perpendicular to the surface, it does not matter at what angle the electron beam strikes the surface. Diffraction will still occur.

In many cases the surface is reconstructed (see Section 10.8) which breaks the periodicity in the surface plane. For example, for the (2x1) surface reconstruction in Figure 10.13, the repeat unit for the surface is one atom spacing along the rows of dimers and two atom spacings perpendicular to the rows. Doubling the periodicity in the surface plane halves the reciprocal lattice rod spacing. This leads to the appearance of extra spots in the diffraction pattern. Counting the spots shows the surface reconstruction. Thus, a Si (111) surface that reconstructs with a 7x7 structure

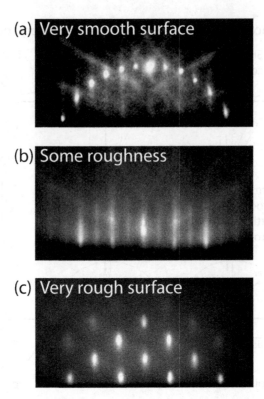

Figure 11.11: Examples of RHEED patterns obtained during non-equilibrium MBE growth of GeSn alloys. (a) Pattern for a very smooth surface as in left side of Figure 11.10 showing points on an arc. The diagonal streaks are multiple diffraction effects known as Kikuchi bands. (b) A pattern for a somewhat rougher surface corresponding to the middle drawing in Figure 11.10. Both (a) and (b) show a half-order reflection between the bright integer-order spots that results from the surface reconstruction. (b) Shows some evidence of residual Kikuchi bands. (c) A pattern for a very rough surface corresponding to the right hand drawing in Figure 11.10 where the electron beam is passing through three dimensional islands. Adapted with permission from O. Gurdal, P. Desjardins, J. R. A. Carlsson, N. Taylor, H. H. Radamson, J.-E. Sundgren, and J. E. Greene, Journal of Applied Physics, 83, 162 (1998). Copyright 1998, American Institute of Physics.

yields six diffraction spots between the bright integral lattice spots of the unreconstructed lattice spacing.

Examples of real RHEED patterns are shown in Figure 11.11.

Measuring the surface reconstruction is important to control of sensitive epitaxial growth processes. In general the ideal growth condition for GaAs includes a higher flux of As to the surface than the flux of Ga. This is because the As evaporates much faster than Ga from the surface. One must therefore supply As in excess. However, too much As degrades the growing film. The surface reconstruction changes when one transitions from too little As to at least enough (Section 10.12.2). The surface reconstruction observed in a RHEED diffraction pattern is therefore a useful tool for monitoring growth processes.

There are additional useful features of the basic diffraction pattern that are important to bear in mind. First, moving the substrate out of the way results in the electron beam striking the phosphor screen directly. This is known as the straight-through beam. The distance on the screen between the straight through beam and the electron beam simply reflected from the surface (the so-called "specular" beam) can be used to calculate the angle at which the electron beam strikes the surface.

Typically the substrate is not cut exactly on a low-index plane. When the diffraction spots are sharp enough, the spots may be split along the reciprocal lattice rod direction (vertically on the pattern) into two spots when the beam is aligned up or down the staircase of surface steps causing the miscut. The separation of these spots reflects the miscut of the sample surface relative to the low index planes. This is useful to know because miscut is used in some cases to enhance heteroepitaxy.

When the surface consists of two layers both contribute to diffraction and interfere destructively at some angles. This leads to one sinusoidal term in the Fourier series modulating the reciprocal lattice normal to the surface. Now it matters at what angle the electron beam hits the sample surface. At some angles electrons reflected from the two layers interfere constructively (the Bragg condition) and at other angles the electrons interfere destructively (the "off Bragg" condition) – see Figure 11.12. In addition, breaking the surface up into islands of atoms on a base layer disrupts the periodicity of the surface and increases the width of the reciprocal lattice in the plane of the surface. When the Ewald sphere cuts the reciprocal lattice rods one gets ellipses rather than points. This results in a streaked diffraction pattern on the screen. The finer the distribution of the islands of atoms on the surface the broader the reciprocal lattice rods and the more elliptical the diffraction spots appear. Reconstruction spots broaden in a similar manner.

Finally, as the surface roughens it eventually becomes possible for the electron beam to strike and pass through the rough spots as if they were three-dimensional islands. This gives rise to conventional diffraction spot patterns, usually a rectangular grid of spots on the screen.

The off-Bragg condition is exceptionally useful for measuring and controlling the film growth process. A perfect flat surface, even in the off-Bragg condition gives a strong diffracted or reflected beam (Figure 11.12). Beginning growth on such a surface has no choice but to add atoms to a second layer. This causes destructive interference in the reflected and diffracted beams and reduces their intensity, which goes on until the surface is as rough as it is likely to get. At that point growth begins

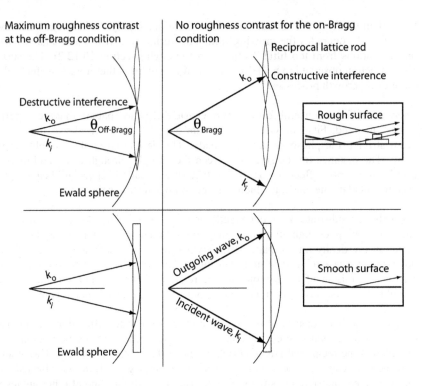

Figure 11.12: RHEED diffraction Ewald sphere constructions for the on-Bragg and off-Bragg conditions. Note that in the on-Bragg condition there is no difference in the diffracted intensity for the rough and smooth surfaces while for the off-Bragg condition alternate layers of the film interfere destructively. Therefore there is strong contrast for the intensity from rough and smooth surfaces. Refer to Figure 4.6 and related discussion for background on this construction.

to smooth the surface or generate a third layer. The reflected intensity increases again because the first and third layers interfere constructively. The result is an oscillation in the specular beam intensity with a period of exactly one monolayer of atoms with maximum intensity at the completion of each monolayer and a minimum at the half-monolayer point. The amplitude of the oscillations tells how rough the surface is as measured by the oscillation envelope (see Figure 11.13). Usually within a few monolayers or tens of monolayers the oscillations disappear because the roughness reaches steady state. However, in some cases the oscillations continue indefinitely. Because the oscillations are gradual and regular, it is possible to obtain a highly accurate growth rate (to better than 1%). A typical RHEED oscillation behavior is shown in Figure 11.13.

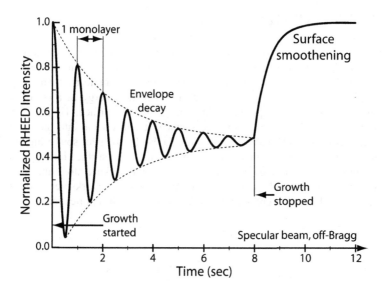

Figure 11.13: A typical example of (hypothetical) RHEED oscillation data during MBE growth at a rate of 1 monolayer/sec. Note that the surface is roughening during the 8 mono-layers of deposition as shown by the decay of the oscillations but upon termination of growth the surface becomes smooth and the intensity recovers. To obtain this type of behavior the experiment is configured in the off-Bragg condition so that alternate monolayers of the film interfere destructively. This type of oscillation is observed in the intensity of all reflected and diffracted beams. The most intense is the specularly-reflected beam so that is the one normally used.

RHEED oscillations have another advantage. When growth is terminated at the maximum of an oscillation (a monolayer is just completed on average) then with time the surface will smooth out as small islands fill in remaining holes. During this smoothing the RHEED intensity increases, eventually reaching a steady state. Thus, one may monitor the surface annealing process and end it after the shortest possible time.

When a multilayer superlattice film is grown with alternating chemistries, one of the layers will inevitably increase the surface energy. From the discussion in Section 10.9, the layer that increases the surface energy will grow in a rough three-dimensional manner. This is bad, especially for very sensitive structures such as laser diodes. Fortunately, once the surface is covered completely with the material that nucleates and grows in three-dimensional islands there is a strong motivation to smooth the surface to reduce surface energy. Hence, if one interrupts growth and waits for the RHEED intensity to recover and the surface to smooth, a superlattice in which all layers are smooth may be grown. The use of RHEED to interrupt growth and smooth the surface, leads to a method known as migration-enhanced epitaxy.

11.3 SPUTTERING

The second common class of physical vapor deposition method is sputtering. This describes all deposition processes in which the vapor of atoms to be deposited is created by a series collisions transferring energy and momentum from an accelerated primary atom or ion to the atoms in a target. When the energy of a target atom is sufficient to overcome the potential binding it in the surface, U, and when its momentum is directed outward from the surface, it will escape into the gas phase. This process is shown schematically in Figure 11.14. In order for sputtering to occur, the energy of the arriving particle must be many times the surface binding potential because each collision divides the energy between the striking atom and the struck atom. The classical hard-sphere Rutherford scattering theory works very well for this type of process, although there are many complications in its application.

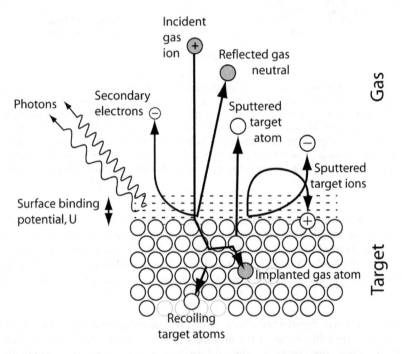

Figure 11.14: A schematic of the sputtering process. Ions impacting the target produce sputtered species and may be reflected from the surface as neutrals through elastic collision effects. In addition, secondary electron emission, photon emission, and heating of the target surface may result from inelastic processes. The surface binding potential results from the electronic bonding of the solid, symbolized by the dashes above the surface.

In addition to sputtering of atoms, a number of other processes may occur when a fast particle (atom or ion) hits the surface, as implied by Figure 11.14. Some of the sputtered particles may be emitted as ions. Typically this is ~1% of the sputtered particles, although the probability varies by several orders of magnitude depending upon the surface composition of the target. These charged ions are what are detected in the secondary-ion mass spectrometry materials analysis technique. Because of the field accelerating the positive primary ions (those from the gas that hit the target), negatively-charged sputtered species are accelerated from the target surface, while positively-charged ions are pushed back into the target.

If the target atoms are significantly heavier than the primary ions, the primary ion may also be reflected as a neutral from the surface. The probability of this event and the energy of the reflected species are determined by classical hard-sphere collision physics. In general, the reflection probability is not high until the target atomic mass is roughly twice the mass of the incident particle. Reflection probabilities are generally of the order of a percent of the impacting particles at the most. However, they may have a very significant effect on the resulting films. Reflected particles are generally thought to be a major source of the dependence of film properties on the sputtering gas pressure as higher pressures result in more scattering of fast reflected particles before they hit the growing film. Reflection ion effects are discussed in Section 11.4.

All of the above effects are due to elastic scattering events in which elastic collision theory works well. In addition, inelastic energy loss effects and neutralization of the incoming ion give rise to emission of secondary electrons and low and high energy photons from the target surface. The photons are normally innocuous but can cause problems for high energy-gap insulators such as SiO_2. Secondary electrons are crucial to maintaining the glow discharge that produces the primary ions, as discussed in Section 11.3.4.

11.3.1 Sputtering yield

When the energy of the arriving particle is well above the bond energy in the solid, the collision cascade can be effectively modeled. The sputtering yield was calculated based on classical collision theory by Sigmund [10] to be:

$$Y(E) = 0.042 \ \alpha(M_T / M_I) \frac{S_n(E)}{U} \qquad 11.2$$

where $S_n(E)$ is the nuclear stopping power of the target, U is the surface binding energy of target atoms, M_T is the average mass of the target atoms, M_I is the mass of the incident high energy atom or ion, E is the incident atom energy, and $\alpha(M_T/M_I)$ is a correction function to account for deviations from a simple single-event hard sphere collision behavior. The nuclear stopping power is generally approximated with a Thomas-Fermi cross-section for impacts. Ignoring energy losses due to excitations of electrons in the target (electronic stopping) the sputtering yield is: [11]

$$S_n(E) = 4\pi\, a\, Z_t Z_i\, q^2\, \frac{M_i}{M_i + M_t}\, s_n(\varepsilon) \qquad\qquad 11.3$$

where $s_n(\varepsilon)$ is a "universal" stopping power and is a function of the reduced energy ε, but not dependent upon the colliding particle masses. The variable **a** is a screened interaction radius, shown by Lindhard to be $a = 0.8853\, a_0\, \sqrt{Z_i^{2/3} + Z_t^{2/3}}$ and where a_0 is the Bohr radius, 0.0529 nm. $Z_t Z_i q^2$ is the product of the nuclear charges of the colliding particles. The universal stopping power function has been written empirically as: [12]

$$s_n(\varepsilon) \approx \frac{3.441\sqrt{\varepsilon}\, \log_{10}(\varepsilon + 2.718)}{1 + 6.35\sqrt{\varepsilon} + \varepsilon(-1.708 + 6.882\sqrt{\varepsilon})}. \qquad\qquad 11.4$$

The reduced energy, ε, was calculated explicitly by Lindhard to be: [11]

$$\varepsilon = \frac{M_t\, a}{(M_i + M_t) Z_i Z_t e^2}\, E \qquad\qquad 11.5$$

Details of the functional form of $\alpha(M_T/M_I)$ and the other considerations in estimating sputtering yields may be found in Sigmund [10] and in Bohdansky [13] and references therein. The behavior of the Sigmund result as a function of particle energy is shown in Figure 11.14.

As the energy of the incident particle decreases one eventually comes to the point where it is highly unlikely that any atom will escape the surface. At this point the sputtering yield falls rapidly below the Sigmund curve near a threshold energy (see Figure 11.15). This threshold has been modeled by a number of workers including in the original work by Sigmund. Following the treatment of Yamamura and Mizuo, the threshold energy may be estimated to be approximately: [14]

$$E_{th} = U\left[0.21 + 4.8\left(\frac{M_T}{M_I}\right)^{0.57} + 0.26\left(\frac{M_I}{M_T}\right) \right] \qquad\qquad 11.6$$

The general behavior of this equation as a function of target mass for a given ion is shown in Figure 11.16. Note that it is generally straightforward to estimate the sputtering threshold from experimental data (c.f. Figure 11.15) but it is more difficult to determine the surface binding potential, U. Therefore, this equation is more useful for estimation of U from E_{th} data than the reverse.

The effect of this threshold on the sputtering yield Y has been modeled by Bohdansky, resulting in the functional form: [13]

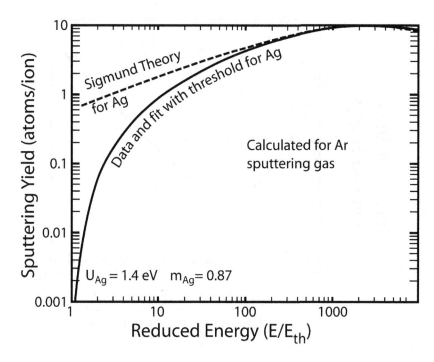

Figure 11.15: A comparison of sputter yield data with both the Sigmund theory and Equation 11.4-8. Equation 11.8 fits the data almost exactly over the range shown and for the values given for fitting parameters in the figure. The higher energy range outside of the range of the figure fits the Sigmund theory well, although the surface binding potential is not the same as when fit to Equation 11.8. Similar fits can be obtained for other elemental sputtering yields.

$$Y(E) \approx 0.042\alpha(M_T/M_I)S_n(E)\left[1 - \left(\frac{E_{th}}{E}\right)^{1-m}\right]\left[1 - \frac{E_{th}}{E}\right]^2 \qquad 11.7$$

In other words, it is the Sigmund yield expression modified by the last two terms in square brackets, which account for the threshold. The basis for terms in this correction are discussed in detail in Bohdansky with additional description in Sigmund. The value of the variable m determines how the threshold behavior is interpolated into the Sigmund type behavior. Bohdansky suggests a value of ~1/3 for m. However, values may vary depending upon the atoms involved. Fitting the experimental sputtering data (Figure 11.15) suggests a value between 0.7 and 0.97 is not uncommon under sputtering conditions. Although Equation 11.7 is not a perfect simulation of the experimental data the general trend is correct and the values estimated give the correct order of magnitude for the result with optimized values for material variables in the equations.

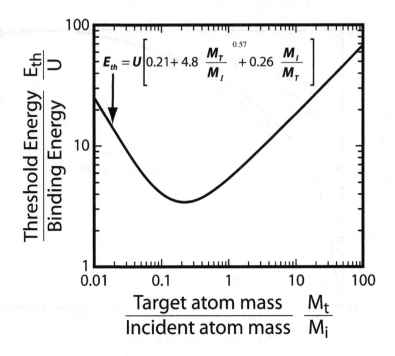

Figure 11.16: The empirical relation for threshold energy for sputtering reduced by surface binding potential as a function of target to ion mass ratio.

Much of this theory has been incorporated into computer simulations of sputtering and ion damage processes. Thus, it is rare to actually employ these expressions directly given the availability of programs such as TRIM, a Monte Carlo type simulation of the collision cascade in solids struck with ions. [15]

Considering the data in Figure 11.15 one may see without reliance on theoretical models that the sputtering yields are generally modest even when the incident ion has thousands of eV of energy, hundreds of times the surface binding potential. Thus, most of the energy of an impacting ion is dissipated in the target as heat rather than ejecting more atoms from the surface. Turning the ion energy up does not greatly aid the situation and eventually the sputtering yield even begins to decrease as the impacting ion travels much farther into the solid before striking a target atom. In other words, the collision probability decreases with increasing ion energy above the peak in the Sigmund curve.

Much can be understood empirically about sputtering when one realizes **that it is a process requiring the turning of the inward-moving kinetic energy and**

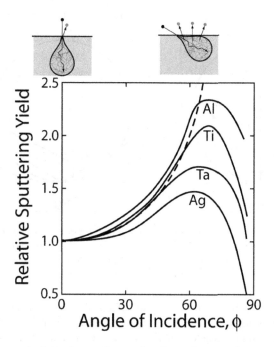

Figure 11.17: The angle-dependence of sputtering for various species as a function of their angle of impact on the surface. The dashed curve is a 1/cosφ dependence. Figure used with permission from Oechsner, H., "Sputtering – a review of some recent experimental and theoretical aspects." *Applied Physics A: Materials Science and Processing*, 1975; 3: 185-196. Copyright Springer, 1975.

momentum of the incident ion into an outward-directed energy and momentum of an atom escaping the surface. Thus, below the peak in the sputtering yield the energy that is turned toward the surface is insufficient to permit atoms to escape easily, while at higher impact energies the transfer of energy to target species occurs too deep in the sample to ultimately reach the surface.

Likewise, the dependence of sputtering yield on angle of impact of the ion with the surface can be understood as related to the turning of the momentum cascade within the solid. The smaller the angle through which this turning must occur, the higher the sputtering yield. This behavior is directly observable and one finds that the yield increases approximately as 1/cosφ, where φ is the angle between the incident ion and the surface normal (see Figure 11.17). One may also understand this behavior by noting that the collision cascade initiated by an incident ion is teardrop shaped. As one increases φ, more of this teardrop intersects the surface, indicating that there are energetic outwardly-directed atoms in the target at that location.

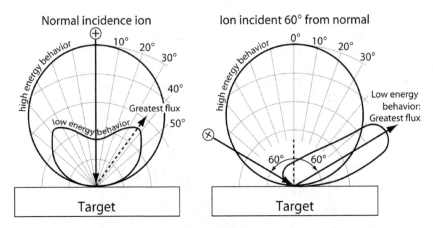

Figure 11.18: Polar plots showing the flux of sputtered particles at two different incidence angles for ions striking a target. The limit of high-energy behavior produces a cosine distribution. This figure was adapted from Yamamura, Y. and Itoh, N. in *Ion Beam Assisted Film Growth*, ed. by T. Itoh, Chapter 4, Copyright Elsevier 1989.

The nature of the collision process also affects the direction at which sputtered species leave the sample surface. When a very high energy particle strikes the target the energy/momentum cascade reaches surface atoms essentially isotropically and atoms leave the target surface in a random distribution. This means that the flux of atoms leaving the target surface is proportional to the cosine of the angle relative to the surface normal at which they leave. This is the same behavior as for evaporation from a full effusion cell (Figure 11.4).

An important but perhaps apparently trivial detail of the previous paragraph deserves special note. **Sputtered atoms are atoms that were at/near the surface of the target and which received energy/momentum from those below them.** Thus, sputtered atoms do not come from deep within the target. The low energy of the sputtered atoms makes escape directly from deep within the target essentially impossible. Thus, *the atoms in the sputtered flux reflect the composition of the target surface multiplied by the sputtering yield of that species. It does not necessarily reflect the bulk composition of the target.*

When an ion strikes the target with a moderate energy the sputtered flux distribution becomes more strongly directed away from the normal to the sample surface, as shown in Figure 11.18. This distribution represents a compromise between turning the momentum cascade through the smallest possible angle, and maximizing the chance of an atom escaping the surface. As the ion energy is reduced the difficulty of turning the collision cascade becomes more significant. Thus the maximum flux leaves the target at ~40° relative to the normal, varying somewhat as the ion mass and energy changes. Likewise, as the ion strikes the target at an increasingly oblique angle the momentum cascade is effectively reflected from the underlying planes of atoms and sputtered particles tend to exit the target primarily along a line representing reflection of the incident particle momentum (Figure 11.18).

These behaviors apply for a smooth target. Real target surfaces are far from smooth on a microscopic scale and often develop shaped surfaces when sputtering is non-uniform across the sample surface. Thus, the actual sputtered flux may vary somewhat. The final flux distribution is a weighted average over the surface orientations. Ions generally strike the surface roughly perpendicular to the average surface of the target in normal thin film deposition sputtering processes as opposed to processes using an ion gun.

One might expect that the forward sputtering effect (right hand side of Figure 11.18) would tend to smooth local roughness on a target because particles sputtered from a projecting rough feature would tend to be sputtered downward into low spots. Nonetheless, targets do become increasingly rough as they are sputtered. Apparently, the angle dependence of sputtering is not the cause of target roughening. Rather, it is the result of local areas where low-sputtering-rate species such as carbides are present on the surface.

11.3.2 Energetic particles

The collision cascade leads to transfer of energy to particles leaving the sample surface. As noted above, this energy must exceed the energy binding atoms into the solid for sputtering to occur. However, we have also seen that the incident particle causing sputtering must normally have far more than this amount of energy. Sometimes considerably more of the incident particle energy than needed is given to the sputtered atom, allowing it to leave the surface with considerable kinetic energy. This may permit the sputtered atom to damage the growing film. This energy dependence has been discussed and modeled by Grapperhaus et al. [18] and Heberlein et al. [19] who proposed a particle distribution probability as a function of sputtered particle energy, p(E), based on Monte Carlo simulations, to be a cascade distribution:

$$p(E) \, \alpha \, \frac{2E_{ion}}{E^2 \left[1 + (U/E) \right]^3} \qquad\qquad 11.8$$

where E_{ion} is the incident particle energy. This behavior scales linearly with ion energy as shown in Figure 11.19. Some interesting conclusions of this behavior are that >90% of the energy of the incident particle is dissipated as heat. Very little is carried away with sputtered particles. The particles that are sputtered may have a few eV of energy, but typically this is no more than what would be needed to break a single bond to the surface. It is far too little for most sputtered atoms to have any significant effect on the film growth process. Nonetheless, the sputtered particles are sufficiently energetic that they may cause chemical reactions on the surface or

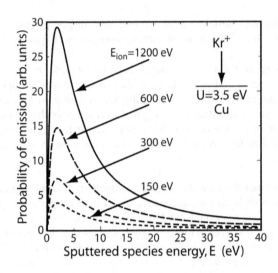

Figure 11.19: The energy distribution of sputtered species leaving a target surface based on Equation 11.8.

enhance surface diffusion. In any case, many of the relatively energetic sputtered species are likely to collide with another atom in the gas phase before reaching the growth surface. This further decelerates them.

Although the sputtered particles themselves typically do not have a significant amount of energy, the energy distribution of particles striking a growing film surface can have a dramatic effect on the properties of that film. These effects and their causes are discussed in detail in Section 11.4. For now we are concerned about the sources of such energetic particles (see Figure 11.20). In addition to the effect in Figure 11.19, these are as follows. First, a sputtering gas atom may be reflected from the target surface if it is lighter than the target atoms. Second, a sputtered particle may be ionized in one of various ways and, if negative and located near the target surface, this can lead to acceleration away from the target. Third, sputtered species and other gas particles may be ionized and accelerated intentionally or (in the case of rf sputtering, unintentionally) to the substrate. We will take these three mechanisms in turn and consider how they work and when.

Reflection of sputtering gas particles from a substrate occurs when the target atoms are heavier than the sputtering gas. Classical inelastic collision theory says that the heavier a struck atom is relative to the striking atom the farther off center a collision can be and still turn the striking atom back in the direction it came (greater than 90° of deflection). When the striking atom is heavier than the struck atom the momentum must continue forward regardless of the point of impact. Reflection of a heavy particle therefore must involve it striking more than one target atom or that the target atom must be moving toward it with significant momentum. Multiple collisions or

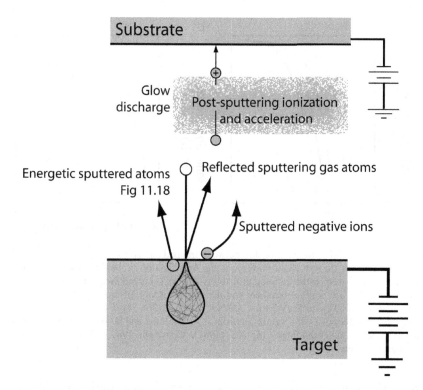

Figure 11.20: shows schematically the processes that may lead to energetic particles leaving the target surface and potentially striking the substrate as the film grows.

processes of this type are so rare that reflection of heavy particles from a light target is effectively impossible.

A reflected particle is nearly always a neutral atom because as a positive ion approaches a surface it will pick up an electron from among the electrons above a surface. To see how this should happen note that the electronic wave functions discussed in Chapters 2 and 5 do not end abruptly at the last atomic nucleus. Therefore the probability of finding an electron outside of the last atomic nucleus is significant. It gives rise to a net electron density over the surface from which the arriving ion can be neutralized. It also gives rise to the work function that binds electrons in a solid. They must cross the resulting surface dipole charge field before leaving.

The reflection probability and average reflected particle energy increases rapidly as the striking particle becomes heavier than struck atom. The reflection probability distribution for a typical process is shown in Figure 11.21. The data for reflection

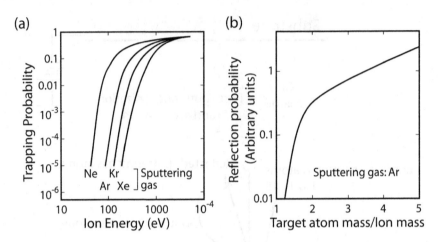

Figure 11.21. (a) Shows the relationship between trapped gas concentration in thin films and ion energy and (b) the estimated corresponding Ar reflection probability for various target materials judged based on the trapping probability. Figure (a) is redrawn with permission from Kornelsen, E.V., "The ionic entrapment and thermal desorption of inert gases in tungsten for kinetic energies of 40 eV to 5keV." *Can. J. Phys.*, 1964; 42: 364-381. Copyright NRC Research Press, 1964. (b) Redrawn with permission from J. A. Thornton and D. W. Hoffman, Journal of Vacuum Science & Technology, 18, 203 (1981). Copyright 1981, AVS The Science & Technology Society.

[Figure 11.21(b)] is derived from trapping of inert gas in a growing thin film as in the data shown in Figure 11.21(a). For a single ion mass and a single film behavior the trapping probability and the reflection probability are assumed to be directly related to each other. This is a reasonable assumption as long as particles reflected from the target and striking the film do not modify its microstructure too much or cause too much resputtering of trapped gas from the film. However, when the gas species is changed as in Figure 11.21(a) no simple assumptions may be made because the energy distribution of reflected particles changes with ion mass, their ability to penetrate a growing film is altered, and their ability to diffuse in the film and escape changes. Therefore, while the data in Figure 11.21(a) certainly include a reflection term, a direct linear relationship cannot be expected.

Reflected sputtering gas atoms leave the target surface with significant energies. Monte Carlo simulations based on the TRIM computer code by Ray and Greene showed that the average energy of a reflected particle increases roughly in proportion to the increase in sputtering ion energy. [20] In the simulation of Ar impacts on an amorphous Ge target the mean reflected Ar energy was approximately 10% of the target impact energy. Approximately half of the reflected particles leave the surface with this energy while 1% of the particles leave with more than ~40% of the target energy. The energy distribution was Gaussian in shape with a maximum at a few eV energy. Although the details will change for different experimental conditions, these general trends should be typical. For a typical target energy of 600 eV about one quarter of the reflected Ar atoms (about 0.5% of the Ar hitting the target) would have enough energy to significantly damage or modify the growing film (about 100 eV).

The energy of the reflected particles is only barely enough to modify the film. Therefore, even one or two collisions between the reflected atom and other gas atoms can eliminate the effect. From Equations 10.1 and 10.2 we know that the mean free path is of the order of 0.7 cm Pascals of pressure but that the path length should increase as particle kinetic energy increases. If these ideal gas equations held at high energies one might expect the mean free path to be thousands of times longer. However, the mean free path does not increase this rapidly. On the other hand, the common assumption that mean free path remains of the order of 0.7 cm divided by pressure in Pa is also incorrect. The fact that gas scattering is important may be seen in the strong effect of pressure on the residual stress, morphology, and other effects in films. An example of this relationship is shown in Figure 11.22. Note however, that the effect shown in this data, while typical, indicates that the simple concepts outlined here are insufficient for a complete explanation.

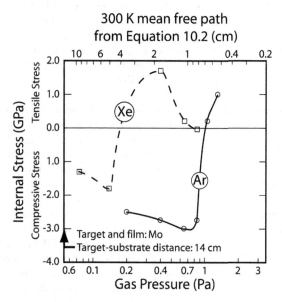

Figure 11.22: An example of the effect of sputtering system pressure on residual stress in thin films. Redrawn with permission from D. W. Hoffman and John A. Thornton, Journal of Vacuum Science & Technology, 17, 380 (1980). Copyright 1980, AVS The Science & Technology Society.

There should, according to the above, be no reflection of Xe ions from the target and no effect of gas pressure on the film. However, Xe does show a pressure effect (although at much lower pressures than for Ar) indicating that there are some energetic particles leaving the target surface even in this case. The low pressure required to show the transition to compressive stress indicates that there are very few energetic particles, thus even very limited scattering in the gas phase eliminates the change in film properties. The effect in the case of a heavy particle striking a light target can be shown explicitly using Compton scattering theory to compute the reflected momentum but this is beyond the scope of this text. For both types of gas the compressive-to-tensile transition is at three to 14 mean free paths for Xe and Ar, respectively. This shows clearly that the high energy of these particles has increased the mean free path well above that of a 300 K atom.

In some cases an atom leaving the target surface may pick up an electron from the surface states. Rather than neutralizing an incoming ion as described above, this results in negative ionization of an outgoing atom. Because the negative ion is created very close to the target surface, it experiences the entire accelerating field resulting from the target bias. Unlike the sputtering gas ions, which are positive and accelerated inward, negative ions are accelerated outward and pass into the gas with very high kinetic energies. Furthermore, because of their very high kinetic energy these atoms may travel a substantial distance through the gas phase and are likely under most sputtering conditions to strike the substrate with most of their energy remaining. It should be apparent from the discussion above that this will cause very great changes to the properties of a growing film.

Highly electronegative species (metalloids from the upper right portion of the periodic table, excluding the noble gasses) are most likely to pick up an electron upon sputtering. Other less electronegative particles may form negative ions at the surface of a target containing highly electropositive species such as one of the alkali metals. Thus sputtering of oxides, chalcogenides, or halides will very likely result in negative ion formation.

It is worth re-emphasizing that negative ions form at the target surface and not in the gas phase in almost all cases. To form a negative ion in the gas requires that an electron be transferred to or be captured by a neutral atom in the presence of another particle that takes away the excess energy with which that electron is bound to the ion. In the gas this requires a "three-body" collision and is unlikely as a result. On a surface the other atoms and electrons in the surface can easily absorb this energy. Therefore, one may generally ignore negative ion formation in the gas phase in sputtering processes.

The final common mechanism for creation of energetic particles is ionization in the gas phase with subsequent acceleration to the substrate. Following the preceding discussion, it should not be surprising that the ions formed in the gas will be positive and will result from an energetic collision of another particle with an atom, resulting in knocking an electron off of the atom. Positive ion formation is what creates and sustains a glow discharge during sputtering. We will see that the details of this process are essential to determining sputtering conditions. What concerns us at the moment is the fact that some of these ions may be accelerated and may strike the substrate.

The simplest way in which ions from the glow discharge may be accelerated and strike the growing film is to apply a negative bias voltage to the substrate. The number of particles leaving the discharge determines the number of impacts that could affect the film properties, while the applied bias voltage determines their energy. This is convenient as it allows, at least in principle, independent control of the current and energy of ions striking the film. Critical energy ranges for modification of the film properties are, with broad generalization, 0-15 eV where atoms are displaced only on surfaces, 15-50 eV where displacement cascades can inject an interstitial atom into a solid, and greater than 50 eV where many other phenomena including resputtering of atoms previously deposited on the film or implantation of the impacting particle may occur. One may see from these energy ranges that careful control of the substrate bias is important to minimizing the damage to the film. Ideally one would generally prefer a high flux of low energy (often less than 25 eV) particles striking the film surface. The effects of these processes on film growth are discussed in Section 11.4.

The other common way in which ions are accelerated toward a growing film is in rf sputtering. Here the glow discharge is the most positively charged portion of the gas. Hence positive ions are pushed out to all of the other surfaces in the system. The details of the formation of electrode potentials that give rise to this phenomenon are discussed in Section 11.3.4.4, below.

11.3.3 Sputtering systems

We have alluded to "dc" and "rf" sputtering above. It is useful to briefly summarize the forms that these systems take before returning to details of how sputtering works. In general, sputtering systems are categorized by the way in which energy is supplied to the gas to form ions and bombard the target (see Figure 11.23). Power supply options are broadly divided into direct current (dc) and radio frequency (rf). We will omit the many other details of power supply design. The glow discharge behaves as a diode. Thus, it conducts current easily in one direction and is resistant to current flow in the other. It has two contacts to the gas, an anode and a cathode. The target normally functions as the cathode and supplies electrons to the gas while the remainder of the system is typically at ground potential and acts as the anode.

Some sputtering systems include additional electrodes that may act as alternate cathodes or anodes. When there is one additional contact (either cathode or anode) the sputtering process is called "triode" sputtering. A separate cathode is usually heated so that it emits electrons spontaneously. This reduces the constraints on the target voltage, typically for the purpose of allowing a lower voltage than would otherwise be needed to maintain the discharge. When using a heated cathode it is

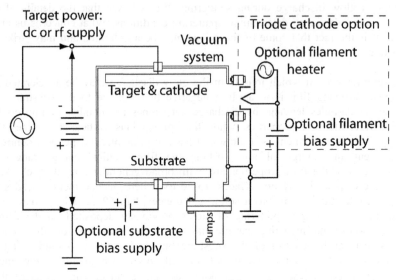

Figure 11.23: A schematic diagram showing some of the options for sputtering system configurations including dc or rf power supplies (but not both), and the possible addition of a substrate bias and a filament or auxiliary anode for a triode configuration.

wise to locate it out of direct line of sight of the substrates because it will typically be sputtered by the gas ions. A separate anode is used when one wants the substrates to be at a potential other than anode potential. Typically this is used to bias the substrates negatively relative to ground so that some low energy ions bombard the substrates. Sputtering when the substrates have a different potential than ground is known as "bias" sputtering.

The sputtering rate is directly related to the flux of ions striking the target. This, in turn, is directly related to the density of ions in the gas. Therefore if this density can be increased the sputtering rate can be increased. The most popular method for increasing the ion density is to place magnets behind the target such that the magnetic field penetrates the target surface forming loops (see Figure 11.24). Electrons leaving the target are deflected by the magnetic field and travel a longer path to reach the anode. In the process they are much more efficient at creating ions. The result is a very high ion density in the gas near the target where electrons are trapped and a very low ion density elsewhere. This has the advantage that sputtering only occurs under the field traps. Avoiding unintentional sputtering of other surfaces

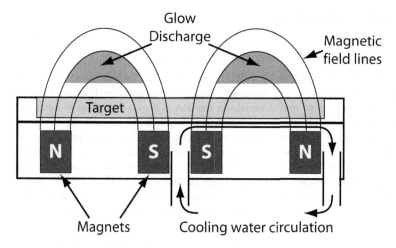

Figure 11.24: A schematic diagram of the design of a magnetron sputtering source.

greatly reduces the chances of contamination of the growing film. Therefore there are many advantages to the magnetron technique. The use of a triode with a heated cathode also increases ion density in the discharge somewhat but is not commonly used. Finally, one can use electron cyclotron resonance in which electrons are driven in circles by an alternating electric field in a constant magnetic field. This is capable of producing over 90% ionization of the gas but is complex and restricts the geometry of the system considerably. Here we will only consider magnetron sputtering for enhancing ion density.

The shape of the target also names sputtering processes. Thus, planar sputtering implies a planar target, cylindrical post sputtering a cylindrical post target, hollow cathode sputtering refers to sputtering off the inside of a tube, etc. These various options in sputtering are shown schematically in Figure 11.25. One may add words to the name of the process to describe the details of the power supply used as in "pulsed dc sputtering". In this process the target bias is switched on and off periodically to reduce the danger of forming an arc. This is particularly important when operating the process near the limits of stability or when sputtering with a reactive gas ("reactive sputtering") where a compound is being formed at the substrate from a metallic target. Pulsed, magnetron, or reactive sputtering can be performed with any target geometry.

11.3.4 Glow discharge basics

11.3.4.1 Types of glow discharge

To complete our brief tour of sputtering methods it is helpful to review briefly how ionized gases are formed in sputtering environments and how they are maintained. It is the formation and maintenance of the ionized gas that largely determines the voltage that must be applied to the target. We will begin with the behavior of dc discharges and consider the more complex case of rf discharges later. The electrical circuit and electrostatic potential across a dc diode system are shown in Figure 11.26. The basic rule of the electric circuit is that the same amount of current must flow in the circuit everywhere. There must also be the same positive charge flow in one direction as there is negative charge flow in the other (to avoid accumulation of electrons or ions, positive or negative charges) in the glow discharge.

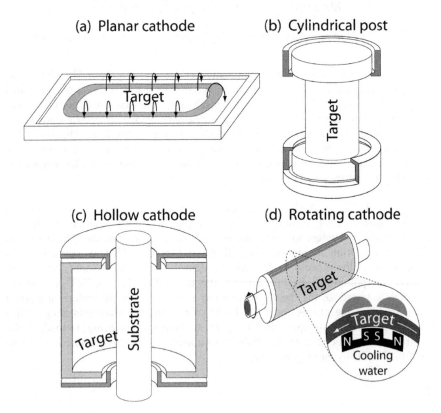

Figure 11.25: Sketches of four magnetron geometries. Note that the rotating target geometry functions in the same was as a planar geometry but the target is not flat. Rotation of the target allows the target to cool between sputtering phases. This allows higher power sputtering and hence faster sputtering rates.

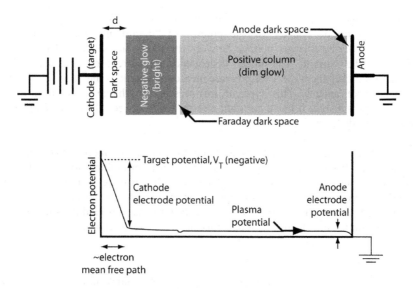

Figure 11.26: The structure of a dc glow discharge and the electron potential across the device from cathode to anode.

The ionized gas normally present in a sputtering process is technically a "glow discharge" not a true plasma because it does not contain an equal number of electrons and ions (required in a plasma). Specifically the glow discharge is positive with respect to ground. This occurs because the total current of electrons and ions leaving the discharge must balance or charge would continue to accumulate one way or the other. Because electrons are much lighter than ions it is necessary to push the ions harder to get them out of the discharge at the same current. Therefore the potential drop between anode and cathode is not divided equally and the dc discharge potential is much closer to that of the anode (Figure 11.26).

Sputtering discharges are divided up according to the way in which the discharge current responds to a change in voltage and pressure. This, in turn, is determined by the chance that an electron leaving the cathode will ionize one or more gas atoms before reaching the anode. Some of the more important types are shown in Figure 11.27.

At low pressures an electron generally hits at most one gas atom as it transits between the cathode and anode. This makes ionization of the gas weak and dependent entirely on gas pressure and the number of electrons leaving the cathode. For a given emission current, ionization depends linearly on pressure. Every ion created generates another electron striking the anode so current also generally depends linearly on pressure. Because very few ions strike the cathode, electron emission is driven entirely by the process of "thermionic emission" (see next section). Applying a larger voltage between the target and substrate has very little effect on the current. The linear dependence of current on pressure in these discharges makes them ideal for pressure measurement and is the basis for ionization type pressure gauges. This type of low-pressure discharge is referred to as a Townsend discharge. It has a constant electric field between the anode and cathode and no well-developed glow.

As the pressure increases electrons are likely to interact with more than one gas atom on their way from anode to cathode, and thus can form multiple ions. Furthermore, electrons generated by ionization events near the cathode can go on to cause ionizations themselves. Eventually ionization events become so common that ions are formed in the gas as fast as they are lost to the surrounding surfaces or to recombination with electrons. When this happens the discharge has a significant

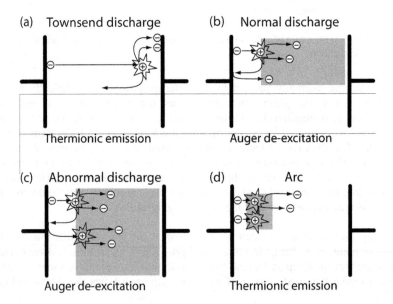

Figure 11.27: A schematic of the four primary types of glow discharge. Shaded areas represent the glow discharge (a) the Townsend discharge occurs at low pressures, results in low ion current densities less than 100 nA/cm² and has a high resistance. (b) At intermediate pressures a normal discharge creates a typical glow discharge as illustrated in Figure 11.25, but covering only part of the cathode area. (c) Higher pressures still lead to an abnormal glow covering the entire cathode and having a relatively low resistance. This is the type of discharge used in typical sputtering processes. (d) At very high pressures an arc occurs where the discharge collapses to a single small area on the cathode. It has a very low resistance.

steady-state ionization (several percent of the gas is ionized). The ion and electron density in the gas is now so great that the gas is strongly conducting of electric current. An electric conductor does not support an electric field greater than Ohm's law would allow for a given conductivity. Consequently the field is no longer evenly distributed from cathode to anode but becomes localized near the electrodes with virtually no field in the discharge itself. (See Figures 11.26 and 11.27.) The increase in field near the cathode accelerates electrons much more quickly and enhances their chance of causing ionizations. In lower pressure ranges the current carrying capacity of the gas is not sufficient to allow high field everywhere across the target. Rather one develops local areas of high field and a local strongly-ionized discharge. Further increases in gas density increase the fraction of the cathode covered by the discharge.

Note that it is the cathode that matters because that is where electrons are emitted and where ionization takes place. No ion emission occurs from the anode, which is simply a sink for electrons. Note also that the electric field is much stronger at the cathode than at the anode (Figure 11.26) to balance the ion and electron current. This has the further advantage of accelerating the electrons emitted by the cathode strongly and enhances their ability to cause ionizations.

As gas density increases with increasing pressure the fraction of the target covered by a well-defined glow discharge with a strong local electric field increases. Eventually the cathode is completely covered by the discharge and further increases in pressure can now no longer increase current by increasing discharge area. This discharge is known as "abnormal". Here the discharge current density increases with applied voltage (low positive resistance), while in the "normal" regime the resistance of the gas is high and increasing voltage does not cause an increase in current density. Sputtering processes are carried out in the abnormal glow discharge pressure regime. In both the normal and abnormal glow discharge regimes the electron emission from the target is governed by ions approaching the target surface (see next section).

Finally, at very high pressures the discharge converts to an arc. Arcs occur when the ion bombardment at the target becomes so energetic that it heats the cathode strongly. As cathode heating increases, thermionic emission of electrons rises exponentially. Eventually this rate exceeds that of ion-induced secondary electron emission. At the hottest point on the target one gets the highest electron emission and hence the discharge above this point becomes the strongest. This further increases discharge density at the highest temperature region of the target, causing further increase in electron emission and further increase in discharge ionization. Quickly the discharge current becomes localized at the target hot spot and the discharge goes out elsewhere. The result is an arc at one point above the target surface. Arcs are so strongly ionized and cathode electron emission currents are so high that they are limited only by the resistance of the external circuit. This is why arcs cause problems in circuits. Arcs are only used in one very high rate deposition process that is part sputtering and part evaporation and is known as cathodic arc deposition.

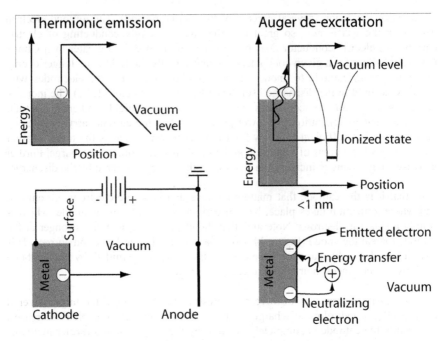

Figure 11.28: The two primary mechanisms for electron emission from a surface. Left: thermionic emission in which heat is used to emit electrons from a surface. Right: Auger deexcitation of the incoming ion can release enough energy to kick another electron out of the solid.

11.3.4.2 Electron emission and gas ionization

As noted above, "secondary" electrons from the cathode (target) are responsible for ionizing the sputtering gas. Therefore, it is important to know how electrons are emitted from the cathode and how they cause ionization. Secondary electrons are emitted by from target as a result of gas ions being neutralized near the target surface. Direct impact of ions on the target does not play a major role in secondary electron emission in normal sputtering processes.

It seems counterintuitive, but under normal sputtering conditions the velocity of ions striking the target and hence their kinetic energy has little or no effect on secondary electron emission from the sputtering target. This is because the ion kinetic energy is modest and not easily coupled to the electrons. Most of this energy is dissipated in the form of heat resulting from "nuclear stopping" due to collisions of the ion with atoms in the target. Electronic stopping of ions is relatively small until the ion kinetic energy is tens of thousands of electron volts.

The velocity of ions approaching the target surface even at 1000 eV of energy is sufficiently small from the point of view of an electron that the ion seems hardly to be moving, and the electronic charge on the target surface is free to adjust to the ion's presence continuously during the approach. During the approach process there is an excellent chance that the ion will capture an electron from the electronic states that lie outside of the outer layer of atomic nuclei. (See Figure 11.28.) It is these states that give rise to a slight negative charge outside of the surface and result in the material having a positive work function (energy is required to remove an electron from the surface.

An ion capturing an electron thus must give up the binding energy of that electron to the gas atom. Typically this is tens of electron volts. There are two ways that such energy can be released, light emission, and the Auger process. The latter occurs when an electron falls into a low energy state in an atom and transfers the resulting energy to another electron, which is then free to escape. This process is shown schematically in Figure 11.28. Likewise, light emitted by the target may be absorbed by another target atom and may ionize that atom. The secondary electrons released from the target surface are accelerated outward by the target electric field.

Accelerated secondary electrons cause ionization of the gas atoms if they have sufficient energy. Figure 11.29 shows the cross section for various processes by which an electron may interact with an Ar atom causing excitation of the atom or ionization. Above a threshold energy the secondary electron begins to ionize the gas. If the electron only loses part of its energy in a given ionization event it may proceed to cause other ionizations. It is necessary for the electron to ionize as many gas atoms as possible as this is generally the limiting factor determining the minimum target voltage to maintain the discharge.

The probability that secondary electrons will cause more than enough ionization events to ignite the glow discharge depends upon their chance of striking a gas atom, which increases with gas pressure, and the energy they will have when that impact occurs. The secondary electron kinetic energy is that portion of the target voltage through which they have dropped before the impact occurs. Therefore there is a negative effect of pressure when sufficiently high – the electron may strike an atom before it gains enough energy to cause ionization. At very low pressures a very high target voltage is necessary to make up for the low probability of impact and ionization of a gas atom. As the pressure increases the required voltage drops because more collisions occur. At some point the voltage required reaches a minimum and begins to increase because collisions occur too soon so a higher field is needed to accumulate the ionization energy within the first mean free path. The result is the Paschen equation for the minimum voltage V_{br} to ignite a glow discharge (also the breakdown voltage of a gas):

$$V_{br} = \frac{APd}{C + \ln(Pd)},$$
11.9

where A and C are constants, P is the pressure, and d is the cathode-anode distance.

Once the glow discharge is ignited the electric field becomes quickly localized near the target because the discharge is electrically conductive as noted above. This means that much less voltage is needed to maintain a discharge than to ignite one. The voltage needed to maintain the discharge is observed to have the following form:

$$V_n = \frac{1}{\eta} \ln\left(1 + \frac{1}{\gamma}\right),$$

         11.10

where η is the number of ions created per electron per eV of energy of the ionizing electron and γ is the probability of secondary electron emission from the target. The number of ions created is related to the ionization cross section (Figure 11.29) and

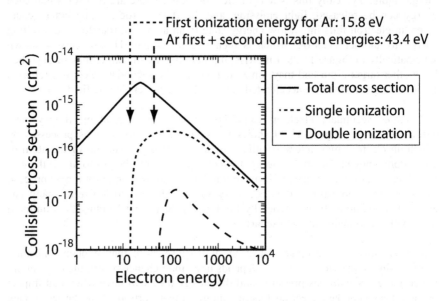

Figure 11.29: Cross sections for collisions events between an electron and a gas atom. An elastic collision results in energy loss and possible excitation of the struck atom but no ionization. Single ionization produces a single-charged ion and a second electron. A double ionization event results in a doubly-charged ion and two additional electrons. The threshold energies for the ionization events are roughly the first ionization energy and the sum of first and second ionization energies for the gas atoms. Events that do not cause ionization (the remainder of the cross-section) results in elastic energy losses. Curves after Thornton, John A. and Penfold, Alan S., "Cylindrical Magentron Sputtering" in Vossen, John L. and Kern, Werner, *Thin Film Processes* volumes I and II. New York, Academic: 1978; pp. 75-113. Copyright Elsevier 1978.

the secondary electron emission probability is related to the Auger de-excitation process (Figure 11.28). Both of these are determined by the chemistry of the surface and the working gas. Therefore, a change in the surface, for example a chemical reaction to form a compound on the target, can change the voltage necessary to maintain a discharge. Likewise, sputtering the target adds target atoms to the gas phase, which changes its chemistry, as does any intentional change in the gas through what is fed into the sputtering system.

The relationship between target voltage and target and gas chemistry is an important diagnostic tool. When doing reactive sputtering in which a reactive gas and a metallic target are used to produce a compound, the reaction of the sputtering gas with the target surface causes control problems. This target reaction is an example of a change that is detectable through the target voltage at constant current.

11.3.4.3 Ion flux

The voltage necessary to initiate and maintain a glow discharge is important but more critical to a sputtering process is the ion current, because the ions do the work of sputtering. As we saw above, the voltage plays only a moderate role in controlling the sputtering yield once one is well above the sputtering threshold energy.

The simplest way to estimate the ion current density striking the target is to ask how many ions cross a given plane, generally taken as the edge of the dark space above the target where the electric field begins, per second. Perhaps surprisingly, this turns out to be a sufficient estimate because once the ions enter the region of high electric field they accelerate toward the target rapidly. The distance is so short that they do not strike another atom on the way to the target so no charge can be transferred.

We saw in Equation 10.3 that the flux of all gas species crossing a plane per second can be written in terms of the partial pressure as:

$$F = \frac{P}{\sqrt{2\pi m k_B T}},$$

The flux of ionized gas species crossing a plane is simply this flux multiplied by the fraction of the gas that is ionized, η. However, there is a complication. In order that the charge on the gas in the glow discharge not change, in other words, to maintain steady state conditions, the ion flux must match the electron flux. It turns out that the electrons gain and lose energy in the glow discharge differently from ions because the types of collisions that the two particles undergo, their relative velocities, and so on are quite different. Therefore in order that the ion current match the electron current at the dark space sheath surface, the ions must behave as if they have the same temperature as the electrons for purposes of calculating their flux. In other words, they have a somewhat different behavior than would be the case for neutral species. The electron temperature is much greater than the neutral gas temperature and therefore the flux of ions from a glow discharge is significantly lower than the result if one simply multipled the neutral gas flux by the fraction of particles ionized. The current density for ions at the target, J_i, is

$$J_i = \frac{P\eta}{\sqrt{2\pi M k_B T_e}},$$ 11.11

where η is the fraction of sputtering gas species ionized, P is their partial pressure in the gas, M is their mass, k_B is Boltzmann's constant, and T_e is the *electron* temperature in the glow discharge.

Equation 11.11 is a true equation but not generally terribly useful because we typically do not know the electron temperature very well. It can be measured with some effort but because it depends on many process parameters, a measured value is not very helpful in general.

More commonly one uses the Child Langmuir law. It was observed that the current density scales with the inverse square of the dark space sheath width and other parameters as:

$$J_i = C \frac{V^{3/2}}{M^{1/2} d^2},$$ 11.12

where C is a constant, d is the dark space sheath width, M is the ion mass, and V is the target voltage. This equation works well at low pressures. However, at high pressures the dark space sheath width becomes dependent upon the current density as:

$$d = \frac{A}{P} + \frac{B}{J_i^{1/2}},$$ 11.13

where A and B are constants. It is easily seen that substituting Equation 11.13 into 11.12 causes the current density to drop out of the equation. Therefore, at high pressures the Child Langmuir law is of little value. Where this transition occurs depends somewhat on the type of sputtering process used. Both equations 11.11 and 11.12 can be useful in estimating the sputtering current dependence on process parameters. In the end, however, it is generally simplest to measure the target current and assume that it is equal to the sputtering ion flux.

11.3.4.4 rf sputtering

The deposition electrode arrangement and the electrical circuit driving an rf sputtering process are shown schematically in Figure 11.30. In all cases where sputtering of a target is the desired outcome the two electrodes making contact to the glow discharge are asymmetric in size, one being relatively small and the other relatively large. The large electrode typically consists of the walls of the vacuum chamber, the substrate, and the other fixtures in the system, and is typically at ground potential for dc voltages. To allow the average dc potential on the target to rise relative to ground, the target electrode is capacitively coupled to the output transformer of the power supply. Note that the reasons why the power supply transfers its energy to the electrodes through a transformer are beyond the scope of this text. Details may be found in any book describing the theory of radio transmitters or rf power supplies.

A critical aspect of understanding rf sputtering is the relationships among the currents at the electrodes during each voltage cycle at steady state. (Refer to Figure 11.31 for an illustration.) The requirement of steady state means that the average dc voltage on each electrode is not changing and that the time scale considered during the rf cycle is long relative to the ability of electrons to respond to the potentials. The conditions of steady state for a typical rf sputtering system operating at a frequency of ~13 MHz are established very rapidly and the system may be considered to be in steady state at all points during the rf cycle.

To avoid a change in the net electron density on any electrode (a necessary condition of steady state), the electron and ion currents at each electrode must balance during the course of that cycle. Because the ions are so much more massive than the

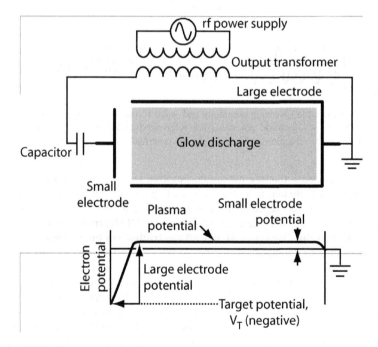

Figure 11.30: The experimental configuration and potential energy diagram for an rf sputtering process. Note that the tuning and matching network that couples the rf energy from the output transformer to the electrodes is not shown.

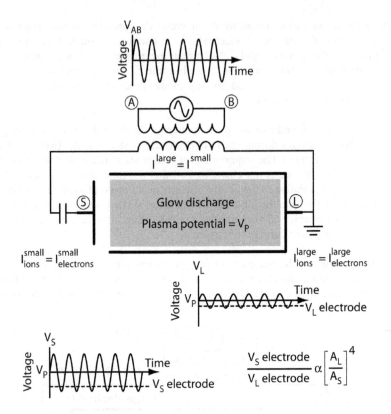

Figure 11.31: Shows the electrode potentials and wave forms at the various points in the rf sputtering circuit. The electrode potentials are referenced to the plasma potential and are negative with respect to the plasma. V_p is the plasma potential.

electrons it is necessary that the maximum negative voltage on the electrode during the cycle be much greater than the maximum positive voltage. This forces the electrode to have a net negative dc voltage relative to the glow discharge. The negative voltage is known as the electrode potential. **All electrodes have a negative voltage relative to the discharge. Hence all electrodes will be bombarded by ions having the electrode potential.**

Second, because the electrical current in the rf power circuit must be constant everywhere along the circuit path, the current at the large electrode must always balance the current at the small electrode. Note that it is the *current* that must balance, not the *current density*. Because one electrode is small, the current density must be significantly greater than at the large electrode. The current density that can be extracted from a glow discharge varies with voltage at low pressures but to make up for the difference in electrode areas the voltage ratio must be much larger than the area ratio. Specifically, the relative dc voltages on the two electrodes scales as the fourth power of the relative areas:

$$\frac{V_{small}}{V_{large}} = \left(\frac{A_{large}}{A_{small}}\right)^4.$$

11.14

Because one generally includes the entire vacuum system and fixtures in the large electrode, the area ratio is quite large. Therefore the vast majority of the applied rf voltage (peak to peak) appears on the small electrode. The first criterion of current balance at that electrode means that the dc electrode potential is also much greater for the target, meaning that ions strike the target with much larger energies than for those that strike the substrate. The electrostatic potential in an rf sputtering system is shown schematically in Figure 11.31.

An important point to note is that the area ratio in Equation 11.14 can be adjusted by controlling how the bulk of the conducting surfaces in the system are coupled to the target and substrate. Indeed, it is possible to electrically couple these surfaces to the target, which leads to sputtering of the substrate. This is useful to clean the substrate surface prior to film growth. It is also interesting to note that because the effective electrical connection between different parts of the vacuum system and the rf power supply can be regulated externally, no physical modification of the deposition system is required to switch from sputtering target or substrate or to adjust the relative electrode potentials and rf power supplies to the two surfaces.

There is much more that could be said about rf sputtering. However, as this is a book about semiconductors and as rf sputtering is little used for their processing due to the ion damage that is likely to occur, we will leave the remainder of the topic to others.

11.4 FAST PARTICLE MODIFICATION OF FILMS

A major advantage of sputtering and related techniques is the typical bombardment of the growing film with energetic particles. The sources of these particles are discussed in Section 11.3.2. Bombardment of the growing film by fast particles allows controlled (and sometimes uncontrolled) modification of the film. The effects of this bombardment include altered solid solubility of alloys, enhanced incorporation of impurities into the film, a reduced tendency of one species to segregate to the surface, modified residual film stress, improved adhesion to the substrate, smoothing of the film on a patterned surface, altered and typically increased grain size, and altered nucleation and growth. Fast particle bombardment typically enhances atomic diffusion, at least on the surface of the growing film. It is even possible in some extreme cases to observe reduced dislocation density, although an increase in dislocations is more common. Changes such as these will certainly affect the mechanical and electrical properties of the resulting film dramatically. Therefore understanding and controlling them are essential to applying them for optimization of film properties in a given application.

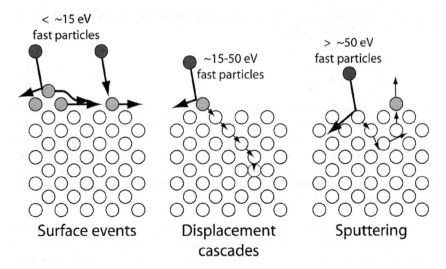

Figure 11.32: A schematic diagram of the various processes that occur as the energy of a fast particle impacting a surface increases.

The effects of fast particles on growing films can be divided roughly into energy regimes where specific effects can occur. These effects and the corresponding energies are shown in Figure 11.32 and were mentioned briefly in Section 11.3.2. Below ~15 eV no displacement of atoms in the bulk of the film can occur and insufficient energy is available upon collision to implant the impacting particle or recoil implant a surface atom into the film. (Recoil implantation is where an atom on a surface is struck and based on the transferred momentum it penetrates into a film. Direct implantation is where the incident fast particle simply penetrates the film deeply itself.) Above about 15 eV implantation and recoil implantation become possible but the energy transferred to the solid is still too low for sputtering. The latter requires energies above about 50 eV. The 15 and 50 eV energies are example values. The specific threshold energies vary with impacting particle and surface atom masses. Rather than repeating this caveat we will take these values as representative with the understanding that your threshold energies may vary. The thresholds are important controls on what types of effects an energetic particle may have on the surface and determine how much a fast particle must be accelerated or decelerated to initiate or terminate a specific effect.

At very low energies (less than ~15 eV), no bulk damage to the film can occur. Therefore, this is an important energy range to consider when designing a process to take advantage of fast particle effects, especially when depositing semiconductors. In this energy range one may enhance diffusion of atoms across the surface of the sample. A simple expectation is that a direct collision would cause a surface atom to bounce across the surface, potentially a long distance. However, this type of single-atom event is limited by several events including rapid transfer of the moving atom momentum to the surface and scattering and energy loss by the surface atom due to impact with surface steps and adatom islands. If there were enough of these events one could imagine that the average diffusivity of atoms on the surface would be significantly enhanced. However, considering the rate of surface diffusion in general it would be necessary to have a very high flux of bombarding particles for the direct impact momentum transfer process to dominate surface diffusion. Such high fluxes can be generated intentionally by intensely ionizing the gas above the substrate and biasing the substrate negatively with respect to the gas, but this type of process is not generally dominant for more conventional sputtering environments.

A second effect due to low energy impacts is decomposition of small clusters of atoms on a surface. Just as when a billiard ball strikes a group of other balls and they all fly apart, a fast particle striking a small cluster of atoms on a surface may completely or partially disrupt that cluster. Because atoms in the cluster are effectively immobile while the individual atoms resulting from the cluster disruption may be quite mobile, this has the effect of greatly increasing the average atomic diffusivity on a surface and is probably much more important than a direct collision transfer of energy. The ability of a collision to disrupt a cluster increases as the impacting particle energy increases. Unlike the general picture in Figure 11.32, there is no specific and clear threshold for this type of disruption. It simply becomes increasingly significant at higher energies and higher particle fluxes. An example of a simulation of fast particle disruption of small clusters may be found in Münger et al. [25] and Charita et al. [26].

Finally, low energy fast particles may stimulate chemical reactions or desorption of atoms from a surface. The activation energy for desorption or reaction is generally much lower than for sputtering processes. Often it only requires breaking one or two bonds and the range of resulting particle movemnt is not critical. As such, the activation of reactions may require less energy or may occur at a lower threshold energy even than for surface diffusion. Likewise a collision cascade dissipating the energy of a much faster impact may stimulate one or multiple reaction or desorption events on a surface. Therefore the distribution of the energy of a fast particle to atoms on the surface may cause events that normally depend upon thermal energy to occur more rapidly. This explains why a high flux of low-energy particles may allow a significant reduction in process temperature. See also the discussion of plasma-enhanced chemical vapor deposition in Chapter 12.

An important intermediate range of energies is from the so-called displacement cascade threshold to the sputtering threshold. Displacement cascades occur when a surface atom strikes an atom below, which strikes one below that, etc. until one atom ends up on an interstitial site. This sequence is shown in Figure 11.32. A displacement cascade is the lowest-energy process that results in a real defect in the solid. Interstitial defects move rapidly. As long as they are close enough to the surface, many interstitials may escape from the film. However, in a film under tensile residual stress the interstitial defects may migrate to regions of high stress and reduce the tension in the film. In a film with vacancies the interstitials may fill the vacancies. If nothing else, an interstitial defect returning to the surface causes diffusion. An important result of displacement cascades is that they mix surface atoms with second layer atoms, second layer atoms with third layer atoms, and so on. This can have a major effect on solubility of alloys. To initiate phase separation it is necessary to nucleate the separated phases. A phase nucleus is a cluster rich in one type of atoms separated from those of another composition. Displacement cascades, coupled with the low-energy cluster disruption effects can mix phase regions and prevent phase separation. This is the basis for increased solid solubility of phases under fast particle bombardment. In summary, fast particles in the intermediate energy range influence alloy miscibility and film stress as well as having the effects described for low-energy impacts.

The third important energy range for fast particles is when sputtering begins. Sputtering changes the composition of the films when the sputtering yield of one atom differs from the other. It changes the surface structure of the deposited film by displacing atoms downward into trenches and low points, it is more effective sputtering the top surface of the film and therefore smooths the surface by removing high points and filling low points. It also removes material from the film. Fast particles in this energy range also produce longer-range direct implantation of the fast species into the film and can recoil implant surface atoms relatively far into the film (several atomic layers) rather than displacement cascades which only move individual atoms one atomic layer downward.

Returning now to the list of effects of fast particles on growth of films, we may trace the sources of the modifications. Enhanced incorporation of impurities, reduction in surface segregation, and enhanced solid solubility all stem from mixing of the surface atomic layer with underlying layers. Segregation is driven primarily by the energy gained by swapping an atom in the surface atomic layer with a free adatom on the surface. Thus, the segregating species "floats" on the growing surface. Mixing the surface adatoms and the surface atoms as well as mixing surface layers with underlying layers opposes the ability of a segregating species to remain on the surface of the growing film. Incorporation of impurities that do not wish normally to be in the film is enhanced by the same mechanism.

Enhanced solubility in alloys results from inhibition of nucleation of second phases by disruption of small nuclei on the surface and by mixing atoms in the near surface layers. Because surface diffusion and nucleation of second phases are so much easier than equivalent processes in the bulk, preventing nucleation on the surface is often sufficient to prevent nucleation anywhere. An example of the effect of fast particles on solid solubility of the semiconductors Ge and GaSb is shown in Figure 11.33. There is almost no solubility for Ge in GaSb or vice versa under normal thermal conditions such as prevail in processes such as MBE, even though these are not in

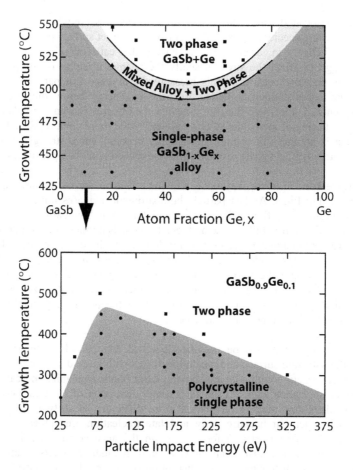

Figure 11.33: The metastable solid solubility obtained by growth of alloys of Ge and GaSb by sputtering. (Based on data in Cadien et al. [26] (top) and Shah (bottom) [27]). Square data points represent two-phase material while circular points represent single phase material. Triangles represent two-phase and single phase mixtures.

true equilibrium. Ion bombardment mixes the materials during growth and prevents second phase nucleation.

Residual stress changes such as in Figure 11.22 result from implantation, recoil implantation and forward sputtering of atoms into a film or into intergranular voids or grain boundaries. At a minimum this requires a displacement cascade. However, because the atoms thus transferred to the bulk of the layer may not be able to return rapidly to the surface or otherwise escape the film, a relatively small number of impacts may have a large effect on film stress. The change in residual stress is most commonly associated with changes in sputtering gas pressure when operating under conditions where species may be reflected from the target surface (see Figure 11.22). Even a small change in gas pressure may change the scattering of particles in the gas phase enough to affect the stress. Usually the change in stress is abrupt around a critical gas pressure because at this pressure the average fast particle leaving the target experiences enough collisions to bring its energy below the threshold for displacements of atoms into the film. Because the effect is related to implantation or forward sputtering of atoms the stress is always made more compressive. Therefore, a film that is naturally in compression simply becomes more strongly compressed while the more typical tensile residual stress is reduced upon fast-particle impact.

Fast particle bombardment may have a major effect on the morphology and grain structure of thin films. As fast particle bombardment increases initially the size of columnar grains in a film will increase and the grain boundaries will gradually densify until the film is fully dense and tensile stress is eliminated. Further increases in bombardment create damage in the film that eventually causes enough defects to generate new grain boundaries. This eliminates the columnar behavior of the films and leads to a so-called equiaxed grain structure in which the grains are roughly the same thickness as they are in diameter. Increases in grain size are usually accompanied by increases in the size of surface facets and a generally rougher surface in terms of total peak-to-valley height.

Film morphology can also be modified because fast particle bombardment can damage the substrate surface and create local sites with more dangling bonds. These are relatively reactive with adatoms, having extra bonds available, and can serve as preferential nucleation sites. At the same time, disruption of clusters increases the critical nucleus size making nucleation more difficult. These two effects do not cancel each other out because early and late nucleation are subject to different constraints. The first nuclei to form do so in the absence of sinks for adatoms. Therefore the adatom density can increase until the adatom supersaturation favors nucleation enough to counteract the disruption of clusters by fast particles. At the same time, the creation of favored nucleation sites also contributes to favoring nucleation. On balance, fast particle bombardment may accelerate the formation of the early nuclei.

Once nucleation has occurred sinks for adatoms are available. Thus disruption of later nuclei allows the associated adatoms to escape to previously formed nuclei rather than allowing secondary nucleation to occur. This is equivalent to increasing the capture radius of a nucleus (see Section 10.6). At the same time the effective increase in critical nucleus size due to disruption of small clusters also works against cluster formation. In the presence of existing stable nuclei the formation of preferential nucleation sites is important because not enough adatoms are present to allow nucleation under conditions of fast particle bombardment.

The combination of accelerated primary nucleation and suppressed secondary nucleation described here has the general effect of increasing the size of crystals on a surface and hence to increasing grain size in a thin film.

Often fast particle bombardment will also change the preferred orientation of the film. When a thin film nucleates on a substrate with which it has no epitaxial relationship the most common arrangement is that the grains will have their close-packed planes perpendicular to the surface on which the grain nucleated. This is simply a result of the atoms in the nucleus trying to be as close together as possible as they form the first layer of the film. However, when fast particles bombard this nucleus they are much more likely to disrupt it if the layer is close packed because, as with billiard balls, energy is transferred much more efficiently among the atoms if they are closer together. Therefore, there is a tendency to favor less close packed planes for surfaces in the presence of fast particles.

One of the most critical aspects of many thin film crystal growth processes is adhesion of one layer of a film to another. A classic case is architectural window glass coatings that are responsible for the blue-green, gold or other colors of glass often present on large buildings. Such a coating must withstand heating and cooling cycles, corrosion, and other consequences of being out in the environment for decades with no failures large enough to be visible, which is not very large. Adhesion is most strongly degraded by two factors, stress in the coating and contamination of the substrate surface. Stress in the film places a force on the film-substrate interface that can cause delamination. Likewise, contamination at the interface can produce a local weak spot where the coating can fail.

Fast particle bombardment can improve adhesion in a number of ways. First, it can be used to reduce residual stress in the film by tuning the deposition parameters, as in Figure 11.22. Second, presputtering the substrate with fast particles before deposition of the film removes contamination and leaves a nearly pristine surface. Sputtering may also break bonds on the surface and give the film more to hang on to. Third, recoil implantation may mix the film and substrate constituents on a small distance scale, which grades the interface chemically. Chemically graded interfaces tend to enhance adhesion because the interface is chemically more similar to both surrounding materials than an abrupt junction would be. Fourth, fast particle sputtering may roughen the substrate surface due to local variations in sputtering rate. This is normally the result of carbon surface contamination that does not sputter well. Roughness may permit mechanical interlocking of the film and substrate and increases the contact area between the film and substrate, both of which enhance adhesion.

Finally, fast particle bombardment may alter the defect density in the material. For semiconductors this is generally a bad thing because the most common effect is to introduce point defects into the growing film. For example, interstitial defects may be created by displacement cascades. High energy impacts may lead to formation of vacancies and antisite defects. Higher fluxes of fast particles may cause enough damage that point defects may coalesce to form dislocation loops and intrinsic or extrinsic stacking faults. In rare cases the increased presence of vacancies can allow dislocations to climb and escape from the film but this is an exception. For semiconductors the formation of defects is generally so unacceptable as to prevent the use of sputter deposition.

11.5 APPLICATION

The primary applications of sputtering are in the deposition of metals, although it has been applied to semiconductors and semiconductor alloys (see for example, Figure 11.33). We will consider an example of sputter deposition of a semiconductor and of a metal alloy here. The two processes are tied together because they involved a project in my laboratory joining the two materials. In particular the deposition of the metal alloy is described because it shows so many of the aspects of fast-particle modification of film properties.

A story from the author's laboratory...

To begin this example we need to set the stage. My group was depositing a semiconductor, $CuInSe_2$ by a sputtering and evaporation technique described below from which we were making solar cells. The phase diagram for $CuInSe_2$ was presented in Chapter 4 as an example. When the material is Cu-rich the solar cells do not function due to the formation of conductive Cu_2Se second phases, which short-circuit the junction. However, we found that when the film was In-rich we suffered delamination from the substrate and our devices also failed. We had an adhesion problem! The surface of the substrate was coated with a layer of Mo metal as a back contact. This is the standard back contact material for these solar cells and we did not know of a better material.

We surmised that if we could mix some Cu with the Mo back contact that we might be able to get a modest amount of Cu diffusion into the $CuInSe_2$ during deposition that would be sufficient to keep the very back side of the material Cu-rich and permit good adhesion while allowing the surface to be In-rich and capable of forming a good device.

This section briefly describes the deposition of the $CuInSe_2$ and then how formation of a Cu-Mo alloy greatly improved adhesion of the film to the substrate.

We are able to sputter-deposit the semiconductor $CuInSe_2$ and fabricate good diodes from it because it is very accepting of point defects, as shown by its phase diagram in Chapter 4. One might expect a large number of fast particles to bombard the growing film because we are working with Se, a relatively electronegative species that should form negative ions at the sputtering target surfaces. Nonetheless, we are able to produce good epitaxial films with the highest recorded majority carrier (hole)

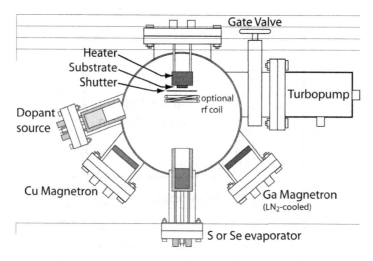

Figure 11.34: A schematic diagram of a sputtering and evaporation system used to produce epitaxial layers of CuInSe$_2$.

mobilities to date. The process we use is to sputter the metals (Cu and In) from two targets and we evaporate Se at the same time. Because Se has a high vapor pressure any excess Se evaporates from the growing film surface. Thus, we supply the Se in excess. The growth rate is determined by the metal arrival rates and the Cu-to-In ratio is determined by the relative fluxes of In and Cu. The deposition system is shown schematically in Figure 11.34.

This growth process has the advantage that the metal fluxes are controlled roughly linearly by control of the target current. The sputtering targets last for months without replacement. Se is very difficult to sputter because of its high vapor pressure. This leads to arcing because any local heating of a Se or other high vapor pressure target material evaporates extra gas there, consequently increases the glow discharge density over the hot spot, and this, in turn, increases sputtering of the surface and heating; quickly, the system arcs. Hence evaporation is a more straightforward method for depositing Se. The low temperature required also makes the evaporation system relatively reliable.

Although the resulting material appears to be of high quality we are obliged to deposit our films at higher temperatures than materials deposited by other physical or chemical vapor deposition methods. One possible explanation is that negative Se ions are damaging the film and that higher temperatures are required to reduce this damage. This example and the example of sputtered alloys in Figure 11.33 show that sputtering can be used to produce semiconductors.

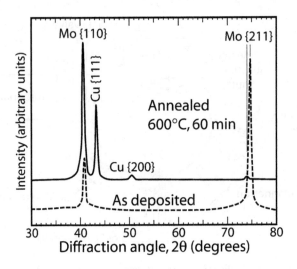

Figure 11.35: Shows the x-ray diffraction spectra for an as-deposited nanograined Cu-Mo alloy film and the separation into Cu and Mo phases resulting from a 60 minute anneal at 600°C. Note the peak shift visible in the {211} peak for the as-deposited film compared to the alloy film. The {110} peak also shifted but the scale of the shift is too small to be visible on these axes. Note also the absence of either the Cu {111} or the Cu {200} peaks in the as-deposited film. The combination of shifted Mo bcc peaks and the absence of Cu peaks shows that a single phase alloy was created. The texture of the film can be observed in the much greater size of the {211} peak in the as deposited film compared to the normally-stronger {110} peak. After annealing the behavior is reversed. [29,30]

With that introduction we now turn our attention to the Cu-Mo back contact and the adhesion problem. To put Cu into our Mo back contacts we soldered Cu strips onto a Mo sputtering target such that both the Cu and Mo would be sputtered together. Our original expectation was that we would produce a fine-grained Mo film with Cu in the Mo grain boundaries. Our expectation was based on the equilibrium phase diagram for Cu and Mo, which shows no significant solid solubility at all. Our deposited films were quite different – single phase nanocrystalline alloys of Cu dissolved in bcc-structure Mo. [29,30] The x-ray diffraction pattern for a film with the highest Cu content studied, $Cu_{0.3}Mo_{0.7}$, is shown in Figure 11.35.

For a bcc structure material the close packed planes are {110} type so we would have expected to observe this texture in the film. However, the as-deposited layers showed a predominantly {211} texture (Figure 11.35, dashed curve). The diffraction peaks were shifted with respect to what one would expect for pure Mo to higher angle, corresponding to a smaller lattice constant. Cu being a smaller atom than Mo, it was not surprising that dissolving Cu in Mo reduced the lattice constant.

The films were deposited in a magnetron sputtering system with a distance between target and substrate of ~5 cm. The sputtering pressure was ~0.53 Pa, corresponding to a thermal mean-free path (Equation 10.2) of 1.3 cm, roughly ¼ of the substrate-to-target distance. As we saw in Figure 11.22, fast Ar particles generated by reflection from a sputtering target can affect film properties significantly over 10 thermal mean free paths from the target. That was certainly the case in my laboratory where we were able to dissolve up to 30% Cu in Mo in spite of the absence of equilibrium solubility. The solution was maintained to temperatures in excess of 500°C.

We prepared solar cells by growth of $CuInSe_2$ on the Cu-Mo alloys and observed large increases in adhesion between the semiconductor film and the metal alloy. In tape tests in which a piece of common office tape was applied to the semiconductor surface and then peeled off virtually all of the semiconductor was removed from a pure Mo metal layer by the tape, while virtually all of the semiconductor remained in place in the base of the metastable alloy contact. Investingating the mechanism showed that the residual stress in the film was almost completely relieved as one heated the sample to deposit the semiconductor. Our solar cells were correspondingly improved by the reduction in the density of defects due to adhesion failures.

The above example shows the dramatic effect of fast particle bombardment on a growing film and of stress on adhesion.

11.6 SUMMARY POINTS

- Evaporation is a process of heating a solid or liquid until its vapor pressure is sufficiently high to produce a gas flux of that species.
- Molecular beam epitaxy is an evaporation process carried out under ultrahigh vacuum to keep contamination of the growing film to a minimum.
- Evaporation processes can be carried out at very high rates under the right conditions and with the right evaporant.
- Evaporation sources include open boats, wire coils, effusion cells, and electron beam evaporators. Sources also used in evaporation instruments include gas sources in conjunction with a gas cracking system.
- Evaporation sources are typically used for evaporant temperatures below ~1100°C while electron beam sources are used for temperatures above ~700°C.
- The flux from an effusion cell depends upon the shape of the crucible and how full it is. The flux pattern is determined geometrically because atoms in the high vacuums used in evaporation systems travel without making collisions in the gas phase. Therefore, only collisions with the crucible sides are important in determining the flux pattern.
- The temperature of effusion cells can be controlled to within better than 1°C, permitting excellent control of the flux of atoms leaving the cell.
- Cracking effusion sources break down multimer clusters leaving an effusion cell such as As_4 or S_8 into smaller clusters such as As_2 or S_2.
- The vapor pressure of a material is the equilibrium pressure of gas above a solid or liquid of the same material at a given temperature. It does not depend upon melting point. The boiling point of the material is where the vapor pressure equals the local atmospheric pressure.
- Vapor pressure changes approximately exponentially with temperature. The energy in the exponent is roughly the latent heat of vaporization.
- Sputtering is a process in which a fast particle, typically an ion, strikes a solid or liquid surface and transfers enough energy and momentum to an atom within that solid that that atom escapes the surface, becoming a gas.
- Sputtering requires the turning of the inward-moving kinetic energy and momentum of the incident particle into an outward directed energy and momentum of an atom escaping the surface.
- Sputtered atoms are atoms that were at/near the surface of the target and received energy and momentum from atoms below them.
- The sputtering probability depends upon the surface binding potential holding an atom in the solid, the mass of target atoms, the mass of the sputtering ions, and the nuclear stopping power of the target.
- There is a threshold energy for sputtering below which not enough energy and momentum are transferred to surface atoms to allow them to escape into the gas. The threshold depends upon the surface binding potential, the mass of the target, and the mass of the substrate.

- The atoms in the sputtered flux reflect the composition of the target surface multiplied by the sputtering yield of that species.
- Sputtering probability increases as the angle of incidence with respect to the surface normal increases up to an angle of ~70° with respect to the normal. Beyond that the probability decreases rapidly.
- The distribution of sputtered flux depends upon the angle at which the sputtered atom emerges from the solid, the energy of the incident particle, and the angle at which the incident particle strikes the solid.
- While most energy in a sputtering collision cascade is dissipated as heat, some energetic particles may result.
- Other sources of energetic particles include reflection of sputtering gas species from the target (if the target atoms are more massive), by formation of negative ions at the target surface, and by formation of an accelerating potential between the glow discharge and the substrate.
- Fast particles striking the substrate cause surface atomic diffusion and disrupt clusters even at energies above a few volts. Higher energy particles may cause displacement cascades, while still higher energy particles may begin to sputter the deposited film.
- The effects of energetic particle bombardment include altered solid solubility of alloys, enhanced incorporation of impurities into the film, a reduced tendency of one species to segregate to the surface of the growing film, modified residual stress in the film, improved adhesion to the substrate, smoothing of the film on a patterned surface, altered and typically increased grain size, and altered nucleation and growth of the film.
- Sputtering systems may use rf or dc power to energize the glow discharge and may have one of several target geometries, usually chosen based on the geometry of the surface to be coated.
- In magnetron sputtering magnets are placed behind the target to enhance the glow discharge ionization locally.
- A glow discharge is an ionized gas with a moderate excess of ions relative to electrons such that it has a net positive charge. This tends to reduce loss of light electrons and push out heavy ions such that the net loss of the two charged species is balanced.
- Glow discharges are maintained by secondary electrons emitted by the cathode.
- In normal sputtering processes the electron emission is by Auger type transfer of energy from one electron to another causing ionization.
- In rf sputtering the glow discharge is the most positive part of the electronic circuit causing ion bombardment of all surfaces.
- Rf sputtering systems must balance electron and ion current at all electrodes and requires the same current at the large and small electrodes. Thus the small electrode must have a much higher current density and therefore a much higher voltage.
- Deposition rate monitors include devices such as quartz crystal monitors and RHEED tools that measure the rate of deposition of a film on a surface and

devices such as EIES and mass spectrometry that measure the partial pressure of a species or all gas molecules in the gas phase.

- RHEED can measure film stress state, surface reconstruction, surface roughness, can count monolayers deposited, surface miscut angle, and other phenomena. For deposition monitoring one sets up the diffraction in the off-Bragg condition such that alternate monolayers on the surface interfere destructively.

11.7 HOMEWORK PROBLEMS

As a member of the MegaJoule Industries research and development team you have been assigned to design a molecular beam epitaxy process for growth of GaAs. Your MBE machine has the following geometry:

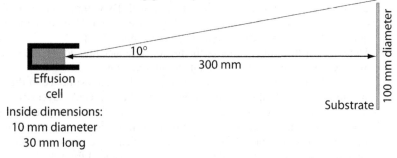

The flux of Ga sticking to the surface of the substrate is 5×10^{14} cm^{-2} s^{-1} for the desired growth rate of 1 monolayer s^{-1}. The sticking coefficient for Ga is 0.5.

1. Assuming that the effusion cell is full of Ga to the brim:

   a. What is the function describing the angular spread of the evaporating Ga flux?

   b. Estimate the fraction of the Ga leaving the effusion cell that hits the substrate. (Hint: integrate the flux distribution from $-90°$ to $+90°$ and compare it with the distribution integrated from $-10°$ to $+10°$.)

   c. Based on the flux of Ga required for growth, the sticking coefficient, the ratio of the area at the effusion cell face to the area of the substrate, and the result in part (b), calculate the flux of Ga required at the effusion cell to achieve the required flux at the substrate.

   d. With reference to Equation 10.4, calculate the pressure required in the effusion cell to achieve this flux. Note that you will need to estimate the temperature, recalculate a pressure, and correct your temperature estimate as necessary since the flux depends upon both pressure and temperature.

   e. Give the temperature for the effusion cell based on the estimate made in part d.

   f. As the effusion cell material is used up, does the fraction of Ga that hits the substrate relative to the amount that hits the walls of the system increase or decrease? Briefly explain why you know this.

   g. If you can tolerate up to a 20% variation in film thickness across the substrate (ie: net growth rate) and if you want to use all of the Ga in the effusion cell before reaching this point, what is the maximum depth of the cell you can use?

   h. For the effusion cell dimensions given above, what is the fraction of the Ga you can expect to use before having to refill the cell?

2. Having established adequate growth conditions for GaAs (and AlAs), you begin growth of GaAs/AlAs superlattices [alternating GaAs and AlAs layers]. Briefly explain the following observations:

    a. Initially the AlAs grows with a very rough surface, although after the GaAs is completely covered by AlAs, the layer smooths out. This can be enhanced if growth is halted and the surface is annealed. By contrast, growth of the GaAs layer is smooth from the start. [Hint: GaAs and AlAs have the same lattice constant so there is no epitaxial lattice mismatch.]

    b. The AlAs surface roughening problem can be reduced by dosing the surface of the growing layer with ~1 monolayer of Te atoms.

    c. If you dose the surface of AlAs with Te, you find almost no Te in the film and all of the Te remains on the surface when the next GaAs layer occurs.

    d. If you grow a very thick layer of AlAs on the GaAs substrate you find problems with adhesion failures when you cool from the growth temperature. Why do you think this happens? (Hint: observing the substrate/film combination, you find that the wafer is convex when viewed from the film side – it bends toward the substrate side at the edges.)

    e. How would you measure the progress of smoothing of the AlAs surface referred to in Part (a), above. Explain briefly how you must set up the experiment used to provide a clear measure of the surface roughness.

As a recent hire on the process development line of MegaJoule Industries you have been assigned to design processing conditions for, to operate and to maintain a sputtering machine. The process design team meets and has been discussing how to handle certain steps in the process. Of particular interest to the team is the following problem. You need to coat a substrate with an Al-Cu-In alloy (5% Cu, 5% In). The In was added to control the microstructure and etching properties as well as the annealing conditions required to obtain the ideal electromigration resistance. This works well except for depositions on this feature on the chips:

Starting Substrate

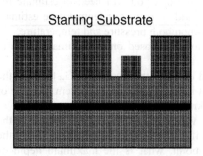

Typical Deposition Result

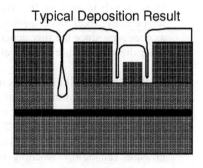

The substrates are eight inch diameter round wafers.

a) You have been instructed to develop a sputtering process for forming the alloy. Recommend a process. Include in your answer with a brief justification:

   i)   the type of sputtering process (reactive, inert gas, magnetron, non-magnetron, dc, rf, single target, multiple target,...)
   ii) the suggested target geometry (planar, cylincrical,...)
   iii) the composition of the target(s)
   iv) Explain how the composition of the resulting film is controlled for the proposed process.

b) You are asked to explain the cause and solution of the "keyhole" defect on the left side of the figure.

   i)   Explain why this forms.
   ii) Propose a possible solution to the problem.

c) You observe that the microstructure of the metal coating along the sides of the holes is very porous and is under tensile stress while the coating elsewhere is in compression and has a very dense microstructure.

   i)   Explain why the microstructures are different in the two areas.
   ii) Propose a method for improving the microstructure on the side walls.

d) You are asked to deposit a thin TiN layer prior to the metal deposition as a reaction barrier.

   i)   What would you use as a target and as sputtering gas?
   ii) How would you control instabilities in the target voltage/sputtering rate?

11.8 SUGGESTED READINGS & REFERENCES

Suggested Readings:

Glocker, David A. and Shah, S. Ismat, *Handbook of Thin Film Process Technology*. Bristol, Institute of Physics Publishing: 2002 (two volumes).

Herman, Marian A. and Sitter, Helmut, *Molecular Beam Epitaxy: Fundamentals and Current Status*. Berlin: Springer, 1996.

Kasper, Erich and Bean, John C., *Silicon Molecular Beam Epitaxy*. Boca Raton, Florida, CRC Press: 1988 (two volumes).

Matthews, J.W., *Epitaxial Growth*. New York, Academic Press: 1975 (two volumes).

Tu, King-Ning; Mayer, James W.; and Feldman, Leonard C.; *Electronic Thin Film Science for Electrical Engineers and Materials Scientists*. New York: Macmillan, 1992.

Vossen, John L. and Kern, Werner, *Thin Film Processes* volumes I and II. New York, Academic: 1978.

References:

[1]   Luscher, P.E. and Collins, D.M.; "Design considerations for molecular beam epitaxy systems", Prog. Crystal Growth and Characterization, 1979; 2: 15-32.

[2]   Dayton, B.B., "Gas flow patterns at entrance and exit of cylindrical tubes." *In 1956 National Symposium on Vacuum Technology Trans.*, ed. by E.S. Perry and J.H. Durant, Oxford: Pergamon; 1957: 5.

[3]   Herman, Marian A. and Sitter, Helmut, *Molecular Beam Epitaxy: Fundamentals and Current Status*. Berlin: Springer, 1996.

[4]   Yamashita, T.; Tomita, T.; and Sakurai, T.; Calculations of molecular beam flux from liquid source." *Jpn. J. Appl. Phys*, 1987; 26: 1192-1193.

[5]   Hoing, R.E., "Vapor pressure data for the more common elements." *RCA Review*, 1957; 18: 195-204.

[6]   Hoing, R.E. and Hook, H.O., "Vapor pressure data for some common gases." *RCA Review*, 1960; 21: 360-368.

[7]   Hoing, R.E., "Vapor pressure data for the solid and liquid elements." *RCA Review*, 1962; 23: 567-586.

[8]   Hoing, R.E. and Kramer, D.A., "Vapor pressure data for the solid and liquid elements." *RCA Review*, 1969; 30: 285-305.

[9]   Gurdal, O.; Desjardins, P.; Carlsson, J.R.A.; Taylor, N.; Radamson, H.H.; Sundgren, J.-E.; and Greene, J.E.; "Low-temperature growth and critical epitaxial thicknesses of fully strained metastable $Ge_{1-x}Sn_x$ (x≤0.26) alloys on Ge(001) 2x1." *J. Appl. Phys*, 1998; 83: 162-170.

[10] Sigmund, Peter, "Theory of sputtering. I. Sputtering yield of amorphous and polycrystalline targets." *Phys. Rev.*, 1969; 184: 383-416.

[11] Lindhard, J. "Energy dissipation by ions in the keV region." *Phys. Rev.*, 1961; 124: 128-130 and references therein, especially Lindhard, J.; Nielsen, V.; Scharff, M. *Kungl. Dan. Vid. Selsk. Mat. Phys. Medd.*, 1958; 36: 10.

[12] Matsunami, N.; Yamamura, Y.; Itakawa, N.; Itoh, Y.; Kazumata, S.; Mayagawa, K.; Morita, K.; and Shimizu, R. *Rad Eff. Lett.*, 1980; 57: 15.

[13] Bohdansky, J., "A universal relation for the sputtering yield of monatomic solids at normal ion incidence." *Nucl. Inst. and Meth. in Phys. Res. B*, 1984; 2: 587-591.

[14] Yamamura, Y. and Mizuo, Y., *J. Nucl. Mater.*, 1984; 128-9: 559.

[15] The methods used in the TRIM computer code are described in Ziegler, J.F.; Biersack, J.P.; and Littmark, U., *The stopping and range of ions in solids*. New York, Pergamon: 1985, latest edition: 2003.

[16] Oechsner, H., "Sputtering – a review of some recent experimental and theoretical aspects." *Applied Physics A: Materials Science and Processing*, 1975; 3: 185-196.

[17] Yamamura, Y. and Itoh, N. in *Ion Beam Assisted Film Growth*, ed. by T. Itoh. Amsterdam, Elsevier: 1989, Chapter 4.

[18] Grapperhaus, M.J.; Krivokapic, Z.; and Kushner, M.J., "Design issues in ionized metal physical vapor deposition of copper." *J. Appl. Phys.*, 1998; 83: 35-43.

[19] Heberlein, T.; Krautheim, G.; and Wuttke, W., *Vacuum*, 1991; 42: 47.

[20] Ray, M.A. and Greene, J.E., unpublished.

[21] Kornelsen, E.V., "The ionic entrapment and thermal desorption of inert gases in tungsten for kinetic energies of 40 eV to 5keV." *Can. J. Phys.*, 1964; 42: 364-381.

[22] Thornton, J.A. and Hoffman, D.W., "Internal stresses in amorphous silicon films deposited by cylindrical magnetron sputtering using Ne, Ar, Kr, Xe, and Ar+H." *J. Vac. Sci. Technol.*, 1981; 18: 203-207.

[23] Hoffman, D.W. and Thornton, J.A., "Compressive stress and inert gas in Mo films sputtered from a cylindrical-post magnetron with Ne, Ar, Kr, and Xe." *J. Vac. Sci. Technol.*, 1980; 17: 380-383.

[24] Thornton, John A. and Penfold, Alan S., "Cylindrical Magentron Sputtering" in Vossen, John L. and Kern, Werner, *Thin Film Processes* volumes I and II. New York, Academic: 1978; pp. 75-113.

[25] Münger, E.P.; Chirita, V.; Greene, J.E.; Sundgren J.-E. "Adatom-induced diffusion of two-dimensional close-packed Pt_7 clusters on Pt(111)." *Surf. Sci.*, 1996; 355: L325-330.

[26] Chirita, V.; Münger, E.P.; Sundgren J.-E.; Greene, J.E., "Enhanced cluster mobilities on Pt(111) during film growth from the vapor phase." *Appl. Phys. Lett.*, 1998; 72: 127-129.

[27] Cadien, K.C., Elthouky, A.H., and Greene, J.E., "Growth of single-crystal metastable semiconducting $(GaSb)_{1-x}Ge_x$ films." *Appl. Phys. Lett.*, 1981; 38: 773-775.

[28] Shah, Syed Ismat Ullah, "Crystal growth, atomic ordering, phase transitions in pseudobinary constituents of the metastable quaternary $(GaSb)_{1-x}(Ge_{2(1-y)}Sn_{2y})_x$." Thesis, Ph.D. – University of Illinois at Urbana-Champaign, 1986.

[29] L. Chung Yang and A. Rockett, "Cu-Mo Contacts to $CuInSe_2$ For Improved Adhesion in Photovoltaic Devices," J. Appl. Phys **75**(2), 1185 (1994).

[30] G. Ramanath, H. Z. Xiao, L. C. Yang, A. Rockett, and L. H. Allen, "Evolution of Microstructure in Nanocrystalline Mo-Cu Thin Films During Thermal Annealing", J. Appl. Phys. **78**(4), 2435 (1995).

Chapter 12

CHEMICAL VAPOR DEPOSITION

Chemical vapor deposition (CVD) refers to a class of methods in which a solid is grown by reaction of gaseous source materials and yielding a product effluent gas. There are a number of variants on the process based on the pressure range at which it is conducted, the type of reactants, and whether some method to activate the reaction is used. CVD can also be conducted in an atomic layer deposition (ALD) mode in which single layers of atoms are produced one at a time. CVD has a number of advantages over physical vapor deposition. For example, the reaction can often be arranged to be selective more easily, depositing material only in certain regions of the substrate rather than covering it with a blanket layer. CVD is generally more conformal than physical vapor deposition, meaning that it covers a rough surface relatively uniformly, tracking the morphology rather than resulting in thin, low-quality coatings on vertical walls of the substrate, as is the case for physical vapor deposition methods. Other advantages include that CVD uses source materials that flow into the process chamber from external reservoirs that can be refilled without contamination of the growth environment, it does not require very high vacuum levels, it can generally process substrates in larger batches than evaporation, and is more forgiving in terms of its tolerance for precision in the process conditions. Counterbalancing these advantages, CVD source materials are generally highly toxic or flammable, requiring great care in the design and operation of a CVD process system. CVD also frequently requires high temperatures. For microelectronics manufacturing the benefits generally outweigh the problems. Thus, most device makers use CVD when possible rather than, for example, MBE.

12.1 OVERVIEW

Chemical vapor deposition (CVD) occurs when molecules from a moderate density gas condense on a surface and react to form a solid. The reaction also creates product molecules that are released into the gas phase. "Moderate" gas density usually means $> \sim 0.1$ Pa. The general features of the process are shown schematically in Figure 12.1. A (simplified) typical CVD reaction for formation of GaAs by reaction of arsine (AsH_3) with triethyl gallium [$Ga(C_2H_5)_3$] might be:

$$Ga(C_2H_5)_3\downarrow + AsH_3\downarrow \rightarrow GaAs + 3C_2H_6\uparrow \qquad 12.1$$

Where $\downarrow$ indicates a gas that adsorbs on the substrate surface before reaction and $\uparrow$ indicates a product molecule that desorbs after reaction. Portions of the reaction may occur on the gas phase as well as on the surface as discussed below (and which is the case for the example above.) A detailed analysis of the kinetics of the above reaction including both gas-phase and surface reactions and all kinetic rate parameters is given, for example, in Ingle et al. [1] The reaction may be limited by any of the required steps including: supply of reactant from the gas phase, precursor reactions in the gas phase, reactant adsorption and desorption rates, surface reaction rates, and product desorption rates. Some of these we have discussed sufficiently in Chapter 10 but others require more attention. As can be seen in Figure 12.1, the extraction of part but not all of the reactant from a moving gas stream makes CVD inherently inefficient as, in some cases, much of the reactant simply passes through the reactor and into the pumps without contributing to growth.

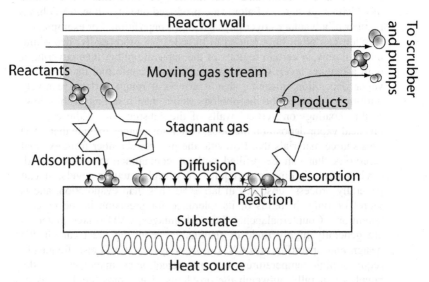

Figure 12.1: A schematic of the events in a CVD growth process.

The process rate limitations may be divided into two groups, gas phase limitations and surface reaction rate limitations. The significance of this division is that **gas phase transport limitations are controlled by reactant concentration in the gas and the viscosity, turbulence, flow rate, and other properties of the gas, while surface reaction rate limitations are controlled by surface temperature and the other processes typical of physical vapor deposition growth,** described in Chapter 10, and of reaction kinetics, described in Chapter 4. In general, operating parameters of the deposition system that control gas phase transport have little impact on surface reaction rate while variables affecting the surface reaction rate have only a small impact on gas phase transport. For example, the substrate temperature generally affects surface reaction rate exponentially while heating of the gas increases the transport rate only modestly. The overall rate of a CVD reaction, in which the details of the gas phase and surface rate limitations are grouped into two parameters, can be written as:

$$R = \frac{C_g}{N_s} \frac{k_f h_g}{k_f + h_g},$$

12.2

where C_g is the concentration of the most critical reactant (the one which is most rate limiting) in the gas phase, k_f is the forward reaction rate constant for the surface reaction, h_g is the gas phase transport rate, and N_s is the concentration of atoms in the solid supplied by the critical reactant species. In general k_f is, as in Equation 4.15:

$$k_f = k_0 \, e^{-\Delta H / k_B T}.$$

12.3

This assumes no reverse reaction, which is the case if the reaction products are rapidly removed from the surface and swept away in the gas phase. Surface coverage dependent terms may also enter into this equation if the reaction rate is limited by available adsorption sites. If a reverse reaction can occur one must also consider the law of mass action behavior as in Equation 4.14, while if there are multiple critical steps one may need to account for this in the reaction rate, as in Equation 4.16. When $k_f \ll h_g$ the rate is surface reaction rate limited, while $h_g \ll k_f$ implies a gas phase transport limitation.

A CVD process that is reaction rate limited has no difficulty supplying reactants to the surface or removing the products once they desorb. The concentration of reactants and products in the gas immediately above the substrate is nearly the same as the concentrations far from the surface. Parameters such as gas pressure have little effect on the reaction rate, other than through the saturation surface coverage of reactants, which may modify the reaction rate as in Equation 12.2, above. Reaction rate limitation produces no significant concentration gradient of reactants or products in the gas phase. This means that many substrates may be placed side by side with only moderate spaces between them, allowing batch processing. This also results in good surface uniformity in the grown layer because the reactant distribution will be uniform across the substrate. Only process variables affecting the reaction rate need be constant across the substrate surface, for example temperature. In typical CVD process implementations, see next section, it is not difficult to obtain excellent temperature uniformity. The major problems with reaction rate limited CVD are that it is relatively slow and it makes inefficient use of the reactant gas.

CVD reactions that are gas phase transport limited depend upon diffusion of species in the gas; which, in turn, depends upon the concentration gradient. The relative concentrations of species in the gas phase for transport limited vs. reaction limited behaviors are given schematically in Figure 12.2. One may approximate the gas phase transport rate as:

$$h_g = D_g/\delta(x),$$
12.4

where D_g is the gas phase diffusivity of the reactant and $\delta(x)$ is the thickness of the boundary layer, discussed in Section 12.3, below. The gas phase diffusivity is proportional to the ratio of gas viscosity, η, to gas density, ρ, and varies moderately with temperature, typically as T^ξ, where $0.5 < \xi < 1.75$. We will return to the implications of gas-phase transport limited reactions below. For now, it is sufficient to note that while transport-limited CVD reactions are more sensitive to the design of the CVD reactor they are also much faster than surface limited reactions.

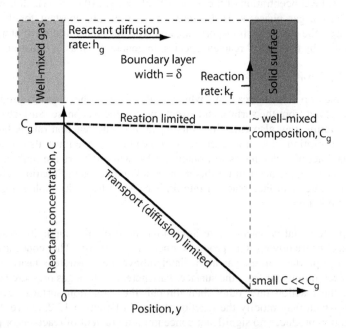

Figure 12.2: A schematic of the transport geometry and reactant concentration between the well mixed gas phase outside of the boundary layer and the substrate surface where growth reactions occur. The behaviors are simplified and approximated as linear. In reality the behaviors are more complex but the result is the same.

The net result of the competition between reaction rate and gas transport rate in Equation 12.2 as a function of temperature is shown schematically in Figure 12.3. At low temperatures the exponential dependence of reaction rate on temperature reduces the reaction rate below the gas transport rate sufficiently that the process is surface reaction rate limited. At higher temperatures diffusion is the dominant rate-limiting step.

The actual CVD reaction process can be significantly more complex than the simplified view presented above. In general, to maintain a constant pressure in the reactor at various reactant flow rates, a significant part of the reactor gas is H_2. Thus, the H_2 along with the temperature allows independent control of the reaction rate, gas density, and reactor pressure. The H_2 also enhances the quality of the deposited material by reducing the importance of atomic vacancies in the resulting semiconductor.

H_2 is often also a reactant. For example, using metal organic sources such as trimethyl gallium [TMG], $Ga(CH_3)_3$ or TEG reacting with arsine (AsH_3) generally involves decomposition steps *in the gas phase* as the molecules diffuse through the stagnant layer (see Figure 12.4). [1,2] Here H replaces the organic ligand attached to the Ga. The result of a reaction sequence in the gas phase is that the molecule that adsorbs on and reacts with the substrate may be quite different from that originally introduced into the reactor. Thus, any model of the process needs to account for this type of gas phase reaction. Complex gas phase processes may be favored if they eliminate more difficult adsorption and reaction events on the surface.

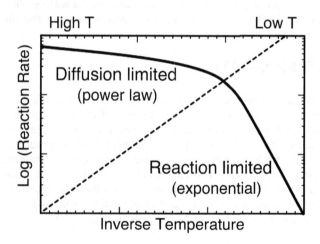

Figure 12.3: The net rate for CVD processes as a function of inverse temperature (solid curve) and the division between diffusion and reaction rate limitation for a variety of reaction rates (dashed line) such as would be varied by concentration of reactant.

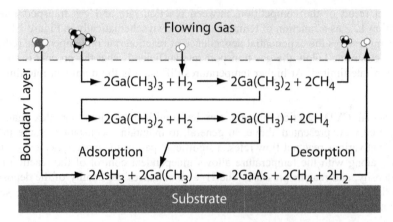

Figure 12.4: A schematic of the pyrolysis processes taking place across the stagnant layer in the gas phase of a cold-wall metal-organic CVD reactor. In a hot-wall reactor the gas may pyrolyze outside of the stagnant layer. Other species present may include completely pyrolyzed TMG resulting in elemental Ga in the gas and pyrolysis products including hydrogen such as $Ga(CH_3)H_2$. Figure based on data in Ingle et al. [1]

It is important to note that the actual reaction forming the compound semiconductor in the case shown in Figure 12.4 still occurs at the substrate, not in the gas (AsH_3 does not react in the gas). Therefore, the process does not produce GaAs dust in the reactor. This is essential to the design of the reaction. **Repeating: the precursors may decompose in the gas phase but the reaction forming the final material should occur on the substrate.**

12.2 CVD APPARATUS

The different reaction rate behaviors introduced briefly above affect the options for reactor design. It is of interest to consider some of these designs and the related systems associated with a CVD machine. A typical apparatus consists of a reactor tube, a heating system, a mechanism for inserting and removing substrates from the reactor, plumbing for introducing reactant gases from storage containers, a pumping system to remove products and unused reactants, a "scrubber" that renders unreacted gases inert and prevents release of toxic species into the atmosphere, a set of safety enclosures and systems, and a control unit incorporating monitors for detecting toxic gases. Such an apparatus is shown schematically in Figure 12.5.

It is an unfortunate feature of most reactants used in CVD processes that they are either highly toxic or pyrophoric (they ignite spontaneously and burn on contact with air) or both. Therefore the safety enclosure and toxic gas detectors are very important. The gases are generally stored in protected and well ventilated locations inside specially-designed safety enclosures and the storage containers generally include mechanisms restricting the rate at which gas can escape even if the valves on

Safety enclosure

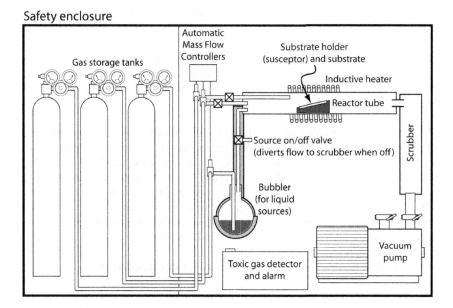

Figure 12.5: A schematic diagram of some of the major elements of a typical low-pressure chemical vapor deposition system. Note that the entire process is typically contained within a safety enclosure because many of the reactants as process gases are either flammable or toxic. A toxic gas detector scans the interior of the enclosure, especially around the gas piping, for leaks. The scrubber reacts the source materials that are left over after passing through the reactor tube to render them harmless. Typically the valves for all sources are operated electronically so that the user does not need to open the safety enclosure except to load substrates on the susceptor. The system shown here includes a single substrate on a tilted susceptor. Batch processes are similar except in the arrangement of substrates in the reactor tube. Other reactor tube designs are also used. More specific reaction examples are given in Section 12.8.

the containers are opened wide or broken off. The safety considerations often make CVD a process that is not used casually or for small-scale low-budget manufacturing. However, for larger scale manufacturing or in laboratory environments where safety systems can be installed, CVD can be used easily and has advantages that make the additional expense and complexity of the safety systems acceptable. With appropriate precautions and apparatus, CVD can be used safely.

Reactants used in CVD can be solid, gaseous, or liquid. Solid and liquid sources will be discussed in more detail in Sections 12.4 and 12.8 but all must be gases by the time they reach the substrate. Normally gaseous sources are stored in high-pressure tanks. The gas pressure is reduced through a regulator as the gas leaves the storage tank and its flow into the reactor is adjusted via a mass flow controller. The latter is essentially an automated needle valve. The flowing gas is either fed directly into the reactor or is diverted into the scrubber (a so-called "vent-run" valving system). This permits continuous flow of the reactant to avoid pressure bursts that would occur if valves were simply opened and closed. If this were done, pressure would accumulate in the supply tubing when switched off. In some cases several reactant gases are pre-mixed in a volume before flowing into the reactor, while in other cases the gases may be introduced at specific locations in the reactor to control the local reactions at that position (see Section 12.8).

CVD reactors are divided into two large classes depending upon whether they operate at atmospheric pressure ("APCVD") or partial vacuum (low pressure CVD or "LPCVD"). In microelectronics applications, virtually all CVD reactors are LPCVD type systems. APCVD is relatively simple because it does not require vacuum pumps and the mechanical design needed to withstand external pressure. APCVD provides high deposition rates (because the reactant concentration is very high) and operates at low temperatures. However, it generally does not provide uniform coverage on rough or patterned surfaces and, most importantly, reactions are likely in the gas phase leading to powder formation. Powder on the substrate is the major reason why APCVD has been abandoned for microelectronic applications.

LPCVD provides excellent uniformity on rough surfaces, excellent purity of the resulting films, and can permit selective deposition under the right conditions. However, it is more complex and expensive, operates at relatively low deposition rates, and often requires higher temperatures. Neither low rates nor high temperatures are a problem if one is trying to produce epitaxial thin films, as both are generally necessary for epitaxy in any case (see Chapter 10).

CVD reactors can also be divided according to how heat is supplied to the substrates, see Figure 12.6. Hot-wall reactor designs place the reactor tube in a furnace that yields exceptional temperature control and uniformity across many substrates. However, hot wall designs heat the gas, encouraging gas phase reactions and reactions on the reactor tube. This can release flakes of film material from the walls, especially upon temperature cycling, which may contaminate the substrate surface. Hot wall designs are useful when large numbers of substrates are to be processed simultaneously as it allows superior temperature uniformity.

Cold wall reactor designs heat only the substrate and its mounting fixture. The most common design for a cold-wall reactor is to place the substrate on a graphite holder and to surround the reactor tube with a radio frequency coil. Graphite is partially conductive so it will absorb radio frequency energy from the coil. Electron currents induced in the graphite produce heat relatively uniformly. The substrate mounted on the graphite susceptor is heated without the need to pass wires into the deposition zone of the system. [A susceptor is a moderately resistive block that is susceptible to absorbing the microwave energy from the rf coil and is thereby heated as its resistance dissipates the absorbed energy by heat.] In cold wall reactors the substrate has much less thermal mass so heating and cooling are more rapid and reactions are

(a)

Thermally-insulating housing

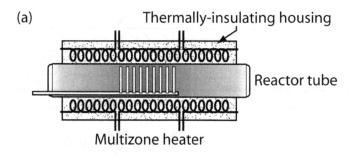

Reactor tube

Multizone heater

(b)

Substrate

rf coil

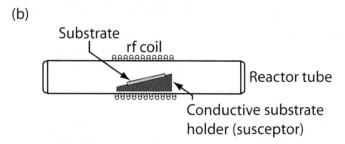

Reactor tube

Conductive substrate
holder (susceptor)

Figure 12.6: Schematic diagrams of (a) the hot wall and (b) the cold wall reactor designs. Hot wall reactors have high termal mass multizone furnaces that provide heating of the entire reactor tube and its contents. Note that in this design one can have many substrates standing upright in the flow inserted using a rod that holds them off of the tube surface. In the cold wall design only the susceptor and substrate get hot due to induction current resulting from the rf field due to the coil.

unlikely on other surfaces, as these are relatively cool. This is why gas phase pyrolysis of reactants (eg: as in Figure 12.4) occurs only in the stagnant gas above the substrate surface in a cold-wall design, as only this gas is heated.

12.3 GAS FLOW IN CVD REACTORS

A critical aspect, especially in the design of cold-wall reactors is the fluid dynamics of the gas flow. Even though the pressure in the reactor is reduced, the gas density is still sufficient to apply conventional fluid mechanics to its flow. Fluid mechanics tells us that a gas flowing down a smooth tube will move without turbulence as long as the Reynolds number, R_e, is less than ~2000. The Reynolds number in a tube is given by:

$$R_e = \frac{\rho v x}{\eta},$$

12.5

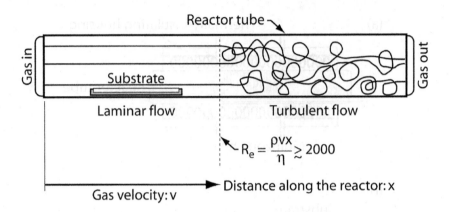

Figure 12.7: A schematic of gas flow in the CVD reactor tube. Because the Reynolds number increases continuously down the tube, it eventually exceeds the critical value for trans-formation to turbulent flow (about 2000). The higher the gas velocity the shorter is this distance. The substrate is best located in the laminar flow region.

where ρ is the gas density, v is the gas velocity, x is the distance traveled along the tube, and η is the gas viscosity. The situation is illustrated schematically in Figure 12.7.

Although one might imagine that the optimal design would be for turbulent flow such that the gas would be well mixed, it turns out that this leads to time-varying local reactant and product concentrations, and yields variable deposition rates and impurity concentrations in the grown material. The optimal design is for laminar (non-turbulent) flow. This has the additional benefit that one can take advantage of the formation of a stagnant boundary layer at the surface of the substrate. In the stagnant layer the gas velocity is reduced as one approaches the wall of the reactor and the substrate surface, and reactants and products must diffuse across this layer to reach the substrate. It is this layer that defines the diffusion distance $\delta(x)$ in Equation 12.4. Therefore, this boundary layer can be used to control the growth rate at different positions along the reactor. The geometry of the boundary layer is shown schematically in Figure 12.8.

The thickness of the stagnant boundary layer can be calculated from fluid mechanics. It is defined as the point at which the viscous force balances inertia under conditions of constant pressure. This may be shown to occur at approximately:

$$\delta(x) = \frac{x}{\sqrt{R_e}} = \sqrt{\frac{\eta x}{\rho v}}, \qquad\qquad 12.6$$

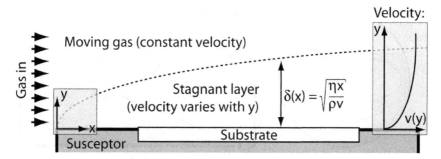

Figure 12.8: A schematic of the stagnant or boundary layer above the substrate. Fluid mechanics results demonstrate that the stagnant layer thickness increases as the square root of the distance along the substrate. Likewise, the velocity of the gas increases from zero at the substrate surface to the average velocity of the moving gas at the boundary layer thickness above the surface. The origin of the coordinate system for these plots is at the leading edge of the susceptor, which is typically upstream of the edge of the substrate such that the stagnant layer is well established before reaching the substrate. The gas velocity above the surface is plotted in the inset to the right of the figure. The stagnant layer thickness, $\delta(x)$ marks the point where the gas velocity reaches the speed of the gas moving down the reactor far from the walls.

where the result on the right was obtained by substituting for R_e from Equation 12.5. A detailed description of the governing equations from which this is derived may be found in most fluid mechanics texts.

From Equation 12.6 one may notice that the boundary layer thickness increases with the square root of the distance along the reactor tube and inversely with the square root of gas velocity. Therefore, if the gas velocity were increased linearly along the tube, $\delta(x)$ would remain constant. To obtain a constant growth rate across a substrate one should maintain an approximately constant boundary layer thickness when operating in a gas-transport-limited growth regime. Therefore, one must arrange that the gas velocity increase linearly along the substrate.

Increasing the gas velocity is achieved in CVD reactors by reducing the tube cross sectional area through which the gas flows. In practice one keeps the tube area constant and inserts a wedge-shaped susceptor into the flow (see Figure 12.9). The flux of gas through any given plane in the tube must be constant at constant pressure. Therefore, as the area of the tube decreases the velocity must increase. A further consideration is that the concentration of reactant decreases gradually in the flowing gas stream along the substrate length while reaction products accumulate. Therefore to achieve a truly constant growth rate it is necessary to reduce the boundary layer thickness gradually along the substrate surface. This is accomplished by increasing the gas velocity more than linearly along the reactor tube. In a round tube, tilting an obstruction in the gas flow not only reduces the distance from the top of the

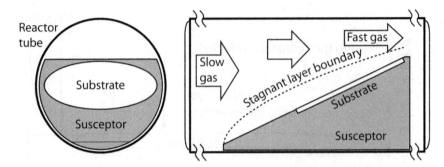

Figure 12.9: A sketch of the use of a tilted susceptor to restrict cross sectional area of the reactor tube and hence to increase gas velocity. End view (left), side view (right). This balances or even reduces the stagnant layer thickness. The gas velocity should not increase to the point of causing turbulence in the gas.

susceptor to the top of the tube along its length but also reduces the width of the channel through which the gas can pass. This automatically accomplishes the super-linear increase in gas velocity along the tube.

Again, in practice reality is more complex than these simple descriptions, although the concepts remain the same – fluid mechanics is used to control the boundary layer thickness such that the growth rate is constant under conditions of gas-transport limited growth. For real CVD reactors a complete fluid dynamics model is coupled to a film growth model to obtain an optimal susceptor design to fit in the reactor tube (which may be rectangular or round or some other shape as desired).

As an example of the complications that must be accounted for in a real reactor design, temperature gradients in the reactant gas occur both along the reactor tube and radially from the center of the tube to the edge. The radial temperature gradient is particularly significant in cold-wall reactors. The complication is in the gas phase transport rate. The diffusivity of gas molecules depends upon the temperature but also on the mass of the gas. Therefore a diffusivity gradient resulting from **a temperature gradient tends to drive large molecules to the cold regions of the reactor**. The result is a temperature dependent gas phase composition gradient that has nothing to do with adsorption/desorption phenomena. It is entirely due to differential diffusivity gradients in the gas that set up differential fluxes of molecules that create composition gradients.

12.4 REACTANT SELECTION AND DESIGN

An additional consideration in the development of a CVD process is design of the reactants such that a desirable growth behavior is obtained. Some of the issues in design and selection of reactants are: the phase of the source material under standard conditions; the ability to produce a vapor from the source material; the stability of the source gas before reaction; gas phase pyrolysis reactions; the thermodynamics and kinetics of its adsorption, desorption, decomposition, and reaction on the substrate; the nature of product species and their desorption from the substrate; the importance of nucleation of the reaction on certain substrate materials compared to others; the presence or absence of undesirable parasitic reactions including reaction with fixtures in the CVD reactor; the purity that may be obtained in the source material; the toxicity and hazard associated with the reactant; and ultimately the quality of the material that can be produced.

Source materials may be solids, liquids or gases under standard laboratory conditions. Solids and liquids are most easily handled, stored, and transported. It is also relatively easy to purify solids and liquids and to keep them pure prior to use, so they are desirable. However, CVD requires gaseous source materials in the reactor. Therefore it is necessary to convert solids and liquids to gases to make use of them.

Solid and low vapor pressure liquid (e.g. Ga) sources may be used in CVD systems by reaction with an etching compound (usually a gas) resulting in a gas phase product. Two examples of etching reactions are:

$$Si + 2I \downarrow \rightarrow SiI_2 \uparrow$$
$$2Ga + 6HCl \downarrow \rightarrow 2GaCl_3 \uparrow + 3H_2 \uparrow$$

12.7

In both cases a vapor containing the initially solid/liquid element is produced, usually upstream from the substrate in the CVD reactor itself. These reactions can run in reverse to redeposit the vaporized species. In some cases it is possible to simply evaporate a high vapor pressure compound such as CdTe in the CVD reactor, although this yields an evaporation-like process more than a CVD process. This case is referred to as closed-space vapor transport and is generally distinct from evaporation as described in Chapter 11 because the substrate and source are in equilibrium with a relatively dense vapor of the source material in a closed environment. The case above for Ga requires an etchant because Ga has a naturally low vapor pressure. In the case of liquids with high vapor pressures a different method can be used.

By heating a high vapor pressure liquid, evaporation will take place, as was the case in Chapter 11 for physical vapor deposition by evaporation. In CVD the liquids involved typically evaporate much more easily than they decompose (at least they should). Such liquids are used by bubbling a carrier gas, usually H_2, through that liquid at a temperature selected to produce a given reactant concentration in the H_2 gas stream. Typical materials vaporized through the use of a bubbler include most metal-organic compounds such as trimethyl gallium, TMG [$Ga(CH_3)_3$] and many chlorides such as $AsCl_3$. The basic bubbler apparatus is shown schematically along with examples of vapor pressures for a variety of gases in Figure 12.10. All piping downstream of the bubbler must be at a temperature at least equal to that of the bubbler to prevent condensation of the source compound but not so hot as to induce decomposition of the reactant.

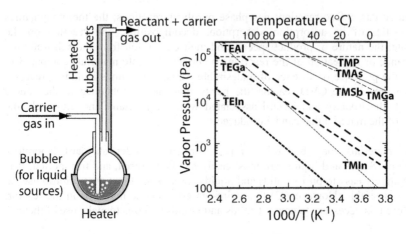

Figure 12.10: Liquid sources are supplied by bubbling a carrier gas such as H_2 or Ar through the liquid. "T" refers indicates "tri", "M" designates "methyl" and "E" is for "ethyl". Al, Ga, and In sources are dashed, P, As, and Sb are solid lines. Ethyl compounds are the heavy lines while methyl compounds are narrow lines. Note that the ethyl compound pressures lie below the methyl vapor pressures. Heavier metals produce lower vapor pressure organic compounds. (TMA is an exception being close to the TEA line.) Vapor pressures of divalent metal organics such as dimethyl tellurium generally lie below the most closely-related group III or group V compound. Note that the In compounds are solids under the conditions shown above and sublime upon heating. Vapor pressures shown here are plotted from data in Ref. 3.

For example, one of the traditional problems in CVD is deposition of Cu. There are few acceptable Cu precursors and most involve large organic groups bound to the Cu molecule. The Cu-to-ligand (organic group) bonding in these precursors is weak and the molecule tends to decompose. Furthermore the vapor pressure of the source molecules is relatively low. Consequently, there is a narrow temperature range where evaporation occurs but decomposition does not. For many precursors this limits the flux of Cu-containing molecules that may be supplied to a substrate. Turning up the source temperature simply leads to decomposition, rather than increased Cu availability. Such problems must be considered in precursor selection.

A primary concern in selection of source materials is the thermodynamics and kinetics of their decomposition and reaction. The source gas must decompose to release the desired element to the growing film and to produce a product species that desorbs harmlessly from the surface. Thus, in an organometallic source one desires relatively stable organic groups and relatively weak but not too weak metal-organic bonds. For example, trimethyl aluminum (TMA) $[Al(CH_3)_3]$ decomposes in the presence of hydrogen to yield both Al and CH_4 and aluminum carbide $[Al_4C_3]$ and H_2. In other words, one may anticipate significant carbon contamination of an

Alcontaining film grown from TMA. Triethyl aluminum (TEA) [Al(C$_2$H$_5$)$_3$] normally decomposes in the presence of H$_2$ to Al and C$_2$H$_6$ and other hydrocarbons. Longer chain hydrocarbons generally bond more weakly to a metal or metalloid atom than do shorter chain hydrocarbons and hydrogen itself. At the same time, heavier metals bond more weakly to hydrocarbons making parasitic reactions less likely. These trends hold true for most group III and group V metalorganic compounds. Likewise the metal-organic bond enthalpy typically increases with valence of the metal, thus TEA has a higher bond enthalpy than does diethyl zinc.

Hydride compounds such as silane (SiH$_4$), arsine (AsH$_3$), and phosphine (PH$_3$) are gases under standard conditions. These are frequently used in CVD reactions as they can be stored conveniently (even though they are generally toxic and pyrophoric) and there are no parasitic reactions to worry about. In some cases hydrides can be created from atomic source materials as needed, eliminating the requirement to store large quantities of toxic compounds. This is highly desirable when an adequate hydride source can be created on demand.

Purity of the source materials is an important consideration. It is much more difficult to purify relatively delicate metalorganic compounds than hydrides. It is easier still to purify molecular gases such as H$_2$ and Cl$_2$ from which, for example, high purity HCl can be produced. Etching purified As or Ga with HCl thereby produces very high purity metal trichlorides. The sensitivity to impurities is also dependent upon the material to be deposited. Thus, deposition of Al compounds is more sensitive to oxygen-containing impurities than for materials less inclined to form oxides.

When depositing an alloy by CVD one must also consider how reactants compete for deposition sites and how precursors interact. For example, the introduction of an Al precursor in a GaAs CVD process might alter the Ga precursor reaction rate in addition to adding Al to the resulting film. The result is that the deposition rate and composition of the resulting film may depend upon the relative reactant flow rates in a complex way. Examples of this situation and a discussion of the kinetic and thermodynamic aspects may be found in References [4-6].

12.5 STIMULATED CVD

The reaction rate limitation presents a practical lower limit to CVD process temperatures in many cases. This is a problem because high temperatures allow reactions among layers in the substrate and permit bulk diffusion. Many substrate materials, especially polymers, are also unable to withstand high process temperatures. Therefore it is of interest to reduce the temperature. Furthermore, some reactants are too stable to be used in conventional CVD, and will never decompose under normal CVD conditions. The classic example is N$_2$. These restrictions have driven the development of methods to stimulate CVD reactions. A stimulated reaction may not require as much heat from the substrate to proceed, therefore permitting lower substrate temperatures. Furthermore, it may be possible to decompose reactant gases that are normally too stable to be used.

The first approach developed for stimulated CVD was photoassisted CVD. Here the plan was to supply energy to either the reactants or the sample surface through a high intensity lamp or laser. The light would be incident either normal to or parallel to the sample surface (see Figure 12.11.) In the normal incidence geometry the light heats the surface locally such that reactions may occur without heating from the back of the substrate. Back heating requires a temperature gradient and hence the back surface must be hotter than the front where the reaction occurs. This is especially important for poor heat conductors such as polymers. Front heating may allow CVD to proceed at the front of the substrate without it melting or decomposing. Although normal-incidence front surface heating is valuable to consider, its benefits and the complications associated with its application are too problematic to make it generally worthwhile. Therefore, this approach is relatively little used.

A second method considered was to pass the light through the reactant gas just above the substrate surface. This has several advantages. First, it puts energy into the gas phase without heating the surface, which can activate the reaction (see further discussion below) and reduce the need for substrate heat. Second, the stimulating wavelength can be tuned to enhance desired reactions without affecting undesirable reactions significantly. This can be used to suppress parasitic reactions preferentially.

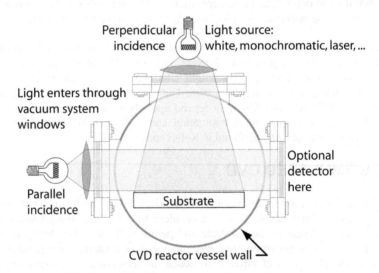

Figure 12.11: A schematic diagram of two configurations for photo-assisted CVD. CVD reactors have windows inserted into the reactor tube (vacuum vessel) wall through which light supporting the reaction enters. A detector in parallel incidence can be included to measure absorption in the gas to determine the gas composition and density.

Plasma-enhanced CVD

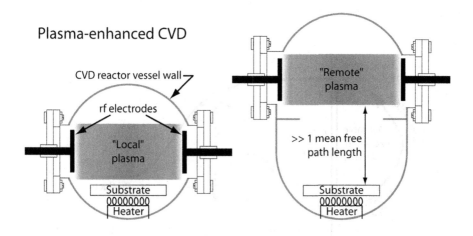

Figure 12.12: Schematic diagrams of typical reactor configurations for PECVD.

Third, optical absorption in the gas phase can be used to probe the reactant concentrations. Unfortunately, this stimulation method is difficult to implement and hard to control. Windows through which the stimulating light passes become coated with a film over time, which reduces intensity. Finding exactly which wavelength to use for selective stimulation requires more experimentation than most process developers care to apply. As with normal incidence photo-assisted CVD, the parallel incidence method has become relatively rare.

The most popular method of stimulating a CVD process is plasma-enhanced CVD (PECVD) [see Figure 12.12]. In this type of approach an rf or dc glow discharge is established in the process chamber. In general an rf discharge is used to minimize the sputtering of electrodes and because higher levels of excitation can be achieved with careful design. There are two implementations of PECVD known by the simple name and as remote-plasma plasma-enhanced CVD (RP-PECVD). The difference is the distance between the glow discharge and the substrate.

The objective of PECVD is to increase the energy of the reactant molecules striking the substrate surface. Surface reactions typically involve an activation energy, which is the energy that appears in Equation 12.3 for the kinetic rate constant of the reaction. Adding energy to a reactant can reduce this barrier and consequently increase the reaction rate dramatically at a given temperature. This situation is illustrated in Figure 12.13. Therefore, one can reduce the surface temperature while maintaining the same or even a higher deposition rate compared to what one would have without the plasma. Activation of the reaction greatly increases the reaction rate constant and effectively eliminates the reaction-rate-limited regime of film growth. The consequence is that PECVD growth processes generally are transport limited

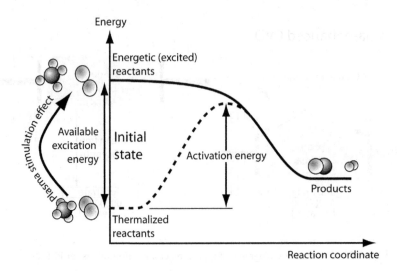

Figure 12.13: Comparison of the reaction energetics for stimulated and conventional CVD. The conventional process is similar to processes described in Chapter 4.

(including being limited by the flux of activated species reaching the substrate) rather than reaction rate limited.

A plasma or glow discharge may have a small number of high-energy particles in it, including electrons, ions, and neutral atoms (see Section 11.3). These are capable of doing damage to certain types of growing film (mostly polymers and dielectrics). When depositing sensitive materials by PECVD it may be necessary to move the plasma away from the film (creating a "remote-plasma"). The plasma excites molecules into high-energy vibrational states as well as accelerating some to high kinetic energies (translational energy). *The critical concept that allows RP-PECVD to work is that vibrational excitations in molecules last longer than their kinetic energy in a gas.* As molecules collide with each other they rapidly transfer kinetic energy and revert to low translational energy within a few impacts – a few mean-free path lengths (Equation 10.1). Transfer of vibrational energy is much less likely so the molecules cool only slowly to an equilibrium excitation level. The distance between the plasma and the substrate in RP-PECVD is therefore beyond the thermalization distance for kinetic energy and below the distance needed to bring the vibrational energy of the molecules to equilibrium.

For PECVD to work the reactant molecules must retain their energy long enough on the surface to reach a potential reaction site and to react. Furthermore, to obtain adequate film properties the surface temperature cannot be reduced below that necessary for atoms to diffuse to acceptable growth sites after reaction and to eliminate crystalline defects. This explains why very large reductions in substrate temperatures can be obtained with PECVD deposition of amorphous SiO_2 compared to conventional CVD processes. The reaction is strongly activated but because the SiO_2 is amorphous there is no significant diffusion of the reacted atoms needed. PECVD is uncommon in deposition of single crystals, especially of semiconductors, because its benefits are largely lost. Single crystal growth is generally more strongly limited by surface diffusion rates than reaction rates. Furthermore, the possibility of damage from energetic particles to the layer, even in RP-PECVD, motivates against application of PECVD to semiconductor growth.

12.6 SELECTIVE CVD

One of the most important advantages of CVD for some applications is the possibility to carry out the deposition selectively. In this case the substrate is presumed to include a surface to be coated with deposited film (which will be referred to here as simply "the substrate") and a masked area that is intended to be devoid of film ("the mask"). Successful selective CVD can eliminate subsequent patterning steps that would be needed if a more conventional deposition blanketed the entire substrate, including the mask, with film. Selective deposition is very helpful for maintaining tight tolerances in current microelectronic devices.

Selectivity is produced by arranging the growth process to strongly favor deposition on the substrate and oppose deposition on the mask. This is obtained through four mechanisms: achieving a high ratio of reactant density on the substrate to the density on the mask, requiring the substrate material in order for the reaction to proceed, large acceleration of the reaction nucleation rate on the substrate compared to that on the mask, and selective etching of any film nuclei on the mask relative to those on the substrate. These mechanisms are outlined schematically in Figure 12.14.

The concentration of reactant on the surface is a dynamic balance between arrival rate, adsorption and desorption. The net surface coverage may be expressed as in Equation 10.8 or as a ratio of the adsorption rate k_a to the desorption rate k_d as:

$$\theta = R\phi\tau = R\frac{k_a}{k_d} \qquad 12.8$$

where all species are assumed atomic, R is the arrival rate per surface site, ϕ is the sticking probability, and τ is the average residence time. Thus, a higher ratio of adsorption to desorption on the substrate to that on the mask will lead to selective deposition. In other words, when a critical reactant adsorbs much better on the substrate than on the mask or desorbs much faster from the mask than from the substrate, the differential reactant concentration on the two surfaces will favor selective growth.

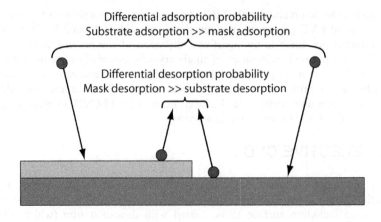

Differential adsorption probability
Substrate adsorption >> mask adsorption

Differential desorption probability
Mask desorption >> substrate desorption

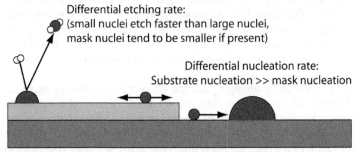

Differential etching rate:
(small nuclei etch faster than large nuclei,
mask nuclei tend to be smaller if present)

Differential nucleation rate:
Substrate nucleation >> mask nucleation

Substrate nucleation promoted by: (1) reaction with the substrate
(2) a nucleation promoting layer on the substrate, (3) naturally favorable
nucleation rate on the substrate or unfavorable rate on the mask.

Figure 12.14: A schematic diagram highlighting some of the mechanisms for achieving selectivity in a CVD process.

The second highly effective method for achieving selectivity is simultaneous etching during growth in a process with a strong nucleation limitation, especially on the mask. In other words, having a process that runs at nearly the same rate forward (growth) as in reverse (etching) results in net deposition but with a strong etch component. As we saw in Chapter 10, nucleation-limited growth processes have a critical nucleus size beyond which the nucleus grows easily and below which the nucleus typically shrinks. Simultaneous etching and growth processes enhance this. One may look at the critical nucleus size as the point at which growth and decomposition balance. Etching increases the nucleus decomposition rate without affecting growth and therefore shifts the balance point to a larger, faster growing, more stable size. Because the average energy binding an atom to a cluster generally increases with cluster size, the tendency to etch a small cluster – even one larger than the critical nucleus size – is greater than the tendency to etch a large cluster.

Therefore the difference in etch rate and growth rate may be closer the smaller a film island is. Hence, nucleation is more strongly suppressed. Because nucleation is designed to be intrinsically more difficult on the mask the nuclei will be smaller and will grow more slowly there. In summary, for many reasons etching suppresses nucleation on the mask relative to the rate on the substrate.

A possibly obvious mechanism for achieving selectivity is to rely on reaction with the substrate material for either nucleation or growth of the desired material. For example, if one were producing a silicide compound such as $NiSi_2$ (see Section 10.1) one might achieve selectivity between a Si substrate and a SiO_2 mask if deposition of Ni everywhere resulted in growth of the silicide where the Si substrate was exposed but without growth on the SiO_2. This is, in fact, the basis for the self-aligned silicide process. In this case the unreacted Ni can be selectively removed later. This example is for a solid phase reaction based on a layer typically deposited by physical vapor deposition. Therefore it is of less relevance to our concerns here. A CVD process that makes use of the substrate to promote reaction is the selective W deposition process described in Section 12.8.

Finally, a similar result enhancing selectivity without relying on reaction with the bulk substrate is achieved by coating the substrate with a nucleation-promoting layer that is chemically different from the substrate material. The nucleation layer is covered with a mask where deposition is not desired. The desired film forms only where the promoter is exposed and continues to grow there because of the existence of the nucleus. This method relies on the reaction being heavily nucleation limited. If nucleation can occur on the mask then selectivity may be lost.

The above methods work best if used together. Simultaneous etching and growth can correct for accidental nucleation on the mask that may result even where there is a difference in reactant supersaturation. The use of a nucleation promoter or an interaction with the substrate to promote nucleation can emphasize nucleation on desired locations enough to allow the etch and growth rates to be more different than would be possible without the nucleation promoter, and so on.

One important point to note about all of the approaches to achieving selectivity relying on differential nucleation rates rather than differential reactant concentrations (which then affect nucleation) is that there will be a large reservoir of reactant on the mask surface. While this may not react on the mask for lack of a nucleus, it is still free to move about. Because the growing film provides an effective reaction site, it may collect additional reactant from the surrounding mask relative to what would have been available far from the mask on a desired deposition region. To make matters worse, the relative deposition rate near the mask edge can therefore depend upon the size of the masked area. Larger areas may provide more reactant to the surrounding unmasked substrate than smaller areas. The situation is shown schematically in Figure 12.15. The non-uniformity of growth rate has become an increasing problem as device dimensions have shrunk and smoothness of the surface

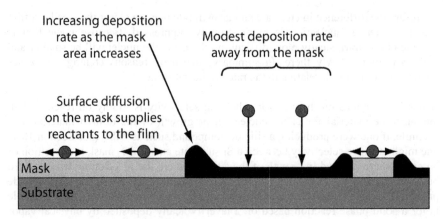

Figure 12.15: Shows the effect of the mask in supplying additional reactant to the substrate area where growth occurs and hence increasing the deposition rate near the mask. The size of the increase is dependent upon the area of the mask.

has become increasingly important. To deal with this a new technique has been developed that is described in the next section.

12.7 ATOMIC LAYER DEPOSITION

Atomic layer deposition (ALD) is a method used to produce materials with highly controlled thicknesses and well-defined compositions, even by comparison to molecular beam epitaxy controlled by reflection high energy electron diffraction. Indeed, one of the earliest forms of ALD was as a form of MBE. More recently it has become primarily a CVD method. It is also used to coat highly textured surfaces or small cavities or crevices uniformly. The basic concept of ALD is outlined in Figure 12.16. Single atomic layers of reactant are adsorbed on the substrate while additional reactant desorbs. In the case of deposition of a compound this would typically involve deposition of a single or partial monolayer of one component atom of the compound in the form of a precursor species. The next element to be added to the compound is then deposited in a single layer that reacts with the species deposited in the preceding step.

ALD source gases must have dramatically different reaction, adsorption and/or desorption rates on surfaces with different chemistries. For example, for deposition of a compound AB, the A precursor should deposit well on the B surface but not to the A surface, while the B precursor is the opposite, depositing well on A and poorly on B. (c.f. Figure 12.16) The difference may be due to much higher sticking of the A precursor to B and vice versa or may be due to much more rapid desorption of precursor A from A than from B or both. Ideally there should at least be a significant

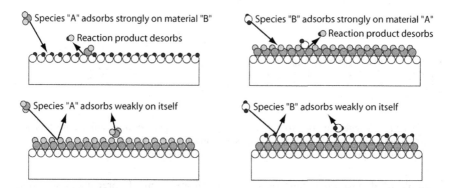

Figure 12.16: A schematic of a typical ALD process for deposition of an AB type compound. A reaction product may also be formed and desorb in one or both half reactions. For example, if the A precursor were a hydride and the B precursor were a chloride, then HCl would be generated one step as a product. In the reaction above the sequence of steps would be top left, bottom left, top right, bottom right, top left, etc.

difference in desorption rate from the surface of the same type to prevent multiple layers of precursor forming. ALD is further enhanced if the surface reaction depositing element A from its precursor is catalyzed by B atoms and/or if decomposition and deposition of B is catalyzed by A or if neither deposition process could occur in the absence of the other species.

The essential element in all of these options is to saturate the B surface with one and only one layer of A atoms in some carrier form followed by coating this A surface with one and only one layer of B.

A major advantage of ALD is that the surface may be exposed to an overdose of precursor (for example A) taking as much time as necessary to fully saturate the surface with one monolayer of that precursor. Having the leisure of time and being automatically limited to one monolayer of deposition means that molecules have plenty of time to find their way into cracks and across complex surfaces without excessively coating the top of the sample. It is also possible to deposit a partial monolayer in each cycle by design of a reactant that is so bulky that a full monolayer equivalent of atoms can not be adsorbed on the surface in the form of precursor. Reconstruction of the surface may also favor partial monolayer deposition. These situations are shown in Figure 12.17.

In ALD the deposition rate is proportional to the reaction cycle time rather than the flux of precursor on the surface and the total thickness of material deposited is exactly related to the number of deposition cycles, making control of growth straightforward at the price, in general, of a slower overall process. However, for a sufficiently reactive precursor the individual steps may be very rapid. A typical

Bulky ligands may limit the number
of adsorbate molecules

Reconstructed surfaces may only
adsorb on certain sites

 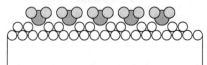

Figure 12.17: A schematic of two options for reducing the number of adsorbate molecules on a surface, attaching large ligands to provide steric hinderance of further adsorption and the use of a species that prefers a specific site on a reconstructed surface.

example of deposition of an AB compound from A and B precursors would be alternation of a hydride with a chloride, yielding hydrogen chloride as one reaction product.

Because it puts layers of atoms down one at a time, less diffusion is needed to transport atoms to desirable reaction sites (these sites are sampled through the gas phase). Therefore the deposition temperature may be significantly lower as compared to other methods. The sequential process also eliminates the possibility of gas phase reactions, allowing ALD to be carried out at higher pressures or with reactants that would be excluded in normal CVD. For example, strongly exothermic reactions can be used more easily in ALD than in conventional CVD. Indeed, this is often desired in order to quickly react a precursor with the active surface and saturate that surface with the species to be deposited. Rapid surface saturation by aggressive reaction also reduces the time of exposure of the surface to a reactant necessary to complete the layer, thus reducing the amount of reactant that flows through the reactor per cycle. Finally, because atoms are put on the surface one layer at a time, one may select the species to be deposited in each layer, allowing excellent control of the sequence of layers produced. One thing ALD does not do easily is to allow variable control of individual layer compositions. However, by producing a superlattice with a changing number of layers per period and subsequently interdiffusing these layers moderately, one may achieve a somewhat similar effect. Overall though, graded layer compositions are not the strength of ALD.

ALD reactions involving more than two elements can also be designed. For the case of deposition of an alloy (for example a pseudobinary alloy), one must rely on a well-characterized competition for surface reaction sites. Thus, to deposit an $Al_xGa_{1-x}As$ alloy one would need to know how an Al precursor would compete with a Ga precursor for a favored metal deposition site. An experimental variable such as surface temperature could alter this competition and give some control of the resulting film alloy composition.

It is even possible to design ALD reactions that deposit a single atomic species one layer at a time, as shown schematically in Figure 12.18. Because the surface is chemically identical in each step it is not possible to use the simple adsorbate to

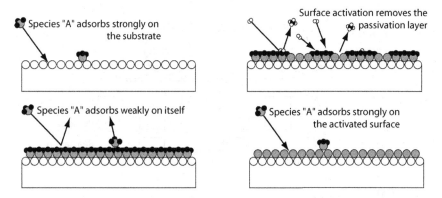

Figure 12.18: A schematic diagram for a simple single species ALD process involving deposition of a monolayer of precursor followed by exposure of the deposited surface to an activator.

control deposition, as it never changes chemistry. Rather, one alternates the depositing species that produces the desired material with an activation step that is necessary for the next monolayer of deposition. This step may involve photolytically decomposing the adsorbate layer or by alternating two precursors, both containing the material to be deposited, but with different ligands that react with each other but not with themselves to deposit the desired material. An example of a single-species ALD process for Cu is given in the next section.

It is frequently the case that a limited range of process temperatures is available for ALD. This may not be surprising as ALD relies on differential adsorption/ desorption/reaction rates of different species. Too high a temperature will lead to reaction or decomposition of the reactant even when the favored partner species is absent, leading to unlimited deposition of a that species. Conversely, too low a temperature can lead to failure of one component to react.

Two sample ALD reactor design options are presented schematically in Figure 12.19. Not shown is a conventional MBE environment such as in Figure 11.1. The CVD-type reactor designs may include a conventional CVD reactor tube (hot or cold wall) with pulsed gas sources sending bursts of one reactant and then the other down the tube. This relies on sufficient physical separation of the pulses in the gas phase as the reactant pulse travels down the tube. A second method is to expose a surface sequentially to reactants supplied at fixed locations by rotating the substrate alternately over each.

12.8 SAMPLE CVD AND ALD PROCESSES

We turn finally to several examples of CVD and ALD processes that illustrate the issues described above. Many CVD processes take advantage of temperature

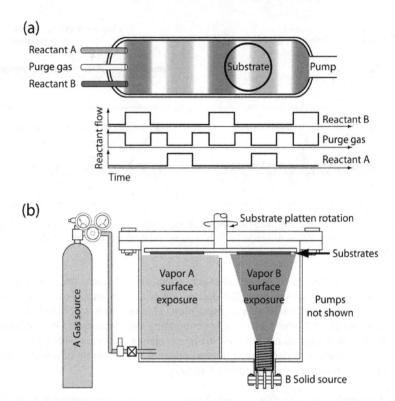

Figure 12.19: Schematic diagrams of two implementations of ALD. (a) Implementation in an essentially normal CVD reactor where pulses of gas of three types flow through the reactor, two reactive species responsible for deposition of the film and a purge gas that separates the two source gases from each other. The purge gas may also flow continuously and the other reactants can be introduced into this stream periodically. (b) A system based on rotation of substrates through different source gases contained in separate portions of the chamber. In the example shown one source might be from a typical CVD type reactor while a second source gas might be supplied by a conventional MBE type evaporation source.

differences in various regions of the reactor to produce net deposition on the substrate. The simplest form of this type of process is the growth of Si by chemical vapor transport using iodine gas. This process is illustrated in Figure 12.20. Iodine reacts with Si to produce SiI_2. [7] The reaction proceeds rapidly at high temperatures, for example at 1100°C. The Si-containing vapor mixes with the iodine in a closed volume and diffuses to a substrate at a lower temperature (for example 950°C). The reverse (exothermic) reaction then deposits Si and returns iodine to the vapor phase. The critical point is the temperature difference. The etching reaction resulting in SiI_2 vapor is endothermic and therefore proceeds more rapidly at high temperatures relative to the deposition reaction.

At lower temperatures the exothermic deposition reaction dominates. The temperature gradient in the sealed reactor tube leads to net transport of Si from a source to the substrate.

A more complex example of a similar process is the growth of GaAs from liquid Ga, $AsCl_3$ and H_2. [8] In this case the $AsCl_3$ reacts with the H_2 to produce As_4 (the same material that is generated by As effusion cells in an MBE process) and HCl vapor. The As_4 then reacts with liquid Ga in a high temperature zone of the CVD reactor tube forming GaAs. This is etched by the HCl vapor yielding $GaCl_3$ and arsenic in various gaseous forms. In a lower temperature zone the gas vapors react to produce GaAs and release HCl. This process is shown schematically in Figure 12.21. Again, the relative temperature of the process affects in which directions the reactions proceed. In this case, the gas phase chemistry is also important because the concentration of HCl has a strong effect on the process via the law of mass action. The effective HCl concentration can be adjusted both through the temperatures in the reactor zones but also through the concentration of H_2. The important point here is that by control of the *non-inert* dilutant gas concentration one may adjust the reaction rate.

The next step up in process complexity is to use gas sources and by separation of the Ga etching reaction and the As reactions [Figure 12.22]. In this case, the Ga is volatilized in the first zone of the reactor at an intermediate temperature by reaction of HCl to produce $GaCl_3$ and H_2. [9] This reaction is controlled by both the temperature of the reactor in the first zone and by the H_2 gas content in this area. Because H_2 is a reaction product, increasing its concentration reduces Ga volatilization. Arsenic is introduced as AsH_3 and decomposes into As_4 as in the previous case. However, because this is downstream of the Ga source, the As_4 reacts directly with $GaCl_3$ to form GaAs at the substrate. The resulting Cl reacts with H to form HCl. By controlling HCl concentration in the gas phase at the substrate, again

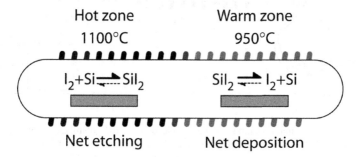

Figure 12.20: A schematic of the iodine-mediated closed-space vapor transport growth process for Si.

the forward and reverse reaction rates can be adjusted independent of the substrate temperature.

A relatively simple but technologically interesting material that can be conveniently deposited by CVD is SiC. SiC is typically produced by reaction of a small organic molecule such as propane (C_3H_8) with silane (SiH_4) in a hydrogen ambient. This reaction has been studied in some detail. For a detailed discussion see, for example, Reference [14]. The reaction runs both forward and in reverse depending upon temperature and H_2 partial pressure.

A single source gas incorporating both Si and C can be used. An example of a high-vapor-pressure liquid containing both Si and C is dimethyl dichloro silane [$(CH_3)_2Cl_2Si$]. Note that because this source gas contains Cl and because HCl vapor can react reversibly with Si, this reaction may be influenced by the partial pressure of H_2 and may yield a reverse reaction that would be controllable by the addition of HCl to the vapor. (See discussion of the purification of metallurgical grade Si in Chapter 4 for a discussion of the related reactions.) It is also interesting to note in conjunction with organo-silane liquid source that the Si to C ratio can be adjusted by control of the dimethyl dichloro silane partial pressure in the reactor. [c.f. Reference 13].

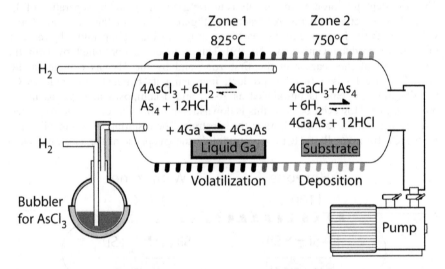

Figure 12.21: A schematic diagram of the halide process for volatilization of Ga using an AsCl$_3$ source and H$_2$ and subsequent deposition of GaAs in the lower temperature zone. Note that the etch reaction 4GaAs + 12 HCl $\rightarrow$ 4GaCl$_3$ + As$_4$ + 6H$_2$ is endothermic and runs forward at 825°C in the presence of excess HCl and in reverse (i.e. as deposition of GaAs) at 750°C in the presence of excess H$_2$, hence the downstream supply of H$_2$.

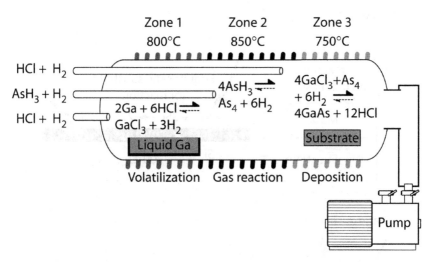

Figure 20.22: A schematic diagram of the growth of GaAs by the halide process. In this approach Ga is volatilized by reaction with HCl to form GaCl₃. The reaction rate is controlled both by the various reactor zone temperatures and by the partial pressure of H₂, a reactant, and HCl, a product.

A typical deposition process would use a tilted graphite susceptor as in Figure 11.1 heated inductively or in a hot-wall geometry, although other susceptor designs also work well. Reactions may be carried out at low pressure or atmospheric pressure depending upon the reactants and growth conditions. Because SiC is a refractory material strong interatomic bonds and a melting point of 2830°C, surface diffusion of atoms is slow. Therefore deposition of SiC is typically carried out at higher temperatures than the other processes discussed above. For example, reaction temperatures between 1100 and 1600°C are common. At 1600°C, H₂ etching of the SiC is rapid, ~1 μm hr⁻¹ at 1.01x10⁵ Pa (1 atm) pressure. [14] Indeed, at higher growth temperatures and reactor pressures (near 1 atm), the deposition rate is controlled primarily by the reverse etching rate. For the case of an organic molecule such as propane or ethane reacting with silane the hydrocarbon tends to have a lower sticking coefficient on the growth surface than does silane. [14] However, gas phase transport is generally the rate-limiting step in the process.

A typical example of selective CVD is the deposition of W on Si using WF₆ gas. [10] This reaction is carried out in two stages, nucleation and growth. The selective W process is shown schematically in Figure 12.23. The nucleation step reaction is

$$2WF_6 + 3Si \xrightarrow{300°C} 2W + 3SiF_4.$$

This can only occur where Si is available as a reactant. An SiO₂ layer does not react with WF₆ as it is too stable. Therefore, it makes a good mask. This means that W

Stage 1: Nucleation

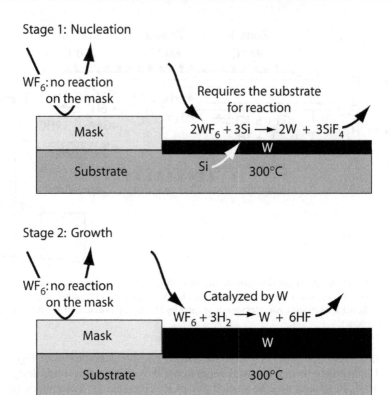

Stage 2: Growth

Figure 12.23: The selective CVD process based on WF_6 decomposition catalyzed during the nucleation step by Si for the initial reaction in the absence of H_2 and the growth reaction catalyzed by W in the presence of H_2.

will form on exposed Si but not on a nearby SiO_2 mask. The requirement for Si as a reactant makes selectivity very high and automatically terminates the process when the W layer becomes a sufficient barrier to prevent Si from reaching the WF_6. This occurs when the W is ~20 nm thick. The growth step begins after the saturated 20 nm of selective W deposition is completed. At this point H_2 is added to the reaction gas and the reaction proceeds in the absence of Si:

$$WF_6 + 3H_2 \longrightarrow W + 6HF$$

The reaction is strongly catalyzed by the presence of W, thus it proceeds most rapidly where a pre-existing W layer is present. Because it does not require Si reactant from the substrate it proceeds without further consumption of the substrate.

The selective W process suggests a method for ALD of W by supplying alternating bursts of a Si-containing precursor and WF_6. This approach was characterized by George et al. and found to operate as follows. WF_6 conformally coats a W surface that includes a flourosilane layer resulting in formation of a fluorinated Si-free W surface. This surface then reacts with disilane (Si_2H_6) or silane (SiH_4) to recover the silanated surface and release hydrogen. George et al.. proposed the two half reactions to be: [11]

$$W \mid SiH_yF_z + WF_6 \downarrow \longrightarrow W \mid WF_x + SiH_aF_b \uparrow$$

$$W \mid WF_x + Si_2H_6 \downarrow \longrightarrow W \mid SiH_yF_z + 2H_2 \uparrow + SiH_aF_b \uparrow$$

where "|" incidates the surface of the deposited layer. The subscripts x, y, z, a, and b are values representing the average stoichiometry of the product species, and having values between 0 and 4. The two half reactions produce well-defined surface coverages of one monolyaer of fluorinated silane-like adsorbate (-SiH_yF_z) or fluorinated W surface (-WF_x) at temperatures above 400K. Note how the basic reaction process is similar to the nucleation step in the selective W process described above. To keep the reaction going indefinitely without the self-limiting behavior in the previous example the Si is supplied from the gas phase as Si_2H_6. Also note that by performing the selective W nucleation stage in Figure 12.23 one can achieve selective W ALD by this method.

Finally, an example of ALD of a single metal species, Cu, can be found in Carlsson et al. [12] In this case the reaction sequence is:

$$2Cu \mid CuCl + H_2O \downarrow \rightarrow Cu_2O + 2HCl \uparrow$$

$$Cu \mid Cu_2O + H_2 \downarrow \rightarrow 2Cu \mid Cu + H_2O \uparrow$$

$$Cu + CuCl \downarrow \rightarrow Cu \mid CuCl$$

where each separate line is a sequential reaction in the ALD process. As above the "|" indicates a surface coated with a monolayer of the material to the right of the |. Note that this reaction involves the selective oxidation and reduction of the surface. It is the self-limiting oxidation and the possibility of reduction of the oxide that permits this process to proceed in an ALD mode.

12.9 SUMMARY POINTS

- Chemical vapor deposition involves supply of gas phase species at moderate pressures to a surface, which react to produce a solid film and a desorbing gaseous product.
- The reaction may be surface reaction rate limited or gas phase transport limited.
- Gas phase transport limitations are controlled by reactant concentration in the gas and the viscosity, turbulence, flow rate, and other properties of the gas, while surface reaction rate limitations are controlled by surface temperature and the other processes typical of physical vapor deposition growth.
- Sufficiently low-temperature CVD processes are normally surface reaction limited while sufficiently high temperature processes are normally gas transport limited.
- Reaction-limited processes produce good uniformity of coverage and may grow material in batch mode in a hot-wall system but are relatively slow.
- Gas-phase transport limited processes are faster than reaction-limited processes.
- A boundary layer forms in the gas phase with a velocity that decreases with decreasing distance from the substrate and with a reactant concentration gradient from high values in the moving gas to low values near the surface.
- Gas phase pyrolysis reactions may permit partial decomposition of a reactant before it lands on the substrate.
- Reactant or product gases may be added to change the reaction rate. Hydrogen is often used to dilute the reactants and may change the reaction process by participation in the pyrolysis process or in the reaction itself.
- CVD process systems generally include a safety enclosure that confines hazardous gases and allows detection of leaks to protect the users.
- In systems operating at lowered pressures the probability of gas phase formation of film material (dust generation) is reduced relative to atmospheric pressure systems.
- Reactants may be solid, liquid or gaseous but all must be vaporized for the CVD process.
- Reactors are generally designed for laminar flow across the substrates.
- The design of the substrate mounting system (susceptor) is used to control gas velocity and hence the stagnant (boundary) layer thickness.
- Conventional CVD reactions are typically endothermic and growth rate is controlled by growth temperature in a transport-limited mode.
- ALD favors use of exothermic reactions to accelerate the reaction and decrease the cycle length.
- Liquid and solid sources are simpler to handle and are often used. High vapor pressure liquids may be vaporized by heating and bubbling a carrier gas through them. Low vapor pressure liquids and and solids are vaporized by reaction with a gas species to produce another gas.

- Stimulated CVD adds energy to the reactants such that the substrate need not be heated as much as in a conventional CVD process. The most common method is plasma-enhanced CVD. The reaction is generally transport limited as a result of the added energy.
- Selective CVD deposits material selectively on exposed substrate rather than on a surrounding masked area.
- Selectivity is usually obtained by 1) achieving a higher reactant concentration on the substrate than on the mask through differential adsorption or desorption rates, 2) catalysis of the reaction by the substrate or where the substrate provides a reactant directly, 3) preferential nucleation on the substrate relative to on the mask, 4) a combined etching and deposition reaction at the surface which enhances the difference in nucleation or growth rate on the substrate relative to the mask.
- Selective deposition may yield increased deposition rates near a masked area.
- ALD deposits single monolayers of reactants on a surface in alternation such that one atomic layer is produced in each cycle.
- For an A/B process cycle the essential element is to saturate the B surface with one and only one layer of A atoms in some carrier form followed by coating this A surface with one and only one layer of B.
- Having the leisure of time and being automatically limited to one monolayer of deposition means that precursor molecules have plenty of time to find their way into cracks and across complex surfaces giving outstanding coverage and uniformity.
- The total amount of material deposited depends upon the total number of cycles only.
- Control of stoichiometry and range of deposition conditions is generally more limited in ALD.

12.10 HOMEWORK PROBLEMS

1. What is the boundary layer in a CVD reactor and what effects does it have (if a boundary layer even forms) on the deposition rate if the growth occurs

    a.   in the reaction rate-limited regime?
    b.   in the diffusion rate-limited regime?

2. A tilted susceptor in the reaction chamber gives control over the boundary layer to keep its thickness constant. In what way does this work? (Hint: an equation for the boundary layer thickness is $\delta(x) = \sqrt{\dfrac{\mu x}{\rho u}}$ where $\delta(x)$ is the boundary layer thickness, μ is the viscosity of the gas, x is the distance along the reactor, ρ is the gas density, and u is the gas velocity.)

3. Your manager at MegaJoule Industries has decided not to buy you that new MBE machine so you will have to make do with the existing CVD process. You have been assigned to deposit some patterned InP thin films on GaAs substrates. The GaAs has been coated with a patterned SiO_2 film.

    You are debating the selection of a growth temperature for the process. Suppose you consider 450°C and 550°C with the reaction:

    $$In(C_2H_5)_3 + PH_3 \rightarrow InP + 3C_2H_6 \text{ (Reaction A)}$$
    or
    $$InCl_3 + PH_3 \rightarrow InP + 3HCl \text{ (Reaction B)}$$
    [Run in the presence of Cl_2 gas there is a reverse etching reaction for Reaction B.]

    a.   At which temperature would you expect the thickness of the boundary layer to be more important to controlling the deposition rate? Explain briefly why this is the case.
    b.   You observe that desorption of TEIn [$In(C_2H_5)_3$] from GaAs has a much higher activation energy than from SiO_2. The two desorption rates are the same at 550°C. Would you expect to achieve better or worse selectivity at 450°C for Reaction A compared to at 550°C. Explain briefly.
    c.   In spite of your best efforts to obtain selectivity with Reaction A, you find nucleation on the SiO2 mask. Once nucleation has occurred the nuclei grow rapidly and you lose the selectivity. How could you reduce this by switching to Reaction B.

d. One of your colleagues recommends replacing the TEIn [In(C2H5)3] with TMIn
   [In(CH3)3] in Reaction A. Would you expect this to increase or decrease the reaction activation barrier (the energy in the Boltzmann factor for the reaction rate) **and** how would it affect the total energy change of the reaction (more or less exothermic/endothermic)?
e. Do you prefer a CVD reaction to be exothermic or endothermic in general? Explain briefly.
f. You decide to run Reaction A at 450°C. However, you find that at this temperature you are strongly reaction rate limited. How might you obtain a higher reaction rate at this temperature?
g. List three major considerations in selection of a source gas for a CVD process.

12.11 SUGGESTED READINGS & REFERENCES

Suggested Readings:

Vossen, John L. and Kern, Werner, *Thin Film Processes* volumes I and II. New York, Academic: 1978.

Tu, King-Ning; Mayer, James W.; and Feldman, Leonard C.; *Electronic Thin Film Science for Electrical Engineers and Materials Scientists*. New York: Macmillan, 1992.

Purcell, Keith F and Kotz, John F., *Inorganic Chemistry*. Philadelphia: W. B. Saunders Co., 1977.

References:

[1] Ingle, N.K.; Theodoropoulos, C.; Mountziaris, T.J.; Wexler, R.M.; and Smith, F.T.J., "Reaction kinetics and transport phenomena underlying the low-pressure metalorganic chemical vapor deposition of GaAs." *J. Cryst. Growth*, 1996; 167: 543-556.

[2] Coleman, James J., "Metalorganic chemical vapor deposition for optoelectronic devices." *Proc. IEEE*, 1997; 85: 1715-1729.

[3] Dupuis, R.D., "Metalorganic chemical vapor deposition (MOCVD)" in *Handbook of Thin Film Process Technology*, ed. by David A Glocker and S. Ismat Shah. Bristol: Institute of Physics, 2002; 1: p. B.1.1:5-6.

[4] Stringfellow, G.B. "A critical appraisal of growth mechanisms in MOVPE." *J. Cryst. Growth*, 1984; 68: 111-22.

[5] Stringfellow, G.B. "Thermodynamic aspects of OMVPE." *J. Cryst. Growth*, 1984; 70: 133-9.

[6] Asai, Toshihiro and Dandy, David S., "Thermodynamic analysis of III-V semiconductor alloys grown by metalorganic vapor phase epitaxy." *J. Appl. Phys.*, 2000; 88: 4407-4416.

[7] See, for example, Bailly, F.; Cohen-Solal, G.; and Mimila-Arroyo, J., "Simplified theory of reactive closed-spaced vapor transport." *J. Electrochem. Soc.*, 1979; 126: 1604-1608.

[8] See, for example, Heyen, M. and Balk, P., "Epitaxial growth of GaAs in chloride transport systems." *Prog. in Cryst. Growth and Characterization*, 1983; 6: 265-303.

[9] See, for example, Putz, N.; Sauerbrey, A.; Veuhoff, E.; Kirchmann, R.; Heyen, M.; Luth, H.; Balk, P., "Effect of total pressure on the uniformity of epitaxial GaAs films grown in the Ga-HCl-AsH$_3$-H$_2$ system." *J. Electronic Materials*, 1985; 14: 645-653.

[10] See, for example, Carlsson, Jan-Otto and Hårsta, Anders, "Thermodynamic investigation of selective tungsten chemical vapour deposition: Influence of growth conditions and gas additives on the selectivity in the fluoride process." *Thin Solid Films*, 1988; 158: 107-122.

[11] George, S.M.; Fabreguette, F.H.; Sechrist, Z.A.; and Elam, J.W., "Quartz crystal microbalance study of tungsten atomic layer deposition using WF$_6$ and Si$_2$H$_6$." *Thin Solid Films*, 2005; 488: 103-110.

[12] Torndahl, T.; Ottosson, M.; Carlsson, J.-O., "Growth of copper metal by atomic layer deposition using copper(I) chloride, water and hydrogen as precursors." *Thin Solid Films,* 2004; 458: 129-136.

[13] Kaneko, T. and Okuno, T., "Growth kinetics of silicon carbide CVD" J. Cryst. Growth: 91 (1988) 599-604.

[14] Danielsson, Ö.; Henry, A.; and Janzén, E.; "Growth rate predictions of chemical vapor deposited silicon carbide epitaxial layers" J. Cryst. Growth: 243 (2002) 170-184.

APPENDIX

Selected Semiconductor Properties

Compound	E_{gap} eV 300K	μ_n cm^2 V^{-1}s^{-1}	μ_p cm^2 V^{-1}s^{-1}	Structure	a_0 nm a/c for wurtzite	ε_r ε_0	N_c cm^{-3}	N_v cm^{-3}
\multicolumn{9}{} Indirect gap semiconductors
Si	1.12	1900	500	d	0.543072	11.8	3.2×10^{19}	1.8×10^{19}
Ge	0.67	3800	1820	d	0.565754	16	1.0×10^{19}	5.0×10^{18}
AlP	2.52	80		z	0.54672			
AlAs	2.16	1200	420	z	0.56622	10.9		
6H SiC	2.39	200	18	w	0.3073/1.0053	10.2	8.9×10^{19}	2.5×10^{19}
\multicolumn{9}{} Direct gap semiconductors
a-Sn	0.08	2500	2400	d	0.64912			
GaAs	1.43	8800	400	z	0.565315	13.2	4.5×10^{17}	9.5×10^{18}
InAs	0.36	33000	460	z	0.605838	14.6		
AlN	6.2	300	14	w	0.498	9.1	6.2×10^{18}	4.9×10^{20}
GaN	3.4	400	150	w	0.3133/0.5187	8.9	2.2×10^{18}	4.6×10^{19}
InN	0.7	20		w	0.3536/0.5701	15.3	9.1×10^{17}	5.2×10^{19}
GaP	2.26	300	150	z	0.54505	11.1	1.8×10^{19}	1.9×10^{19}
InP	1.27	4600	150	z	0.586875	12.4	5.7×10^{17}	1.1×10^{19}
ZnS	3.54	180	4	z	0.54093	8.9		
ZnSe	2.58	540	28	z	0.56676	9.2		
CdS	2.42	250	15	z	0.55818	8.9		
CdSe	1.73	650		z	0.605	10.2		

d: diamond structure, z: fcc zincblende, w: hcp wurtzite

USEFUL CONSTANTS

Avagadro's number, N_A: 6.0222×10^{23} particles mol^{-1} or amu g^{-1}

Boltzmann's constant, k_B: 1.3806×10^{-23} J K^{-1} or 8.617×10^{-5} eV K^{-1}

Permittivity of free space ε_0: 8.85×10^{-14} F cm^{-1}

Permeability of free space, μ_0: 1.2566×10^{-6} N A^{-2} ($= 4\pi \times 10^{-7}$ N A^{-2} exactly)

Planck's constant, h: 6.6262×10^{-34} J s or 4.14×10^{-15} eV s

$\hbar$: 1.0546×10^{-34} J s $= h/2\pi$

Speed of light: 2.997×10^{10} cm sec^{-1} or 29.97 cm ns^{-1}

Electronic charge: 1.60219×10^{-19} C

Electronic mass: 9.1096×10^{-31} kg

Proton mass: $1836 \times$ electron mass (1.67261×10^{-27} kg)

Rydberg: 1.09678×10^5 cm^{-1}

Hydrogen first ionization energy: 13.60 eV or 1312 kJ mol^{-1}

Bohr radius: 0.05292 nm

Bohr magnetron: 9.2741×10^{-24} J T^{-1}

UNITS

Variable:	Common unit	in SI units only
Acceleration, a:	$m\ s^{-2}$	$m\ s^{-2}$
Angular frequency, ω:	$rad\ s^{-1}$	s^{-1}
Angular momentum, L:	$kg\ m^2\ s^{-1}$	$kg\ m^2\ s^{-1}$
Area, A:	m^2	m^2
Capacitance, C:	$F\ (or\ C\ V^{-1})$	$A^2\ s^4\ kg^{-1}\ m^{-2}$
Charge, q:	C	$A\ s$
Conductivity, σ:	$\Omega^{-1}\ m^{-1}\ (0.01\ \Omega^{-1}\ cm^{-1})$	$A^2\ s^3\ kg^{-1}\ m^{-3}$
Current, I:	A	A
Density, ρ:	$kg\ m^{-3}$	$kg\ m^{-3}$
Electric dipole moment, p:	$C\ m$	$A\ s\ m$
Electric field, E:	$V\ m^{-1}$	$kg\ m\ A^{-1}\ s^{-3}$
Electromotive force, V:	V	$kg\ m^2\ A^{-1}\ s^{-3}$
Energy, E:	J	$kg\ m^2\ s^{-2}$
Entropy, S:	$J\ K^{-1}$	$kg\ m^2\ s^{-2}\ K^{-1}$
Force, F:	N	$kg\ m\ s^{-2}$
Frequency, f:	Hz	s^{-1}
Inductance, L:	H	$kg\ m^2\ A^{-2}\ s^{-2}$
Magnetic field, B:	T	$kg\ A^{-1}\ s^{-2}$
Mass, m:	kg	kg
Momentum, k:	$kg\ m\ s^{-1}$	$kg\ m\ s^{-1}$
Power, P:	$W\ (or\ J\ s^{-1})$	$kg\ m^2\ s^{-1}$
Pressure, P:	$Pa\ (N\ m^2)$	$kg\ m^{-1}\ s^{-2}$
Resistance, R:	$\Omega\ (or\ V\ A^{-1})$	$kg\ m^2\ A^{-2}\ s^{-3}$
Temperature, T:	K	K
Time, t:	s	s
Velocity, v:	$m\ s^{-1}$	$m\ s^{-1}$

UNIT CONVERSIONS

Length units:
$10^8\ Å\ cm^{-1}$　　$10^9\ nm\ m^{-1}$　　$10^7\ nm\ cm^{-1}$
$10^6\ \mu m\ m^{-1}$　　$10^4\ \mu m\ cm^{-1}$　　$10^2\ cm\ m^{-1}$

Angle:
$180°/\pi$　　　57.30 degrees per radian

Density:
$10^3\ (kg\ m^{-3})/(g\ cm^{-3})$

Force:
$10^5\ dyne\ N^{-1}$
Pressure:

$0.1\ (N\ m^{-2})/(dyne\ cm^{-2})$
$10^2\ Pa\ mbar^{-1}$
$1013.25\ mbar\ atm^{-1}$
$133.32\ Pa\ Torr^{-1}$
$6.895 \times 10^3\ Pa\ psi^{-1}$

Energy:
$10^7\ erg\ J^{-1}$
$4184\ J\ kcal^{-1}$
$1.60219 \times 10^{-19}\ J\ eV^{-1}$
$23.06\ (kcal\ mol^{-1})/(eV\ molecule^{-1})$
$9.648 \times 10^4\ (J\ mol^{-1})/(eVmolecule^{-1})$

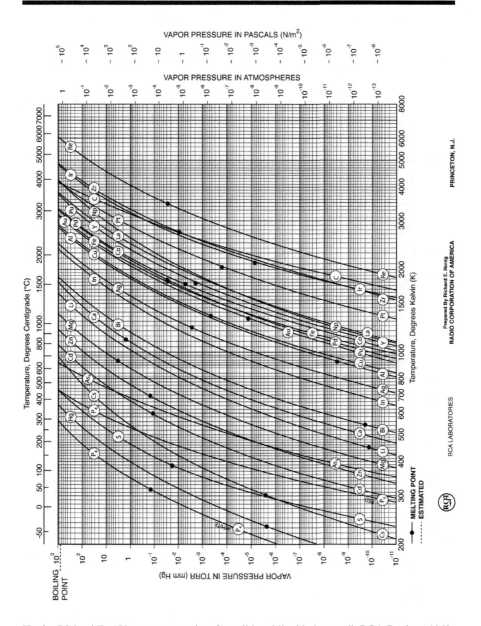

Honig, Richard E., "Vapor pressure data for solid and liquid elements", RCA Review, 1962; 23: 567-586. © 1962 RCA Laboratories. Courtesy of The David Sarnoff Library.

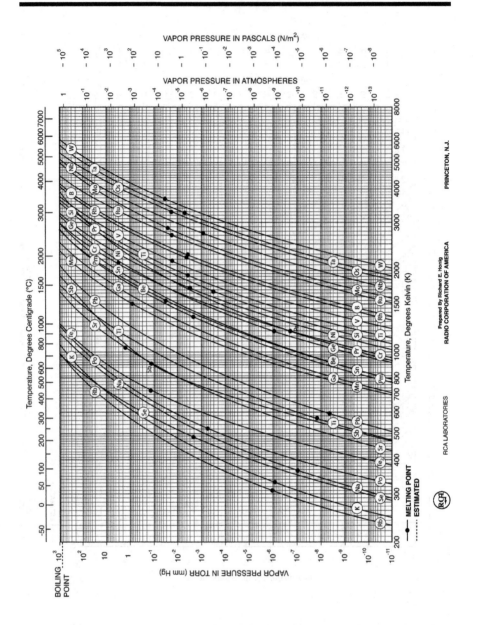

Honig, Richard E., "Vapor pressure data for solid and liquid elements", RCA Review, 1962; 23: 567-586. © 1962 RCA Laboratories. Courtesy of The David Sarnoff Library.

INDEX